Fundamentals of Controlled/Living Radical Polymerization

RSC Polymer Chemistry Series

Series Editors:
Professor Ben Zhong Tang (Editor-in-Chief), *The Hong Kong University of Science and Technology, Hong Kong, China*
Professor Alaa S. Abd-El-Aziz, *University of Prince Edward Island, Canada*
Professor Stephen L. Craig, *Duke University, USA*
Professor Jianhua Dong, *National Natural Science Foundation of China, China*
Professor Toshio Masuda, *Fukui University of Technology, Japan*
Professor Christoph Weder, *University of Fribourg, Switzerland*

Titles in the Series:
 1: Renewable Resources for Functional Polymers and Biomaterials
 2: Molecular Design and Applications of Photofunctional Polymers and Materials
 3: Functional Polymers for Nanomedicine
 4: Fundamentals of Controlled/Living Radical Polymerization

How to obtain future titles on publication:
A standing order plan is available for this series. A standing order will bring delivery of each new volume immediately on publication.

For further information please contact:
Book Sales Department, Royal Society of Chemistry, Thomas Graham House, Science Park, Milton Road, Cambridge, CB4 0WF, UK
Telephone: +44 (0)1223 420066, Fax: +44 (0)1223 420247
Email: booksales@rsc.org
Visit our website at www.rsc.org/books

Fundamentals of Controlled/ Living Radical Polymerization

Edited by

Nicolay V Tsarevsky
Department of Chemistry, Southern Methodist University, Dallas, TX, USA
Email: nvt@smu.edu

Brent S Sumerlin
Department of Chemistry, University of Florida, Gainesville, FL, USA
Email: sumerlin@chem.ufl.edu

RSCPublishing

RSC Polymer Chemistry Series No. 4

ISBN: 978-1-84973-425-7
ISSN: 2044-0790

A catalogue record for this book is available from the British Library

Published by The Royal Society of Chemistry,
Thomas Graham House, Science Park, Milton Road,
Cambridge CB4 0WF, UK

Registered Charity Number 207890

For further information see our web site at www.rsc.org

Printed in the United Kingdom by Henry Ling Limited, Dorchester, DT1 1HD, UK

Preface

There is little doubt the advent of controlled/living radical polymerization (CRP) has revolutionized modern polymer science. Perhaps more than any synthetic development of the last several decades, CRP has unlocked access to a plethora of new functional and well-defined polymers to address a variety of complex applications. Previously considered useful for primarily the facile and rapid preparation of high molecular weight polymers for mostly commodity applications, radical polymerization can now be readily employed to prepare specialty materials based on (co)polymers that are well-defined, have controlled molecular weights and end group functionality, complex topologies, and predefined segment sequences (*e.g.*, block copolymers).

More than any other characteristic, it is the molecular weight of a polymer that controls its properties and potential utility. While conventional radical polymerization had been employed for many years to achieve high molecular weight polymers, precise control of chain length remained difficult. External chain transfer agents proved useful to limit molecular weights, but the obtained polymers were characterized with broad molecular weight distributions. The diverse family of CRP techniques developed over the last 2–3 decades now allows access to polymers of a specific chain length that can be derived from virtually any vinyl monomer polymerizable by a radical method.

Conventional radical polymerization had limited success in the preparation of polymers with specific chain-end functional and/or reactive groups. Although certain functionalized radical initiators showed some potential, the range of functional groups that could be attached efficiently to either end of the macromolecules was relatively narrow. Further, due to the ever-present termination and transfer reactions, which rapidly "kill" the propagating chains, it is virtually impossible to prepare samples in which nearly all macromolecules are chain-end-labeled. In contrast, CRP methods have proved very powerful in the synthesis of polymers containing either identical or different α- and

RSC Polymer Chemistry Series No. 4
Fundamentals of Controlled/Living Radical Polymerization
Edited by Nicolay V Tsarevsky and Brent S Sumerlin
© The Royal Society of Chemistry 2013
Published by the Royal Society of Chemistry, www.rsc.org

ω-chain-end functional groups. The synthesis of composite materials by CRP is also straightforward.

One of the key benefits of classical radical polymerization is the facility with which many monomers can be incorporated into a single chain by copolymerization. Depending on the relative reactivity of the comonomers, random, alternating and blocky sequences can be achieved. However, block copolymers with well-defined junction points linking segments of differing repeat units are nearly impossible to prepare directly by conventional radical polymerization. While it was possible to prepare block copolymers by coupling two end-functional homopolymers or by polymerization from homopolymers terminated with a moiety that allowed a subsequent polymerization step, facile access to block copolymers was not achieved by a radical mechanism prior to the development of controlled radical methods. Since then, an enormous number of new block copolymers have been prepared by CRP, and this has provided access to a variety of new materials with potential applications as drug delivery agents, cosmetics, coatings, adhesives, thermoplastic elastomers, *etc.* Additionally, the copolymerization of monomers with different reactivities under CRP conditions yields gradient copolymers, in which the composition changes along the polymer backbone, instead of from one chain to another (as is the case in conventional radical polymerization). Gradient copolymers were inaccessible through traditional radical polymerization and are very promising materials in various fields, such as surfactants, degradable polymers, *etc.*

Another aspect of chain structure that has become more readily manipulated is chain architecture. By employing new initiators or reversible chain transfer agents with multiple sites capable of initiating chain growth, a wide variety of complex architectures previously considered inaccessible can now be readily prepared by CRP. Stars, combs, and brushes represent only a fraction of branched chain topologies that have been synthesized. This aspect of CRP has enabled both new applications and fundamental insight into polymer structure-property relationships.

Many of the aforementioned characteristics of CRP are in common with those of previously developed (pseudo)living ionic polymerization techniques. However, it is the relative ease with which CRP can be employed to achieve these characteristics that has led to the continued success and growth of the field. Indeed, because most of the methods described in the chapters that follow can be conducted under relatively non-stringent conditions in a variety of solvents and at a wide range of temperatures, the barrier to entry to the field is minimal. Precision polymer synthesis is now possible in laboratories without access to expensive experimental equipment or extensive expertise.

It is important to remember that the CRP techniques are based on the success achieved over many years in the areas of living ionic polymerizations. The success of Szwarc and others in recognizing the far-reaching utility of living anionic polymerization led many to pursue the goal of polymerization control during other chain growth methods. Molecular weight and end group control in cationic polymerization was subsequently achieved by introduction of the concept of dormant-active equilibration (*i.e.*, "reversible deactivation") that kinetically limited the extent of chain breaking reactions by rendering a significant fraction of chains dormant at any given time. Unfortunately, ionic polymerizations are applicable only to a relatively narrow range of monomers and required stringent reaction conditions (*e.g.*, absence of moisture, carbon dioxide, and many other impurities often encountered in commercially available monomers and solvents). Owing to significant differences in the reactivity ratios of the ionically polymerizable monomers, copolymerization reactions are often challenging, which limits the number of accessible materials. For a relatively long time, it was considered that radical polymerizations that were living or at least resembled living ionic process were impossible to design due to fast termination reactions between the propagating radicals. It was thought that living radical polymerization could only be achieved in system where the mobility of the growing radicals was drastically diminished by, for instance, carrying out the polymerizations in very viscous media and preferably at low temperatures (which not only increases the viscosity but also minimizes transfer). Of course, such approaches were not particularly practical. The principle of reversible deactivation of propagating radicals was extended to radical polymerization in the 1980s and proved to be of tremendous utility in the design of living-like radical polymerizations. It is the specifics of the chemistry employed to achieve reversible deactivation that differentiates the various CRP processes. The fundamental aspects such as kinetics, thermo-dynamics, structure-reactivity correlations, solvent effects, *etc.*, of the most important CRP methods are the subject of chapters in this book. It should be noted that radical polymerizations are never truly living (termination reactions always take place), but using the reversible deactivation strategy, these reaction show many of the characteristics of living polymerizations, such as control over molecular weight (determined by the ratio of monomer to initiator), narrow molecular weight distribution (if initiation is fast), high degree of chain-end functionalization, which is responsible for the ability to synthesize block copolymers, *etc.* Owing to the ability to control numerous molecular parameters and due to the fact that radical polymerizations with reversible deactivation resemble closely the classical living ionic polymerizations, we have adopted the term controlled/living radical polymerization in this book.

We hope this book provides insight into the current state of the art in CRP while also giving historical perspective to the field. While many other books have focused on controlled/living polymerizations, including CRP, and have provided extensive insight into the materials that can be prepared therefrom, our intent here is to elucidate the *fundamentals* governing the most common types of CRP. In many cases, particular attention is dedicated to the kinetic and thermodynamic factors that effect polymerization control. We hope this collection will be useful to both experts and newcomers to this rapidly expanding field.

Nicolay V. Tsarevsky
Southern Methodist University, Dallas
Brent S. Sumerlin
University of Florida

Contents

RSC Polymer Chemistry Series No. 4
Fundamentals of Controlled/Living Radical Polymerization
Edited by Nicolay V Tsarevsky and Brent S Sumerlin
© The Royal Society of Chemistry 2013
Published by the Royal Society of Chemistry, www.rsc.org

CHAPTER 1

Kinetics and Thermodynamics of Radical Polymerization

F. EHLERS,* J. BARTH AND P. VANA

Georg-August-University of Göttingen, Institute of Physical Chemistry,
Tammannstr. 6, D-37077 Göttingen, Germany
*Email: Florian.Ehlers@chemie.uni-goettingen.de

1.1 Introduction

Radical polymerization processes are of great scientific and economic importance. Knowledge about their kinetic principles is a prerequisite for the efficient synthesis of a wide range of polymeric products. Since the dawn of macromolecular chemistry in the 1920's, the study of these principles has been a central topic of academic research. Although a radical polymerization process is basically constituted by just four types of reactions, which are initiation, propagation, transfer and termination, the coupled nature of these reactions leads to a complexity that makes it difficult to determine the individual rate constants and to evaluate their effects on the properties of the final polymer, like its molecular weight distribution. Scheme 1.1 shows a set of fundamental reaction equations for a radical polymerization process.

There is a kinetic rate law expression for each of these reactions. Determination of the corresponding rate coefficients is the main task of all kinetic experiments in this field. The employed experimental techniques can roughly be separated into two classes: one approach focuses on accurate measurements of the overall polymerization rate or time-resolved species concentration, while the other one is based on the analysis of the resulting molecular weight distributions. Provided that all relevant rate coefficients for a certain

RSC Polymer Chemistry Series No. 4
Fundamentals of Controlled/Living Radical Polymerization
Edited by Nicolay V Tsarevsky and Brent S Sumerlin
© The Royal Society of Chemistry 2013
Published by the Royal Society of Chemistry, www.rsc.org

Initiator decomposition	$I_2 \xrightarrow{k_d} 2I^\bullet$
Initiation	$I^\bullet + M \xrightarrow{k_i} R_1^\bullet$
Propagation	$R_i^\bullet + M \xrightarrow{k_p} R_{i+1}^\bullet$
Transfer	$R_i^\bullet + S \xrightarrow{k_{tr}} R_1^\bullet + P_i$
Termination	$R_i^\bullet + R_j^\bullet \xrightarrow{k_t} P_{i+j}$ or $P_i + P_j$

Scheme 1.1

polymerizing system are known, it is possible to make precise predictions about the kinetics of the entire process, and therefore also about the molecular weight distribution. Today, computer simulations are an important tool in polymer research, allowing for precise numerical simulations of even very complex polymerizing systems and thus contributing to a deeper understanding of radical polymerization kinetics.

The intention of the following chapter is to give a general overview of our knowledge about the kinetics of conventional radical polymerization and its implications for the process and the formed product. This basic knowledge is also mandatory for the understanding and optimization of controlled polymerization processes.

1.2   Initiation

For a radical polymerization to occur, the first thing needed are free radicals. These are initially provided by some agent, the initiator, during the reaction step called initiation. The initiation step is commonly characterized by two coefficients, the initiation rate coefficient k_i and the initiator efficiency f. Knowledge of these parameters is of crucial importance for both industrial applications and theoretical studies of radical polymerizations.

The vast majority of initiators belong to one of two groups, thermal initiators or photoinitiators. Thermal initiators form radical species upon heating, while photoinitiators decompose when exposed to visible or UV light. While in commercial processes mainly thermal initiators are used, kinetic studies are preferentially performed using photoinitiators. This is because the irradiation can precisely be timed, defining a sharp starting point for the polymerization reaction. A general scheme for the decomposition reaction, regardless of the type of initiator, is given in Scheme 1.2.

The initiator (I) decay, be it caused by light absorption or heating, follows a first order rate law:

$$-\frac{dc_I}{dt} = k_d c_I \tag{1.1}$$

For the polymerization kinetics, the initiator concentration is not the important quantity. More important is the concentration of primary radicals

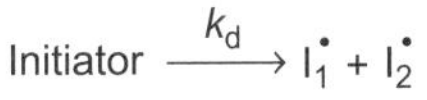

$$\text{Initiator} \xrightarrow{k_d} I_1^{\bullet} + I_2^{\bullet}$$

Scheme 1.2

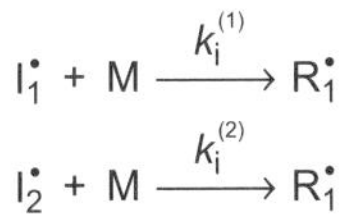

$$I_1^{\bullet} + M \xrightarrow{k_i^{(1)}} R_1^{\bullet}$$

$$I_2^{\bullet} + M \xrightarrow{k_i^{(2)}} R_1^{\bullet}$$

Scheme 1.3

formed by the initiation process. The rate R_d of formation of radicals that can start chain growth can be expressed by the following first order rate law:

$$R_d = \frac{dc_{I^{\bullet}}}{dt} = -2f\frac{dc_I}{dt} = 2fk_dc_I \tag{1.2}$$

where k_d is the rate coefficient for the initiator decomposition reaction, and f is the initiator efficiency. Integration of eqn (1.2) leads to the following expression, showing the exponential decrease of initiator concentration with time:

$$c_I = c_I^0 e^{-k_d t} \tag{1.3}$$

Initiator decay alone is not sufficient to start a new polymer chain. The formed radical has to react with a monomer unit. Right after decay, the (usually two) freshly formed radicals $I_1^{\bullet}$ and $I_2^{\bullet}$ are still in close proximity of each other and surrounded by solvent molecules. The primary radicals' ability to leave the solvent cage unreacted and then react with a monomer is expressed by the initiator efficiency f, with values ranging from zero (no initiation) to unity. In a real system, not every primary radical will actually start a polymer chain. Radicals can recombine before leaving the solvent cage or undergo a side reaction before they encounter a monomer molecule. Typically, f has a value between 0.5 and 0.8 and depends on the viscosity of the system, indicating that the diffusion-controlled escape from the solvent cage is the crucial factor.

If the initiator molecule is asymmetric, *i.e.* $I_1^{\bullet} \neq I_2^{\bullet}$, the formed radical species generally do not show identical reactivity towards the monomer. Thus, the initiation process will take on the form shown in Scheme 1.3, where M indicates a monomer molecule, $R_1^{\bullet}$ refers to a macroradical of chain length 1 and $k_i^{(1)}$ and $k_i^{(2)}$ refer to the rate coefficients of initiation for the respective initiator fragments. The overall rate of initiation, R_i, can be calculated according to eqn (1.4):

$$R_i = \frac{dc_{R_1^{\bullet}}}{dt} = -\frac{dc_{I_1^{\bullet}}}{dt} - \frac{dc_{I_2^{\bullet}}}{dt} = k_i^{(1)}c_M c_{I_1^{\bullet}} + k_i^{(2)}c_M c_{I_2^{\bullet}} \tag{1.4}$$

The rate coefficient of initiation k_i can be expressed as the arithmetic mean of the coefficients for the individual fragments, since $c_{I_1^{\bullet}} = c_{I_2^{\bullet}} = c_{I^{\bullet}}/2$

$$R_i = k_i c_M c_{I^{\bullet}} \quad \text{with} \quad k_i = \frac{k_i^{(1)} + k_i^{(2)}}{2} \tag{1.5}$$

1.2.1 Thermal Initiation

There are mainly two types of thermally activated initiators: azo-type, and peroxy-type. Their general structures are shown in Scheme 1.4.

Thermal initiators decompose in a first order reaction upon heating, displaying a characteristic half life at a certain temperature. It is correlated to the decomposition rate coefficient by eqn (1.6). The half-life, $t_{1/2}$, is the amount of time it takes for half of a sample of initiator to decompose.

$$t_{1/2} = \frac{\ln 2}{k_\mathrm{d}} \qquad (1.6)$$

A large variety of initiators has been described in the literature, and many of them are available commercially, so that the initiator can be chosen according to the desired decomposition rate. For practical use, initiators are often characterized by the temperature where they show a half life of 10 h. Typical values for these temperatures lie in the range from 20 to 120 °C. A large collection of data on initiator properties has been published in the *Polymer Handbook*.[1] However, the decomposition rate is not the only property to consider when selecting an initiator. Possible side reactions with monomer or solvent might also play a role, or the ability of the initiator to act as a transfer agent (compare Section 4.2).

1.2.2 Photoinitiation

Photoinitiators are seldomly used in commercial applications, because large reaction volumes cannot be irradiated easily in a uniform way. Still, there are some applications in the area of UV hardening lacquers and printing inks. In research, on the other hand, photoinitiators are used frequently, for they allow to precisely define the beginning and end of the initiation process by flash photolysis of the initiator. Also, most photoinitiators show almost no temperature dependence of the decomposition rate, but a strong dependence on the (UV) light intensity.

A good photoinitiator for a given polymerization should have the following properties:

1. An irradiation wavelength should be available were the initiator shows strong absorption, but the monomer and the solvent show almost no absorption.

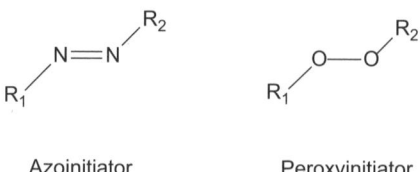

Azoinitiator   Peroxyinitiator

Scheme 1.4

2. The initiator should show a high efficiency.
3. At best it should only generate a single radical species.

There are two groups of photoinitiators that differ in the mechanism of radical formation. Type I photoinitiators generate radicals *via* unimolecular bond cleavage after irradiation, similar to thermal initiators. Type II initiators show no bond cleavage directly after irradiation. Instead, they enter a bound excited state which reacts with a so-called co-initiator molecule to generate free radicals, mostly *via* H-abstraction. Most visible light active photoinitiators belong to type II.

Type I photoinitiators show considerable structural variety, one of the most common motives being the acetophenone group. The general structure of this kind of photoinitiators is shown in Scheme 1.5.

Upon irradiation, the initiator molecules will absorb a certain amount of energy changing from the electronic ground state to an excited state. The accessible electronic states of a molecule are usually shown in a Jablonski diagram. An example of such a diagram is shown in Figure 1.1.

In most cases, absorption will cause the initiator molecule to enter the first singlet excited state, commonly denoted as S_1. In general, more than one

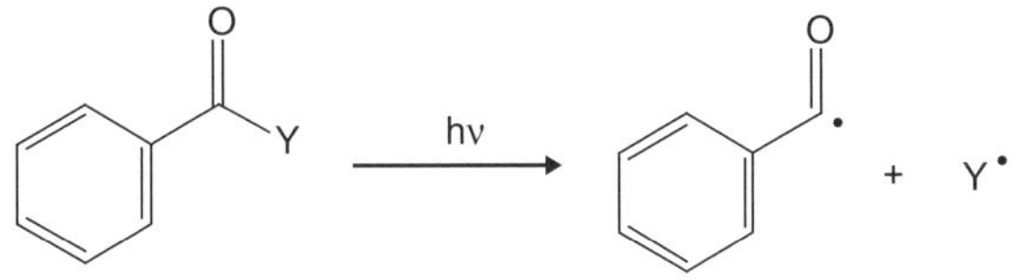

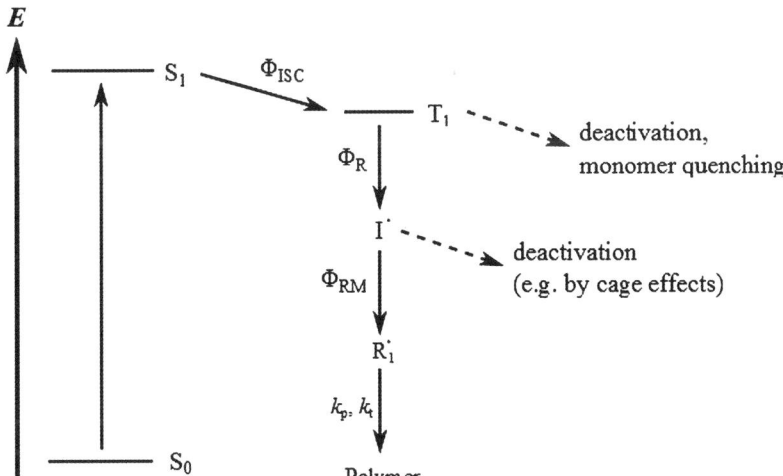

Scheme 1.5

Figure 1.1 Simplified Jablonski diagram of type I photoinitiator decomposition, *i.e.* for an acetophenone type initiator.

deactivation channel will be active for the excited species, and not all of them do necessarily lead to free radical generation. The fraction of the excited molecules that are actually converted into primary radicals is expressed by the overall quatum yield Φ, which is the product of the quantum yields of three successive elementary processes: intersystem crossing from the lowest excited singlet to the lowest triplet state (1.8), bond scission in the triplet state (1.9), and reaction of the formed radical with a monomer molecule (1.10).

$$S_0 \xrightarrow{h\nu} S_1 \tag{1.7}$$

$$S_1 \longrightarrow T_1 \qquad \Phi_{ISC} \tag{1.8}$$

$$T_1 \longrightarrow R^\bullet \qquad \Phi_R \tag{1.9}$$

$$R^\bullet \xrightarrow{M} RM^\bullet \quad \Phi_{RM} \tag{1.10}$$

where $\Phi = \Phi_{ISC}\,\Phi_R\,\Phi_{RM} \leq 1$.

For ketones, the intersystem crossing step usually has a rather high quantum yield,[2] so setting the value of Φ_{ISC} to one is a good approximation in most cases. The overall quantum yield is then determined by the values of Φ_R and Φ_{RM}. Φ_R may be reduced due to alternative deactivation pathways from the T_1 triplet state. The dominating deactivation reactions in free radical polymerizations are the reaction with molecular oxygen and the deactivation by monomer molecules (compare ref. 3). The influence of the former may be reduced by thorough degassing of the polymerization mixture prior to initiation.

A long life time of the triplet state tends to lead to a reduced overall quantum yield, because chances are higher that alternative deactivation routes may successfully compete with radical formation. 2,2-Dimethoxy-2-phenylacetophenone (DMPA) is an example for an initiator with a rather short-lived first triplet state ($\tau < 0.1$ ns),[4] causing a high quantum yield Φ_R. The quantum yield Φ_{RM} for the formation of macroradicals from the initiator fragment radicals is also known as the initiator efficiency, f, which is defined analogously to the efficiency of thermal initiators.

Most acetophenone-type initiators decompose into more than one kind of radical upon irradiation. Typically, two different radical species $R_1^\bullet$ and $R_2^\bullet$ are formed, showing very different reactivities towards the monomer. While the carbonyl radical is efficiently forming macroradicals, the additional radical does not add much to the initiator efficiency. On the contrary, for DMPA the methoxy radical is thought to be involved exclusively in termination steps.[5] In such a case, it is common to say that the carbonyl radical is the "effective" and the methoxy radical the "ineffective" primary radical. Their initial concentrations are equal and denoted as ρ, while the overall initial radical concentration is 2ρ. Polymerization kinetics can strongly be influenced by the formation of two sorts of primary radicals with markedly different reactivities.

Azoinitiators may also be used as photoinitiators. They decompose upon irradiation by a mechanism that is markedly different from the acetophenone type. By example of 2,2-azobisisobutyronitrile (AIBN), this mechanism is shown in Scheme 1.6.

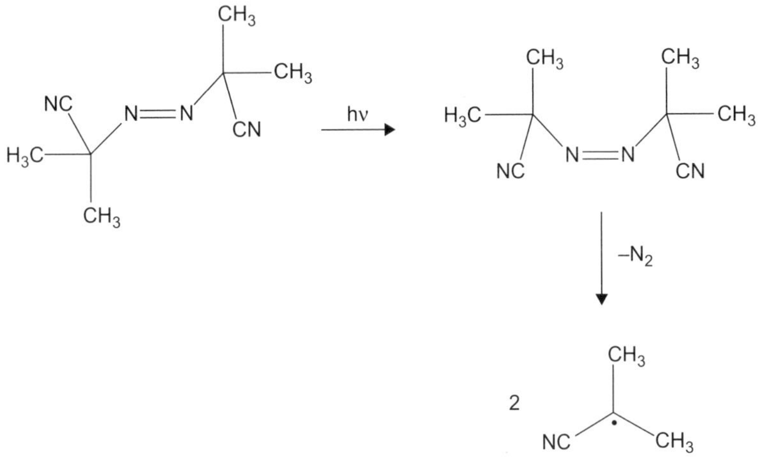

Scheme 1.6

UV irradiation leads to a *cis-trans* isomerization reaction. Since the rate of this isomerization reaction is finite, it is observed in experiments that the time of incidence of the laser pulse is not identical with the time of primary radical formation, but there is a certain delay, usually in the order of microseconds.

The concentration of effective primary free radicals ρ that is generated by irradiation may be calculated by:

$$\rho = 2\Phi \frac{n_{abs}}{V} \tag{1.11}$$

with Φ the primary quantum yield, n_{abs} the number of absorbed photons and V the volume considered.

The number of absorbed photons is given by the Beer–Lambert Law:

$$n_{abs} = \frac{E_p}{E_\lambda}(1 - 10^{-\varepsilon cd}) \tag{1.12}$$

E_p: energy absorbed by the sample
E_λ: molar energy of photons at the irradiation wavelength λ
ε: molar absorption coefficient of the initiator molecule at wavelength λ
c: photoinitiator concentration
d: optical path length

Detailed information about different photoinitiators and their decomposition behaviour has been gathered by Gruber.[6]

1.2.3 Self-initiation

It is not strictly necessary for a radical polymerization to be started by an initiator. It might also be initiated by impurities, by peroxy compounds that are formed in the presence of molecular oxygen, or even by the monomer itself.

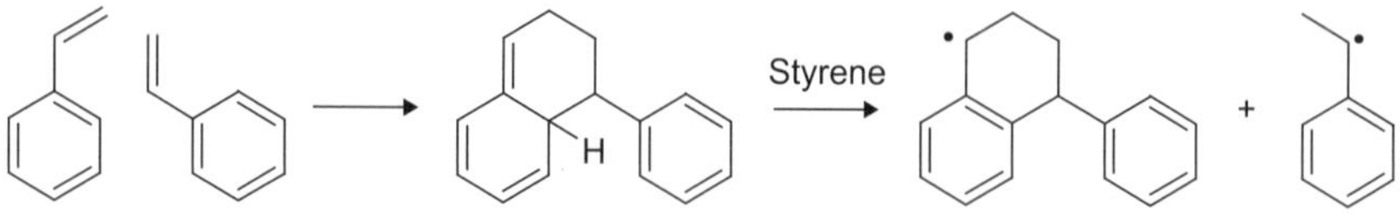

Scheme 1.7

$$Fe^{2+} + RO\!-\!OR \longrightarrow Fe^{3+} + RO^- + RO^\bullet$$

Scheme 1.8

A prominent example for the latter is the self-initiation of styrene, which proceeds *via* Diels–Alder reaction of two monomers, as depicted in Scheme 1.7.[7] Such self-initiated polymerization processes are typically limited to elevated temperatures, and can often be prevented under very pure conditions. Few monomers are capable of self-initiation even under very pure conditions, one of which is styrene, reacting *via* a self Diels–Alder cycloaddition mechanism.[8–10] The self-initiated bulk polymerization of styrene has a substantial activation energy: a 50% monomer conversion needs 400 days at 29 °C, but only 4 h at 127 °C. However, the produced polystyrene is very pure due to the absence of initiators and other additives.

1.2.4 Redox-initiation

Redox-initiation is most frequently used in polymerizations in aqueous systems but may be used in organic solvents as well. A redox-initiator consists of an oxidizing and a reducing agent. In most redox initiators, the redox reaction leads to the formation of only one radical, avoiding cage termination processes and thus enhancing the initiator efficiency. As an example, Scheme 1.8 shows the radical forming reaction in an iron-(II)-peroxide system.

For more information on redox-initiation, the reader is referred to more specialized literature, for example the review article of Sarac[11] and the sources cited therein.

1.3 Propagation

The propagation step, that is the addition of another monomer unit to a macroradical, can be described by a rate law expression as shown in eqn (1.13).

$$-\frac{dc_M}{dt} = k_p\, c_{R^\bullet} c_M \tag{1.13}$$

where k_p is the rate coefficient of propagation of a macroradical. It is well known that up to high monomer conversions, the propagation step is controlled chemically. Therefore, the rate coefficient is independent of monomer conversion as long as it stays lower than about 80%. The chemical control of the propagation reaction becomes obvious when comparing the frequency of propagation reactions, typically on the order of $10^3 \, s^{-1}$, to the average collision frequency in a liquid at room temperature, which is about $10^{12} \, s^{-1}$, meaning that only one in 10^9 collisions is reactive.

1.3.1 Chain Length Dependence

While the rate coefficient of propagation does not depend on conversion (except in the high viscosity regime), it is doubtlessly dependent on chain length. The first few addition steps are particularly fast and may reach rate coefficients many times the long chain limit, as in the case of methyl methacrylates, where the first propagation step at 60 °C exceeds the long chain limit by a factor of about 16.[12–14] It has also been found that the product of monomer concentration and propagation rate coefficient is strongly dependent on chain length up to several hundred monomer units.[15–18] Since currently no method is known to determine the propagation rate coefficient independently of the concentration of monomer, the apparent chain length dependence of k_p might actually reflect a structuring of monomer concentration in the surrounding of the propagating chain end, rather than a chemical effect.

1.3.2 Monomer Effects

The absolute value of the propagation rate coefficient is determined by the reactivity of the propagating radical as well as by the properties of the monomer unit. The relationship between monomer reactivity and reactivity of the propagating radical is roughly reciprocal, *i.e.* a very reactive monomer corresponds to a rather unreactive radical and *vice versa*.

An important condition for chain growth is that the macroradical is sufficiently stable, which means it has to survive a large number of ineffective collisions with monomer or solvent in order to finally react with a monomer molecule. Any possible decomposition or side reaction has to be sufficiently slow so that it will not compete with propagation.

When going from ethene as the simplest possible monomer for radical polymerization to monomers with higher reactivity, one or more hydrogen atoms are formally substituted by groups activating the double bond and stabilizing the macroradical. Of course, steric hindrance is added at the same time, so a complete separation of entropic and enthalpic effects on the propagation rate seems impossible.

Nevertheless, a certain degree of separation is achievable by comparing Arrhenius parameters of different monomers, as shown in Scheme 1.9. It is found that electronic changes are reflected mainly by the activation energy E_A, while steric effects tend to change the pre-exponential factor A. In comparison with methyl methacrylates (MMA), dimethyl itaconate (DMI)[19,20] shows a considerably lower pre-exponential factor, while the activation energy is almost identical.

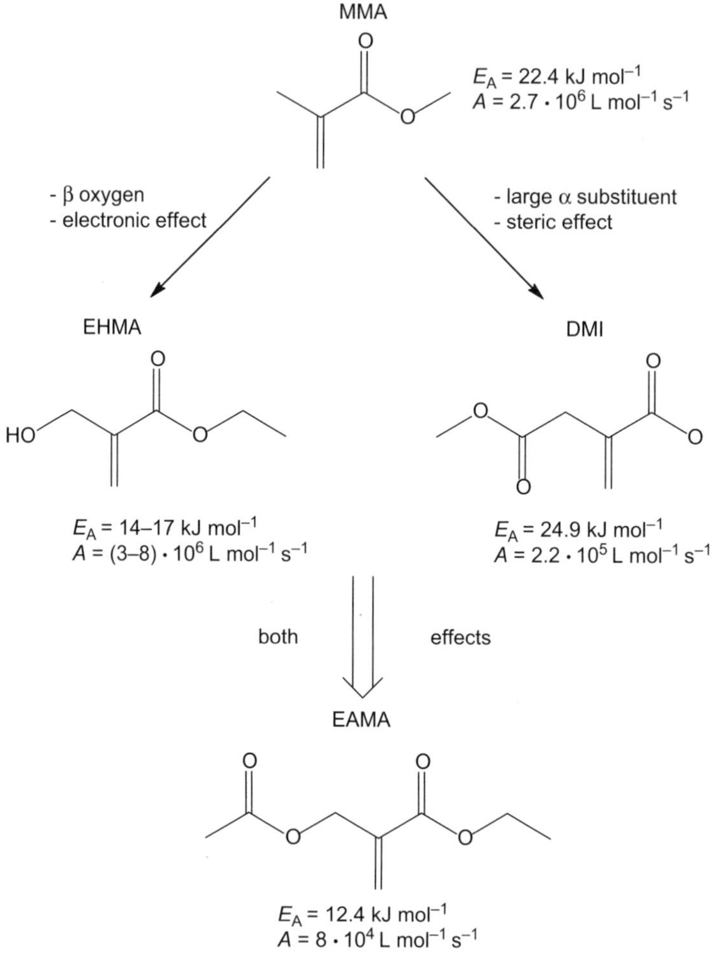

MMA

E_A = 22.4 kJ mol^{-1}
A = 2.7 · 10^6 L mol^{-1} s^{-1}

- β oxygen
- electronic effect

- large α substituent
- steric effect

EHMA

DMI

HO

E_A = 14–17 kJ mol^{-1}
A = (3–8) · 10^6 L mol^{-1} s^{-1}

E_A = 24.9 kJ mol^{-1}
A = 2.2 · 10^5 L mol^{-1} s^{-1}

both   effects

EAMA

E_A = 12.4 kJ mol^{-1}
A = 8 · 10^4 L mol^{-1} s^{-1}

Scheme 1.9

This effect is attributed to the greater sterical hindrance in DMI. If MMA is compared to ethyl α-hydroxy methacrylates (EHMA) instead, the activation energy is reduced due to the electronic effect of the additional β-oxygen atom, while the pre-exponential factor remains unchanged. In a monomer that combines both effects in one molecule, like ethyl α-acetoxymethyl acrylate, a decrease in both Arrhenius parameters is observed (compare Scheme 1.9).

Structurally similar monomers often display similar values for the propagation rate coefficient. For example, acrylate monomers with linearly increasing ester groups, as in methyl, ethyl, propyl acrylate, and so on, have virtually identical k_p values. The same is true for a series of methacrylate systems. Data on rate coefficients and Arrhenius parameters for a large range of common monomers can be found in the literature.[21] Selected values are presented in Table 1.1. These values cover data up to 2000 and were obtained

Table 1.1 Activation parameters for the propagation step for various monomers, obtained *via* PLP-SEC, and according k_p values at 60 °C. All values from ref. 147 unless otherwise indicated.

Monomer	E_A/kJ mol^{-1}	A/L mol^{-1} s^{-1}	k_p at 60 °C/L mol^{-1} s^{-1}
Methyl methacrylate	22.3	2.65×10^6	833
Ethyl methacrylate	23.4	4.07×10^6	873
Butyl methacrylate	22.9	3.80×10^6	976
Isodecyl methacrylate	20.8	2.19×10^6	1590
Dodecyl methacrylate	21.0	2.51×10^6	1280
2-Ethylhexyl methacrylate	20.4	1.87×10^6	1190
Cyclohexyl methacrylate[28]	23.0	6.29×10^6	1560
Benzyl methacrylate[28]	22.9	6.83×10^6	1750
Glycidyl methacrylate[28]	22.9	6.19×10^6	1590
Isobornyl methacrylate[28]	23.1	6.13×10^6	1460
Hydroxyethyl methacrylate	21.9	8.89×10^6	3270
Hydroxypropyl methacrylate	20.8	3.51×10^6	1900
3-[Tris(trimethylsilyloxy)silyl]propyl methacrylate[29]	19.9	1.44×10^6	1092
2-Ethoxyethyl methacrylate[30]	24.1	5.4×10^6	899
Poly(ethylene glycol) ethyl ether methacrylate[30]	24.4	9.3×10^6	1390
Dimethyl itaconate[19]	24.9	2.20×10^5	27
Dicyclohexyl itaconate[31]	22.0	1.74×10^4	6
Methyl acrylate[32]	18.5	2.5×10^7	3.14×10^4
Butyl acrylate[24]	17.9	2.21×10^7	3.45×10^4
Dodecyl acrylate[a]	15.8	1.09×10^7	36400
Isobornyl acrylate[23]	17.0	1.12×10^7	2.42×10^4
tert-Butyl acrylate[23]	17.5	1.90×10^7	3.43×10^4
1-Ethoxyethyl acrylate[23]	13.8	6.3×10^6	4.32×10^4
2-Ethylhexyl acrylate[32]	15.8	9.1×10^6	3.03×10^4
Styrene	32.5	4.27×10^7	341
p-Me-Styrene	32.4	2.84×10^7	236
p-Cl-Styrene	32.1	4.48×10^7	415
p-F-Styrene	32.0	3.50×10^7	336
Vinyl acetate	20.4	1.49×10^7	9460
Acrylonitrile[26]	15.4	1.79×10^6	6890
N-Vinylcarbazole[33]	25.3	1.0×10^8	1.08×10^4
N-Vinylindole[34]	17.5	8.49×10^4	153

[a]Experiments carried out at 100 bar.

via pulsed-laser polymerization size-exclusion chromatography (PLP-SEC),[22] which is the IUPAC-recommended and most accurate kinetic method for determining k_p values. Since then, many other k_p values have been obtained and published, *e.g.*, for bulky acrylates,[23] butyl acrylate,[24] for methacrylic acid,[25] for acrylonitrile[26] and for radical polymerization from solid surfaces,[27] to name but a few.

It is important to note here that there is sometimes more than one type of propagating radicals in one monomer system. These radicals may exhibit significantly different reactivities. This is true, for instance, for acrylates, vinyl acetate or ethylene, where a backbiting reaction (intramolecular transfer for polymer, see below) induces the transformation of secondary propagating radicals (SPR) to tertiary mid-chain radicals (MCR). These two radicals may have very different k_p values, differing by orders of magnitude.[24] The ratio of these two types of radicals depends on the reaction temperature[35,36] and will impact the overall rate of propagation, which is lower than expected for pure terminal radicals. Assuming that backbiting is the only cause of MCR formation, an effective propagation rate k_p^{eff} may be calculated when the backbiting rate k_{bb} (see below) and monomer addition rate of SPR and MCR are known:[24,37]

$$k_p^{eff} = \frac{k_p^{SPR}}{\dfrac{k_{bb}}{k_p^{MCR} c_M} + 1} \qquad (1.14)$$

For most monomers, a strong pressure dependency of the propagation rate coefficient is observed, with k_p increasing at higher pressures. The termination rate coefficients, on the contrary, typically show a decrease with increasing pressure, leading to an increased overall polymerization rate at elevated system pressures (see below).

1.3.3   Solvent Effects

The solvent usually has no large influence on the propagation rate, since the reaction is under chemical control up to high conversions. Studies of the solvent dependence of propagation rates generally found only minor effects.[38–40] These are rationalized by assuming a difference in monomer concentration between the solution and the surrounding of the growing chain end, depending on the difference of molar volumes of monomer and solvent.[41]

In contrast to these relatively weak influences of many solvents, water has a pronounced impact on aqueous radical polymerization. It is frequently found that the polymerization rates of water-soluble monomers in aqueous solutions are higher than in organic solvents.[42] PLP-SEC studies into different systems (methacrylic acid (MAA)[43–45] acrylic acid (AA),[46–48] *N*-isopropyl acrylamide,[49] acrylamide,[50] and *N*-vinyl pyrrolidone[51] revealed a huge solvent effect on k_p. Figure 1.2 demonstrates this effect on the example of *N*-vinyl pyrrolidone.

Nonionized methacrylic acid (MAA), for instance, has been studied within the entire concentration range from bulk to dilute aqueous solution.[44] Hydrogen-bonded interactions between the propagating MAA macroradical and an environment that, depending on the particular MAA concentration, consists of different relative amounts of MAA and water molecules do not significantly affect the activation energy for propagation reaction, $E_A(k_p)$, but largely influence the pre-exponential factor, $A(k_p)$ by about a factor of 10. This is assigned to the increased friction that the relevant degrees of rotational motion of the transition state structure experience[52] upon replacing

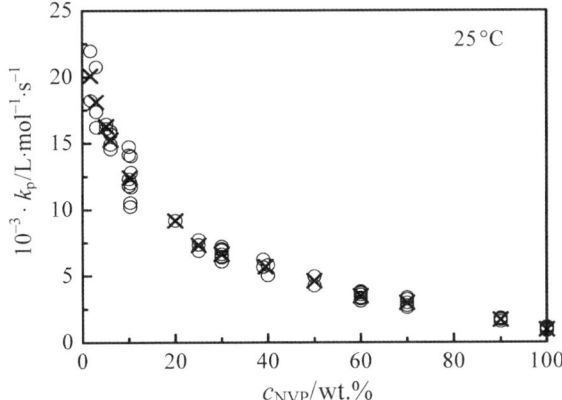

Figure 1.2 Variation of k_p with monomer concentration for polymerizations of *N*-vinyl pyrrolidone at 25 °C in aqueous solution (open circles). Crosses show arithmetic mean values for individual concentrations.
(Reprinted with permission from: M. Stach *et al.*, *Macromolecules*, 2008, **41**, 5174. Copyright 2008 American Chemical Society.)

H_2O by MAA molecules. This finding for MAA appears to be a general kinetic effect for hydrogen-bonded monomers in aqueous solution. It is, however, not consistent with the data that the variation in k_p is due to a deviation in local monomer concentration at the site of the propagating radical. For protic monomers, like MAA and acrylic acid (AA), polymerizations at different degrees of ionic dissociation revealed that ionization also strongly affects k_p, which also can be explained by the above-mentioned entropic effect, which primarily acts on the Arrhenius pre-exponential factor.[53] For both AA and MAA, a decrease of k_p with increasing degree of dissociation is observed. A solution of 5 wt% of monomer in water, at a temperature of 6 °C, shows a decrease of k_p by about an order of magnitude in the range $\alpha = 0–1$.[54]

In the presence of ionic liquids, radical polymerization rates and polymer molecular weights have also been reported to be significantly enhanced as compared to reaction in conventional organic solvents or in bulk.[55–60] This is partly due to an increased k_p value when using this specialty solvents. By applying the PLP-SEC method it was found that k_p for methyl methacrylate (MMA) in mixtures containing 50 vol% 1-butyl-3-methylimidazolium hexafluorophosphate ([bmim]PF$_6$) is about a factor of 2.5 above the bulk polymerization value.[61,62] The k_p in ionic liquid solution containing only 20 vol % MMA was found to be enhanced even by a factor of 4 in comparison to the bulk value.[63] The analysis of the k_p data from PLP-SEC experiments on two methacrylate monomers dissolved in four ionic liquids suggests that the observed, rather pronounced solvent effects are primarily due to a lowering of the activation energy for propagation upon gradually replacing monomer molecules by ionic liquid species (Figure 1.3).

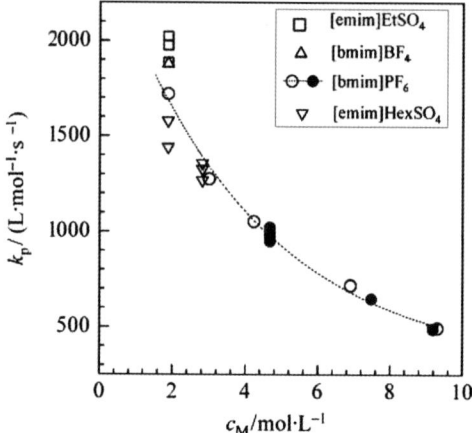

Figure 1.3 Variation of k_p with monomer concentration for MMA polymerizations in several ionic liquids at 40 °C. The dotted line serves as a visual guide. (Reprinted with permission from: I. Woecht *et al.*, *Journal of Polymer Science Part. A: Polymer Chemistry*, 2008, **46**(4), 1460. Copyright 2007 John Wiley and Sons.)

1.4 Transfer

Models that include only initiation, propagation, and termination reactions generally predict higher average molecular weights than experimentally observed. The main reason for this discrepancy is stopped chain growth due to transfer reactions. A transfer reaction is a reaction that transfers the radical function from a chain end to another molecule, as shown in Scheme 1.10.

A transfer agent T contains a weakly bonded atom X that can be abstracted by a propagating macroradical $R_i^{\bullet}$ of chain length i, thus generating a dead polymer molecule and a new radical $T^{\bullet}$. The newly formed radical might initiate new chain growth or undergo a termination reaction with some other radical. The transfer agent might be a purposefully added molecule, but transfer to monomer, solvent or initiator are well-known and mostly undesirable side reactions. The rate law for a general transfer reaction is given by the following equation

$$-\frac{dc_T}{dt} = k_{tr} c_{R^{\bullet}} c_T \tag{1.14}$$

where k_{tr} is the rate coefficient of the transfer reaction and c_T and $C_{R^{\bullet}}$ are the concentrations of the transfer agent and the macroradical, respectively. It is common to express the influence of transfer reactions as the transfer constant C, which is the ratio of the transfer rate coefficient to the propagation rate coefficient:

$$C = \frac{k_{tr}}{k_p} \tag{1.15}$$

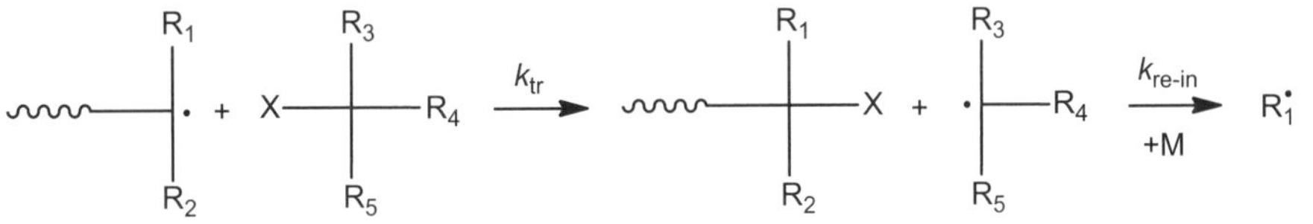

Scheme 1.10

A large collection of data on initiator properties has been published in the *Polymer Handbook*.[1] In special cases, transfer has a measurable impact on the overall reaction kinetics, *i.e.*, the polymerization rate R_p depends on these two rate coefficients as well as the coefficient $k_{re\text{-}in}$ for re-initiation, meaning the initiation of a new polymerizing chain by the transfer agent radical. Four different scenarios may occur, depending on the relative sizes of the rate constants:

(i) $k_p \gg k_{tr}$ and $k_{re\text{-}in} \approx k_p$

This is the most common case. The average molecular weight is reduced, while the transfer reaction has no significant influence on the polymerization rate.

(ii) $k_p \gg k_{tr}$ and $k_{re\text{-}in} < k_p$

In this case, the transfer reaction displays an inhibitory effect, slowing the polymerization reaction and decreasing the average molecular weight.

(iii) $k_p \ll k_{tr}$ and $k_{re\text{-}in} \approx k_p$

Telomerization is observed, the molecular weight is severely reduced while the overall reaction rate remains unaltered.

(iv) $k_p \ll k_{tr}$ and $k_{re\text{-}in} < k_p$

Degradative chain transfer occurs: both the average molecular weight and the overall reaction rate are decreased.

(These scenarios will also be addressed within the context of inhibition and retardation, see below.) Although transfer can be an inconvenient side reaction, it can also be used as a tool to deliberately influence a polymer's molecular weight distribution, for example in a case (i) scenario, where the overall reaction kinetics is not disturbed, because the number of free radicals remains unaltered, but the molecular weight distribution is shifted to a lower average weight. Case (i) is also assumed in the justification of the Mayo equation, which is commonly employed for determination of transfer rate coefficients (see below).

1.4.1 Transfer to Monomer

Only one type of transfer reaction is unavoidable, and that is transfer to monomer. In absence of other transfer reactions, transfer to monomer sets an upper limit to achievable chain lengths. Typical rate coefficients for this process are in the range of 3×10^{-5} and 20×10^{-5} L mol^{-1} s^{-1}, which thankfully is slow

Table 1.2 Transfer constants to monomers, C_M.

Monomer	$\theta/\,°C$	$C_M \times 10^5$	Reference
Acrylamide	60	6.0	64
Acrylonitril	60	3.3–10.2	65
1-Butene	60	73	
Butyl acrylate	60	3.33–12.5	
o-Chlorostyrene	50	2.5–2.8	
Ethyl acrylate	60	5.79	
Ethylene	60	4–42	66
	110	50–53.2	
	130	16–112	
Methacrylonitrile	60	58.1	
Methyl acrylate	60	0.36–3.25	67
Methyl methacrylate	0	1.28–1.48	
	30	1.17–2.6	
	50	5.15	68
	100	3.8	
α-Methylstyrene	50	412	68
Styrene	25	3.5	69
	60	7.8–8.7	
	90	15–16.5	
Vinyl acetate	0	5.0–9.6	
	25	9.0–107	
	40	12.9–13.2	
	60	18	70
Vinylidene chloride	60	380	71

[a]Data from Polymer Handbook,[1] unless other reference is noted.

enough to allow for appreciable polymer growth. It is not possible to reduce the influence of transfer to monomer by lowering the monomer concentration, because both the transfer rate and the propagation rate are proportional to monomer concentration, so it cancels out in the ratio. Nevertheless, the activation energies of propagation and transfer to monomer are generally different, typically being greater for the transfer reaction, so a decrease in temperature typically results in a reduced monomer transfer constant C_M. Table 1.2 lists C_M values for a selection of usual monomers.

For monomers containing aliphatic hydrogen atoms, the transfer reaction usually occurs *via* abstraction of one such hydrogen atom. Ethene is an example for a monomer not containing aliphatic hydrogens. Nevertheless, transfer occurs also by abstraction of a hydrogen atom from a monomer molecule by a propagating radical, as shown in Scheme 1.11.

In the case of styrene, direct hydrogen abstraction is unlikely, since the chain transfer coefficient of ethylbenzene is significantly lower than that of styrene. So the hydrogen is rather abstracted from a Diels–Alder adduct of two styrene molecules, where the carbon–hydrogen bonds are comparatively weak.

In the case of vinyl chloride, transfer proceeds *via* chlorine abstraction: a monomer molecule adds head-to-head to a propagating radical. The resulting radical is highly unstable and reacts with another monomer molecule in a chlorine abstraction step.[72] This special reaction pathway is the reason for C_M

$$H_3C\!-\!CH_2^{\bullet} \; + \; H_2C\!=\!\!CH_2 \longrightarrow H_3C\!-\!CH_3 \; + \; H_2C\!=\!\!CH^{\bullet}$$

Scheme 1.11

Table 1.3 Transfer coefficients to initiators, C_I, at 60 °C unless stated otherwise.

Initiator	C_I styrene	C_I methyl acrylate	C_I MMA
2,2′-Azobis (isobutyronitrile)	0.09–0.14[73]	—	0.02[74]
tert-Butyl peroxide	2.3–6.0 × 10⁻⁴ [75]	4.7 × 10⁻⁴(65 °C)[67]	1×10⁻⁴ (20 °C)[76]
2-Butanone peroxide	0.46 (50 °C)	0.05 (65 °C)	2.5–6.98 × 10⁻³ (65 °C)
tert-Butyl hydroperoxide	0.035	0.01	—
Ethyl peroxide	6.6 × 10⁻⁴	—	—
2,2′-Azobis(2,4,4-trimethyl valeronitrile)	0.59 (25 °C)	—	—
Benzoyl peroxide	0.101[77]	0.0246[67]	0.02[78]

[a] Data from Polymer Handbook,[1] unless other reference is noted.

being much greater for vinyl chloride than for most other industrially important monomers.

Allylic compounds ($CH_2\!=\!CHCH_2X$) are usually reluctant to homopolymerization, due to the activation of the allylic hydrogen atom towards abstraction. The generated allylic radical is highly stabilized by both the substituent X and by delocalization of the free electrons into the double bond. These radicals add to monomer very slowly and perform side reactions that in turn lead to retardation.

1.4.2 Transfer to Initiator

Except in the rare case of self-initiated polymerization, transfer to initiator is also unavoidable, but as long as initiator concentration is kept low, its impact on overall reaction kinetics is small. For some common initiators, transfer constants C_I are given in Table 1.3.

1.4.3 Transfer to Solvents or Transfer Agents

Since the possibilities of influence on transfer to monomer and transfer to initiator are very limited, interest is often focused on the transfer to solvent reaction. In typical industrial polymerization processes, the solvent is the major component of the reaction mixture, so even comparatively slow reactions gain importance, due to the high concentration. For many solvents, transfer rate coefficients are similar to those of common monomers given above, but there are exceptions, like tetrachloromethane, showing considerably higher transfer constants.

Solvents displaying very high transfer constants cannot be used as solvent for a polymerization reaction mixture, but can be useful as a purposefully added transfer agent. For a transfer constant greater than unity, even small amounts of the respective substance can significantly influence molecular weights, making it a useful tool to control viscosity and thus enhance heat flow in an industrial process.

Another means to control the average molecular weight is to increase the initiator concentration. But this has the drawback of a considerable increase in polymerization rate, which might lead to uncontrollable reaction conditions. Examples for transfer agents displaying very high transfer constants are thiols and compounds with high halogen contents, like tetrabromomethane. If large quantities of such efficient transfer agents are applied, this results in telomerization, that is, the formation of extremely short-chained products, mainly consisting of di- and trimers. Table 1.4 gives an overview of some important transfer agents.

1.4.3.1   Thiols

Thiols may be used as transfer agents in a wide variety of free radical polymerization processes.[85] Scheme 1.12 shows the general reaction mechanism for this class of transfer agents. Nucleophilic radicals react more readily with thiols than electrophilic radicals, so transfer coefficients are higher for vinyl esters and styrene than for acrylates and methacrylates. Aromatic thiols react more readily than aliphatic ones, *i.e.*, the chain transfer constant is higher, but they also show a stronger retardation effect as the resulting S-centered radicals are less prone for monomer addition due to their increased stability. The product of the transfer reaction is a thiyl radical, which is electrophilic and will react preferably with the more electron rich monomer in copolymerizations.

It is feasible to use functionalized thiols as transfer agents to prepare end-functional polymers. Such materials have successfully been used in the synthesis of block and graft copolymers.

1.4.3.2   Sulfides and Disulfides

A broad spectrum of organic disulfides has been employed as transfer agents. Among these are dialkyl, diaryl and diaroyl disulfides as well as xanthogens. For monomers such as styrene or methacrylates, transfer coefficients of disulfides are very low, but in the polymerization of vinyl esters they may come close to unity. For xanthogens and thiurams, transfer coefficients are usually higher than for disulfides, which is commonly explained by their *iniferter* capabilities.[86,87]

Monosulfides are less effective transfer agents than disulfides, due to steric reasons and because of the greater strength of the C–S bond compared to the S–S bond.

1.4.3.3   Catalytic Chain Transfer

Some metal complexes show the ability to catalyze transfer to monomer reactions, so they lead to very high transfer coefficients without being consumed

Table 1.4   Transfer coefficients to solvents and additives, C_T, at 60 °C.[a]

Transfer agent	$C_T \times 10^4$ Styrene	Methyl methacrylate
2-Aminoethanthiol hydochloride	—	1100[79]
2-Butanone	4.98	0.45
Acetaldehyde	8.5	6.5
Acetic acid	2.22 (40 °C)	0.24 (80 °C)
Acetone	0.32[80]	0.195
Acetonitrile	0.44	—
Aniline	2.0	4.2
Benzaldehyde	4.5–5.5	0.86–2.5
Benzene	0.01–0.04	0.04–0.83
Benzenesulfonyl chloride	4330	5
Benzenethiol	—	27 000
Carbon tetrabromide	2 500 000[81]	1500–2700
Carbon tetrachloride	69–148	0.5–20.11
Chloroform	0.41[80]	0.45–1.77
Copper(II) chloride	10^8	2×10^7
Cumene	0.8–3.88	1.9–2.56
Cyclohexane	0.024–0.063	0.1–0.2 (80 °C)
Diethyldisulfide	45 (99 °C)	1.3
Dibenzyldisulfide	100	63
Diphenyldisulfide	1500	110
Ethyl acetate	15.5	0.1–0.46
Ethyl ether	5.64	—
Ethyl iodoacetate	8000	—
Ethyl bromoacetate	430	—
Ethyl tribromoacetate	100 000	—
Ethyl trichloroacetate	100	—
Heptane	0.42	1.8 (50 °C)
Iron(III)chloride	5 360 000	4000
Isopropanol	3.05	0.583
Methanol	0.296–0.74	0.2
Mercaptoacetic acid methyl ester	14 000[82]	3000[82]
N,N-dibenzylhydroxylamine	5000	—
N,N-dimethyl acetamide	4.6	—
N,N-dimethyl formamide	4.0	—
n-Butanol	1.6	0.394
n-Butanethiol	220 000	6600
n-Dodecanethiol	150 000	9700–12 300[83]
Pentaphenylethane	20 000	—
Phenyl ether	7.86	9.13
Pyridine	0.6	0.176 (70 °C)
Tetraethylthiuram	3200[84]	—
Tetrahydrofurane	0.5 (50 °C)	—
Toluene	0.105–2.05	0.17–0.45
Trichlorotoluene	57.5	—
Triethylamine	1.4–7.5	8.3
Water	0.006–0.31	—

[a]Data from Polymer Handbook,[1] unless other reference is noted.

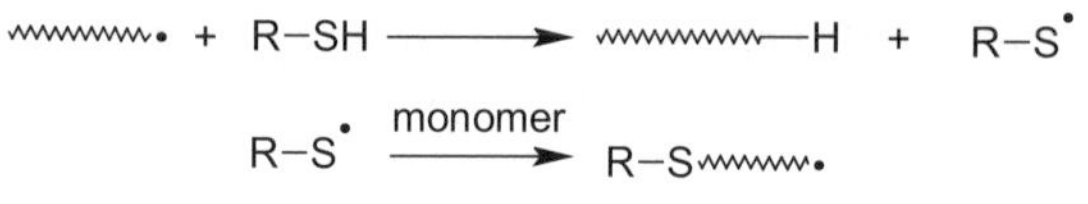

Scheme 1.12

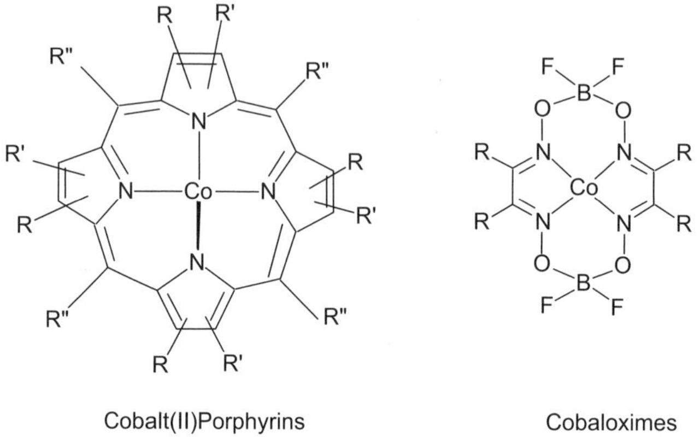

Cobalt(II)Porphyrins   Cobaloximes

Scheme 1.13

in the transfer reaction or built into the polymer chain.[88–90] The transfer catalysts that are most commonly employed are based on low-spin cobalt macrocycles.[91] Scheme 1.13 shows some typical structures of such complexes.

While catalytic chain transfer is very efficient for tertiary radicals, as for example in methyl methacrylate polymerization, its efficiency is much lower for secondary radicals. This is explained by cobalt–carbon bonding as a side reaction, which significantly reduces the concentration of the catalytically active species. Adducts containing cobalt–carbon bonds have been detected *via* matrix-assisted laser desorption ionization (MALDI) mass spectrometry.[92]

The mechanism is thought to operate as shown in Scheme 1.14 for the example of methyl methacrylate. A β-hydrogen atom is abstracted by the cobalt complex, thus forming a terminally unsaturated polymer chain and a cobalt hydride species that re-initiates a monomer unit by hydrogen atom transfer.

Catalytic chain transfer has become a useful tool in commercial synthesis, allowing for the production of low molecular mass polymers with terminal double bonds. Such materials may be used as chain transfer agents for block structure synthesis or as macromonomers in the synthesis of comb- or star-shaped polymers.

1.4.3.4   Alkyl Halides

Alkyl halides may also be used as transfer agents in radical polymerization processes. Their reactivity depends not only on the nature of the halogen atom and the alkyl unit, but also on the character of the propagating radical.

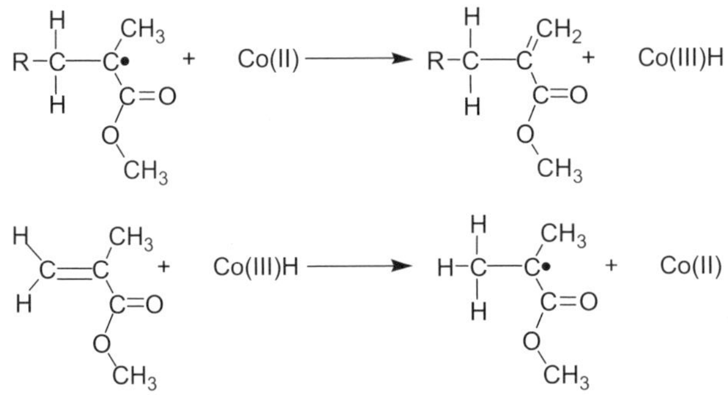

Scheme 1.14

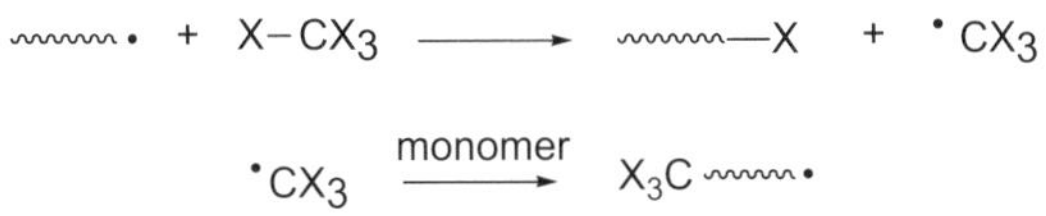

Scheme 1.15

For nucleophilic radicals like styrene, the reactivity is much higher than for electrophilic radicals like acrylates. The haloalkyl radicals that are formed in the transfer reaction are, in turn, electrophilic in nature.

The transfer coefficients increase with increasing period of the halogen, $Cl < Br < I$, and also increases with the number of halogen atoms. For example, in the polymerization of styrene, transfer coefficients for iodoacetic acid, bromoacetic acid and chloroacetic acid are $C_I = 0.8$, $C_{Br} = 0.043$ and $C_{Cl} = 0.020$, respectively, whereas the values for dibromo- and tribromoacetic acid are $C_{Br2} = 0.27$ and $C_{Br3} > 10$. Halogenated organic acids and their esters and halomethanes are the most commonly employed groups of alkyl halides.

The application of high concentrations of effective transfer agents leads to the exclusive formation of oligomeric species, also known as telomerization. Both halomethanes and halogenated esters have found widespread application in the synthesis of telomers, due to their low cost and high transfer efficiencies. The kinetics of 35elomerisation reactions are rather difficult, since the reactivities of the various involved radical species scatter over a broad range. The tetrahalomethanes react according to the mechanism shown in Scheme 1.15, leading to a trihaloalkyl radical that may re-initiate polymerization. Of course, the polymer chain started by the trihalogenomethyl radical may again take part in a transfer reaction, but the reactivity decreases rapidly with the number of halogen atoms, as mentioned above.

In the case of hydrohalomethanes, two alternative routes are possible for the transfer reaction, either halogen- or hydrogen-atom abstraction. The hydrogen transfer is favoured for chloroform, because the C–H bond is weaker than the C–Cl bonds.

1.4.3.5   Transition Metal Halides

Transition metal halides can also act as transfer agents. For example, copper(II) chloride or iron(III) chloride may be applied. Transfer coefficients for these two halides have been determined in DMF at 60 °C. For copper(II) chloride, the transfer coefficients are $C = 10^4$, $C = 10^3$ and $C = 10^2$ for styrene, MMA and acrylonitrile, respectively.[93] Iron(III) chloride is less efficient and gives values of $C = 626$, $C = 306$, $C = 86$, $C = 4$ and $C = 2$ for vinyl acetate, styrene, vinyl chloride, MMA and acrylonitrile, respectively.[94] The transfer process forms alkyl halides and metal species in lower oxidation states, the latter occasionally being capable of activating the former. Such a mechanism of reversible activation and deactivation is utilized in atom transfer radical polymerization (ATRP).

1.4.4   Transfer to Polymer

Transfer to polymer occurs *via* intra- or intermolecular abstraction of a hydrogen from the polymer backbone by the chain-end macroradical. By this transfer reaction, the number of growing chains is not increased. Hence, the average molecular mass of the obtained polymer is not intrinsically lowered by transfer to polymer. A lowering of molecular mass, as may be expected from the ideal kinetic chain-length (ratio between propagation and termination rate) in conjunction with chain-end radical k_p, is generally observed due to the decreased propagation reactivity of radical species (MCR) produced by transfer to polymer (see above). Further, short- and long-chain branches are introduced in the chains. Transfer to polymer is thermodynamically favored as the position of the radical at the chain-end is less stabilized as compared to a radical position along the polymer backbone. In ethene polymerization, the driving force comes from the transformation of a primary chain-end radical into a secondary mid-chain radical.[95,96] Thus, the transfer to polymer rate constants are considerably higher (approximately by a factor of 10) than those observed for the corresponding monomer.

Transfer coefficients to polymers are not as readily determined as other transfer coefficients because the process does not necessarily lead to a reduction of the molecular weight. Polymerization in the presence of polymer yields a mixture of the polymer initially present and the new polymer formed, and thus the decrease in the molecular weight cannot be accurately evaluated. However, chain transfer coefficients to polymers are accessible *via* the structural investigations of the generated branched macromolecules, *e.g. via* NMR.[97,98] An alternative approach utilizes the reversible addition-fragmentation chain transfer (RAFT) mechanism *via* modelling of Z-RAFT star polymerizations, in which long chain branching can be detected selectively *via* RAFT-group absorbance of a star-star couple containing two living cores (see Scheme 1.16)[99] By this approach, the rate coefficient of *inter*molecular transfer to polymer at 60 °C was estimated to be $k_{tr}^{P,inter} = 0.33$ L mol^{-1} s^{-1} in butyl acrylate polymerization and $k_{tr}^{P,inter} = 7.1$ L mol^{-1} s^{-1} in dodecyl acrylate polymerization.

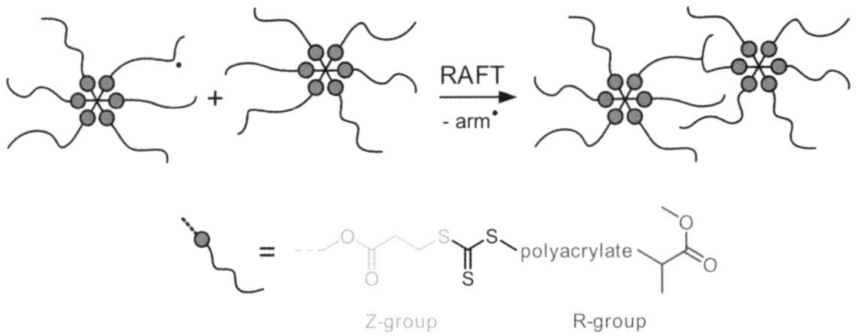

Scheme 1.16

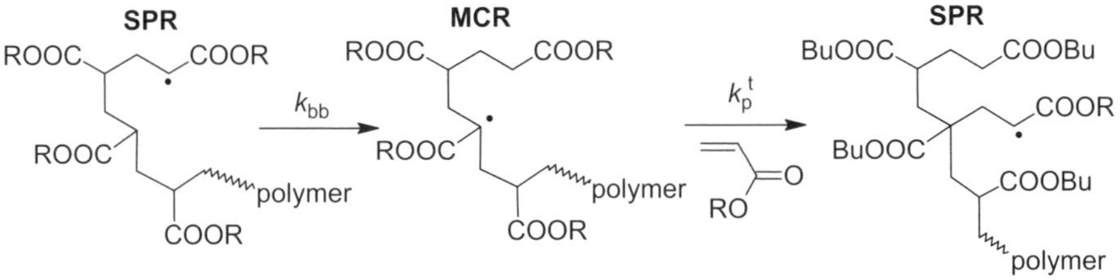

Scheme 1.17

As mentioned above, it is known that *intra*molecular chain transfer, in particular, 1,5-hydrogen shift, does also occur during the polymerization of monomers that yield very reactive macroradicals, such as acrylates[100–103] and acrylic acid.[104] This so-called backbiting reaction, by which a secondary radical (SPR) is transformed into a more stabilized tertiary (MCR) one, proceeds *via* a six-membered cyclic transition state with rate coefficient k_{bb} (see Scheme 1.17). In principle, *intra*molecular chain transfer to a remote chain position and *inter*molecular chain transfer to another polymer molecule may also take place.[105] These latter processes are, however, found to be not significant in butyl acrylate polymerization at low and moderate degrees of monomer conversion and temperature.[106,107]

The overall kinetics is rather complex: tertiary mid-chain radicals, MCRs, are produced by backbiting reactions of secondary propagating chain-end radicals, SPRs. On the other hand, monomer addition to an MCR with rate coefficient k_p^t, again produces an SPR under simultaneous formation of a short-chain branch. SPRs rapidly propagate, undergoing backbiting or termination with another SPR or an MCR with rate coefficients, k_t^{ss} and k_t^{st}, respectively. Propagation from an MCR is approximately two orders of magnitude slower than the one from an SPR. Also MCR homo-termination with rate coefficient k_t^{tt} is slower than SPR homo-termination.

For acrylate and acrylic acid monomers, the polymerization kinetics and polymer properties are affected decisively by the formation of less reactive MCR species. The polymerization rate is significantly lower than would be expected from the ideal process with propagation of SPRs only, as already mentioned above. The fraction of MCRs may be estimated by assuming that dynamic equilibrium has been reached and making a quasi-steady-state assumption on dc_{MCR}/dt.[102]

$$x_{MCR} = \frac{c_{MCR}}{c_{SPR} + c_{MCR}} = \frac{k_{bb}}{k_p^t c_M + 2k_t^{tt} c_{MCR} + 2k_t^{st} c_{SPR} + k_{bb}} \quad (1.15)$$

Derivation of eqn (1.15) neglect other side-reactions such as transfer to monomer and β-scission (see below). By implementing the so-called long-chain hypothesis ($k_p^t c_M \gg 2k_t^{tt} c_{MCR} + 2k_t^{st} c_{SPR}$) into eqn (1.16) yields a simplified expression for the fraction of MCRs:

$$x_{MCR} = \frac{k_{bb}}{k_p^t c_M + k_{bb}} \quad (1.16)$$

Eqn (1.16) is particularly suitable for describing x_{MCR} at high monomer concentrations and higher temperatures, whereas eqn (1.15) is usually a more reliable description for x_{MCR} at low temperatures where high concentrations of SPRs are present. Experimental methods for determination of reliable rate coefficients for k_{bb} are (i) ^{13}C NMR,[108–110] (ii) frequency-tuned (ft)-PLP-SEC[111] and (iii) single-pulse pulsed-laser polymerization coupled with online time-resolved electron-spin resonance spectroscopy (SP-PLP-EPR).[112] The propagation rate coefficient for MCRs may be obtained *via* ft-PLP-SEC[111] and SP-PLP-EPR.[112] Termination rate coefficients $k_t^{s,s}$, $k_t^{s,t}$ and $k_t^{t,t}$ are only accessible from SP-PLP-EPR,[112] in which different types of radicals can simultaneously be traced as a function of time. Remaining kinetic coefficients can then be obtained *via* computer modeling. Table 1.5 collates kinetic coefficients for butyl acrylate polymerization as an example.

Bonds in the β-position to radical functionalities are relatively labile, particularly at higher temperatures. Due to the associated high activation barrier, β-scission reactions may often be neglected below 80 °C. The mechanism of β-scission is illustrated in Scheme 1.18.

Table 1.5   Rate coefficients for BA polymerization in a solution of toluene.[112]

	Pre-exponential factor/ $L\,mol^{-1}\,s^{-1}$ *or* s^{-1}	Activation energy/ $kJ\,mol^{-1}$	k at 50 °C/$L\,mol^{-1}\,s^{-1}$ *or* s^{-1}
$k_t^{ss}(1,1)$	1.3×10^{10}	8.4	5.7×10^8
$k_t^{st}(1)$	4.2×10^9	6.6	3.6×10^8
k_{bb}	1.6×10^8	34.7	3.9×10^2
k_p^t	9.2×10^5	28.3	25

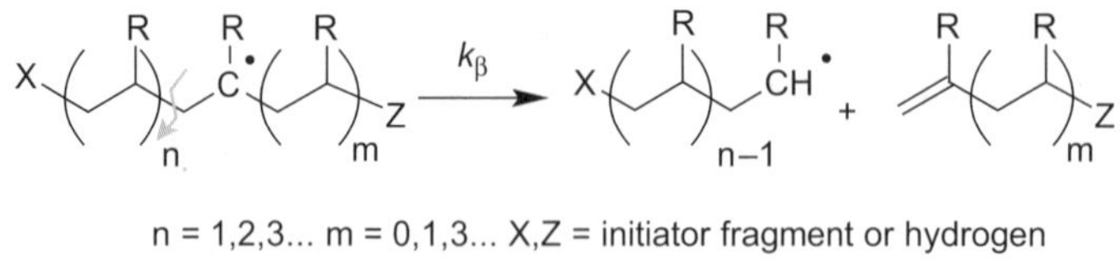

n = 1,2,3... m = 0,1,3... X,Z = initiator fragment or hydrogen

Scheme 1.18

By β-scission, a macroradical is cleaved into a chain-end radical and a double-bond-terminated molecule. In the special case that $m = 0$ and $Z = H$ (see Scheme 1.18), a chain-end radical with degree of polymerization n is cleaved into a similar chain-end radical with degree of polymerization $(n - 1)$ and a monomer molecule. This reaction is often referred to as depropagation, since it is the back reaction of a propagation step. For $m \geq 1$, β-scission of a midchain radical (typically $m \geq 2$ for an MCR produced *via* backbiting) produces an SPR of chain length n and a so-called macromonomer, MM, of chain length m. It needs to be noted that MMs will subsequently add to SPRs during radical polymerizations. Further details on β-scission and macromonomer synthesis are provided in ref. 113, 114 and 115.

1.5 Termination

The termination reaction in free radical polymerization is arguably the most complex reaction step of the polymerization process. The corresponding termination rate coefficient, k_t, is influenced by a variety of different factors, *i.e.*, (i) the system viscosity, (ii) the chain length of the terminating free macroradicals, (iii) the temperature, (iv) the pressure and the (v) monomer conversion. These parameters are very difficult to separate and experimental access to k_t is thus far from being trivial. In addition, many of these parameters vary during the course of a radical polymerization reaction. The enormous scatter of reported termination rate coefficients over several orders of magnitude, as *e.g.* compiled in the *Polymer Handbook*,[1] for the same monomer at the same reaction temperature is a direct manifestation of the influence of these various parameters on the termination event. More recent data of k_t is addressing these influences with greater care and also employ more precise k_p values obtained by PLP-SEC,[22] which are often needed for the determination of k_t. It is for this reason that the quality of k_t data has been improved enormously during the last years.

1.5.1 Combination *versus* Disproportionation

There are two modes of termination: one is the direct coupling (combination) of two free macroradicals yielding a dead polymer chain of chain length $i + j$. The associated rate coefficient is $k_{t,comb}$. The second mode is disproportionation, where a hydrogen atom is transferred from one of the radical chain end to the other radical, giving two dead polymer chains of which one carries a double bond. This reaction is associated with the rate coefficient $k_{t,dis}$. These processes are illustrated in Scheme 1.19 on the example of poly(methyl methacrylate) macroradicals.

The reactions between two carbon-centered radicals generally give a mixture of disproportionation and combination. Which termination mode dominates depends largely on the structure of the monomer unit, but also – however to a lesser extent – on the reaction temperature, pressure and solvent.[116] Disproportionation is (slightly) favoured at higher reaction temperatures. The reasons for this behaviour have yet to be clarified, but there is some evidence pointing

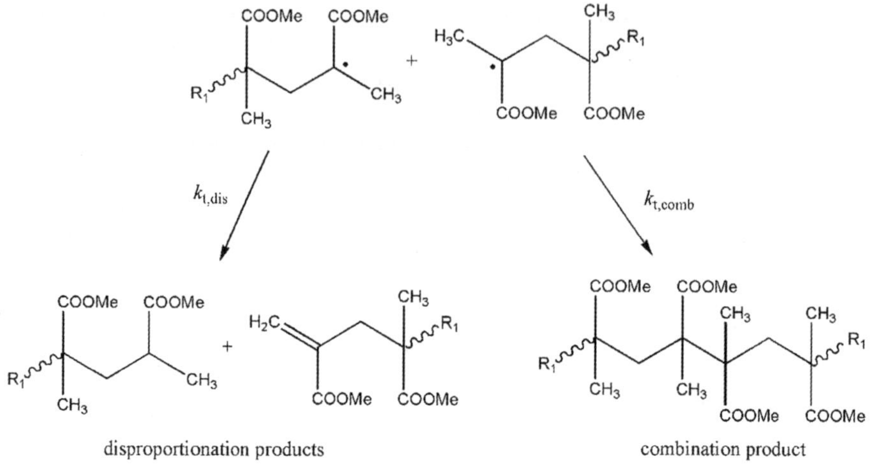

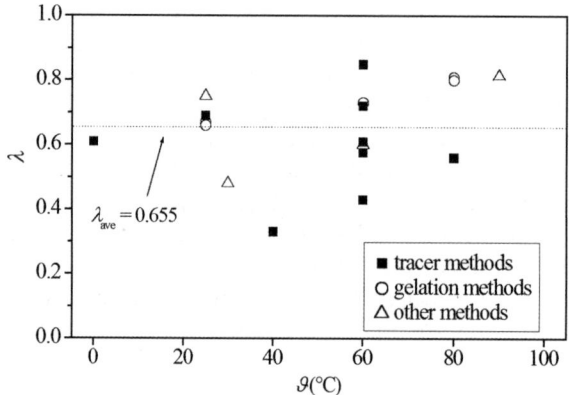

Scheme 1.19

Figure 1.4 graph: λ (y-axis from 0.0 to 1.0) versus $\vartheta(°C)$ (x-axis from 0 to 100). $\lambda_{ave} = 0.655$. Legend: ■ tracer methods, ○ gelation methods, △ other methods.

Figure 1.4   Literature values of λ, the fraction of termination by disproportionation, as a function of temperature for radical polymerization of methyl methacrylate; λ_{ave}, the average of all these values of λ, is also shown. Values were obtained *via* tracer methods, gelation methods and other methods, as indicated.
(Reprinted with permission from ref. 121: M. Buback *et al.*, *Macromolecules*, 2009, **42**, 652. Copyright 2009 American Chemical Society.)

towards a different temperature dependence of the corresponding pre-exponential factors in the Arrhenius expression for $k_{t,comb}$ and $k_{t,dis}$. Combination and disproportionation are clearly two different reactions with two distinct transition states, which is supported by theoretical studies.[117–120] However, the measured effects are rather small and associated with a large experimental scatter, as indicated in Figure 1.4 for the example of the temperature dependence of the contribution of disproportionation to the overall termination process, λ, for a MMA polymerization at ambient pressure.

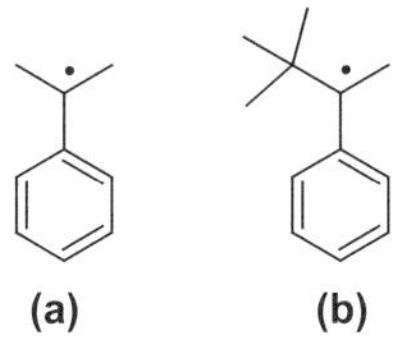

(a)   (b)

Scheme 1.20

$$\lambda = \frac{k_{t,dis}}{(k_{t,dis} + k_{t,comb})} \tag{1.17}$$

This impressively indicates that λ is a very difficult parameter to measure accurately. Recently, the situation was somewhat improved by applying high-resolution mass spectrometry for determining λ, which allows a very precise measurement of end-groups originating from the two termination processes.[121] This method arrived at a value for MMA at 85 °C of $\lambda = 0.63$, which is very close to the average value λ of all data obtained before, which is an indication of this method's accuracy.

For a given series of radicals, λ increases with the number of β-hydrogen atoms. However, there is no direct correlation and other factors are involved.[116,122] In addition, it is generally observed that the extent of disproportionation increases with rising substitution at the radical centre. Steric effects play a very important role as can be demonstrated on the self-reaction of cumyl radicals (Scheme 1.20(a)) and the *tert*-butyl substituted equivalent (Scheme 1.20(b)). The termination reaction of radical (a) shows predominantly combination, whereas radical (b) gives predominantly disproportionation, although there are less β-hydrogen atoms. This finding may imply that the combination reaction is more suppressed by the steric hindrance than the disproportionation reaction. The statistical effect that favours disproportionation when more β-hydrogen atoms are available is hence to be considered less pronounced than the steric effect.[123] The steric crowding can lead in extreme cases to persistent radicals (*e.g.* di-*tert*-butyl methyl radical[124] and triisopropylmethyl radical[125] that are relatively reluctant to perform radical–radical reactions.

The mode of termination has no direct impact on the rate of the free radical polymerization. However, the generated molecular weight distributions are influenced by the termination mode. Since some methods for the determination of the termination rate coefficient rely on the analysis of the full molecular weight distributions, it is mandatory to have reliable data on λ. The termination mode also determines whether one (disproportionation) or two (combination) of the end groups are initiator-derived. λ also determines whether dead polymer produced during a controlled radical polymerization occurs at the same chain length as the "living" polymer (and remains invisible) or occurs as hump at the doubled chain length. Table 1.6 presents selective data on termination modes for various monomers at ambient pressure and various temperatures.

Table 1.6   Contribution of disproportionation to the overall termination, λ, for different monomers.

Monomer	$\theta/\,^\circ C$	λ	Ref.
Acrylonitirile	10–90	0	126–128
Butyl methacrylate	80	0.54	129
Dicyclohexyl itaconate	25	0.8–1.0	31, 130
Ethyl methacrylate	80	0.42	129
Methacrylonitrile	25	0.65	131
Methylacrylate	−34	~0.25	130
Methyl methacrylate	0	0.61	132
	25	0.67	133
	60	0.73	133
	85	0.63	121
	90	0.81	134
α-Methyl styrene	55	0.091	135
Styrene	20–50	0.17	136
	30	0.14	137, 138
	50	0.2	137
	60	0.1–0.2	139
	90	0.054	134

A more detailed review of the literature known data for various monomers and model systems can be found in ref. 140. However, unambiguous numbers for λ are scarce and there is in most cases only a qualitative agreement between different literature values. Despite these divergences, the following statements generally hold:

1. Polymerizations of vinyl monomers predominantly terminate *via* combination.
2. Polymerizations of α-methyl-vinyl monomers always show a contribution of disproportionation.
3. The hydrogen atoms of the α-methyl-group are more prone to abstraction during the disproportionation reaction than the methylene hydrogen atoms.
4. Within a series of vinyl or α-methyl-vinyl monomers, λ apparently decreases according to the radical stabilization ability of the substituent.

1.5.2   Termination Rate

The rate law expression for termination in its simplest form is

$$R_{\text{term}} = -\frac{dc_R}{dt} = 2k_t c_{R\bullet}^2. \qquad (1.18)$$

with the radical concentration $c_{R\bullet}$ and the termination rate coefficient, k_t. There sometimes is confusion whether to incorporate the factor 2 of the rate law expression into the termination rate coefficient.[141] The factor 2 is needed if the rate law describes the rate of macroradical loss, but unnecessary if termination events are considered. Nevertheless, the IUPAC ruling[142] on this is clear: termination rate coefficients are to be reported without the factor two. All termination rate coefficients given in this chapter are in accordance with this IUPAC guideline.

It is, however, very difficult to tabulate single values for k_t, because of the above-mentioned chain-length and monomer conversion dependence of the termination rate. It is however possible to give chain-length averaged k_t value for a specific monomer conversion, $<k_t>$, which may be sufficient for rough estimations of the polymerization process. Many simulations of technical processes indeed rely on this simplified concept, as the average chain-length of the radical population is not changing drastically in conventional polymerization. This simplified approach, however, cannot be used in controlled radical polymerization, where chain lengths systematically increase during polymerization. The impact of monomer conversion, on the other hand, has to be considered with both types of radical polymerizations in case it changes drastically during the reaction. In conclusion, k_t is best given as functional dependence both of chain-length and monomer conversion. This function, however, is to date not available for more than one or two systems (see below) and simpler approximations are still in use.

1.5.3 Monomer Conversion Dependence

For low and intermediate monomer conversion, the termination reaction itself can be broken down into three consecutive stages (see Scheme 1.21):[143–146]

1. Translational centre-of-mass diffusion (TD) of both species towards each other travelling through the reaction medium.
2. Segmental diffusion (SD) of the radical chain ends towards each other. This occurs when a coil pair is penetrating each other in an entangled state. Segmental diffusion brings the chain ends into a position that enables them to react.
3. The chemical reaction (CR) of the two radical sites yielding the polymeric product(s).

The diffusion-controlled termination rate coefficient for low and intermediate monomer conversion, $k_{t,D}$, is expressed *via* eqn (1.19) were k_{TD}, k_{SD} and k_{CR} denote the rate coefficients associated with the reaction steps in Scheme 1.21.

$$\frac{1}{k_{t,D}} = \frac{1}{k_{TD}} + \frac{1}{k_{SD}} + \frac{1}{k_{CR}} \qquad (1.19)$$

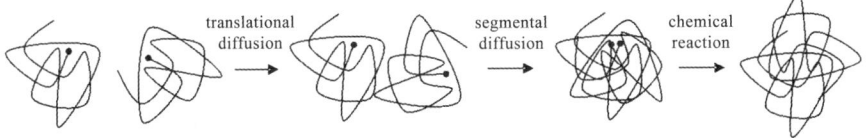

translational diffusion   segmental diffusion   chemical reaction

Scheme 1.21

At low and moderate conversions, overall k_t is adequately represented by $k_{t,D}$, *i.e.* $k_t = k_{t,D}$. Since the chemical reaction between two (macro)radical functionalities is extremely fast (associated with a high value of k_{CR}), termination is diffusion controlled from the initial phase of polymerization on. Termination usually depends on the rate-determining diffusion step, denoted by SD- or TD-controlled k_t.

At high degrees of conversion, termination may in parallel occur to a significant extent *via* so-called reaction diffusion (RD). Termination *via* RD, with rate coefficient $k_{t,RD}$, occurs by chain-end encounter after successive addition of monomer units. This mechanism plays a major role when macroradicals are immobilized (trapped) in a polymer network. The termination rate coefficient k_t is thus given by eqn (1.20).

$$k_t = k_{t,D} + k_{t,RD} \qquad (1.20)$$

The conversion dependence of termination was experimentally investigated in detail *via* the single-pulse pulsed-laser polymerization coupled with online time-resolved near infrared spectroscopy (SP-PLP-NIR) technique.[142,147] This method provides access to chain-length-averaged termination rate coefficients, $\langle k_t \rangle$. The variation of $\langle k_t \rangle$ towards increasing monomer-to-polymer conversion, X, is exemplified in Figure 1.5 for bulk polymerization of MMA.

The plot of $\langle k_t \rangle$ *vs.* X in Figure 1.5 reveals distinct regimes of rather different conversion dependencies, which can be assigned to $\langle k_t \rangle$ being controlled by specific termination mechanisms, denoted as SD, TD or RD control.

The SD control of $\langle k_t \rangle$ in the initial stage of a polymerization is often called plateau-regime, since $\langle k_t \rangle$ remains more or less constant with increasing conversion. The plateau level depends on (mostly the viscosity of) monomer and solvent. SD control is characterized by fast centre-of-mass diffusion of macroradicals through the environment of mostly monomer and solvent and subsequent segmental re-orientation, which also occurs against the friction of

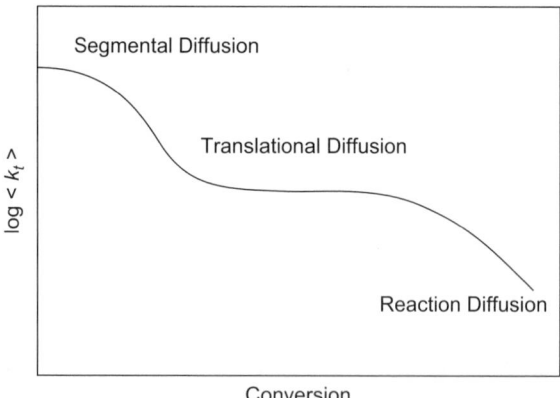

Figure 1.5   Conversion dependence of the termination rate coefficient $\langle k_t \rangle$ typical for bulk polymerizations.

monomer and solvent environment. The formation of polymer induces an increase in bulk viscosity, even though this does not to a major extent influence the mobility of the terminating radicals, since the large mesh-size of the polymer chains allows for macroradical diffusion essentially controlled by monomer and solvent fluidity. Center-of-mass diffusion in the SD regimes is not correlated with bulk viscosity, but rather determined by a so-called "microviscosity" of the monomer–solvent mixture. The past section is of particular relevance in view of chain-length dependent termination.

The pronounced decrease of $\langle k_t \rangle$ in the TD regime is associated with the occurrence of the so-called gel-effect.[148] Also known as the 'Trommsdorff', 'Norrish–Smith' or 'Norrish–Trommsdorff' effect, this effect can cause problems within both an industrial and scientific context ranging from a product mixture to reactor explosion, due to its exothermic nature.[148,149] Increasing polymer content induces overlap of polymer chains and decreases the mesh-size in between the polymer chains beyond a critical limit. As a consequence, TD may become the rate-determining step in Scheme 1.21 for the majority of macroradicals, thus $\langle k_t \rangle$ decreases by orders of magnitude in some cases. It is important not to confuse the gel effect with the auto-acceleration that is observed when a polymerization is carried out under non-isothermal conditions, so that the reaction temperature increases with increasing monomer conversion, due to the exothermic nature of the polymerization reaction. The gel effect is observed under isothermal reaction conditions. The cause of the gel effect has been discussed extensively and various theories have emerged which can explain all or part of the experimental data (excellent reviews on the topic can be found in ref. 150 and 151).

The RD regime is indicated by a less pronounced dependence of $\langle k_t \rangle$ on conversion. As termination depends on propagation rate, $k_{t,RD}$ is directly proportional to (constant) k_p and to monomer concentration. Decreasing monomer concentration with ongoing conversion essentially explains the decay of $\langle k_t \rangle$ in the RD region. The stronger decrease of termination beyond conversions of *ca.* 90% is explained by strong deceleration of propagation rate. Buback has developed a model for the dependence of the termination rate on the monomer conversion, which considers segmental, translational and reaction diffusion processes.[152] This model has been successful in describing a large set of data up to high monomer conversions.[153–155]

1.5.4 Chain-length Dependent Termination (CLD-T)

Caused by the diffusion-controlled termination steps in the full course of polymerization, k_t depends on the chain lengths i and j of the associated terminating radicals. During radical polymerization carried out under continuous initiation, termination generally occurs between macroradicals of different chain lengths, thus, termination rate coefficients $k_t(i,j)$ need to be considered. During the course of a radical polymerization, the average chain-lengths of terminating macroradicals may be altered, even under continuous initiation. This is especially true for controlled radical polymerizations, in which the radical population continuously grows in chain length. Consideration of CLD-T may thus significantly improve kinetic models used for modeling and simulation.

Three averaging models are commonly used to describe $k_t(i,j)$ as a function of the individual chain lengths i and j, of $k_t(1,1)$ associated with termination of two monomeric radicals and of the exponent value α, in which the strength of CLD-T is expressed. Values for $k_t(1,1)$ and α can be obtained in a reliable way by experimental techniques (see below).

$$k_t(i,j) = k_t(1,1)\left(\frac{2\,ij}{i+j}\right)^{-\alpha} \quad \text{(hm)} \qquad (1.21)$$

$$k_t(i,j) = \frac{1}{2}k_t(1,1)(i^{-\alpha}+j^{-\alpha}) \quad \text{(dm)} \qquad (1.22)$$

$$k_t(i,j) = k_t(1,1)(\sqrt{ij})^{-\alpha} \quad \text{(gm)} \qquad (1.23)$$

The individual models: harmonic-mean (hm), diffusion-mean (dm) and geometric-mean (gm) include different weighting of the contribution of shorter and longer chains. For example, the dm-model is directly based on the Smoluchowski equation, eqn (1.27), *i.e.* the extent of contribution to $k_t(i,j)$ of the individual macroradical refers to the size of diffusion coefficients associated with i and j. Simulation of polymerization processes by implementation of one of these models is however extremely complex and time-consuming, since the chain-length distribution of macroradicals present at any stage during the radical polymerization has to be implemented into the model in addition to an adequate function for $k_t(i,j)$. Averaging-model-free determination of $k_t(i,j)$ is in principle possible but also rather difficult from an experimental point of view.

Chain-length averaged termination rate coefficients $\langle k_t \rangle$ may be estimated from eqn (1.21)–(1.23) *via* eqn (1.24), provided that data for the concentration of macroradicals as a function of chain-length, $C_{R(i)}$, is available.

$$\langle k_t \rangle = \frac{\displaystyle\sum_i \sum_j k_t(i,j)c_{R(i)}c_{R(j)}}{(c_R)^2} \qquad (1.24)$$

Enormous progress has been made in the past decade in determining chain-length-dependent termination rate coefficients for two macroradicals of almost identical chain length, $k_t(i,i)$, which is somewhat easier to handle. Barner-Kowollik and Gregory have recently presented a comprehensive review on this.[156] Termination between radicals of identical chain-length plays an important role in controlled radical polymerization, since the chain-length distributions of active chains are narrowly distributed and increase constantly with conversion. Hence, rational modeling of controlled radical polymerization intrinsically relies on the availability of chain-length dependent k_t data. Practical approaches to implement experimental data for $k_t(i,i)$ into the kinetic schemes used for simulations of technical relevant processes have also been made *e.g.* by eqn (1.25).

$$\langle k_t \rangle = a\,k_t(i,i); \quad 0 < a < 1 \tag{1.25}$$

The chain length, i, in eqn (1.25) refers to the number average degree of polymerization of macroradicals. The correction factor a is found to be much smaller than unity which empirically expresses the impact of short-long termination.

Specially designed techniques for determination of $k_t(i,i)$ are based on controlling the radical chain length either by laser single-pulse initiation[157–159] or by RAFT polymerization.[160,161] All being well, these techniques induce a narrow size distribution of radicals with degree of polymerization increasing linearly with time and with monomer conversion. Thus, the obtained termination rate coefficients, $k_t(i,i)$, vary with time and refer to the length, i, of radicals present at each instant.

It is found that the so-called composite model for termination, eqn (1.26),[162] seems to be obeyed by all monomers:

$$\begin{aligned}
k_t(i,i) &= k_t(1,1)i^{-\alpha_s}; \quad i \leq i_c \\
k_t(i,i) &= k_t(1,1)(i_c)^{-\alpha_s+\alpha_1}i^{-\alpha_1} = k_t^0 i^{-\alpha_1}; \quad i > i_c
\end{aligned} \tag{1.26}$$

This model postulates that there are two distinct regimes of chain-length dependence. For short radicals, $k_t(i,i)$ strongly decreases with i, and the exponent α_s is found to be between 0.50 and 0.65 for styrene, methacrylates and some other monomers. This is consistent with termination being controlled by centre-of-mass diffusion. These values of α_s are consistent with the power-law exponents found in measurements of diffusion coefficients, D_i, as a function of chain length i of deliberately synthesized oligomers.[163,164] For pure translational diffusion, theory predicts the exponent α_s to be 0.5 or 0.6 for the diffusion of coiled chains in theta and athermic solvents, respectively, and to be 1.0 for stiff, rod-like chains.[165–167] Figure 1.6 shows the function of k_t *vs.* chain-length for two acrylates, which was obtained *via* pulsed-laser RAFT polymerization.[168] It can easily be seen that k_t is indeed changing systematically within two distinct regimes, which supports the concept of the composite model. This model, however, is somewhat semi-empirical, and not all aspects of the differences in chain-length dependencies of k_t in different systems are fully understood yet. Table 1.7 collates chain-length dependent k_t data for some important monomers for the segmental diffusion regime, *i.e.*, for low monomer conversion roughly up to 20 to 30%.

For radicals of size above a certain crossover chain length i_c of around 50, the dependency becomes much weaker, with observed values of α_l mostly falling in the range 0.15–0.30.[156] Such values are in accord with theoretical predictions of $\alpha_l = 0.16$ for control of (long-chain) termination by segmental diffusion in a good solvent.[169]

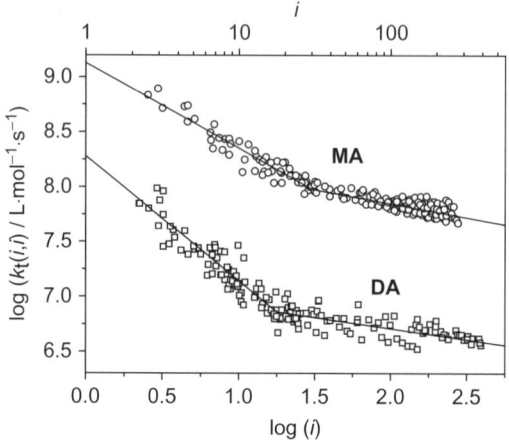

Figure 1.6 Chain-length dependent termination rate coefficients determined *via* the
SP-PLP-NIR-RAFT technique for methyl acrylate and dodecyl acrylate
bulk polymerization at 60 °C and 1000 bar.
(Reprinted with permission from ref 180: M. Buback *et al.*, *Australian
Journal of Chemistry*, 2007, **60**, 779. Copyright 2007 CSIRO Publishing.)

Table 1.7 Chain-length dependent termination rate coefficients of some
important monomers for the segmental diffusion regime.

Monomer	θ/ °C	$k_t(1,1)$	α_S	i_c	α_L	Ref.
MMA	80	$>10^{10}$	0.65	100	0.15	170
n-Dodecyl MA	0	1.1×10^7	0.64	50	0.18	171
Cyclohexyl MA	0	3.7×10^7	0.50	90	0.22	171
Benzyl MA	0	2.4×10^7	0.51	90	0.21	171
Benzyl MA	−10	1.3×10^7	0.45	90	0.16	171
Benzyl MA	−20	2.3×10^7	0.55	90	0.18	171
n-Butyl MA	−30–60	1.3×10^8	0.65	50	0.20	172
t-Butyl MA	−30–60	9.1×10^7	0.56	70	0.20	172
Styrene	80	2×10^9	≈ 0.77	≈ 15	0.14	160
Styrene	90	5×10^8	0.53	≈ 30	0.15	173, 174
Methyl acrylate	25	$\approx 1 \times 10^9$	0.41–1.15	30..50	0.35–0.36	175, 176
Ethyl acrylate	25	$\approx 7 \times 10^8$	0.41–1.15	30..50	0.35–0.37	175, 176
n-Butyl acrylate	25	$\approx 3 \times 10^8$	0.41–1.15	30..50	0.17–0.19	175, 176
Methyl acrylate	50	1×10^9	0.78	≈ 18	0.15	174
Methyl acrylate	80	$>1 \times 10^9$	>0.36	≈ 5	0.36	177
n-Butyl acrylate	80	3×10^9	1.04	40	0.20	178
n-Dodecyl acrylate	60	$>1 \times 10^9$	1.20	≈ 20	0.28	179
n-Dodecyl acrylate	80	$>4 \times 10^8$	1.15	≈ 15	0.22	179
Methyl acrylate 1000 bar	60	1×10^9	0.78	30	0.26	180
n-Butyl acrylate 1000 bar	60	4×10^9	≈ 1	≈ 10	0.22	181
n-Butyl acrylate	80	1×10^9	1.25	27	0.22	178
n-Dodecyl acrylate 1000 bar	60	2×10^8	1.12	20	0.20	180
Dibutyl itaconate	0–60	7.2×10^5	0.5	≈ 100	≤ 0.16	182
n-Butyl MA	−30–60	1.3×10^8	0.65	50	0.20	172
n-Dodecyl MA	−20–0	1×10^7	0.64	50	0.18	171
Tridecafluorooctyl MA	80–100	4.3×10^7	0.65	58	0.20	183
Methyl MA	80	4.9×10^8	0.65	100	0.15	170, 184
Butyl acrylate 77% toluene	−40	3.2×10^8	0.85	30	0.22	185

1.5.5 Combined Chain-Length and Monomer Conversion Dependence

When measuring $k_t(i,i)$ during the course of a controlled radical polymerization – which *e.g.* may be realized by online rate measurement *via* DSC during a RAFT polymerization (RAFT-CLD-T method)[160] – one obtains k_t data that is also inherently linked to a distinct monomer conversion as it is impossible to determine $k_t(i,i)$ at fixed conversion. This difficulty can be turned into a virtue by using different RAFT agent concentrations during a set of experiments on the same monomer system. This approach was introduced by Barner-Kowollik and co-workers and leads to different paths of k_t through the (i,X)-space. By combining all such data and fitting it, one may map out $k_t(i,X)$, as illustrated in Figure 1.7 on the example of methyl acrylate polymerization.[186]

This method was applied to MA,[186] vinyl acetate,[187] and to MMA in order to study the gel regime.[188,189] Such 3D-plots are naturally difficult to tabulate here and the interested reader is referred to the original literature. It is, however, without doubt the wish of all scientists working in this field that such generalized data become available for many other monomers and systems at various reaction parameters in the future and that computer-based models for k_t become available to apply the correct values of k_t for every i,X-point that is passed during a (controlled) radical polymerization.

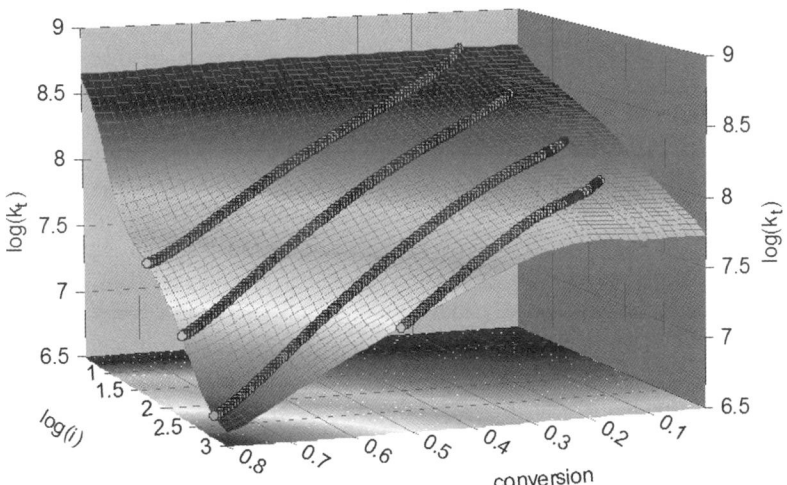

Figure 1.7 3-Dimensional plot of $k_t(i, i)/(L\ mol^{-1}\ s^{-1})$, for bulk polymerization of methyl acrylate (MA) at 80 °C as a function of chain length, i, and fractional conversion of monomer into polymer. The curves are the results from four RAFT experiments with different initial RAFT agent concentration.
(Reprinted with permission from ref. 186: A. Theis *et al.*, *Macromolecules*, 2005, **38**, 10323. Copyright 2005 American Chemical Society.)

1.5.6 Temperature and Pressure Dependence

Because monomeric radicals are so small, their termination $k_t(1,1)$ must be *via* centre-of-mass diffusion. This situation can adequately be described by the Smoluchowski equation (1.27)

$$k_t(1,1) = 2\pi P_{spin}(D_1 + D_1)R_c N_A \qquad (1.27)$$

where N_A is the Avogadro number, D_1 is the self-diffusion coefficient of the monomer, *i.e.* radical of chain length unity, R_c is the capture radius for termination, and P_{spin} is the probability of encounter involving a singlet pair: on straight statistical grounds this value will be 0.25.[190,191] The most important quantity in eqn (1.27) is D_1. Its behavior should be captured by the well-known Stokes–Einstein equation:

$$D_1 = \frac{k_B T}{6\pi r_1 \eta} \qquad (1.28)$$

Here k_B is the Boltzmann constant, T is (absolute) temperature, r_1 is the hydrodynamic radius of monomer, and η is the viscosity of the reaction mixture. For polymerization systems, η should be understood as the micro-viscosity (or solvent viscosity), because it is well known that termination rate coefficients do not vary according to bulk viscosity (see above). From the above considerations one expects that $k_t(1,1) \sim (r_1\eta)^{-1}$. The additional expectation is that $E_A(k_t(1,1)) \approx E_A(\eta^{-1})$, where E_A denotes activation energy. Table 1.8 gives activation energies and activation volumes ($\Delta V^{\neq} = -RT(\partial \ln k/\partial p)_T$) for selected acrylate and methacrylate termination rate coefficients.

Inspection of Table 1.8 shows that the activation energies are rather low which is consistent with the diffusion (either segmental or translational) controlled nature of the termination reaction, *i.e.* the fact that these activation energies correspond to temperature dependence of the inverse system viscosity. Currently, there is no data on the activation energy of the actual termination reaction itself. However, it is very likely that the termination reaction itself has a close to zero activation energy. This conclusion may be deduced from the activation parameters observed for small radical termination.[195,190] Importantly, the activation volume of the termination reaction is positive, *i.e.* the value of k_t decreases with increasing pressure. This observation can also be connected with the pressure dependency of the viscosity. The reaction medium

Table 1.8 Activation energies and volumes for the termination rate co-efficients of selected acrylates and methacrylates.

Monomer	E_A/kJ mol^{-1}	$\Delta V^{\neq}$/cm^3 mol^{-1}	*Reference*
MA	8.0 (1000 bar)	16.0 (30 °C)	192
BA	6.0 (1000 bar)	16.0 (40 °C)	193
DA	3.0 (1000 bar)	20.0 (40 °C)	192
MMA	11.0 (1 bar)	15.0[a] (40 °C)	194

[a]This activation volume is pressure dependent, the value is valid for the pressure range from 1000 to 1500 bar.

tends to be more viscous at higher reaction pressures, thus slowing the rate of termination, *i.e.* the activation volume of the termination rate coefficient is very close to the corresponding activation volume that characterizes the pressure dependence of the inverse of the monomer viscosity.[196] It is important to note that the pressure dependencies of the termination and propagation rate coefficients display opposite behaviour, *i.e.* allowing for increased rates of polymerization at elevated pressures.

1.5.7 Solvent Effects

In general, solvent effects on k_t are rather small. There are, like with k_p, two important exceptions to this rule, namely ionic liquids and water. When using ionic liquids as solvent, it is observed that polymerization rate and polymer molecular weight are enhanced compared to polymerizations in conventional organic solvents or in bulk.[197–200] One reason for this is the enhancement of the propagation rate coefficient, k_p (see above).[201,202] In addition to an increase in k_p, k_t decreases in the highly viscous solution containing ionic liquid.[201] Both rate coefficients, k_p and k_t, thus contribute to an enhancement of polymerization rates in the presence of ionic liquids. Barth and Buback[203] studied this effect by time-resolved EPR spectroscopy after single laser pulse initiation in detail and found a decrease in k_t by around one order of magnitude (see Table 1.9).

k_t in both ionic liquids are well below the value measured for MMA bulk polymerization. The direction of change agrees with the inverse of the bulk viscosities at 10 °C: 0.67 cP for MMA,[204] 55.9 cP for dry [emim] NTf$_2$ and 171 cP for dry [bmim] BF$_4$.[205] This indicates that the increased microviscosity of the solution phase is the cause of the reduced termination rate. Chain length dependency does not change with solvent.

The situation is somewhat more complex when using water as the solvent. Data on k_t in aqueous solution, however, is scarce[206] and not all effects are fully understood. Studies into k_t in polymerization of 1-vinylpyrrolidin-2-one in a solution of water, for instance, has revealed that the full mechanism of termination is shifted with increasing content of water (see Figure 1.8)[207] Termination is being enhanced with increasing concentration of water and the transition between segmental and translational diffusion control is shifting simultaneously.

Table 1.9 Chain-length averaged termination rate coefficients for MMA polymerization in ionic liquids at 10 °C referring to the regime of short chains ($i \leq 100$) and $\langle k_t \rangle$ of MMA in bulk for $i < 1000$.

Solvent	$\langle k_t \rangle / \text{L mol}^{-1}\text{s}^{-1}$	Ref.
[bmim] BF$_4$	$(2.4 \pm 0.1) \times 10^6$	203
[emim] NTf$_2$	$(7.2 \pm 0.3) \times 10^6$	203
Bulk	2×10^7	142

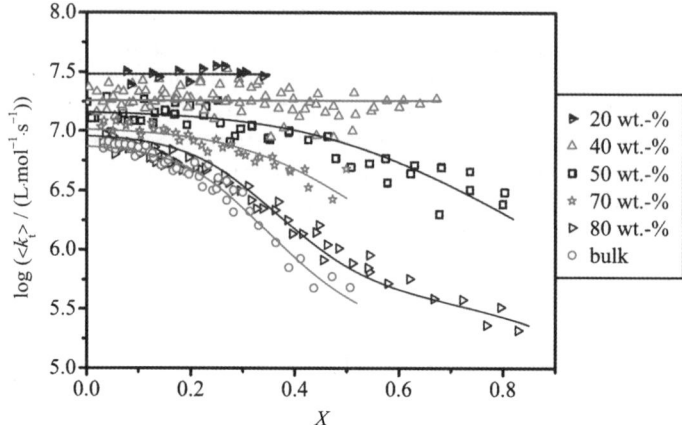

Figure 1.8 Conversion dependence of the chain-length-averaged termination rate coefficient, $<k_t>$, for aqueous 1-vinylpyrrolidin-2-one polymerizations at 40 °C and 2000 bar for various initial contents of 1-vinylpyrrolidin-2-one in water.[207]
(Reprinted with permission from ref. 207: J. Schrooten, M. Buback, P. Hesse, R. A. Hutchinson, I. Lacik, *Macromol. Chem. Phys.*, 2011, **212**, 1400. Copyright 2011 John Wiley and Sons.)

1.6 Rate of Polymerization

A conventional radical polymerization is in nearly all cases a steady-state polymerization system, which is characterized by a constant free radical concentration over time as given by eqn (1.28)

$$\frac{dc_{R\bullet}}{dt} = 0 \tag{1.28}$$

A simple but rather general expression for the rate of polymerization, R_p, can be derived when assuming the following approximations:

- All reactions are irreversible.
- Monomer species are consumed only by propagation of radical species.
- All macroradicals are of identical reactivity, regardless of their chain length and the degree of monomer-to-polymer conversion.
- Termination of macroradicals takes place either by bimolecular combination or disproportionation.

Under steady state conditions, the rate of initiation, R_i, is equal to the rate of termination eqn (1.29), which is an assumption necessary for the establishment of a constant free radical concentration

$$2fk_dc_I = 2k_tc_{R\bullet}^2 \tag{1.29}$$

Rearrangement of eqn 1.29 and insertion into the simplified form of eqn 1.30

$$R_p = -\frac{dc_M}{dt} = k_p c_M c_{R\bullet} \tag{1.30}$$

yields the final expression for the rate of polymerization, R_p:

$$R_p = -\frac{dc_M}{dt} = k_p \left(f c_I \frac{k_d}{k_t} \right)^{0.5} c_M \tag{1.31}$$

Eqn 1.31 indicates a reaction order of one of monomer concentration on the rate of polymerization and a reaction order of 0.5 of the initiator concentration. These dependencies have been confirmed experimentally on the example of many polymerizing systems, although it is important to note that deviations from ideality, such as chain length dependent rate coefficients, back-biting, and primary radical termination, lead to a change in the exponents associated with the initiator and monomer concentrations.[208,209] In such cases, the rate of polymerization will scale with a weaker than square root dependence on c_I and a stronger than linear dependence on c_M. Very low monomer concentrations can also alter the exponents of monomer and initiator concentration. For many conventional polymerization systems, however, this equation holds and can be integrated to yield an expression which correlates the monomer conversion with an observed overall kinetic rate coefficient, k_{obs}.

$$\ln\left(\frac{1}{1-X}\right) = \ln\left(\frac{c_M^0}{c_M}\right) = k_{obs}\, t \quad \text{where} \quad k_{obs} = k_p \left(f c_I \frac{k_d}{k_t} \right)^{0.5} \tag{1.32}$$

X is the fractional monomer conversion. This equation is the famous "pseudo first-order rate law" of radical polymerization, which is very often used to characterize the kinetics of a radical polymerization process. The finding that a plot according to eqn (1.32) is linear indicates that the assumptions made above about steady-state polymerization are fulfilled. A linearity of a first order rate plot primarily indicates that the concentration of active sites – radicals in the case of radical polymerization – is constant over time. Deviations from linearity are mainly due to depletion of the initiator. It must be stressed here, that every ideal conventional radical polymerization gives a linear pseudo first-order rate plot. Very often, a linear pseudo first-order rate plot is taken as proof for a successful controlled radical polymerization, which is absolutely *not correct*. In contrast, controlled radical polymerizations based on the persistent radical effect, such as NMP and ATRP, do in principle *not* give a linear pseudo first-order rate plot. RAFT polymerization – as it rests on degenerative chain transfer – does show linear pseudo first-order rate plots, on the other hand. It is thus not possible to evaluate the quality of a controlled radical polymerization by inspecting pseudo first-order rate plots. This misconception originates from living anionic polymerization, where the linearity of the pseudo first-order rate plots is indication of the fact that active species do not terminate. This concept cannot be transferred to controlled radical polymerization. Unfortunately, this misconception can still be found very often in literature.

The temperature dependence of the overall polymerization rate is given by the temperature dependencies of the individual rate coefficients. Each rate coefficient follows its own Arrhenius law, $k = A\exp(-E_A/RT)$, where A is the pre-exponential factor and E_A denotes the activation energy. The overall activation energy of the rate of polymerization, $E_A^{R_p}$, p, equals the sum of the weighted activation energies of the elementary reactions, propagation (E_A^p), initiation (E_A^i) and termination (E_A^t).

$$E_A^{R_p} = E_A^p + \frac{1}{2}E_A^i - \frac{1}{2}E_A^t \qquad (1.33)$$

Activation energies for commonly used thermally decomposing initiators, E_A^i, are in the order of 120 to 150 kJ mol^{-1}. The E_A^p values for most common monomers lie within the range of 20 to 40 kJ mol^{-1}, and E_A^t is generally in the range of 4 to 10 kJ mol^{-1}. Hence, typical values for overall activation energies for the rate of polymerization initiated by a thermally decomposing initiator are close to 80 kJ mol^{-1}. This corresponds to a two or threefold rate increase in rate for a 10 °C temperature increase. Photochemical polymerization rates have a much lower activation energy of about 20 kJ mol^{-1}, according to close to zero activation energy of the photoinitiation process.

A deviation from ideality in conventional radical polymerization, which is often observed and which also occurs frequently in controlled radical polymerization, is the effect of inhibition and retardation. These effects are due to reactions of macroradicals with transferring or terminating species. In order to derive the kinetics of retardation and inhibition, these effects are described in terms of chain transfer, and termination is considered as transfer to species with negligible re-initiation ability. The chain transfer process stops the chain growth and the transfer agent itself becomes a radical. If the re-initiation ability of the generated radical is in the order of that of the macroradical and if the time necessary for the chain transfer process is within the range of one propagation step, 'normal' chain transfer occurs (see above). This normal or conventional chain transfer does not lead to any change in the polymerization rate. If the generated radical is less reactive than the propagating radical, retardation (degradative chain transfer) takes place, which slows down the rate of polymerization. This effect is *e.g.* observed when using phenol as retarder in MMA polymerization. If the retardation is very effective, the polymerization process is completely suppressed and this is referred to as inhibition. This can for instance be observed with oxygen in MA polymerization. Inhibition leads to an induction period where no polymerization takes place at all, until the inhibitor is consumed. The reaction then starts with the conventional rate of polymerization (see Figure 1.9).

The kinetics of the inhibition effect for a steady-state polymerization, *i.e.* the scenario that the generated radical does not reinitiate, can be analyzed

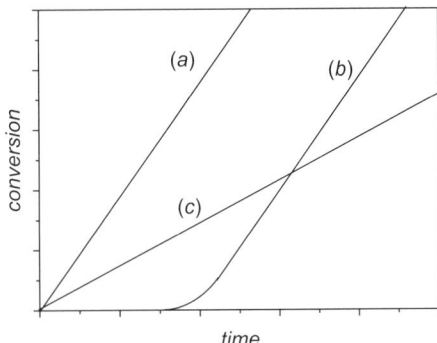

Figure 1.9 Typical conversion *vs.* time profiles for conventional radical polymerization (*a*) with addition of inhibitor (*b*) and retarder (*c*).

by adding an additional reaction step to the basic scheme of polymerization.[210,211]

$$R_i^\bullet + Q \xrightarrow{k_Q} P_i + Q^\bullet \tag{1.34}$$

where Q is the retarder or inhibitor and k_Q is the associated rate coefficient of the inhibition reaction. The generated radical, $Q^\bullet$, is assumed to not reinitiate, but remains persistent. This scenario may *e.g.* be realized by addition of typical "stabilizers" such as substituted phenols (*e.g.* *p*-methoxyphenol, 2,6-di-*tert*-butyl-4-methyl phenol) used as stabilizer for styrene[212] or as benzoquinone, which gives a phenoxy radical as a result of a rapid addition of the radicals present in the system to the C=O double bond.[213-215] The steady state assumption, which is only a very rough approximation until all inhibitor is consumed,[216] can then be written as

$$-\frac{dc_{R^\bullet}}{dt} = R_i - 2k_t c_{R^\bullet}^2 - k_Q c_{R^\bullet} \cdot c_{Q^\bullet} = 0 \tag{1.35}$$

which yields in combination with the expression (1.30), assuming that the rate of monomer loss, $-dc_M/dt$, equals the rate of polymerization, R_p

$$R_i - \frac{2k_t R_p^2}{k_p^2 c_M^2} - \frac{k_Q R_p c_{Q^\bullet}}{k_p c_M} = 0 \tag{1.36}$$

The ratio of the rate coefficients for retardation, k_Q, and propagation, k_p, is often referred to as the inhibition constant, $z = k_Q/k_p$, which reflects the ability of a molecule to cause inhibition. If the inhibition constant is large ($z \gg 1$), the second term of the l.h.s. of eqn 1.36 will become much smaller than the third one. In this case, the rate of inhibition is much larger than the rate of termination. Eqn (1.36) then reads

$$R_p = \frac{k_p c_M R_i}{k_Q c_{Q^\bullet}} \tag{1.37}$$

Eqn 1.37 shows that the polymerization rate is inversely proportional to the inhibitor concentration. It should be kept in mind that the inhibitor concentration will decrease during the reaction, since each radical generated by the initiation process will consume one inhibitor molecule. Propagation can become competitive with the inhibition reaction, when the inhibitor concentration becomes low.

Dividing the rate law for the loss of inhibitor, $-dc_{Q\bullet}/dt = k_Q c_{R\bullet} c_{Q\bullet}$, by the rate law of propagation, $-dc_M/dt = k_p c_{R\bullet} c_M$, leads to

$$\frac{dc_{Q\bullet}}{dc_M} = \frac{z c_{Q\bullet}}{c_M} \tag{1.38}$$

and subsequent integration with $c_{Q\bullet}^0$ and c_M^0 being the concentration of inhibitor and monomer at the beginning of the reaction.

$$\ln\left(\frac{c_{Q\bullet}}{c_{Q\bullet}^0}\right) = z \cdot \ln\left(\frac{c_M}{c_M^0}\right) \tag{1.39}$$

It is apparent from eqn 1.39 that, if z is large, the monomer conversion remains nearly zero until the inhibitor is consumed.

1.7   The Chain Length Distribution

The rate coefficients of the individual reaction steps of radical polymerization are determining the rate of polymerization, R_p. The same kinetic parameters may be employed to calculate the lengths of macroradicals and the final polymer. The chain length distribution of a polymer is defined as the fraction of molecules x_P that contains P basic monomer units. The degree of polymerization P is equivalent to the chain length i. The propagating macroradicals by which the dead polymer is generated through any chain-stopping event exhibit a chain length distribution, too. Both distributions are closely related to each other and the chain length distribution of the dead polymer can be calculated *via* the derivative of the distribution of the living macroradicals.

Like any other distribution function, the chain length distribution may be described by its statistical moments, which are defined as

$$m^{(k)} = \sum_{P=1}^{\infty} P^k x_P \tag{1.40}$$

The combination of such moments define mean values for the degree of polymerization, $\bar{P}$. In practice, there are two mean values calculated by the first three statistical moments, which are extensively used. The number average degree of polymerization, $\bar{P}_n$, and the weight average degree of polymerization, $\bar{P}_w$.

$$\bar{P}_n = \frac{m^{(1)}}{m^{(0)}} = \frac{\sum_{P=1}^{\infty} P x_P}{\sum_{P=1}^{\infty} x_P} \qquad \bar{P}_w = \frac{m^{(2)}}{m^{(1)}} = \frac{\sum_{P=1}^{\infty} P^2 x_P}{\sum_{P=1}^{\infty} P x_P} \tag{1.41}$$

To calculate the number average degree of polymerization, $\bar{P}_n$, of a polymer produced by a steady-state polymerization, it is mandatory to know how many propagation steps occur before the chain mechanism is stopped. It has to be distinguished between the term 'chain' used in a molecular sense and 'chain' used as a kinetic concept. The kinetic chain length, ν, is defined as

$$\nu = \frac{\text{total number of polymerized monomer units}}{\text{total number of initiation steps}} \tag{1.42}$$

With the assumption of chain length independent rate coefficients, eqn (1.42) can be rewritten as

$$\nu = \frac{R_p}{R_i} = \frac{k_p c_{R\bullet} c_M}{2 f k_d c_I} \tag{1.43}$$

Elimination of c_R by means of eqn (1.29) leads to an expression for the kinetic chain length, ν, which shows the dependence on the various kinetic parameters.

$$\nu = \frac{k_p c_M}{2(f k_d k_t c_I)^{0.5}} \tag{1.44}$$

This equation illustrates one important characteristic of the free radical polymerization: the sizes of the macromolecules are inversely proportional to the square root of the initiator concentration. Increasing the initiator concentration leads to shorter chains. Disregarding any transfer effect, as a first approximation, correlates the kinetic chain length with the number average degree of polymerization, $\bar{P}_n$. In the case of termination by disproportionation one polymer molecule is produced per every kinetic chain

$$\bar{P}_n = \nu \tag{1.45}$$

Termination by combination leads to one polymer molecule per two kinetic chains, reflecting the combination mechanism.

$$\bar{P}_n = 2\nu \tag{1.46}$$

Any mixture of these both mechanisms can be described by using the value λ (see Section 5.1), the contribution of disproportionation to the overall termination process.

$$\bar{P}_n = \frac{2}{1 + \lambda} \nu \tag{1.46}$$

The transfer process does not change the radical concentration, but the chain length of the polymer. Without changing the free radical concentration, chain transfer processes remain hidden in any experiment measuring the rate of polymerization alone. The kinetic chain length is also unaffected by transfer, because the growing free radical centre stays active after the transfer, although more than one polymer chains are produced. For this reason eqn (1.46) does not hold true if chain transfer occurs. Taking chain transfer into account, the number average degree of polymerization, $\bar{P}_n$, can be described as

$$\bar{P}_n = \frac{\text{total number of polymerized monomer units}}{\text{half the number of formed end groups}} \qquad (1.47)$$

"End group" here means any group at the end of a polymer chain that is not propagating. The various reactions within the polymerization process generate different amounts of end groups per initiation step:

initiation	1 end group
propagation	0 end groups
transfer	2 end groups
termination by disproportionation	1 end group
termination by combination	0 end groups

At steady-state, the number of polymerized monomer units can be substituted in eqn (1.47) with the rate of polymerization and the numbers of end groups by the rate of their formation.

$$\bar{P}_n = \frac{R_p}{\frac{1}{2}(R_i + R_{t,d} + 2R_{tr})} \qquad (1.48)$$

Insertion of the simplified rate laws of the different processes

initiation $R_i = 2fk_d c_I = 2(k_{t,d} + k_{t,c})c_{R\bullet}^2$ (1.49)

propagation $R_p = k_p c_M c_{R\bullet}$ (1.50)

termination (disproportionation) $R_{t,d} = 2k_{t,d} c_{R\bullet}^2$ (1.51)

chain transfer $R_{tr} = \sum_k k_{tr}^{T_k} c_{T_k} c_{R\bullet} + k_{tr}^M c_M c_{R\bullet}$ (1.52)

and subsequent inversion leads to

$$\frac{1}{\bar{P}_n} = \frac{2k_{t,d} + k_{t,c}}{k_p^2 c_M^2} R_p + \frac{k_{tr}^M}{k_p} + \sum_k \frac{k_{tr}^{T_k}}{k_p} \frac{c_{T_k}}{c_M} \qquad (1.53)$$

c_{T_k} is the concentration of any molecule that is capable of taking part in a chain transfer reaction, including solvent S, initiator I, polymer P and added chain transfer agent T. It is usual to define chain transfer constants for the different molecules

$$C_M = \frac{k_{tr}^M}{k_p} \quad C_S = \frac{k_{tr}^S}{k_p} \quad C_I = \frac{k_{tr}^I}{k_p} \quad C_P = \frac{k_{tr}^P}{k_p} \quad C_T = \frac{k_{tr}^T}{k_p} \qquad (1.54)$$

Thus, eqn (1.53) becomes

$$\frac{1}{\bar{P}_n} = \frac{2k_{t,d} + k_{t,c}}{k_p^2 c_M^2} R_p + C_M + C_S \frac{c_S}{c_M} + C_I \frac{c_I}{c_M} + C_P \frac{c_P}{c_M} + C_T \frac{c_T}{c_M} \qquad (1.55)$$

This equation gives the fundamental correlation of the number average degree of polymerization with the rate of polymerization and the various chain transfer constants. Performing a polymerization experiment with only low conversion of monomer to polymer, the concentration of polymer is often too low to show significant chain transfer. The same holds true for the initiator, which is mainly used in the range of low concentrations. Without addition of solvent and additional chain transfer agent, eqn (1.55) reads after introduction of eqn (1.17)

$$\frac{1}{\bar{P}_n} = \frac{(1+\lambda)k_t}{k_p^2 c_M^2} R_p + C_M \qquad (1.56)$$

Hence, a plot of the inverse number average degree of polymerization, $\bar{P}_n$ against the rate of polymerization R_p, – the rate of polymerization can be easily varied by the concentration of the initiator – yields the monomer chain transfer constant C_M as intercept of a linear plot. The value of C_M^{-1} constitutes a limit for the maximum number average degree of polymerization, $\bar{P}_n^{max}$, as transfer to monomer cannot be avoided. Methyl methacrylate, for instance, has a monomer chain transfer constant of about $C_M = 5 \times 10^{-5}$ at 60 °C, leading to a maximal chain length of about 20 000, whereas in a radical polymerization of vinyl acetate with $C_M = 2 \times 10^{-4}$ at 60 °C, the limit is already reached at 5000.

So far, only the average degree of polymerization has been considered. To calculate the distribution function itself for a steady state polymerization it is convenient to choose a statistical approach based on kinetic parameters. A probability factor α of propagation is defined as the probability that a radical will propagate rather than terminate. The factor α is the ratio of the rate of propagation over the sum of the rates of all possible reactions the macroradical can undergo.

$$\alpha = \frac{R_p}{R_p + R_{tr} + R_t} \qquad (1.57)$$

First, we assume that termination occurs only by disproportionation and that the propagation probability factor is equal for each chain length. The probability for the occurrence of a polymer chain – hence its distribution function – with the length P is given by the probability of $P - 1$ propagation steps and the probability of one chain stopping event (termination *or* transfer).

$$x_{P,d} = \alpha^{P-1}(1 - \alpha) \qquad (1.58)$$

The molecular weight averages can be evaluated by calculating the moments of this distribution function by insertion of eqn (1.58) into eqn (1.59–1.61),

$$m^{(0)} = \sum_{P=1}^{\infty} x_{p,d} = 1 \tag{1.59}$$

$$m^{(1)} = \sum_{P=1}^{\infty} P\, x_{p,d} = (1 - \alpha)^{-1} \tag{1.60}$$

$$m^{(2)} = \sum_{P=1}^{\infty} P^2\, x_{p,d} = (1 + \alpha)(1 - \alpha)^{-2} \tag{1.61}$$

and subsequent insertion into eqn (1.62).

$$\bar{P}_{n,d} = m^{(1)}/m^{(0)} = (1 - \alpha)^{-1} \quad \bar{P}_{w,d} = m^{(2)}/m^{(1)} = (1 + \alpha)(1 - \alpha)^{-1} \tag{1.62}$$

The ratio of the weight average and the number average degree of polymerization, $\bar{P}_w / \bar{P}_n$, equals the dispersity index, $Đ$, of a chain length distribution and can be expressed by

$$\frac{\bar{P}_{w,d}}{\bar{P}_{n,d}} = Đ = \frac{m^{(2)}m^{(0)}}{m^{(1)}m^{(1)}} = 1 + \alpha \tag{1.63}$$

It should be noted that the propagation step must be highly favoured over chain stopping events to produce polymer with a significant chain length and the value of α must be near to one. Hence, eqn (1.63) shows that for a chain-length distribution of a polymer produced in a steady-state experiment, where chain stopping events are termination by disproportionation or transfer, the dispersity becomes nearly 2.

Expressions may also be derived for the chain length distribution produced, when the termination process is by combination. The expression for the probability of the occurrence of a chain with the chain length P is now given by the contributions of two chains with the chain length n and m, which form the desired molecule by combination. Hence, the auxiliary condition $n + m = P$ must hold true.

$$x_{p,c} = \sum_{n=1}^{P-1} \alpha^{n-1}(1 - \alpha)\, \alpha^{m-1}(1 - \alpha) = (P - 1)\, \alpha^{P-2}(1 - \alpha)^2 \tag{1.64}$$

Evaluating the moments of this distribution function by insertion of eqn (1.64) into eqn (1.65–1.67) as above

$$m^{(0)} = \sum_{P=1}^{\infty} x_{p,c} = 1 \tag{1.65}$$

$$m^{(1)} = \sum_{P=1}^{\infty} P\, x_{p,c} = 2(1 - \alpha)^{-1} \tag{1.66}$$

$$m^{(2)} = \sum_{P=1}^{\infty} P^2 x_{p,c} = (4 + 2\alpha)(1 - \alpha)^{-2} \quad (1.67)$$

leads to

$$\bar{P}_{n,c} = m^{(1)}/m^{(0)} = 2(1 - \alpha)^{-1} \quad \bar{P}_{w,c} = m^{(2)}/m^{(1)} = (2 + \alpha)(1 - \alpha)^{-1} \quad (1.68)$$

The dispersity can then be given by

$$\frac{\bar{P}_{w,c}}{\bar{P}_{n,c}} = D = \frac{m^{(2)}m^{(0)}}{m^{(1)}m^{(1)}} = 1 + \frac{\alpha}{2} \quad (1.69)$$

Keeping in mind that α has a value close to one, 1.69 leads to a dispersity of 1.5 for a polymer produced in a polymerization process where termination is by combination. The corresponding chain length distribution is somewhat narrower than that generated by disproportionation, because of the statistical coupling of two chains with different sizes.

Almost every polymerization system shows both disproportionation and combination modes. In order to combine the two modes the general expression for the dispersity of any given termination controlled chain length distributions reads

$$D = \frac{1}{2}(3 - \lambda)(1 + \lambda) \quad (1.70)$$

Because the value of α is close to one, the expression $\ln(\alpha) \approx \alpha - 1$ holds, leading to $\alpha \approx \exp[-(1 - \alpha)]$. With this in mind, the combination of eqn (1.58) with the l.h.s of eqn (1.62) gives

$$x_P = \frac{1}{\bar{P}_n}\alpha^{P-1} \approx \frac{1}{\bar{P}_n}\exp\left[-\frac{(P - 1)}{\bar{P}_n}\right] \approx \frac{1}{\bar{P}_n}\exp\left[-\frac{P}{\bar{P}_n}\right] \quad (1.71)$$

with the factor $(P - 1)$ substituted by P, since $P \gg 1$. eqn (1.71) demonstrates that the chain length distribution of the polymer formed by disproportionation or chain transfer follows an exponential function in the limit of infinite chain length.

The same calculation procedure, starting with eqn (1.58), also leads to an exponential expression for the chain length distribution for termination by combination.

$$x_P = \frac{4P}{\bar{P}_n^2}\exp\left[-\frac{2P}{\bar{P}_n}\right] \quad (1.72)$$

However, eqn (1.72) exhibits the chain length P additionally in the pre-exponential factor.

Evaluation of eqn (1.56) immediately leads to

$$\alpha = \frac{k_p c_M}{k_p c_M + 2(f k_d k_t c_I)^{0.5} + k_{tr} c_T} \quad (1.73)$$

All derived distribution functions and average degrees of polymerization may now be expressed *via* the kinetic coefficients.

For further reading on the topic of molecular weight distributions of polymers the reader is referred to more specialized literature, *e.g.* the works of Peebles,[217] or Bamford *et al.*[218]

1.8 Thermodynamics

The propagation step in radical polymerization cannot be seen as being irreversible at higher temperatures, where it is reversible leading to a thermodynamic equilibrium. This equilibrium can be described by the free energy difference, ΔG_p, between polymer and monomer. The polymerization process is thermodynamically favoured if ΔG_p is negative. The value of the free energy difference is given by the fundamental equation,

$$\Delta G_p = \Delta H_p - T \Delta S_p \tag{1.74}$$

The enthalpy and entropy changes in the propagation reaction are effectively those of the overall polymerization reaction for long polymer chains.[219,220] The polymerization enthalpy, ΔH_p, of radical polymerizations are mostly negative with typical values of –30 to –100 kJ mol^{-1}. The values for the standard polymerization entropies are negative, too, because the monomer is losing degrees of freedom when becoming part of the polymer chain. Typical values for the polymerization entropies are –100 to –120 J K^{-1} mol^{-1}. The two terms on the r.h.s. of eqn (1.74) are thus antagonistic. The exothermicity of the reaction exceeds the entropic term at normal temperatures and ΔG_p becomes negative. At elevated temperatures, however, the entropic term becomes significantly larger and finally equals the enthalpic term at the so-called ceiling temperature, T_c. At the ceiling temperature, the free energy difference becomes zero and polymerization does not occur any longer to a significant extent.[221] Only a few systems are known in which both the enthalpy and entropy change of the polymerization are positive. The polymerization of sulfur of the eight-membered ring conformation is one example:[222–224] the entropy increases during polymerization as the S_8-ring is rigid and the ring opening reaction makes additional conformations available. It follows from eqn (1.74) that in such cases there exists a floor temperature, T_f, with polymerization being feasible only above a certain temperature.

Polymerization reactions can also be described in view of kinetics with propagation being reversible:[225,226]

$$R_i^{\bullet} + M \xrightarrow{k_p} R_{i+1}^{\bullet} \tag{1.75}$$

$$R_{i+1}^{\bullet} \xrightarrow{k_{dp}} R_i^{\bullet} + M \tag{1.76}$$

with the rate coefficient of depropagation, k_{dp}. Since depropagation is a β-scission reaction (see above), its activation energy, $E_{A,dp}$, is much higher than that of the propagation reaction, $E_{A,p}$. The difference between these two activation energies is equal to the enthalpy change of the polymerization reaction (see eqn 1.77):

$$E_{A,p} - E_{A,dp} = \Delta H_p \qquad (1.77)$$

At standard conditions, the standard free energy difference of polymerization can be related to the equilibrium constant, K, of polymerization according to eqn (1.78)

$$\Delta G_p^0 = -RT \ln K \qquad (1.78)$$

where the kinetic expression for K is the ratio of the rate coefficients of the forward reaction to the rate coefficient of the backward reaction. This can be set equal to the thermodynamic definition of K. If the degree of polymerization is very large, the concentrations of growing chains i and $i+1$ can be considered as being nearly identical, which leads to

$$K = \frac{\overrightarrow{k}}{\overleftarrow{k}} = \frac{k_p}{k_{dp}} = \frac{[R_{i+1}^\bullet]}{[R_i^\bullet][M]_e} \approx \frac{1}{[M]_e} \qquad (1.79)$$

where $[M]_e$ is the equilibrium monomer concentration (*e.g.* 10^{-6} mol^{-1} L^{-1} for styrene at 25 °C). With the standard reaction enthalpy, ΔH_p^0, and the standard reaction entropy change ΔS_p^0 for $[M] = 1$ mol L^{-1}, it follows that

$$\Delta G_p^0 = \Delta H_p^0 - T\Delta S_p^0 = R \ln[M]_e \qquad (1.80)$$

With the assumption that the reaction enthalpy is independent of temperature, the reaction enthalpy equals the standard reaction enthalpy, $\Delta H_p = \Delta H_p^0$. Inserting the fact that $\Delta G_p = 0$ at T_c then leads to

$$T_c = \frac{\Delta H_p}{\Delta S_p} = \frac{\Delta H_p^0}{\Delta S_p^0 + R \ln[M]_e} \qquad (1.81)$$

Standard polymerization enthalpies, ΔH_p^0, and standard polymerization entropies, ΔS_p^0, of various monomers are collated in Table 1.10. Eqn (1.81) indicates that the ceiling temperature is a function of the equilibrium monomer concentration. This implies that there exists a specific ceiling temperature for every given monomer concentration. The maximum ceiling temperature is reached for the bulk polymerization system. It is determined by thermodynamic parameters and is independent of the polymerization mechanism. The applicability of the above relations depends on depropagation being the exact reverse of propagation. Hence, the observation of a ceiling temperature

Table 1.10 Standard polymerization enthalpies, ΔH_p^0, and polymerization entropies, ΔS_p^0, of various monomers for the reaction of liquid monomers to condensed polymers.

Monomer	$-\Delta H_p^0/\text{kJ mol}^{-1}$	$-\Delta S_p^0/\text{J K}^{-1}\text{mol}^{-1}$
α-methylstyrene	35	110
α-vinyl naphtalene	36	—
Acrylamide	79	—
Acrylonitrile	76	109
methyl acrylate	78	—
methyl methacrylate	54	112
styrene	70	105
sulfur, S_8	−19	−31
tetrafluoroethylene	138.1	112
vinyl acetate	89	—
vinyl chloride	108.8	—
vinylidene chloride	60	106

[a]Data from *Polymer Handbook*.[1]

requires the presence of active centres. In their absence, polymers can exist at temperatures above the ceiling temperatures. The introduction of active centres, *e.g.* *via* UV-radiation and/or high temperatures, may however open up the pathway for depropagation, which is one pathway for the degradation of polymers.

Depropagation hardly occurs in most of the typical radical polymerization systems. In some 1,1-disubstituted monomer systems, however, the effects of the reverse reaction cannot be ignored at some conditions. α-Methyl styrene, for instance, has a low ceiling temperature of around $60\,^\circ\text{C}$ in bulk polymerization, which originates from its relatively low heat of polymerization.[227] Methacrylate and styrene monomers also exhibit depropagation, although at much higher temperatures ($220\,^\circ\text{C}$ and $310\,^\circ\text{C}$, respectively, for bulk polymerizations). The depropagation process lowers the rate of polymerization according to

$$R_p = k_p[\text{M}][\text{R}^\bullet] - k_{dp}[\text{R}^\bullet] = \left[k_p - \frac{k_{dp}}{[\text{M}]}\right][\text{M}][\text{R}^\bullet] = k_p^{\text{eff}}[\text{M}][\text{R}^\bullet] \qquad (1.82)$$

The effective rate coefficient of propagation can therefore be defined as

$$k_p^{\text{eff}} = k_p - \frac{k_{dp}}{[\text{M}]} \qquad (1.83)$$

It can be seen that the impact of depropagation[228–230] (right term in eqn (1.83)) is inversely proportional to the monomer concentration, which is part of the thermodynamic equilibrium. The effective propagation rate coefficient can be directly measured by PLP-SEC. The deviation of k_p^{eff} from the linear slope of an Arrhenius plot at higher temperatures is impressively demonstrated in Figure 1.10.[231]

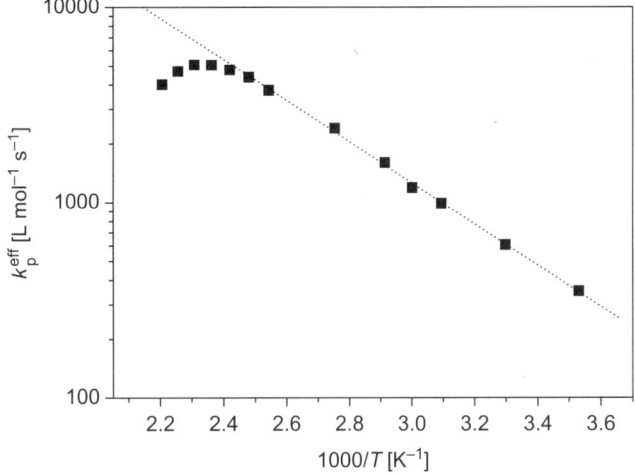

Figure 1.10   Temperature dependence of the effective propagation rate coefficient, k_p^{eff}, in bulk polymerization of dodecyl methacrylate.
(Reprinted with permission from ref. 231: R. A. Hutchinson *et al.*, *Industrial & Engineering Chemical Research*, 1998, **37**, 3567. Copyright 1998 American Chemical Society.)

References

1. A. Brandrup, E. H. Immergut and E. A. Grulke, *Polymer Handbook*, Wiley-Interscience, New York, 1999.
2. N. J. Turro, *Macromolecular Photochemistry*, The Benjamin-Cummings Publ. Co. Inc, San Francisco, 1978.
3. N. S. Allen (ed.), *Photopolymerization and Photoimaging Science and Technology*, Elsevier Science Publisher Ltd., Amsterdam, 1989.
4. J. P. Fouassier, P. Jacques, D. J. Lougnot and T. Pilot, *Polymer Photochem.*, 1984, **5**, 57.
5. H. Fischer, R. Baer, R. Hany, I. Verhoolen and M. J. Walbiner, *Chem. Soc., Perkin Trans.*, 1990, **2**, 787.
6. H. F. Gruber, *Prog. Polym. Sci.*, 1992, **17**, 953.
7. H. F. Kauffmann, O. F. Olaj and J. W. Breitenbach, *Makromol. Chem.*, 1976, **177**, 939.
8. O. F. Olaj, H. F. Kauffmann and J. W. Breitenbach, *Makromol. Chem.*, 1977, **178**, 2707.
9. O. F. Olaj, H. F. Kauffmann, J. W. Breitenbach and H. Bieringer, *J. Polym. Sci., Polym. Lett. Ed.*, 1977, **15**, 229.
10. H. F. Kauffmann, O. F. Olaj and J. W. Breitenbach, *Makromol. Chem.*, 1976, **177**, 939.
11. A. S. Sarac, *Prog. Polym. Sci.*, 1999, **24**, 1149.
12. A. A. Gridnev and S. D. Ittel, *Macromolecules*, 1996, **29**, 5864.

13. J. P. A. Heuts, R. G. Gilbert and L. Radom, *Macromolecules*, 1995, **28**, 8771.

14. M. Deady, A. W. H. Mau, G. Moad and T. H. Spurling, *Makromol. Chem.*, 1993, **194**, 1691.

15. O. F. Olaj, P. Vana, A. Kornherr and G. Zifferer, *Macromol. Rapid Commun.*, 2000, **19**, 913.

16. O. F. Olaj, P. Vana and M. Zoder, *Macromolecules*, 2002, **35**, 1208.

17. O. F. Olaj, M. Zoder, P. Vana, A. Kornherr, I. Schnöll-Bitai and G. Zifferer, *Macromolecules*, 2005, **38**, 1944.

18. J. P. A. Heuts, G. T. Russell, G. B. Smith and A. M. van Herk, *Macromol. Symp.*, 2007, **248**, 12.

19. L. H. Yee, M. L. Coote, T. P. Davis and R. P. Chaplin, *J. Polym. Sci. Part A: Polym. Chem.*, 2000, **38**, 2192.

20. P. Vana, L. H. Yee, C. Barner-Kowollik, J. P. A. Heuts and T. P. Davis, *Macromolecules*, 2002, **35**, 1651.

21. A. M. van Herk, *Macromol. Theory Simul.*, 2000, **9**, 433.

22. O. F. Olaj, I. Bitai and F. Hinkelmann, *Makromol. Chem.*, 1987, **188**, 1689.

23. B. Dervaux, T. Junkers, M. Schneider-Baumann, F. E. Du Prez and C. Barner-Kowollik, *J. Polym. Sci. Part A: Polym. Chem.*, 2009, **47**, 6641.

24. J. M. Asua, S. Beuermann, M. Buback, P. Castignolles, B. Charleux, R. G. Gilbert, R. A. Hutchinson, J. R. Leiza, A. N. Nikitin, J.-P. Vairon and A. M. van Herk, *Macromol. Chem. Phys.*, 2004, **205**, 2151.

25. S. Beuermann, M. Buback, P. Hesse, F.-D. Kuchta, I. Lacík and A. M. Van Herk, *Pure Appl. Chem.*, 2007, **79**, 1463.

26. T. Junkers, S. P. S. Koo and C. Barner-Kowollik, *Polym. Chem.*, 2010, **1**, 438.

27. R. Rotzoll and P. Vana, *Macromolecular Rapid Communications*, 2009, **30**, 1989.

28. S. Beuermann, M. Buback, T. P. Davis, N. García, R. G. Gilbert, R. A. Hutchinson, A. Kajiwara, M. Kamachi, I. Lacíkand and G. T. Russell, *Macromol. Chem. Phys.*, 2003, **204**, 1338.

29. L. M. Muratore, M. L. Coote and T. P. Davis, *Polymer*, 2000, **41**, 1441.

30. R. Siegmann, A. Jeličić and S. Beuermann, *Macromol. Chem. Phys.*, 2010, **211**, 546.

31. P. Vana, L. H. Yee and T. P. Davis, *Macromolecules*, 2002, **35**, 3008.

32. T. Junkers, M. Schneider-Baumann, S. S. P. Koo, P. Castignolles and C. Barner-Kowollik, *Macromolecules*, 2010, **43**, 10427.

33. M. Yin, T. P. Davis, J. P. A. Heuts and C. Barner-Kowollik, *Macromol. Rapid Commun.*, 2003, **24**, 408.

34. M. Ohoka, H. Ohkita, S. Ito and M. Yamamoto, *Polym. Bull.*, 2004, **51**, 373.

35. R. X. E. Willemse, A. van Herk, E. Panchenko, T. Junkers and M. Buback, *Macromolecules*, 2005, **38**, 5098.

36. M. Buback, P. Hesse, T. Junkers, T. Sergeeva and T. Theis, *Macromolecules*, 2008, **41**, 288.

37. T. Junkers and C. Barner-Kowollik, *J. Polym. Sci. Part A: Polym. Chem.*, 2008, **46**, 7585.

38. O. F. Olaj and I. Schnöll-Bitai, *Monatshefte für Chemie*, 1999, **130**, 731.
39. M. D. Zammit, T. P. Davis, G. D. Willett and K. F. O'Driscoll, *J. Pol. Sc. Part A: Polym. Chem.*, 1997, **35**, 2311.
40. M. L. Coote, T. P. Davis, B. Klumperman and M. J. Monteiro, *M. Macromol. Sci., Rev.. Macromol. Chem. Phys.*, 1998, **C38**, 567.
41. S. Beuermann, *Macromol. Rapid Commun.*, 2009, **30**, 1066.
42. V. F. Gromov, E. V. Bune and E. N. Teleshov, *Russ. Chem. Rev.*, 1994, **63**, 507.
43. F. D. Kuchta, A. M. van Herk and A. L. German, *Macromolecules*, 2000, **33**, 3641.
44. S. Beuermann, M. Buback, P. Hesse and I. Lacík, *Macromolecules*, 2006, **39**, 184.
45. S. Beuermann, M. Buback, P. Hesse, F.-D. Kuchta, I. Lacík and A. M. van Herk, *Pure Appl. Chem.*, 2007, **79**, 1463.
46. I. Lacík, S. Beuermann and M. Buback, *Macromolecules*, 2001, **34**, 6224.
47. I. Lacík, S. Beuermann and M. Buback, *Macromolecules*, 2003, **36**, 9355.
48. I. Lacík, S. Beuermann and M. Buback, *Macromol. Chem. Phys.*, 2004, **205**, 1080.
49. F. Ganachaud, R. Balic, M. J. Monteiro and R. G. Gilbert, *Macromolecules*, 2000, **33**, 8589.
50. S. A. Seabrook, M. P. Tonge and R. G. Gilbert, *J. Polym. Sci., Part A: Polym. Chem.*, 2005, **43**, 1357.
51. M. Stach, I. Lacík, D. Chorvát Jr., M. Buback, P. Hesse, R. A. Hutchinson and L. Tang, *Macromolecules*, 2008, **41**, 5174.
52. J. P. A. Heuts, R. G. Gilbert and L. Radom, *Macromolecules*, 1995, **28**, 8771.
53. S. Beuermann, M. Buback, P. Hesse, S. Kukuckova and I. Lacík, *Macromol. Symp.*, 2007, **248**, 23.
54. L. Lacik, L. Ucnova, S. Kukuckova, M. Buback, P. Hesse and S. Beuermann, *Macromolecules*, 2009, **42**, 7753.
55. P. J. Kubisa, *Polym. Sci. Part A: Polym. Chem.*, 2005, **43**, 4675.
56. P. Kubisa, *Prog. Polym. Sci.*, 2004, **29**, 3.
57. K. Hong, H. Zhang, J. W. Mays, A. E. Visser, C. S. Brazel, J. D. Holbrey, W. M. Reichert and R. D. Rogers, *Chem. Commun.*, 2002, 1368.
58. V. Strehmel, A. Laschewsky, H. Wetzel and E. Görnitz, *Macromolecules*, 2006, **39**, 923.
59. H. Zhang, K. Hong, M. Jablonsky and J. W. Mays, *Chem. Commun.*, 2003, 1356.
60. M. G. Benton and C. S. Brazel, *Polym. Int.*, 2004, **53**, 1113.
61. S. Harrisson, S. R. Mackenzie and D. M. Haddleton, *Chem. Commun.*, 2002, 2850.
62. S. Harrisson, S. R. Mackenzie and D. M. Haddleton, *Macromolecules*, 2003, **36**, 5072.
63. I. Woecht, G. Schmidt-Naake, S. Beuermann, M. Buback and N. García, *J. Polym. Sci. Part A: Polym. Chem.*, 2008, **46**, 1460.
64. C. F. Jasso, E. Mendizabal and M. E. Hernandez, *Rev. Plast. Mod.*, 1991, **62**, 823.

65. I. Capek, *Collect. Czech. Chem. Commun.*, 1986, **51**, 2546.
66. B. Erussalimsky, N. Tumarkin, F. Duntoff, S. Lyubetzky and A. Goldenberg, *Makromol. Chem.*, 1967, **104**, 288.
67. V. Mahadevan and M. Santhappa, *Makromol. Chem.*, 1955, **16**, 119.
68. D. Kukulj, T. P. Davis and R. G. Gilbert, *Macromolecules*, 1998, **31**, 994.
69. H. Kapfenstein-Doak, C. Barner-Kowollik, T. P. Davis and J. Schweer, *Macromolecules*, 2001, **34**, 2822.
70. S. P. Potnis and A. M. Deshpande, *Makromol. Chem.*, 1972, **153**, 139.
71. K. Matsuo, G. W. Nelb, R. G. Nelb and W. H. Stockmayer, *Macromolecules*, 1977, **10**, 654.
72. W. H. Starnes, F. C. Schilling, K. B. Abbas, R. E. Cais and F. A. Bovey, *Macromolecules*, 1979, **12**, 556.
73. J. G. Braks and R. Y. M Huang, *J. Appl. Polym. Sci.*, 1978, **22**, 3111.
74. G. Ayrey and A. C. Haynes, *Makromol. Chem.*, 1974, **175**, 1463.
75. W. A. Pryor, A. Lee and C. E. Witt, *J. Am. Chem. Soc.*, 1964, **86**, 4229.
76. I. M. Bel'govskii, L. S. Sakhonenko and N. S. Enikolopyan, *Vysokomolekul. Soedin.*, 1966, **8**, 369.
77. J. A. May and W. B. Smith, *J. Phys. Chem.*, 1968, **72**, 216.
78. S. Henrici-Olive and S. Olive, *Fortschr. Hochpolym. Forsch.*, 1961, **2**, 496.
79. C. P. R. Nair, M. C. Richou, P. Chaumont and G. Clouet, *Eur. Polym. J.*, 1990, **26**, 811.
80. N. Y. Kaloforov and E. Borsig, *J. Polym. Sci. Polym. Chem.*, 1973, **11**, 2665.
81. G. Gleixner, J. W. Breitenbach and O. F. Olaj, *Makromol. Chem.*, 1977, **178**, 2249.
82. L. Businelli, H. Deleuze, Y. Gnanou and B. Maillard, *Macromol. Chem. Phys.*, 2000, **201**, 1833.
83. J. P. A. Heuts, T. P. Davis and G. T. Russell, *Macromolecules*, 1999, **32**, 6019.
84. C. P. R. Nair, G. Clouet and P. Chaumont, *J. Polym. Sci. Polym. Chem.*, 1989, **27**, 1795.
85. A. H. Krause, *Rubber Age*, 1954, **75**, 217.
86. T. Otsu and M. Yoshida, *Makromol. Chem. Rapid Commun.*, 1982, **29**, 127.
87. G. Gans and D. Duesentrieb, *Enth. J. Polym. Res.*, 1954, **206**, 503.
88. N. S. Enikolopyan, B. R. Smirnov, G. V. Ponomarev and I. M. Belgovski, *J. Polym. Sci. Polym. Chem.*, 1981, **19**, 879.
89. A. Debuignea, R. Polib, C. Jérômea, R. Jérômea and C. Detrembleura, *Progress in Polymer Science*, 2009, **34**, 211.
90. J. P. A. Heuts and N. M. B. Smeets, *Polym Chem.*, 2011, **2**, 2407.
91. A. A. Gridnev and S. D. Ittel, *Chem. Rev.*, 2001, **101**, 3611.
92. G. E. Roberts, J. P. A. Heuts and T. P. Davis, *Macromolecules*, 2000, **33**, 7765.
93. W. I. Bengough and T. O'Neill, *Trans. Faraday Soc.*, 1968, **64**, 1014.
94. W. I. Bengough and N. M. Chawdry, *J. Chem. Soc., Faraday Trans. 1*, 1972, **68**, 1807.

95. M. J. Roedel, *J. Am. Chem. Soc.*, 1953, **75**, 6110.
96. J. S. S. Toh, D. M. Huang, P. A. Lovell and R. G. Gilbert, *Polymer*, 2001, **42**, 1915.
97. D. Britton, F. Heatley and P. A. Lovell, *Macromolecules*, 2001, **34**, 817.
98. D. Britton, F. Heatley and P. A. Lovell, *Macromolecules*, 1998, **31**, 2828.
99. D. Boschmann and P. Vana, *Macromolecules*, 2007, **40**, 2683.
100. R. X. E. Willemse, A. M. van Herk, E. Panchenko, T. Junkers and M. Buback, *Macromolecules*, 2005, **38**, 5098.
101. J. M. Asua, S. Beuermann, M. Buback, P. Castignolles, B. Charleux, R. G. Gilbert, R. A. Hutchinson, J. R. Leiza, A. N. Nikitin, J. P. Vairon and A. M. van Herk, *Macromol. Chem. Phys.*, 2004, **205**, 2151.
102. A. N. Nikitin and R. A. Hutchinson, *Macromolecules*, 2005, **38**, 1581.
103. K. F. O'Driscoll and H. K. Mahabadi, *J. Polym. Sci. Polym. Chem.*, 1976, **14**, 869.
104. B. C. Gilbert, J. R. L. Smith, E. C. Milne, A. C. Whitwood and P. Taylor, *J. Chem. Soc. Perk. T. 2*, 1994, **1759**.
105. T. Junkers and C. Barner-Kowollik, *J. Polym. Sci. Polym. Chem.*, 2008, **46**, 7585.
106. J. Barth, M. Buback, P. Hesse and T. Sergeeva, *Macromol. Rapid Commun.*, 2009, **30**, 1969.
107. C. Farcet, J. Belleney, B. Charleux and R. Pirri, *Macromolecules*, 2002, **35**, 4912.
108. C. Plessis, G. Arzamendi, J. M. Alberdi, A. M. van Herk, J. R. Leiza and J. M. Asua, *Macromol. Rapid Commun.*, 2003, **24**, 173.
109. P. A. Lovell, N. M. Ahmad, B. Charleux, C. Farcet, C. J. Ferguson, S. G. Gaynor, B. S. Hawkett, F. Heatley, B. Klumperman, D. Konkolewicz, K. Matyjaszewski and R. Venkatesh, *Macromol. Rapid Commun.*, 2009, **30**, 2002.
110. M. Gaborieau, P. Castignolles, R. Graf, M. Parkinson and M. Wilhelm, *Polymer*, 2009, **50**, 2373.
111. A. N. Nikitin, R. A. Hutchinson, M. Buback and P. Hesse, *Macromolecules*, 2007, **40**, 8631.
112. J. Barth, M. Buback, P. Hesse and T. Sergeeva, *Macromolecules*, 2010, **43**, 4023.
113. A. M. Zorn, T. Junkers and C. Barner-Kowollik, *Macromol. Rapid Commun.*, 2009, **30**, 2028.
114. T. Junkers and C. Barner-Kowollik, *Macromol. Theory Simul.*, 2009, **18**, 421.
115. T. Junkers, F. Bennet, S. P. S. Koo and C. Barner-Kowollik, *J. Polym. Sci. Polym. Chem.*, 2008, **46**, 3433.
116. M. J. Gibian and R. C. Corley, *Chem. Rev.*, 1973, **73**, 441.
117. S. W. Benson, *Acc. Chem. Res.*, 1986, **19**, 335.
118. J. J. Dannenberg and B. Baer, *J. Am. Chem. Soc.*, 1987, **109**, 292.
119. M. Imoto, S. Sakai and T. Ouchi, *J. Chem. Soc. Jpn.*, 1985, **1**, 97.
120. T. Minato, S. Yamabe, H. Fujimoto and K. Fukui, *Bull. Chem. Soc. Jpn.*, 1978, **51**, 1.

121. M. Buback, F. Günzler, G. T. Russell and P. Vana, *Macromolecules*, 2009, **42**, 652.

122. C. Reichardt, *Top. Curr. Chem.*, 1980, **88**, 1.

123. G. Fraenkel and M. J. Geckle, *J. Chem. Soc., Chem. Commun.*, 1980, 55.

124. D. Griller and K. U. Ingold, *Acc. Chem. Res.*, 1976, **9**, 13.

125. D. Griller and P.R. Marriott, *Int. J. Chem. Kinet.*, 1979, **11**, 1163.

126. J. C. Bevington and D. E. Eaves, *Trans. Faraday. Soc.*, 1959, **55**, 1777.

127. B. E. Bailey and A. D. Jenkins, *Trans. Faraday Soc.*, 1960, **56**, 903.

128. C. H. Bamford, A. D. Jenkins and R. Johnston, *Trans. Faraday Soc.*, 1959, **55**, 179.

129. S. Bizilj, D. P. Kelly, A. K. Serelis, D. H. Solomon and K. E. White, *Aust. J. Chem.*, 1985, **38**, 1657.

130. P. Vana, T. P. Davis and C. Barner-Kowollik, *J. Aust. Chem.*, 2002, **55**, 315.

131. C. H. Bamford, R. W. Dyson and G. C. Eastmond, *Polymer*, 1969, **10**, 885.

132. J. C. Bevington, H. W. Melville and R. P. Taylor, *J. Polymer Sci.*, 1954, **14**, 463.

133. C. H. Bamford, G. C. Eastmond and D. Whittle, *Polymer*, 1969, **10**, 771.

134. M. D. Zammit, T. P. Davis, D. M. Haddleton and K. G. Suddaby, *Macromolecules*, 1997, **30**, 1915.

135. R. C. Neuman Jr. and M. J. Amrich Jr., *J. Org. Chem.*, 1980, **45**, 4629.

136. O. F. Olaj, J. W. Breitenbach and B. Wolf, *Monatsh. Chem.*, 1964, **95**, 1646.

137. K. C. Berger and G. Meyerhoff, *Makromol. Chem.*, 1975, **176**, 1983.

138. K. C. Berger, *Makromol. Chem.*, 1975, **176**, 3575.

139. G. Moad, D. H. Solomon, S. R. Johns and R. I. Willing, *Macromolecules*, 1984, **17**, 1094.

140. G. Moad and D. H. Solomon, *The Chemistry of Radical Polymerization*, 2nd edn, Elsevier, Oxford, 2006, pp. 251.

141. R. Szymanski, *Macromol. Theory Simul.*, 2011, **20**, 8.

142. M. Buback, M. Egorov, R. G. Gilbert, V. Kaminsky, O. F. Olaj, G. T. Russell, P. Vana and G. Zifferer, *Macromol. Chem. Phys.*, 2002, **203**, 2570.

143. A. M. North and G. A. Reed, *Trans. Faraday Soc.*, 1961, **57**, 859.

144. S. W. Benson and A. M. North, *J. Am. Chem. Soc.*, 1962, **84**, 935.

145. A. M. North and G. A. Reed, *J. Polym. Sci. Part A*, 1963, **1**, 1311.

146. A. M. North and G. A. Reed, *Trans. Faraday Soc.*, 1961, **57**, 859.

147. S. Beuermann and M. Buback, *Prog. Polym. Sci.*, 2002, **27**, 191.

148. E. Trommsdorff, H. Kohle and P. Lagally, *Makromol. Chem.*, 1948, **1**, 169.

149. R. G. W. Norrish and R. R. Smith, *Nature*, 1942, **150**, 336.

150. G. A. O'Neil, M. B. Wisnudel and J. M. Torkelson, *Macromolecules*, 1996, **29**, 7477.

151. J. Gao and A. Penlidis, *J. Macromol. Sci., Rev. Macromol. Chem. Phys.*, 1996, **C36**, 199.

152. M. Buback, *Makromol. Chem.*, 1990, **191**, 1575.

153. M. Buback and J. Schweer, *Z. Phys. Chem.*, 1989, **161**, 153.

154. M. Buback, B. Degener and B. Huckestein, *Makromol. Chem., Rapid Commun.*, 1989, **10**, 311.

155. M. Buback and B. Degener, *Makromol. Chem.*, 1993, **194**, 2875.

156. C. Barner-Kowollik and R. T. Gregory, *Prog. Polym. Sci.*, 2009, **34**, 1211.

157. M. Buback, M. Egorov, T. Junkers and E. Panchenko, *Macromol. Chem. Phys.*, 2005, **206**, 333.

158. O. F. Olaj, P. Vana, A. Kornherr and G. Zifferer, *Macromol. Chem. Phys.*, 1999, **200**, 2031.

159. M. Buback and T. Junkers, *Macromol. Chem. Phys.*, 2006, **207**, 1640.

160. P. Vana, T. P. Davis and C. Barner-Kowollik, *Macromol. Rapid Commun.*, 2002, **23**, 952.

161. T. Junkers, A. Theis, M. Buback, T. P. Davis, M. H. Stenzel, P. Vana and C. Barner-Kowollik, *Macromolecules*, 2005, **38**, 9497.

162. G. B. Smith, G. T. Russell and J. P. A. Heuts, *Macromol. Theory Simul.*, 2003, **12**, 299.

163. M. C. Griffiths, J. Strauch, M. J. Monteiro and R. G. Gilbert, *Macromolecules*, 1998, **31**, 7835.

164. M. C. Piton, R. G. Gilbert, B. E. Chapman and P. W. Kuchel, *Macromolecules*, 1993, **26**, 4472.

165. H. K. Mahabadi and K. F. O'Driscoll, *Macromolecules*, 1977, **10**, 55.

166. H. K. Mahabadi and K. F. O'Driscoll, *J. Polym. Sci. Polym. Chem.*, 1977, **15**, 283.

167. P. J. Flory, *Principles of Polymer Chemistry*, Cornell University Press, Ithaca, NY, 1953.

168. M. Buback, P. Hesse, T. Junkers, T. Theis and P. Vana, *Aust. J. Chem.*, 2007, **60**, 779.

169. B. Friedman and B. O'Shaughnessy, *Macromolecules*, 1993, **26**, 5726.

170. G. Johnston-Hall, A. Theis, M. J. Monteiro, T. P. Davis, M. H. Stenzel and C. Barner-Kowollik, *Macromol. Chem. Phys.*, 2005, **206**, 2047.

171. M. Buback, E. Müller and G. T. Russell, *J. Phys. Chem. A*, 2006, **110**, 3222.

172. J. Barth, M. Buback, P. Hesse and T. Sergeeva, *Macromolecules*, 2009, **42**, 481.

173. G. Johnston-Hall and M. J. Monteiro, *Macromolecules*, 2008, **41**, 727.

174. G. Johnston-Hall and M. J. Monteiro, *J. Polym. Sci., Polym. Chem. Ed.*, 2008, **46**, 3155.

175. J. B. L. de Kock, A. M. van Herk and A. L. German, *J. Macromol. Sci., Polym. Rev.*, 2001, **C41**, 199.

176. J. B. L. de Kock, *PhD thesis*, Technical University of Eindhoven, Eindhoven, The Netherlands, 1999.

177. A. Theis, A. Feldermann, N. Charton, M. H. Stenzel, T. P. Davis and C. Barner-Kowollik, *Macromolecules*, 2005, **38**, 2595.

178. T. Junkers, A. Theis, M. Buback, T. P. Davis, M. H. Stenzel, P. Vana and C. Barner-Kowollik, *Macromolecules*, 2005, **38**, 9497.

179. A. Theis, A. Feldermann, N. Charton, T. P. Davis, M. H. Stenzel and C. Barner-Kowollik, *Polymer*, 2005, **46**, 6797.
180. M. Buback, P. Hesse, T. Junkers, T. Theis and P. Vana, *Aust. J. Chem.*, 2007, **60**, 779.
181. M. Buback, T. Junkers and P. Vana, *Macromol. Rapid Commun.*, 2005, **26**, 796.
182. M. Buback, M. Egorov, T. Junkers and E. Panchenko, *Macromol. Chem. Phys.*, 2005, **206**, 333.
183. J. Barth, *PhD thesis*, University of Göttingen, Göttingen, Germany, 2011.
184. M. Stickler, D. Panke and W. Wunderlich, *Makromol. Chem.*, 1987, **188**, 2651.
185. J. Barth, M. Buback, P. Hesse and T. Sergeeva, *Macromolecules*, 2010, **43**, 4023.
186. A. Theis, T. P. Davis, M. H. Stenzel and C. Barner-Kowollik, *Macromolecules*, 2005, **38**, 10323.
187. A. Theis, T. P. Davis, M. H. Stenzel and C. Barner-Kowollik, *Polymer*, 2006, **47**, 999.
188. G. Johnston-Hall, M. H. Stenzel, T. P. Davis, C. Barner-Kowollik and M. J. Monteiro, *Macromolecules*, 2007, **40**, 2730.
189. G. Johnston-Hall and M. J. Monteiro, *Macromolecules*, 2007, **40**, 7171.
190. H. Fischer and H. Paul, *Acc. Chem. Res.*, 1987, **20**, 200.
191. G. T. Russell, R. G. Gilbert and D. H. Napper, *Macromolecules*, 1993, **26**, 3538.
192. C. H. Kurz, *Ph.D. Thesis*, University of Göttingen, Göttingen, 1995.
193. S. Beuermann, M. Buback and C. Schmaltz, *Ind. Eng. Chem. Res.*, 1999, **38**, 3338.
194. J. Barth and M. Buback, *Macromol. Rapid Commun.*, 2009, **30**, 1805.
195. A. J. L. Beckwith, D. Griller, and J. P. Lorand, in *Landolt-Börnstein, Numerical Data and Functional Relationships in Science and Technology*, ed. H. Fischer, Springer-Verlag, Berlin, Heidelberg, New York, Tokyo, 1984, pt A, ch. 1.
196. M. Buback and F.-D. Kuchta, *Macromol. Chem. Phys.*, 1997, **198**, 1445.
197. H. L. Hong, H. W. Zhang, J. W. Mays, A. E. Visser, C. S. Brazel, J. D. Holbrey, W. M. Reichert and R. D. Rogers, *Chem. Commun.*, 2002, 1368.
198. V. Strehmel, A. Laschewsky, H. Wetzel and E. Gornitz, *Macromolecules*, 2006, **39**, 923.
199. H. W. Zhang, K. L. Hong, M. Jablonsky and J. W. Mays, *Chem. Commun.*, 2003, 1356.
200. M. G. Benton and C. S. Brazel, *Polym. Int.*, 2004, **53**, 1113.
201. S. Harrisson, S. R. Mackenzie and D. M. Haddleton, *Macromolecules*, 2003, **36**, 5072.
202. S. Beuermann, I. Woecht, G. Schmidt-Naake, M. Buback and N. Garcia, *J. Polym. Sci. Polym. Chem.*, 2008, **46**, 1460.
203. J. Barth, M. Buback, G. Schmidt-Naake and I. Woecht, *Polymer*, 2009, **50**, 5708.

204. M. Stickler, D. Panke and W. Wunderlich, *Makromol. Chem.*, 1987, **188**, 2651.
205. J. Jacquemin, P. Husson, A. A. H. Padua and V. Majer, *Green Chem.*, 2006, **8**, 172.
206. J. Barth and M. Buback, *Macromolecules*, 2011, **44**, 1292.
207. J. Schrooten, M. Buback, P. Hesse, R. A. Hutchinson and I. Lacik, *Macromol. Chem. Phys.*, 2011, **212**, 1400.
208. G. T. Russell, *Macromol. Theory Simul.*, 1995, **4**, 519.
209. F. Tudos and T. Foldes-Berezsnich, *Magy. Kem. Foly.*, 1982, **88**, 213.
210. M. D. Goldfein, G. P. Gladyshev and A. V. Trubnikov, *Polymer Yearbook*, 1996, **13**, 163.
211. G. C. Eastmond, in *Comprehensive Chemical Kinetics, Volume 14A, Free-Radical Polymerisation*, ed. C. H. Bamford and C. F. H. Tipper, Elsevier, Amsterdam, 1976, p 125.
212. F. Lartigue-Peyrou, *Ind. Chem. Libr.*, 1996, **8**, 489.
213. T. Zytowski and H. Fischer, *J. Am. Chem. Soc.*, 1997, **119**, 12869.
214. T. L. Simandi, A. Rockenbauer and L. I. Simandi, *Eur. Polym. J.*, 1995, **31**, 555.
215. T. L. Simandi and A. Rockenbauer, *Eur. Polym. J.*, 1991, **27**, 523.
216. P. E. M. Allen and C. R. Patrick, *Kinetics and Mechanisms of Polymerization Reactions*, John Wiley & Sons, New York, 1974.
217. L. H. Peebles, *Molecular Weight Distributions in Polymers*, Interscience, New York, 1971.
218. C. H. Bamford, W. G. Barb, A. D. Jenkins and P. F. Onyon, *The Kinetics of Vinyl Polymerization by Radical Mechanisms*, Academic Press, New York, 1958.
219. H. Sawada, *J. Macromol. and Rev. Sci., Macromol. Chem.*, 1969, **C3**, 313.
220. D. G. Hawthorne and D. H. Solomon, *J. Polym. Sci. Polym. Symp.*, 1976, **55**, 211.
221. E. W. Merrill, *Plast. Eng.*, 1996, **33**, 13.
222. F. Fairbrother, G. Gee and G. T. Merrall, *J. Polymer Sci.*, 1955, **16**, 459.
223. A. V. Tobolsky, *Polym. Prepr., Amer. Chem. Soc., Div. Polym. Chem.*, 1970, **11**, 165.
224. R. Steudel, S. Passlack-Stephan and G. Holdt, *Z. Anorg. Allg. Chem.*, 1984, **517**, 7.
225. H. Sawada, *Encycl. Polym. Sci. Eng.*, 1985, **4**, 719.
226. Z. Szablan, M. H. Stenzel, T. P. Davis, L. Barner and C. Barner-Kowollik, *Macromolecules*, 2005, **38**, 5944.
227. H. W. McCormick, *J. Polymer Sci.*, 1957, **25**, 488.
228. W. Wang and R. A. Hutchinson, *Chem. Eng. Technol.*, 2010, **33**, 1745.
229. Z. Szablan, H. Ming and M. Adler, *J. Polym. Sci. A, Polym. Chem.*, 2007, **45**, 1931.
230. W. Wang, R. A. Hutchinson and M. C. Grady, *Ind. Eng. Chem. Res.*, 2009, **48**, 4810.
231. R. A. Hutchinson, D. A. Paquet, S. Beuermann and J. H. McMinn, *Ind. Eng. Chem. Res.*, 1998, **37**, 3567.

CHAPTER 2

Fundamental Aspects of Living Polymerization

The University of Southern Mississippi, School of Polymers and High
Performance Materials, Hattiesburg, MS 39406, USA
Email: Robson.Storey@usm.edu

2.1 Introduction

The term living polymerization describes any chain growth polymerization that
proceeds in the absence of chain termination and chain transfer reactions.
This condition of uninterrupted chain growth confers great synthetic utility,
enabling the synthesis of sophisticated macromolecules such as block copolymers,
star polymers, end-functional (telechelic) polymers, uniform graft copolymers, *etc.*

Living polymerization was first observed by Ziegler[1] and Abkin and
Medvedev,[2] but its profound significance upon polymer synthetic chemistry
was first caused to be generally recognized by Szwarc *et al.*[3,4] who successfully
polymerized styrene and butadiene using sodium naphthalide in tetrahy-
drofuran and suggested the utility of living polymerization for the synthesis of
block copolymers. In the original system of Szwarc *et al.*, which was carried out
in a good solvating medium for ions, essentially all of the chains were active in
propagation simultaneously. Species possessing different states of association,
e.g., aggregated paired ions, single paired ions, solvent-separated ion pairs and
free dissociated ions, existed in equilibrium with one another, but in general
their interconversion was fast, and the positions of the various equilibria were

RSC Polymer Chemistry Series No. 4
Fundamentals of Controlled/Living Radical Polymerization
Edited by Nicolay V Tsarevsky and Brent S Sumerlin
© The Royal Society of Chemistry 2013
Published by the Royal Society of Chemistry, www.rsc.org

not perturbed by propagation. This fact, in conjunction with a rate of initiation comparable to or faster than propagation and an absence of significant depropagation (temperature well below the ceiling temperature), enabled the synthesis of polymers with very narrow molecular weight distributions. It has been suggested that polymerizations, which are both living and of narrow molecular weight distribution by virtue of fast initiation, rapid interconversion of multiple species, and absence of significant depropagation, be termed controlled/living.[5] Such narrow molecular weight distribution polymers are nowadays the hallmark of living polymerization, but the three latter conditions upon which they additionally depend were not part of the original definition of living polymerization given by Szwarc.

Since the original discovery of living anionic polymerization, many different polymerization processes have appeared that, although not strictly living according to the original Szwarc definition, exhibit characteristics of living polymerizations such as molecular weight control, narrow polydispersity, end group control, and enablement of block copolymer synthesis. These newer processes include group transfer polymerization (GTP),[6–9] living carbocationic polymerization (LCP),[10–12] and controlled/living radical polymerizations including stable-radical-mediated polymerization (SRMP),[13,14] atom transfer radical polymerization (ATRP),[15,16] and reversible addition-fragmentation chain transfer (RAFT) polymerization.[17,18] Nearly all of these newer processes differ from the original living polymerization system introduced by Szwarc *et al.* in one important respect, and that is the existence of an equilibrium between active and dormant (reversibly terminated) chains. It has been recently suggested that such systems be referred to as reversible-deactivation polymerizations (RDP).[19] As pointed out by Iván,[20] living polymerizations consisting of active and dormant chains are the more general phenomenon, and the original system of Szwarc *et al.*, with all the chains active all of the time, was actually a special case of the more general classification.

Although the subject of this book is controlled/living radical polymerization, this chapter will discuss those fundamental features common to all living polymerizations, and the discussion will generally concern a generic active species that could be anionic, cationic, or free radical. In some cases, features that are particularly relevant to a specific type of living radical polymerization will be addressed. Several excellent reviews specifically directed to controlled/living radical polymerizations have been published by Matyjaszewski.[21–23]

2.2 Diagnostic Criteria for Livingness of Polymerizations

In the absence of chain transfer, and if initiation is complete prior to significant conversion of the monomer, then the degree of polymerization, $\overline{X}_n$, of the formed polymer molecules in a living polymerization will increase linearly with monomer conversion, p, and $\overline{X}_n$ at any conversion will be predetermined by the concentration of consumed monomer, $[M]_o - [M]$, divided by the initiator concentration, $[I]_o$:

$$\overline{X}_n = \frac{[M]_o - [M]}{[I]_o} = p\frac{[M]_o}{[I]_o} \tag{2.1}$$

Eqn (2.1) provides a diagnostic criterion for absence of chain transfer in a chain propagation process. If an experimental polymerization obeys eqn (2.1), then it is free of chain transfer within the limits of detection. As will be discussed later, the ease of detection of chain transfer is related to $[I]_o$; the higher is $[I]_o$, the more difficult it is to detect chain transfer, and *vice versa*. Figure 2.1 shows a plot of $\overline{X}_n$ *vs. p* for an ideal living polymerization (Theoretical), for a polymerization in which chain transfer is operable (Chain Transfer), and for two polymerizations that are free of transfer but characterized by slow initiation. For all four plots, the target degree of polymerization is $[M]_o/[I]_o = 100$; for the Slow Initiation plots, $[I]_o = 0.01\ mol\ L^{-1}$.

The $\overline{X}_n$ *vs. p* plot of Figure 2.1 is not affected by termination, provided that 100% of the chains do not become terminated prior to complete monomer conversion.[24] With complete initiation, and in the absence of chain transfer, the number of polymer molecules is determined solely by the quantity $[I]_o$; however, the analysis of degree of polymerization, upon which Figure 2.1 is based, cannot distinguish between chains that were living *vs.* chains that were already terminated at the time of analysis.

To demonstrate that a polymerization is free from termination, one must show that the rate of polymerization remains proportional to monomer

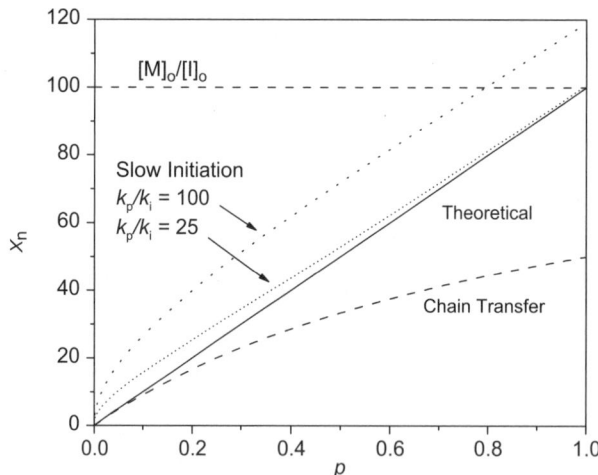

Figure 2.1 Number average degree of polymerization *vs.* monomer conversion with $[M]_o/[I]_o = 100$. Theoretical curve is ideal living and assumes instantaneous initiation and no transfer. Slow initiation curves assume no transfer and $[I]_o = 0.01\ mol\ L^{-1}$. Chain transfer curve assumes instantaneous initiation and a transfer rate constant that is 100 times smaller than the propagation rate constant.

concentration throughout the course of reaction, *i.e.*, that the polymerization is characterized by a constant number of growing chains. This is most conveniently done by constructing a plot of $\ln([M]_o/[M])$ *vs.* time (semi-logarithmic first-order kinetic plot), according to eqn (2.2):

$$\ln \frac{[M]_o}{[M]} = k_p[I]_o t \qquad (2.2)$$

where, k_p = rate constant for propagation (L mol^{-1} s^{-1}) (as discussed later, k_p is often actually an apparent rate constant.)

If the polymerization is free from termination, and initiation is fast, then the resulting plot will be linear and pass through the origin. The inverse statement is not necessarily true; if the first-order plot is linear, then the polymerization may be free of termination or it could be characterized by a stationary-state condition, such as for many radical polymerizations. However, if linearity of the first-order plot is due to stationary-state kinetics, then the $\overline{X}_n$ *vs.* p plot will not be linear. We may therefore state that, if a polymerization is characterized by a linear first-order plot that passes through the origin, and a plot of $\overline{X}_n$ *vs.* p is also linear and passes through the origin, the polymerization is free of both termination and chain transfer, and is thus living. Penczek *et al.* have suggested that the chain transfer and termination criteria be incorporated into a single equation, which alone provides both a necessary and sufficient condition for livingness:[24]

$$\ln\left(1 - \frac{[I]_o}{[M]_o}\overline{X}_n\right) = -k_p[I]_o t \qquad (2.3)$$

Figure 2.2 shows a hypothetical first-order plot for an ideal living polymerization (Theoretical), for a polymerization in which unimolecular chain termination is operable (Termination), and for two polymerizations that are free of termination but characterized by different degrees of slow initiation. For all four plots, $[I]_o = 0.01$ mol L^{-1}.

2.3 Quantifying Degree of Livingness

A living polymerization is a theoretical model or idealized system (akin to an ideal gas in thermodynamics) that is approached more or less closely by real systems. Several authors including Matyjaszewski[25] and Penczek and Duda[26] have suggested that the degree to which a given polymerization process approaches the ideal may be quantified by the ratio of the propagation rate constant to that of the chain-breaking reaction responsible for non-ideality, *i.e.*, termination, chain transfer to monomer, chain transfer to solvent, *etc.* Thus, the livingness of a polymerization with regard to absence of termination may be quantified by the ratio of the rate constant for propagation to the rate constant for termination, k_p/k_t. Likewise, the livingness of a system with regard to its freedom from chain transfer to monomer may be quantified by the ratio of the rate constant for propagation to the rate constant for chain transfer to monomer, $k_p/k_{tr,M}$. This ratio

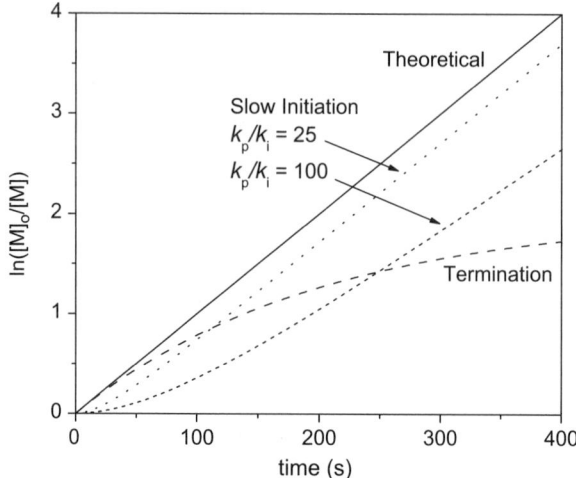

Figure 2.2 First-order kinetic plots for $[I]_o = 0.01 \, \text{mol L}^{-1}$. Theoretical curve is ideal living and assumes instantaneous initiation and no termination. Slow initiation curves assume no termination and $[M]_o = 1 \, \text{mol L}^{-1}$. Termination curve assumes instantaneous initiation and unimolecular (spontaneous) termination with $k_{t,sp}/k_p = 0.005 \, \text{mol L}^{-1}$.

is, of course, the inverse of the chain transfer constant, C_M. The higher these numbers, the greater the livingness of the particular polymerization process. The greater the livingness, the higher may be the molecular weight of well-defined polymers produced from that particular process. The livingness of a polymerization, quantified by rate constant ratios in this way, is an intrinsic or absolute livingness. Polymerizations also manifest a quality termed apparent livingness, which will be discussed in a later section. The meaning of apparent livingness is how well a polymerization system meets some targeted polymer molecular characteristic (*e.g.*, molecular weight), regardless of how easily met is that target.

2.4   Types of Living Systems

Living polymerization systems may generally be classified as either belonging to the reversible-deactivation or degenerative-transfer type (Figure 2.3). The former class includes classical anionic polymerization of olefins (in the limit of $k_d \cong 0$), anionic polymerization of epoxides, GTP (associative mechanism[27]), carbocationic polymerization of olefins, cationic ring opening polymerizations involving tertiary oxonium ion centers, SRMP, and ATRP. For ionic polymerizations, G is a counterion; for SRMP, G is a persistent radical. Activation may be catalyzed, such as for ATRP. Polymerization systems of the degenerative-transfer type include GTP (dissociative mechanism[8,9]) and RAFT. In Figure 2.3, if $\sim$P* represents a radical species, irreversible

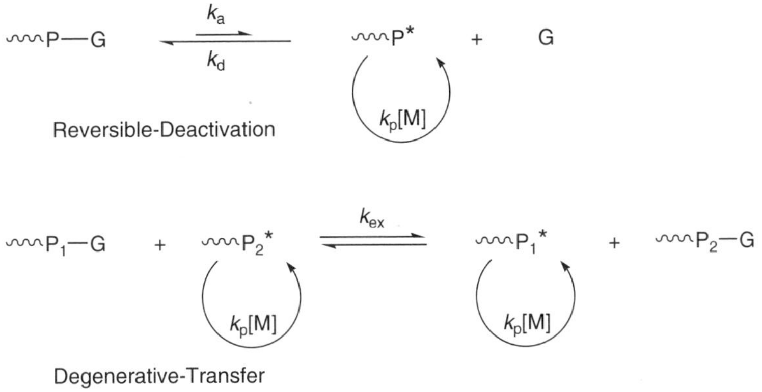

Figure 2.3 Reversible-deactivation and degenerative-transfer type polymerization systems.

Table 2.1 Typical identity of G (see Figure 2.3) for various polymerization systems.

Reversible Termination System	*G*
Anionic polymerization	Metal cation
Cationic polymerization	Halogen, ester
Nitroxide mediated polymerization	Nitroxide radical
Atom transfer radical polymerization	Halogen

Degenerative Transfer System	*G*
Group transfer polymerization	Silyl ketene acetal
Reversible addition fragmentation-transfer polymerization	Dithioester, trithiocarbonate

bimolecular termination can also take place, but when the radical concentration is maintained very low, as in most controlled/living radical polymerizations, the rate of termination can be negligible. The identity of G in Figure 3 varies from system to system; typical examples are listed in Table 2.1.

2.5 Propagation Kinetics of Reversible-Deactivation Polymerizations

The kinetics of living polymerizations with reversible termination may be treated as follows. The rate of propagation is given by the following:

$$\frac{-d[M]}{dt} = k_p[\sim P^*][M] \qquad (2.4)$$

where, $[M]$ = concentration of monomer and $[\sim P^*]$ = concentration of instantaneously active chains.

Since $[\sim P^*]$ is often unknown and difficult to measure, whereas the total concentration of growing chains, $[CE]$ ($= [\sim P - G] + [\sim P^*]$), is generally known from the stoichiometry of the polymerization (*i.e.* $[CE] =$ initiator concentration, $[I]_o$), eqn (2.4) is typically recast in terms of $[CE]$:

$$\frac{-d[M]}{dt} = k_{app}[CE][M] \tag{2.5}$$

where, $k_{app} =$ apparent rate constant for propagation (L mol^{-1} s^{-1}).

Integration of eqn (2.5) yields:

$$\ln\frac{[M]_o}{[M]} = k_{app}[CE]t \tag{2.6}$$

Eqn (2.6) is the same as eqn (2.2) introduced earlier, except that the rate constant for propagation is now shown as an apparent rate constant, which recognizes the existence of reversible deactivation. $[CE]$ is related to the concentration of active species through the activation-deactivation equilibrium, $K_{eq} = k_a/k_d$ as follows:

$$[CE] = \frac{K_{eq} + 1}{K_{eq}}[\sim P^*] \tag{2.7}$$

Obviously, if K_{eq} is small, the term $(K_{eq} + 1)/K_{eq}$ reduces to $1/K_{eq}$. Eqn (2.7) can assume a number of different forms depending on whether activation is catalyzed and the nature and concentration of G and its state of association with the active chain end. The form presented here applies to an uncatalyzed (thermal) activation process and a unimolecular deactivation process characteristic of, for example, ionic polymerizations in which the active species is a paired ion. This represents the simplest case; alternative forms can be more complicated, but the principles to be demonstrated are essentially the same. For example, if the activation process is catalyzed, the following approximation, derived by Müller,[28] holds for $[\sim P^*] \ll [CE]$:

$$[CE] = \frac{[\sim P^*]}{K_{eq}([C]_o - [\sim P^*])} \tag{2.8}$$

where, $[C]_o =$ initial catalyst concentration.

It is noteworthy that the same proportionality seen in eqn (2.7) exists between k_p, the true bimolecular rate constant for propagation, and k_{app}:

$$k_p = \frac{K_{eq} + 1}{K_{eq}}k_{app} \tag{2.9}$$

Substituting eqn (2.7) or (2.9) into the integrated rate expression (eqn (2.6)) yields:

$$\ln\frac{[M]_o}{[M]} = k_{app}\frac{K_{eq} + 1}{K_{eq}}[\sim P^*]t = k_p\frac{K_{eq}}{K_{eq} + 1}[CE]t = k_p[\sim P^*]t \tag{2.10}$$

For classical living anionic polymerizations and other systems in which all of the chains are active all of the time, $K_{eq} = \infty$, and thus $[CE] = [\sim P^*]$, and $k_{app} = k_p =$ true bimolecular rate constant for propagation.

2.6 Chain Breaking Reactions in Reversible-Deactivation Polymerizations

Chain transfer reactions are bimolecular or unimolecular (spontaneous). Typical bimolecular chain transfer reactions are transfer to monomer, initiator, and external chain transfer agents (especially impurities), and intermolecular chain transfer to polymer; typical unimolecular chain transfer reactions are transfer to counterion in ionic polymerizations, intramolecular chain transfer to polymer, and transfer to solvent (pseudo unimolecular).

Chain transfer to monomer is a fundamentally important bimolecular chain transfer reaction because it usually sets the upper limit of attainable molecular weight in a chain polymerization system that has been rigorously purified of all other chain transfer agents. It consists of reaction of the propagating chain with monomer *via* some reaction other than the normal propagation reaction. As such, its rate equation is identical in form to that for propagation:

$$\frac{d[P_{tr,M}]}{dt} = k_{tr,M}[\sim P^*][M] \tag{2.11}$$

where, $[P_{tr,M}] =$ concentration of terminated chains formed by chain transfer to monomer and $k_{tr,M} =$ rate constant for chain transfer to monomer.

Terminated chains, $P_{tr,M}$, are formed by transfer of a hydrogen atom, proton, or other reactive species from the monomer to the active chain end, or *vice versa*; for example, in radical polymerization systems this would normally be transfer of a hydrogen atom from monomer to the active chain end; in carbocationic polymerization systems, it would be transfer of a proton from the active chain end to monomer. In certain systems, for example anionic polymerization of propylene oxide,[29] $P_{tr,M}$ is indistinguishable from the dormant form of the chain end, and thus is not actually terminated in an absolute sense, but simply adds to the reservoir of reversibly terminated chains.

Dividing eqn (2.11) by eqn (2.4) and integrating yields the following equation, which indicates that the concentration of terminated chains, $P_{tr,M}$, is a linear function of monomer conversion, $p = ([M]_o - [M])/[M]_o$:

$$[P_{tr,M}] = \frac{k_{tr,M}}{k_p} p[M]_o \tag{2.12}$$

Often the transferred reactive species (*i.e.*, radical, cation, anion, *etc.*) rapidly re-initiates monomer. In this case, the polymerization kinetics are unaffected, and each chain transfer event creates one additional polymer chain. Thus, the total number of chains in the system, $[P]_{tot}$, is given by:

$$[P]_{tot} = [CE] + [P_{tr,M}] = [I]_o + [P_{tr,M}] \tag{2.13}$$

The "Chain Transfer" curve shown earlier in Figure 2.1 was constructed using $\overline{X}_n = p[M]_o/[P]_{tot}$, with $k_{tr,M}/k_p = 0.01$.

Two important observations can be made regarding eqn (2.12) and (2.13). As discussed earlier, the absolute degree of livingness of a polymerization with regard to chain transfer may be characterized by the ratio $k_{tr,M}/k_p$. The smaller this number (or the greater its reciprocal), the greater is the degree of livingness. It should also be noted that, the higher the initiator concentration, $[I]_o$ (*i.e.*, the lower the target molecular weight), the less significant becomes the fraction of terminated chains relative to the total number of chains. Therefore, for a given value of $k_{tr,M}/k_p$, the apparent livingness of a system increases as the target molecular weight decreases.

Chain transfer to a chain transfer agent, S, (*e.g.* an impurity) is handled mathematically similarly to the above and results in the following expression:

$$[P_{tr,S}] = [S]_o\left[1-(1-p)^{C_S}\right] \quad (p < 1) \tag{2.14}$$

where, $[P_{tr,S}]$ = concentration of terminated chains formed by chain transfer to impurity S, $[S]_o$ = initial concentration of S and C_S = chain transfer constant for S ($k_{tr,S}/k_p$).

Unimolecular (spontaneous) chain breaking processes include unimolecular chain transfer and unimolecular termination. Both cause a first-order decay of the original living chains (*i.e.*, those chains initiated from the initiator, I), and hence negatively impact such synthetic objectives as block copolymer preparation and end group control. In the case of spontaneous transfer, the transferred reactive species (*e.g.*, radical, cation, anion, *etc.*) re-initiates a new chain, and therefore, the polymerization kinetics remain the same but the degree of polymerization is reduced. In the case of spontaneous termination, the polymerization kinetics are affected, but the average degree of poly-merization is unaffected at a given monomer conversion.

The rate expression for spontaneous chain transfer is given by:

$$\frac{d[P_{tr,sp}]}{dt} = k_{tr,sp}[\sim P^*] \tag{2.15}$$

where, $[P_{tr,sp}]$ = concentration of terminated chains formed by spontaneous chain transfer and $k_{tr,sp}$ = rate constant for spontaneous chain transfer.

Dividing eqn (2.15) by eqn (2.4) and integrating yields the following equation, which indicates that the concentration of terminated chains, $[P_{tr,sp}]$, increases exponentially with monomer conversion:

$$[P_{tr,sp}] = -\frac{k_{tr,sp}}{k_p}\ln(1-p) \tag{2.16}$$

The livingness of a polymerization with regard to spontaneous chain transfer is characterized by the ratio $k_{tr,sp}/k_p$. The smaller this number (or the greater its reciprocal), the greater is the degree of absolute livingness. Since the transferred reactive species is assumed to rapidly re-initiate monomer, polymerization kinetics are unaffected and described by eqn (2.10), and $[P]_{tot}$ increases according to eqn (2.13). Thus, the higher is $[I]_o$ (lower is the target molecular weight) the less noticeable is the effect of spontaneous chain transfer (greater apparent livingness). Eqn (2.16) clearly shows that the incidence of spontaneous chain transfer increases dramatically at high monomer conversion.

If the transferred reactive species does not re-initiate monomer, because for example, it is too unreactive or it is purposefully trapped, or if some purely terminative chain breaking process is involved, then the process is termed unimolecular or spontaneous termination. In this case $[P]_{tot}$ remains constant and equal to $[I]_o$, and we must revise eqn (2.13) as follows:

$$[P]_{tot} = [I]_o = [CE]_o = [CE] + [P_{t,sp}] = [\sim P - G] + [\sim P*] + [P_{t,sp}] \quad (2.17)$$

where, $[CE]_o$ = original concentration of growing chains (dormant plus active) prior to unimolecular termination and $[P_{t,sp}]$ = concentration of terminated chains formed by spontaneous termination.

The rate of spontaneous termination is written as:

$$\frac{d[P_{t,sp}]}{dt} = k_{t,sp}[\sim P*] \quad (2.18)$$

where, $k_{t,sp}$ = rate constant for spontaneous termination.

The concentration of active chains, $[\sim P*]$, can be expressed as a function of $[P_{t,sp}]$ through combination with eqn (2.7) and (2.17):

$$\frac{d[P_{t,sp}]}{dt} = k_{t,sp} \frac{K_{eq}}{K_{eq} + 1} \left([I]_o - [P_{t,sp}] \right) \quad (2.19)$$

The solution of differential eqn (2.19) is the following expression,[30] which yields the fraction of chains that have terminated as a function of polymerization time:

$$\frac{[P_{t,sp}]}{[I]_o} = \left(1 - e^{-k_{t,sp}\frac{K_{eq}}{K_{eq}+1}t} \right) \quad (2.20)$$

Eqn (2.20) can also be used to calculate the fraction of original chains that have terminated *via* spontaneous chain transfer, by substituting $[P_{tr,sp}]$ in place of $[P_{t,sp}]$ and $k_{tr,sp}$ in place of $k_{t,sp}$.

Assuming that the dormant-active chain end equilibrium is established instantaneously, spontaneous termination causes a first-order decay of the active species, according to the following expression:

$$[\sim P*] = [\sim P*]_o e^{-k_{t,sp}\frac{K_{eq}}{K_{eq}+1}t} \quad (2.21)$$

Substituting eqn (2.21) into the expression for the rate of propagation (eqn (2.4)), and noting that $[\sim P*]_o = \frac{K_{eq}}{K_{eq}+1}[I]_o$, yields:

$$\frac{-d[M]}{dt} = k_p \frac{K_{eq}}{K_{eq} + 1} [I]_o e^{-k_{t,sp}\frac{K_{eq}}{K_{eq}+1}t}[M] \quad (2.22)$$

Integration of eqn (2.22) results in the following expression:

$$\ln\frac{[M]_o}{[M]} = \frac{k_p}{k_{t,sp}}[I]_o\left(1 - e^{-k_{t,sp}\frac{K_{eq}}{1+K_{eq}}t}\right)$$ (2.23)

The "Termination" curve shown earlier in Figure 2.2 was constructed using eqn (2.23), with $k_{t,sp}/k_p = 0.005$ mol L^{-1}, $[I]_o = 0.01$ mol L^{-1}, and $K_{eq} = \infty$ (all of the chains active all of the time).

The form of eqn (2.23) predicts a maximum conversion, p_{MAX}, beyond which the system cannot progress; thus, at long times, the exponential approaches zero, and eqn (2.23) reduces to:

$$p_{MAX} = 1 - e^{-\frac{k_p}{k_{t,sp}}[I]_o}$$ (2.24)

2.7   Rate of Initiation

Szwarc's original definition of living polymerization did not explicitly address the rate of initiation relative to the rate of propagation,[3,4] but the qualitative description of their sodium naphthalene complex/styrene/THF system implied rapid initiation. The effect of initiation rate on degree of polymerization and kinetics of a living polymerization may be treated generally as follows. The rate of initiation is given by:

$$\frac{-d[I]}{dt} = k_i[I][M]$$ (2.25)

where, $[I] =$ concentration of initiator and $k_i =$ rate constant for initiation.

The rate of monomer consumption is written as the sum of the rate consumed by initiation and propagation, with the growing chain end concentration, $[CE]$, expressed as the concentration of consumed initiator, $[I]_o - [I]$.

$$\frac{-d[M]}{dt} = k_i[I][M] + k_p([I]_o - [I])[M]$$ (2.26)

Combination of eqn (2.25) and (2.26) to remove the time variable, followed by integration, yields the following expression:

$$p[M]_o = [M]_o - [M] = \frac{k_p}{k_i}[I]_o\ln\frac{[I]_o}{[I]} + \left(1 - \frac{k_p}{k_i}\right)([I]_o - [I])$$ (2.27)

The "Slow Initiation" curves shown in Figure 2.1 were constructed as $\overline{X}_n = p[M]_o/([I]_o - [I])$ *vs.* p, with $p[M]_o$ defined by eqn (2.27), and with $[M]_o = 1$ mol L^{-1} and $[I]_o = 0.01$ mol L^{-1}. Examination of Figure 2.1 shows that as k_p/k_i approaches unity, the "Slow Initiation" curves approach the "Theoretical" curve. Indeed for $k_p/k_i = 1$, eqn (2.26) reduces to the classical rate equation for a living polymerization with instantaneous initiation. Thus, "instantaneous" initiation should be understood to simply mean $k_i \geq k_p$.

The effect of slow initiation on kinetics of living polymerizations is treated in the following way. Rearranging eqn (2.27) to yield [M] as a function of [I], and then substituting into eqn (2.25), yields the following non-linear differential equation:

$$\int_{[I]_o}^{[I]} \frac{-d[I]}{[I](C_1 + C_2 \ln[I] + C_3[I])} = k_i t \tag{2.28}$$

where, $C_1 = [M]_o + [I]_o \left[\frac{k_p}{k_i}(1 - \ln[I]_o) - 1\right]$, $C_2 = \frac{k_p}{k_i}[I]_o$, and $C_3 = 1 - \frac{k_p}{k_i}$.

Eqn (2.28) may be solved numerically[31] to yield [I] as a function of t, and then through the use of eqn (2.27), [M] as a function of t can be obtained. The latter was used to construct the "Slow Initiation" curves shown in Figure 2.2, with $[I]_o = 0.01$ mol L^{-1} and $[M]_o = 1$ mol L^{-1}.

2.8 Polydispersity of Living Polymers

The absence of termination and chain transfer in living polymerizations potentially allows all chains to experience the same monomer concentration at equivalent stages of growth and to result in polymers with remarkably narrow size distribution. For this to happen, the initiation rate must be approximately equal to or greater than the propagation rate ($k_i \geq k_p$), and interconversion of chain-end species of different reactivities must be rapid relative to the rate of propagation. The theoretical distribution for such a system was shown by Flory[32] to be characterized by the Poisson distribution:

$$\frac{[P_x]}{[I]_o} = \frac{e^{-\nu}\nu^{x-1}}{(x-1)!} \tag{2.29}$$

where, $[P_x]$ = concentration of growing chains that have added x monomer units and ν = average kinetic chain length, *i.e.*, $([M]_o - [M])/[I]_o$, equivalent to the average degree of polymerization for monofunctional initiators.

Figure 2.4 shows a plot of the Poisson distribution represented by eqn (2.29) for $\nu = 100$. For comparison, the analogous plot is shown for a conventional radical polymerization with $\nu = 100$ and termination strictly by combination of radicals.[33] The comparison is striking and demonstrates how relatively narrow is the distribution that is obtained from an ideal living polymerization with instantaneous initiation.

The weight fraction distribution can be obtained from eqn (2.29) by making use of the monomer unit molecular weight. The number and weight average molecular weights can then be calculated, leading to the following theoretical molecular weight distribution (polydispersity):

$$\frac{\overline{M}_w}{\overline{M}_n} = 1 + \frac{\nu}{(\nu+1)^2} \tag{2.30}$$

Eqn (2.30) predicts that the molecular weight distribution will asymptotically approach 1 as the degree of polymerization approaches infinity.

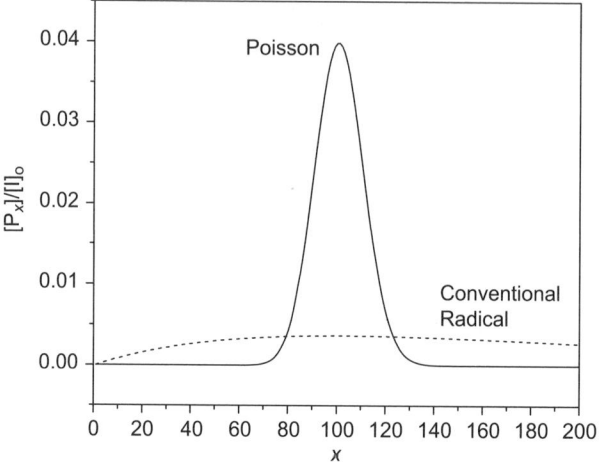

Figure 2.4   Number fraction of polymer chains containing x repeat units *versus* x for a living polymerization displaying a Poisson distribution compared to a conventional radical polymerization with termination by combination.

For reversible-deactivation polymerizations, the distribution predicted by eqn (2.29) and (2.30) will theoretically be obtained under certain circumstances, namely if the exchange between dormant and active chains is rapid relative to propagation. In this regard, it is useful to introduce the concept of run number (RN), which here is defined as the average number of monomer units added per activation-deactivation cycle, and may be less than one.[34] For example, in a reversible-deactivation system with an uncatalyzed activation process and a unimolecular deactivation, $RN = k_p[M]/k_d$.[35] For a degenerative-transfer system, it is given by $RN = k_p[M]/k_{ex}[P-G]$, where [P-G] is the concentration of dormant chains. If a chain undergoes many activation-deactivation cycles per monomer addition, *i.e.* $RN < 1$, then the distribution of the resulting polymer chains will be Poisson, as given by eqn (2.29) and (2.30). For higher run numbers, in general the distribution broadens, but the precise mathematical form of the distribution equation depends on the characteristics of the particular system. For example, activation-deactivation may be of the reversible-deactivation or degenerative-transfer type; activation may be catalyzed or uncatalyzed, and deactivation may be unimolecular or bimolecular involving a deactivator. Litvinenko and Muller[36] have published a review of the various distributions obtained from detailed kinetic analyses of the common systems.

For reversible-deactivation systems that proceed according to the top mechanism pictured in Figure 2.3, Müller *et al.*[37,38] derived the following approximate expression for the polydispersity as a function of monomer conversion, p:

$$\frac{\overline{M}_w}{\overline{M}_n} = 1 + \frac{1}{\beta}\left(\frac{2}{p} - 1\right) \tag{2.31}$$

Two cases apply to eqn (2.31). If deactivation is unimolecular (*e.g.* ion pairs in ionic polymerization, associative mechanism of GTP) the parameter β is given by $\beta = \frac{k_d}{k_p[I]_o}$, and if deactivation is bimolecular involving a deactivating species G (*e.g.* free ions in ionic polymerization, dissociative mechanism of GTP, SRMP), then the parameter β is given by $\beta = \frac{k_d}{k_p[I]_o}[G]$. For degenerative transfer systems that proceed according to the bottom mechanism pictured in Figure 2.3, eqn (2.31) also holds if the parameter β is redefined as follows:[37] $\beta = \frac{k_{ex}}{k_p}$

Eqn (2.31) is valid only for $\beta \geq 10$ (fast exchange, low RN), $[\sim P^*]/[I]_o \leq 10^{-2}$ (equilibrium strongly shifted toward dormant species), long chains ($p[M]_o/[I]_o \gg 1$), and complete initiation. Under conditions of incomplete initiation, a correction must be applied, resulting in the following:[21,36]

$$\frac{\overline{M}_w}{\overline{M}_n} = \left[1 + \frac{1}{\beta}\left(\frac{2}{p} - 1\right)\right]\left[1 - (1-p)^{\beta}\right] \qquad (2.32)$$

Figure 2.5 shows plots of $\overline{M}_w/\overline{M}_n$ *vs.* conversion for polymerization systems possessing a Poisson molecular weight distribution (eqn (2.30)) and three distributions predicted by eqn (2.31) ($\beta = k_d/k_p[I]_o$) with $k_p/k_d = 5$, 2, and 0.5 L mol^{-1}. For the system with slow exchange ($k_p/k_d = 5$ L mol^{-1}), the final distribution at $p = 1.0$ is 1.05, significantly broader than Poisson. For $k_p/k_d = 0.5$ L mol^{-1}, the distribution of the reversible-deactivation polymerization is very nearly Poisson at high conversions but deviates from Poisson at low conversions. For $k_p/k_d < 0.5$ L mol^{-1}, eqn (2.31) predicts a distribution that is more narrow than Poisson, especially at low conversions.

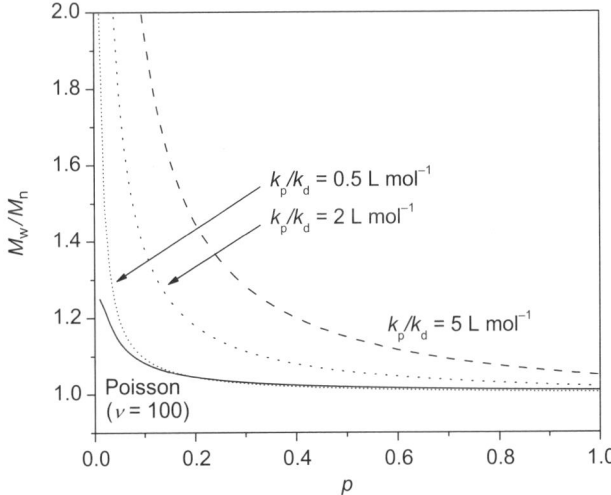

Figure 2.5 $\overline{M}_w/\overline{M}_n$ *vs.* conversion for polymerization systems possessing a Poisson molecular weight distribution (eqn (2.30)) and three distributions predicted by eqn (2.31) ($\beta = k_d/k_p[I]_o$) with $k_d/k_p = 5$, 2, and 0.5 mol L^{-1}. For all systems, $[I]_o = 0.01$ mol L^{-1}.

2.9   Shelf Life of Living Polymerizations

The concept of shelf life (shelf time) of a chain polymerization has been discussed by Szwarc.[39] Shelf life in this context is defined as the time available to the operator to complete a given synthetic task, for example, block copolymer synthesis by sequential monomer addition, or end-capping (functionalization) of the polymer chains by reaction with a quenching agent. The equations dealing with chain breaking in the previous section may be used to demonstrate the concept of shelf life and how it is favorably affected by the operation of a reversible-deactivation equilibrium.

The discussion below follows closely to that of Szwarc[39] and applies to a cationic chain polymerization, but the principle is general. A traditional non-living carbocation polymerization system might be characterized by the following typical kinetic parameters:

$$k_p = 10^5 \, L \, mol^{-1} \, s^{-1}$$
$$k_{tr,sp} = 0.1 \, s^{-1}$$
$$k_{tr,M} = 0$$
$$K_{eq} = \infty$$
$$[CE]_o = [\sim P^*] = 5 \, mmol \, L^{-1}$$

Thus, in the particular system chosen, chain breaking is by spontaneous chain transfer, and there is no dormant-active equilibrium (all centers are active all of the time). The rate constant for propagation represents the composite for paired and free ions. Eqn (2.10) (with $K = \infty$ and $k_{app} = k_p$) shows that for this system, 99% conversion is reached within about 9 ms. Moreover, eqn (2.20) shows that 99% of the original chains have decomposed via spontaneous chain transfer within about 46 s. Clearly, such a system would not be considered living and it would offer no control from the point of view of end group fidelity or block copolymer synthesis.

A similar analysis may be applied to this system, but now assuming that the active chains are in equilibrium with a reservoir of dormant chains with $K_{eq} = 10^{-6}$. Eqn (2.20) shows that now only 1% of the original chains have decomposed in 28 h. However, 95% of the monomer is consumed by these reversibly-deactivated chains in about 100 min. The reversible-deactivation equilibrium has thus transformed a poorly controlled, apparently non-living system into a controlled/living system with practical synthetic possibilities. This analysis also assumes that the ratio $k_{tr,sp}/k_p$ was unaffected by whatever system change (solvent, counterion, temperature, *etc.*) that was carried out to create the large reservoir of dormant chain ends. Very likely, the decrease in ionicity (*e.g.*, elimination of free ions in the system) also caused this number to become smaller such that the intrinsic livingness of the system was also improved.

Another important point should be noted from this example. For many practical applications, a monomer conversion higher than 95% is desired. However, rarely would any synthetic objective require greater than 10 monomer half-lives, which achieves 99.9% conversion. Under the conditions given, this would require about 230 min and would suffer loss of less than 0.01% of the original chains. If the operator were to carelessly leave the polymerization to run over the weekend (~ 64 h), the increase in conversion is of course negligible, but the loss of original chains has increased to 2.3%. Although this degree of operator negligence is extreme, it illustrates the general rule that carrying a living polymerization to excessively long reaction times, say, beyond 10 monomer half-lives, is not good practice since there is little to be gained but much to lose if any spontaneous chain breaking process with significant $k_{t,sp}/k_p$ is operable.

2.10   Conclusion

Since the concept of living polymerization was first introduced by Szwarc, great progress has been made in extending this concept to other classes of chain polymerizations. This was driven chiefly by practical synthetic goals, but also by the elegance of the living polymerization process. As new systems were introduced, it became apparent that the classical anionic systems are actually a special case of the more general process, in which a small concentration of polymerization-active species is spread over a much larger concentration of growing chains, *via* a dynamic exchange process. This circumstance allows the lifetime of a given growing chain to be greatly extended, making possible synthetic manipulations such as sequential monomer addition, coupling reactions, chain end functionalizations, *etc.* The application of this concept to radical chain polymerizations, the topic of this book, is of great practical and theoretical importance due to the broad applicability of radical polymerization to many classes of monomers.

References

1. K. Ziegler, *Angew. Chem.*, 1936, **49**, 499–502.
2. A. Abkin and S. Medvedev, *Trans. Faraday Soc.*, 1936, **32**, 286–296.
3. M. Szwarc, M. Levy and R. Milkovich, *J. Am. Chem. Soc.*, 1956, **78**, 2656–2657.
4. M. Szwarc, *Nature*, 1956, **178**, 1168–1169.
5. K. Matyjaszewski and A. H. E. Müller, *ACS Div. Polym. Chem., Polym. Prepr.*, 1997, **38**(1), 6–9.
6. O. W. Webster, W. R. Hertler, D. Y. Sogah, W. B. Farnham and T. V. RajanBubu, *ACS Div. Polym. Chem., Polym. Prepr.*, 1983, **24**(2), 52–53.
7. O. W. Webster, W. R. Hertler, D. Y. Sogah, W. B. Farnham and T. V. RajanBubu, *J. Am. Chem. Soc.*, 1983, **105**, 5706–5708.
8. R. P. Quirk and G. P. Bidinger, *Polym. Bull.*, 1989, **22**, 63–70.
9. R. P. Quirk and J. Ren, *Macromolecules*, 1992, **25**, 6612–6620.
10. M. Miyamoto, M. Sawamoto and T. Higashimura, *Macromolecules*, 1984, **17**, 265–268.

11. R. Faust and J. P. Kennedy, *J. Polym. Sci., Part A: Polym. Chem.*, 1987, **25**, 1847–1869.
12. J. P. Kennedy and B. Iván, *Designed Polymers by Carbocationic Macro-molecular Engineering: Theory and Practice*, Hanser Publishers, Munich, Germany, 1991, pp. 96–113.
13. D. H. Solomon, E. Rizzardo and P. Cacioli, U.S. Patent 4 581 429, 1985.
14. G. Moad, E. Rizzardo and D. H. Solomon, in *Comprehensive Polymer Science*, ed. G. C. Eastmond, A. Ledwith, S. Russo and P. Sigwalt, P., Pergamon, London, 1989, vol. 3, p. 145.
15. M. Kato, M. Kamigaito and M. Sawamoto, *Macromolecules*, 1995, **28**, 1721–1723.
16. J. S. Wang and K. Matyjaszewski, *J. Am. Chem. Soc.*, 1995, **117**, 5614–5615.
17. J. Chiefari, Y. K. Chong, F. Ercole, J. Krstina, J. Jeffery, T. P. T. Le, R. T. A. Mayadunne, G. F. Meijs, C. L. Moad, G. Moad, E. Rizzardo and S. H. Thang, *Macromolecules*, 1998, **31**, 5559–5562.
18. G. Moad, E. Rizzardo and S. H. Thang, *Aust. J. Chem.*, 2009, **62**, 1402–1472.
19. A. D. Jenkins, R. G. Jones and G. Moad, *Pure Appl. Chem.*, 2010, 483–491.
20. B. Iván, *Makromol. Chem., Macromol. Symp.*, 1993, **67**, 311–324.
21. K. Matyjaszewski, *ACS Symp. Ser.*, 2000, **768**, 2–26.
22. K. Matyjaszewski, in *Handbook of Radical Polymerization*, ed. K. Matyjaszewski, and T. P. David, John Wiley & Sons, Hoboken, 2002, ch. 8, pp. 361–406.
23. W. A. Braunecker and K. Matyjaszewski, *Prog. Polym. Sci.*, 2007, **32**, 93–146.
24. S. Penczek, P. Kubisa and R. Szymanski, *Makromol. Chem., Rapid Comm.*, 1991, **12**, 77–80.
25. K. Matyjaszewski, *Macromolecules*, 1993, **26**, 1787–1788.
26. S. Penczek and A. Duda, *Macromol Symp.*, 1996, **107**, 1–15.
27. W. B. Farnham and D. Y. Sogah, *ACS Div. Polym. Chem., Polym. Prepr.*, 1986, **27**(1), 167–168.
28. A. H. E. Müller, *Macromolecules*, 1994, **27**, 1685–1690.
29. R. W. Body and V. L. Kyllingstad, in *Encyclopedia of Polymer Science and Engineering*, ed. H. F. Mark, N. M. Bikales, C. G. Overberger and G. Menges, Wiley-Interscience, New York, 1986, vol. 6, p. 280.
30. R. F. Storey, C. L. Curry and L. B. Brister, *Macromolecules*, 1998, **31**(4), 1058–1063.
31. The numerical solutions to eqn (2.28) were obtained using the program Mathcad® 15.0, Parametric Technology Corporation.
32. P. J. Flory, *J. Am. Chem. Soc.*, 1940, **62**, 1561–1565.
33. A. M. North, *The Kinetics of Free Radical Polymerization*, Pergamon Press, Oxford, 1966, pp. 14–16.
34. "Run number" was defined as the average number of monomer units per productive activation ($\bar{l} = 1 + \text{RN}$) in the following reference: J. E. Pukas, G. Kaszas and M. Litt, *Macromolecules*, 1991, **24**, 5278–5282.

35. L. K. Breland, Q. A. Smith and R. F. Storey, *Macromolecules*, 2005, **38**, 3026–3028.
36. G. Litvinenko and A. H. E. Müller, *Macromolecules*, 1997, **30**, 1253–1266.
37. A. H. E. Müller, G. Litvinenko and D. Yan, *Macromolecules*, 1996, **29**, 2339–2345.
38. A. H. E. Müller, G. Litvinenko and D. Yan, *Macromolecules*, 1996, **29**, 2346–2353.
39. M. Szwarc, *Makromol. Chem., Rapid Commun*, 1992, **13**, 141–145.

CHAPTER 3

Controlled/Living Radical Polymerization in the Presence of Iniferters

MEHMET ATILLA TASDELEN*[a,b] AND YUSUF YAGCI*[a]

[a] Istanbul Technical University, Faculty of Science and Letters, Chemistry Department, Maslak, TR-34469, Istanbul, Turkey; [b] Yalova University, Faculty of Engineering, Department of Polymer Engineering, TR-77100, Yalova, Turkey
*Email: tasdelen@yalova.edu.tr, yusuf@itu.edu.tr

3.1 Introduction

In an attempt to combine the advantages of living polymerization with the flexibility of free radical polymerization, there has been a significant amount of research into the possibility of controlled/living radical polymerization (C/LRP). These polymerization methods[1-4] allow nearly unlimited control of the polymer's composition, architecture and functionality.[5] It seems possible nowadays to prepare various novel architectures from virtually all kinds of vinyl monomers by C/LRP methods in common mass, suspension or even emulsion processes.

Control over radical polymerization of vinyl monomers was firstly attempted in 1955 by Ferington and Tobolsky.[6] These authors observed that dithiuram disulfides in radical polymerizations lead to higher transfer constants, which resulted in retardation of the polymerization. Borsig et al.[7] reported more evidence of C/LRP; the use of diaryl and triaryl protected groups in the vinyl polymerization increased the molecular weights with conversion and also the

RSC Polymer Chemistry Series No. 4
Fundamentals of Controlled/Living Radical Polymerization
Edited by Nicolay V Tsarevsky and Brent S Sumerlin
© The Royal Society of Chemistry 2013
Published by the Royal Society of Chemistry, www.rsc.org

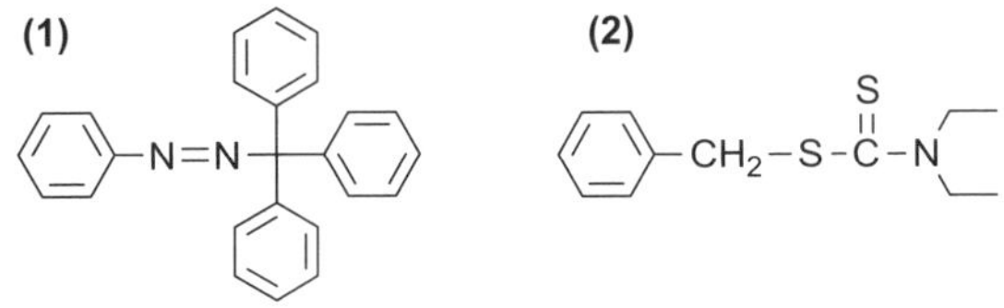

(1) **(2)**

Chart 3.1 Examples of thermal iniferter (1) and photoiniferter (2).

formation of a block copolymer was observed. This system was later extensively studied by Braun.[8] However, initiation efficiencies were low, molar mass and polydispersities were relatively high and molecular weight did not evolve linearly with conversion. A possible reason for these deviations from living characteristics was a slow, but continuous, initiation by the bulky organic radicals. In some cases, these systems could resemble dead-end systems, in which new chains were not generated and transfer did not take place. The ideas of Borsig were later refined *via* iniferter systems and nitroxides, which do not initiate polymerization themselves and reversibly reactivate the growing chains.

In 1982 Otsu,[9] for the first time, provided a model for C/LRP to describe styrene (St) and methyl methacrylate (MMA) polymerization in the presence of phenylazotriphenylmethane and benzyl-*N*,*N*-diethyldithiocarbamate of the following structures (Chart 3.1).

By analogy with the inifers used in carbocationic polymerization,[10] he proposed that dithiocarbamates act as iniferters, namely, agents that initiate, transfer, and terminate.[11] The model that Otsu proposed is very similar to modern C/LRP methods. Mechanistically, C/LRP is distinguished from conventional free radical polymerization by the existence of a reversible activation process. C/LRP methods additionally require fast initiation and relatively slow propagation in order to be able to control the chain-length distribution. The dormant species are turned on to the active chain by thermal, photochemical, and/or chemical stimuli in the presence of monomer and propagates until it is deactivated to dormant species. If a living chain experiences the activation-deactivation cycles frequently enough over a period of polymerization time, all living chains will have a nearly equal chance to grow, yielding a low-polydispersity product.

3.2 Kinetics

Assuming perfect ideal living polymerization in which all polymer chains are induced to initiate the simultaneous propagation of chains without any termination and chain transfer reactions, livingness can be validated by analyzing the linear first order (semilogaritmic) kinetic plot of monomer consumption, conversion-molecular weight and molecular weight-inverse iniferter concentration relationships.[12,13] However, such an interpretation appears to be too simple to describe the whole process of iniferter polymerization, which is far more complex than expected.[14] Iniferter polymerizations, like conventional free radical polymerizations, involve initiation,

Reaction Step	Reaction Details	Reaction Rate Expression
Initiation	i) $\quad A-B \xrightarrow[h\nu]{k_d} A^\bullet + {}^\bullet B$	Iniferter decomposition: $k_d[AB]$
	ii) $\quad A^\bullet + M \xrightarrow{k_i} P_1^\bullet$	Primary radical addition: $k_i[\dot{A}][M]$
Reversible Activation	iii) $\quad P_n B \underset{k_{deact}}{\overset{k_{act}}{\rightleftharpoons}} P_n^\bullet + {}^\bullet B$	Activation: $k_{act}[P_n B]$ Deactivation: $k_{act}[P_n^\bullet][B]$
Propagation	iv) $\quad P_n^\bullet + M \xrightarrow{k_p} P_{n+1}^\bullet$	Propagation: $k_p[P_n^\bullet][M]$
Termination	v) $\quad P_n^\bullet + P_m^\bullet \longrightarrow P_{n+m}$	Combination: $k_{tc}[P^\bullet]^2$
	vi) $\quad P_n^\bullet + P_m^\bullet \xrightarrow{k_{td}} P_n + P_m$	Disproportination: $k_{td}[P^\bullet]^2$

Scheme 3.1   The general mechanism steps of the iniferter polymerization.

propagation, and termination steps (Scheme 3.1).[15] The initiation step relies on the photochemical or thermal dissociation of the iniferter molecule (A–B) into a reactive carbon-centered radical (A$^\bullet$) and a relatively stable diethylthiocarbamyl-thiyl radical (B$^\bullet$) (i). The carbon radicals produced in these reactions are extremely reactive and initiate the polymerization by reacting with a monomeric double bond to generate a macroradical (P$_n{}^\bullet$) whereas the other is less or nonreactive and cannot enter initiation. After the addition of a certain number of monomer species, the macroradical recombines with the stable radical (B$^\bullet$) and is converted into a polymeric macroiniferter molecule (P$_n$B) (ii–iv). The unique feature of this iniferter polymerization is that it proceeds in a controlled manner, in which "active" and "dormant" propagating chain ends are reversibly equilibrated throughout the course of polymerization (iii). The irreversible carbon–carbon radical termination results in a dead unreactive polymer by recombination or disproportionation of the growing radicals.[16] A considerable amount of research has been conducted on the kinetics and the role of iniferters in free radical polymerization, as compared to conventional polymerizations.[17–19] The main difference between iniferter polymerization and free radical polymerization is the reversible activation/deactivation step. Usually, for an effective iniferter polymerization, the deactivation is always faster than the activation. Initially, a lower rate of polymerization is observed due to the slow rate of initiation and the only factor that promotes the propagation reaction is the high ratio of monomer to stable radicals. At high monomer conversion this ratio may decrease to the level where the effect of reversible termination becomes the controlling factor compared with the rates of propagation. With increased conversion, the viscosity of the reaction mixture increases which may also be a considerable factor in decreasing the rate of polymerization. The presence of irreversible terminations may also reduce the free radical concentrations with polymerization time and hence rate of polymerization also decreases. Overall, the rates of polymerization are much lower than those of conventional free radical polymerization which is consistent with controlled polymerization methods.[20] Moreover, Kannurpatti *et al.*[17,18] and Ward *et al.*[19] found that the autoacceleration effect observed in classical free radical polymerization is either reduced or eliminated by iniferter molecules. This is clear evidence of the reversible carbon–thiyl radical combination reactions, which are not diffusion limited, dominating compared to the diffusion limited, and irreversible carbon–carbon radical termination reactions. This domination of carbon–thiyl radical combination reactions prevents the irreversible terminations and other side reactions, so the polymerization proceeds through controlled radical polymerization mechanism.

The success of living polymerization is strongly dependent on the mediating radical B$^\bullet$, and a variety of different stabilized radicals have been developed and employed. The reactivities of bond dissociation of the initiation and propagation steps on the molecular level are needed for the complete understanding of this polymerization. Livingness in this polymerization depends on the dissociation ability of the C–S bond of the iniferters. The C–S bond dissociation energy increases in the order tertiary < secondary < primary carbon, and is reduced with electron-withdrawing groups (*e.g.*, nitrile or carbonyl) attached to the carbon atom (Chart 3.2).

According to Ishizu *et al.*[21–23], among various iniferters, the compound possessing nitrile and carboxylic acid groups (7) enables the control of the molecular weight and molecular weight distribution in the polymerization of styrene and methyl methacrylate. Depending on the type of monomer, the initiator efficiency (*f*) was over 0.9, and the polydispersity was close to 1.2. These polymerization characteristics, high initiation efficiency and slow polymerization rate, may be explained in terms of the following statements: (i) the first dissociation of iniferter and followed by monomer addition proceed at a very high rate due to a very low bond dissociation energy (112 kJ mol^{-1}) and (ii) the polymerization rate, determined by the bond dissociation energy of the macroiniferters (dithiocarbamate end-functionalized polystyrene and poly(methyl methacrylic acid)) is low due to the relatively high bond dissociation energy. In contrast, compound (5) has very low initiator efficiency (*f* = 0.32) for styrene monomer and yields polymers with polydispersity around

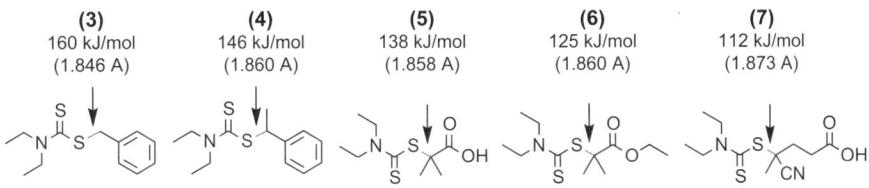

(3)	**(4)**	**(5)**	**(6)**	**(7)**
160 kJ/mol	146 kJ/mol	138 kJ/mol	125 kJ/mol	112 kJ/mol
(1.846 A)	(1.860 A)	(1.858 A)	(1.860 A)	(1.873 A)

Chart 3.2 Several model compounds and their C–S bond lengths and dissociation energies.

2 regardless of the molecular weights. As summary, the following issues should be considered for an ideal C/LRP process:

(i) A fast initiation system where the dissociation rate constant of iniferters is much higher than that of propagation rate constant of polymerization. This means that all chains are initiated at the beginning of the process, grow at a similar rate, and survive throughout the polymerization (there is no irreversible chain transfer or termination).

(ii) A slow initiation system where the rate of propagation is much faster than the rate of dissociation of iniferters leads to a broadening of polydispersity and a loss of molecular weight control.

(iii) The polymerization can be restarted at any time. One important aspect of this condition is that a block copolymer can be synthesized if a different monomer is added to the system before restarting the polymerization.

The polymerizations of St and MMA with dithiocarbamates were found to proceed *via* a living radical mechanism even in a homogeneous system, *i.e.*, both the yield and the molecular weight of the polymers produced increase with reaction time (Figure 3.1).

The isolated polymers could further initiate the polymerization in the presence of second monomer to give block copolymers. Iniferter polymerization has three main characteristics over classical radical polymerization; (i) the end groups of the polymers are iniferter fragments; (ii) both the molecular weight of the polymer and monomer conversion increase with reaction time and; (iii) the polymer prepared through iniferter technique should, if needed, act as polymeric iniferter to produce block copolymers.

However, iniferter polymerization has several drawbacks:

(i) The counter-radicals, dithiocarbamates or trityl radicals, were not really persistent radicals and although at a much slower rate they can initiate new chains.[25]

(ii) Benzyl dithiocarbamates was often used as iniferter for MMA polymerization. According to the current state of art of C/LRP, benzyl derivatives are inefficient initiators for MMA.[26]

(iii) Degradative transfer occurred in some systems.[27] In principally, alkyl dithiocarbamates can act as a degenerative transfer agent, which is employed in the RAFT mechanism. However, the rate of addition of radicals derived from methacrylates and styrene to dithiocarbamates is very slow since those radicals are not reactive enough to exchange with dithiocarbamates.[28,29]

(iv) Activation required irradiation by UV light, thermal activation was relatively inefficient in the system.

(v) Dissociation of dithiocarbamates radicals led to undesired irreversible termination reactions.[30]

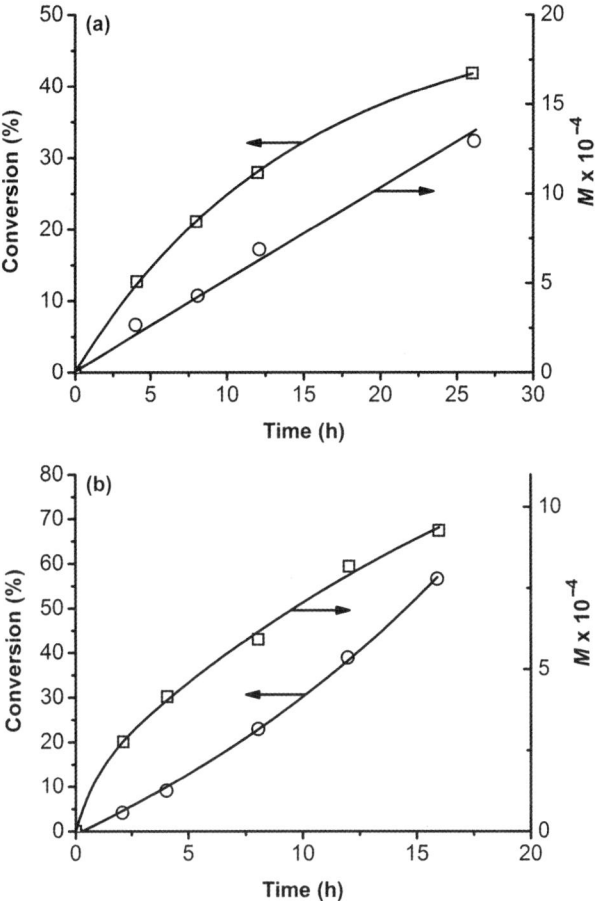

Figure 3.1 Time-conversion and time-molecular weight relationships for photopolymerization of St with [tetraethylthiuram disulfide] = 7.7×10^{-3} mol L^{-1} at 30 °C in bulk (a) and MMA with [tetraethylthiuram disulfide] = 4.6×10^{-3} mol L^{-1} at 30 °C in benzene (b).
Adapted from ref. 9 and 24.

Despite all these deficiencies, iniferter polymerization allows the facile control of the polymerization reaction by means of irradiation time and UV intensity, has a comparatively fast polymerization rate compared to other C/LRP methods, as well as the fact that the polymerization can be easily performed at room temperature or below to avoid thermal polymerization of heat-sensitive monomers. Moreover, the photoiniferter technique is indispensable for surface modification in both organic and aqueous media and it is suitable for micro-patterning.[31,32] Additionally, the polymerization is only initiated at reactive carbon radicals on the surface and not by cleaved dithiocarbamyl radicals, thus preventing homopolymer contamination in the bulk solution, and eliminating the need for extensive cleaning steps after brush formation.

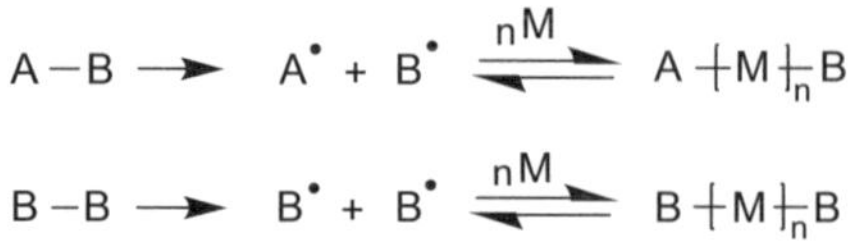

$$A-B \longrightarrow A^\bullet + B^\bullet \overset{nM}{\rightleftharpoons} A\text{-}[M]_n\text{-}B$$

$$B-B \longrightarrow B^\bullet + B^\bullet \overset{nM}{\rightleftharpoons} B\text{-}[M]_n\text{-}B$$

Scheme 3.2   C/LRP of vinyl monomers using unsymmetrical A–B and symmetrical B–B type iniferters.

3.3   Iniferters

Iniferters can be classified into two main types according to the structures of radicals formed (unsymmetrical A–B and symmetrical B–B type iniferters). Unsymmetrical A–B type iniferters decompose into two different radicals where $A^\bullet$ is the reactive radical, which participates only in initiation, and $B^\bullet$ is the less or non-reactive radical which cannot enter initiation and acts as a primary radical terminator. On the other hand, the symmetrical B–B type iniferters dissociate into two identical radicals where $B^\bullet$ is the less reactive radical which can enter into both initiation and primary radical termination (Scheme 3.2).

Several compounds containing sulfides, phenylazo compounds, amines, alkoxyamines, halides, and thiols for the A–B type, and peroxides, disulfides and tetraphenylethanes for the B–B type are used as iniferters.[19] However, the B–B type has several disadvantages compared with A–B type iniferters: the reactivity of $B^\bullet$ radicals towards the monomer is quite low, resulting a broadening of molecular weight distribution and a loss of molecular weight control. Therefore, A–B type iniferters are more versatile than B–B type for the synthesis of polymers with controlled molecular weights and architecture. The functionality of iniferters can be adjusted by changing the number of the A–B bond introduced into an iniferter molecule.

Depending on the stimulus which activates the polymerization, thermal iniferters or photoiniferters can be used. The chemistry of thermal iniferters is mainly based on the thermolysis of triphenylmethane derivatives (Chart 3.1a), whereas photodissociation of carbon-carbon or carbon-sulfur bonds is the main process in photoiniferters (Chart 3.1b).

3.3.1   Thermal Iniferters

Most of the thermal iniferters containing carbon-carbon bonds are symmetrical disubstituted tetraphenylethane derivatives, such as tetraphenylsuccino-dinitrile, tetra(*p*-methoxy phenyl) succinodinitrile, pentaphenylethane, 1,1,2,2-tetraphenyl-1,2-bis(trimethylsilyloxy)-ethane, 1,1,2,2-tetraphenyl-1,2-bis(phenoxy)ethane, phenylazotriphenylmethane, dialkyl disulfides and diphenyl disulfides (Table 3.1). For example, phenylazotriphenylmethane can act as a thermal iniferter where the phenyl radical initiates the polymerization of vinyl-type monomers and the triphenylmethyl radical is a stable radical which couples with the growing chain, exhibiting a kind of living radical nature

Table 3.1 Structures of typical thermal iniferters and their polymerization conditions.

Iniferter	Conditions	Monomers
(Ph)(Ph)C–N=N–C(Ph)(Ph) with pendant phenyl	at 60 °C, in bulk, under inert atmosphere	PMMA,[9,39] PMMA-*b*-PSt,[39] PMAN,[40] PMAN-*b*-PHFMA[40]
NC–C(Ph)(Ph)–C(Ph)(Ph)–CN	at 60 °C, in bulk, under inert atmosphere	PMMA[41,42]
H–C(Ph)(Ph)–C(Ph)(Ph)–Ph	at 60 °C, in bulk, under inert atmosphere	PMMA[41]
(Ph)C(Ph)(Ph)–C(Ph)(Ph)(Ph)	at 80 °C, in bulk, under inert atmosphere	PMMA,[37,38] PSt[37,38]
C_2H_5–C(Ph)(Ph)–C(Ph)(Ph)–C_2H_5	at 40–80 °C, in bulk, under inert atmosphere	PMMA,[43–47] PMMA-*b*-PSt[48]
Br–C(CH₃)₂–C(=O)–O–C(Ph)(Ph)–C(Ph)(Ph)–O–C(=O)–C(CH₃)₂–Br	at 70 °C, in bulk, under inert atmosphere	PMMA,[49] PSt[49]
HO–C(Ph)(Ph)–C(Ph)(Ph)–C(=O)–O–N(H)–C₆H₄–CH₃	at 50 or 80 °C, in DMF, under inert atmosphere	PMMA,[50] PMMA-*b*-PBMA[50]
succinimidyl N–Br (O=C...C=O, N–Br)	at 110 °C, in bulk, under nitrogen atmosphere	PMA,[51] PMMA,[51] PSt[51]
EtOOC–C(CN)(Ph)–C(CN)(Ph)–COOEt	at 50–80 °C, in bulk or toluene, under nitrogen atmosphere	PMMA,[52–55] PSt,[52–55] PMMA-*b*-PSt[52–55]
HO–CH₂CH₂–N(CH₂CH₂OH)–C(=S)–S–S–C(=S)–N(CH₂CH₂OH)–CH₂CH₂–OH	at 110 °C, in bulk, under argon atmosphere	PSt[56]

Table 3.1 (*Continued*)

Iniferter	Conditions	Monomers
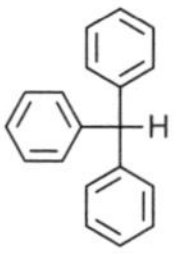	at 85 °C, in cyclohexanone, under nitrogen atmosphere	PMMA,[57] PSt[57]
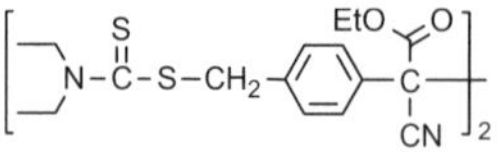	at 60 or 80 °C, at r.t., in bulk, under inert atmosphere	PMMA,[58,59] PSt,[58–60] PIP-*b*-PSt-*b*-PIP,[59] PSt-*b*-P*t*BMA[61]
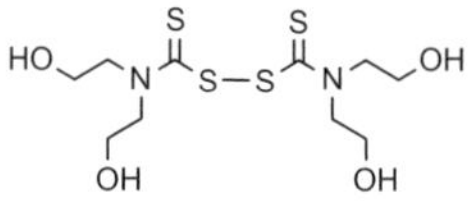	at 70 °C or UV lamp, at r.t., in DMF, under nitrogen atmosphere	PSt[62]
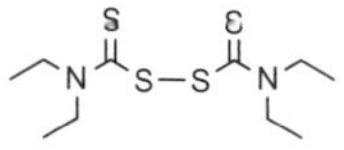	at 85 °C, in bulk, under nitrogen atmosphere	PSt[62]
	at 120 °C, in bulk, under inert atmosphere	PSt[63]
	at 70 °C, in bulk, under nitrogen atmosphere	PMMA[64]
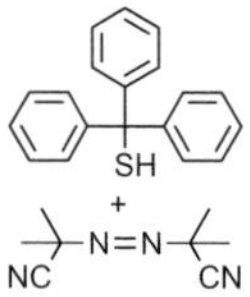	at 70 °C, in bulk, under nitrogen atmosphere	PSt,[65] PMMA,[65] PSt-*b*-PMMA[65]
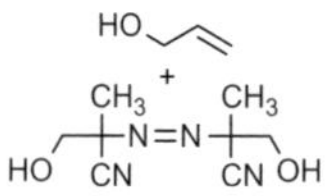	at 100 °C, in THF, under inert atmosphere	P*n*BA,[66] PMMA,[66] PSt,[66] PVAc[66]

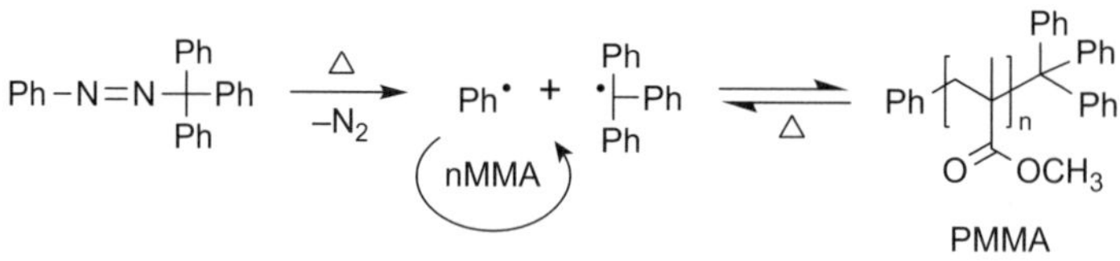

Scheme 3.3 C/LRP of methyl methacrylate using phenylazotriphenylmethane as thermal iniferter.

(Scheme 3.3). In this polymerization, the molecular weight of the polymers produced increased with the reaction time.

In this connection, it should be mentioned that the terminating triphenylmethyl radical can be converted to the corresponding cation by diphenyliodonium hexafluorophosphate in free radical polymerizing systems and, thus, improve the efficiency of phenylazotriphenylmethane as a free radical initiator in conventional radical polymerization.[33,34]

Thermal iniferters have been mainly used for the polymerization of methyl methacrylate. It was reported that polymerization of styrene with some thermal iniferters such as 1,1,2,2-tetraphenyl-1,2-bis(phenoxy)ethane[35] and tetraarylsccinonitriles[36] proceeds *via* a dead-end polymerization mechanism, *i.e.*, it does not follow living radical polymerization conditions. On the other hand, Jongh *et al.*[37] and Chernikova *et al.*[38] reported that some specially designed hexasubstituted-ethane derivatives can act as iniferters for styrene polymerization.

3.3.2 Photoiniferters

Photopolymerizations offer many advantages (such as temporal and spatial control of initiation, cost efficiency, and solvent-free systems) over conventional thermal polymerization. The photochemical polymerization of vinyl monomers with dithiocarbamates, referred to as photoiniferters, is one of the most widely studied photoinitiated C/LRP. UV irradiation of a dithiocarbamate yields a reactive carbon radical and a relatively unreactive dithiocarbamyl radical. The carbon radical, reacting with the monomer molecules, initiate the radical polymerization and propagates the process upon addition of monomer molecules. On the other hand, the dithiocarbamyl radical does reversibly terminate growing polymer chains ("capping/decapping") and, depending on reaction conditions, enables controlled radical polymerization during photoirradiation (Scheme 3.4). Simple evidence used to support the postulated hypothesis included the low polydispersities of the product (typically ranging from 1.2 to 2.0), a linear increasing trend of molecular weight with the conversion, and a formation of a related block copolymer.

In a recent study, Tasdelen *et al.* synthesized a phenacyl morpholine-4-dithiocarbamate, which can act as both a photoiniferter and reversible addition fragmentation chain transfer (RAFT) agent.[67] Polymerization of styrene was carried out in bulk under UV irradiation at above 300 nm at room temperature. The polymerization showed living characteristics up to 50% conversions and produced well-defined polymers with molecular weights close to those predicted from theory and relatively narrow poyldispersities ($M_w/M_n \sim 1.30$). End group determination and block copolymerization

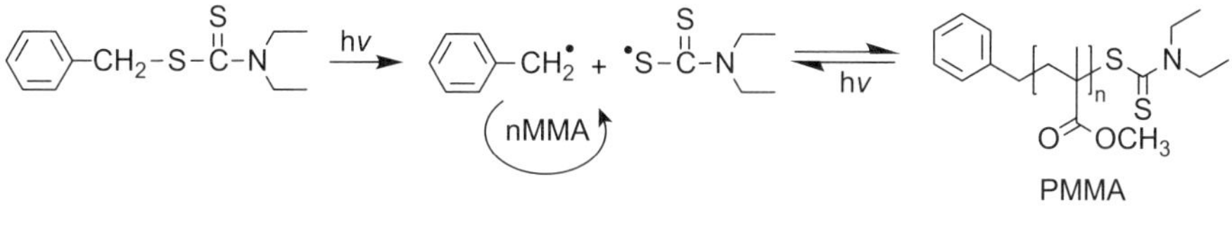

Scheme 3.4 C/LRP of methyl methacrylate using benzyl-*N,N*-diethyldithiocarbamate as photoiniferter.

Table 3.2 Results of photo- and thermal block copolymerization of methyl acrylate using conventional initiators with PSt-MDC (conv: 18%; M_{nth}: 4000 g mol^{-1}, M_{nNMR}: 4600 g mol^{-1}, M_{nGPC}: 5200 g mol^{-1}, M_w/M_n: 1.65) as a macro-RAFT agent.[67]

No	[MA]/[PSt-MDC]/[I]	Time/min	Conv.^{c}(%)	$M_{n,th}$	$M_{n,NMR}{}^d$	$M_{nGPC}{}^e$	$M_w/M_n{}^e$
1^a	200/1/0.1	15	60	14 800	42 100	18 300	1.80
2^a	200/1/0	15	7.2	5700	10 600	6500	1.72
3^b	200/1/0.1	60	17	7400	6900	17 700	1.53
4^b	200/1/0	60	—	—	—^{f}	5250	1.60

^{a} Photopolymerization carried out above 365 nm using (2,4,6-trimethylbenzoyl) diphenylphosphine oxide as photoinitiator in the presence of CH$_2$Cl$_2$ (interference filter: cupric sulfate aqueous solution).
^{b} Thermal polymerization carried out at 70 °C using azobisisobutyronitrile as thermal initiator in the presence of toluene.
^{c} Determined gravimetrically.
^{d} Calculated from ^{1}H NMR of the corresponding polymers.
^{e} Determined by GPC based on polystyrene standard.
^{f} Did not observed -OCH$_3$ protons belongs to PMA in ^{1}H-NMR spectrum.

with methyl acrylate (MA) suggested that morpholino dithiocarbamate groups were attained at the end of the polymer. The occurrence of a RAFT mechanism was proved by polymerization experiments performed at various conditions, which were block copolymerizations of MA using conventional free radical initiators with PSt morpholino dithiocarbamate (PSt-MDC) as a macro-RAFT agent. The block copolymerizations did not proceed in the absence of an added initiator, indicating that PSt-MDC did not act as an iniferter, whereas in the presence of free radical initiators, polymerizations proceeded in a controlled manner (Table 3.2). Photolysis and polymerization studies revealed that both RAFT and photoiniferter mechanisms are operative in the system.

A large number of monomers, such as St, MMA, *n*-butyl acrylate, acrylamide, acrylonitrile, and methacrylonitrile, can be polymerized in a controlled manner with photoiniferters (Table 3.3). However, for other monomers such as vinyl acetate and methyl acrylate, dithiocarbamates provide poor or no control over the polymerization. The living character of the polymerization seems to depend on the nature of the monomer and decreases from styrene to methyl methacrylate, and basically disappears in the case of acrylates.

3.3.3 Redox Iniferters

Compared with other modes of polymerization, redox initiation has the advantage of being applicable in aqueous media and at low temperature. Therefore, the probability of side reactions can be minimized. To conduct iniferter polymerization at room temperature, Otsu and Tazaki proposed a redox iniferter system. The system is based on the reaction of nickel (Ni$^{(0)}$) with an organic halide (R–X) such as benzyl chloride to form the benzyl radical,

Table 3.3 Structures of typical photoiniferters and their polymerization conditions

Iniferter	Conditions	Polymers
	UV light, at r.t., in various solvents, under inert atmosphere	PAAc,[68] PAOA,[69] PnBA,[70] PBMA,[68] PDMAEMA,[71] PEGDMA,[72] PFOMA,[71] PMMA,[73–75], PSt,[73] PTAN,[71] PMMA-b-PEA,[73] PSt-b-PEA,[73] PSt-b-PMMA,[73] PMMA
	UV light, at r.t., in bulk, under argon atmosphere	PMMA[75]
	UV light, at r.t., in bulk, under argon atmosphere	PMMA[75]
	UV light, at r.t., in bulk, under argon atmosphere	PMMA[75]
	UV light, at r.t., in bulk, under nitrogen atmosphere	PSt[76,77]
	UV light, at r.t., in benzene, under inert atmosphere	PMMA,[64] PSt[78]
	UV light, at 60–80 °C, in water	PMMA[79,80]
	UV light, at r.t., in water	PMMA,[81,82] PMOEP,[83] PMMA-co-PSt[81]
	UV light, at r.t., in bulk, under inert atmosphere	PBA,[48,49] PSt,[84] PSt-b-PMMA,[84] PSt-b-PVAc[84]
	UV light, at r.t., in bulk, under inert atmosphere	PBA,[85] MA,[85] PSt[85]
	UV light, at r.t., in THF, under nitrogen atmosphere	PBzMA,[86] PHEMA,[86] PMMA[87]
	UV light, at r.t., in bulk, under argon atmosphere	PMMA[87]

Table 3.3 (*Continued*)

Iniferter	Conditions	Polymers
	UV light, at r.t., in bulk, under argon atmosphere	PMMA[87]
	UV light, at r.t., in bulk, under argon atmosphere	PMMA[87]
	UV light, at r.t., in bulk, under argon atmosphere	PMMA[87]
	UV light, at r.t., in bulk, under argon atmosphere	PMMA[88]
	UV light, at r.t., in DMF, under nitrogen atmosphere	PDMAm,[89] PSt[89]
	UV light, at r.t., in DMF, under nitrogen atmosphere	PDMAm,[89] PSt[89]
	UV light, at r.t., in DMF, under nitrogen atmosphere	PDMAm,[89] PSt[89]
	UV light, at r.t., in DMF, under nitrogen atmosphere	PDMAm,[89] PSt[89]
	UV light, at 40 °C, in THF, under nitrogen atmosphere	PEMA,[90,91] PSt[90,91]
	UV light, at r.t., in acetone, under inert atmosphere	PSt,[90,92] PSt-*co*-PMAnh[93]
	UV light, at r.t., in THF, under nitrogen atmosphere	PMA,[94] PMMA[94]
	UV light, at r.t., in toluene, under inert atmosphere	PMA,[22] PSt,[23] PBA-*b*-PMMA,[95] PSt-*b*-PMA[22,23]

Table 3.3   (*Continued*)

Iniferter	Conditions	Polymers
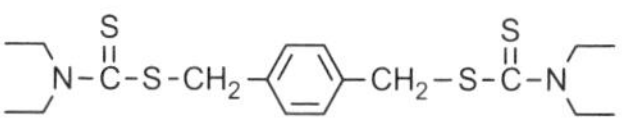	UV light, at 60 °C, in bulk, under nitrogen atmosphere	PHEMA,[18] PTEGDMA[18]
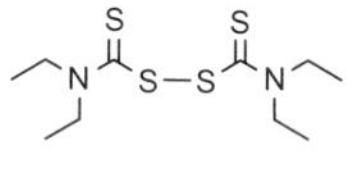	UV light, at r.t., in bulk, under nitrogen atmosphere	PSt,[62] PTEGDMA[96]
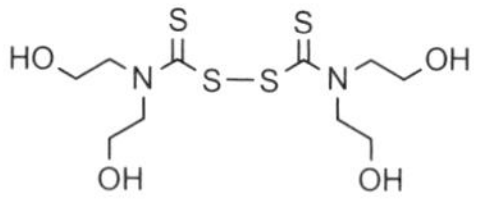	UV light, at r.t., in bulk, under nitrogen atmosphere	PSt[62]
	UV light, at r.t., in bulk and benzene, under nitrogen atmosphere	PSt,[97] PMMA[97]
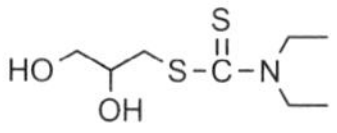	UV light, at r.t., in DMSO, under nitrogen atmosphere	PMMA,[98] PMMA-*b*-PSt[98]
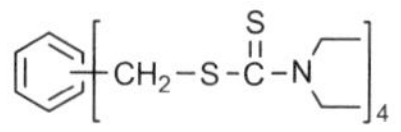	UV light, at r.t., in toluene, under nitrogen atmosphere	PDMAPMA,[99,100] PDMAPAAm-*co*-PDMAm[101]
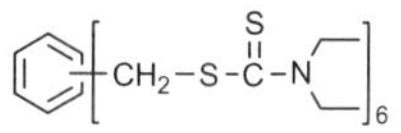	UV light, at r.t., in chloroform, under nitrogen atmosphere	PDMAm[102]

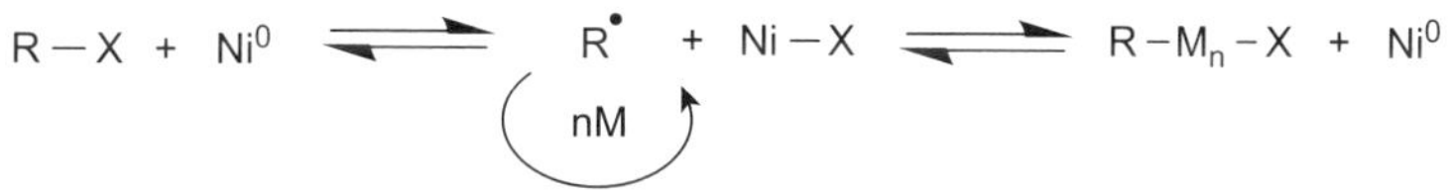

$$R-X \;+\; Ni^0 \;\rightleftharpoons\; R^{\bullet} \;+\; Ni-X \;\rightleftharpoons\; R-M_n-X \;+\; Ni^0$$

$$nM$$

Scheme 3.5   C/LRP of vinyl monomers by redox iniferter.

which can initiate polymerization. The dormant chains are generated by reversible deactivation of the growing chains by the oxidized Ni complex and the C–X bond at the chain end further reacts with Ni$^{(0)}$ *via* redox reaction, the polymerization proceeds *via* a C/LRP mechanism (Scheme 3.5). A successful extension of the redox iniferter system to atom transfer radical polymerization, ATRP, provided a new and efficient method for synthesis of well-defined polymers.

There are a number of factors that affect the progress of the iniferter polymerization, the extent of its livingness, and thereby the control over the end product.[86] These factors are related to iniferter structures,[103,104] types of monomer[105] and solvent,[105] and temperature. Since the iniferter itself is incorporated into the growing/propagating polymer chain during polymerization, a desirable end group functionality can be designed by careful choice of iniferter.[15]

3.4 Applications

3.4.1 Telechelic Polymers by Iniferters

So far, many iniferters have been reported and utilized for the synthesis of telechelic polymers.[106] The preparation of telechelic polymers is based on the concept of locating the required function on the alkyl group of the thiuram disulfide and using it in the photo- or thermal polymerization. Since end groups are introduced *via* initiation and transfer and common bimolecular termination between two growing chains are negligible, bifunctional telechelics are available in a good yield. Several functional disulfides and substituted tetraphenylethylenes have also been used as iniferters in free radical polymerization. The functionalities of the telechelics prepared by the iniferter method were reported to be close to 2, within experimental error. The formation of the nonfunctional polymers was claimed to be negligible because of the triple function of the iniferter. For example, carboxylic acid and amino functionalities were introduced to polystyrene using the corresponding disulfides.[107] Diamino functional poly (*t*-butyl acrylate) was also prepared.[108] In this case, polymers were readily hydrolyzed to polyacrylic acid possessing amino terminal groups, which is a useful material for the application of polyelectrolytes.

The photochemically and thermally induced iniferter properties of the tetraalkylthiuram disulfides during free radical polymerization were also exploited to end functionalize PMMA and PSt (Scheme 3.6).[11,109] Table 3.4 summarizes the functional iniferters used for obtaining telechelic polymers.

3.4.2 Complex Macromolecular Architectures by Iniferters

Block copolymers are generally defined as macromolecules in which chemically different blocks (or segments) are connected together to combine their macroscopic properties and to design hybrid materials. Traditionally, block

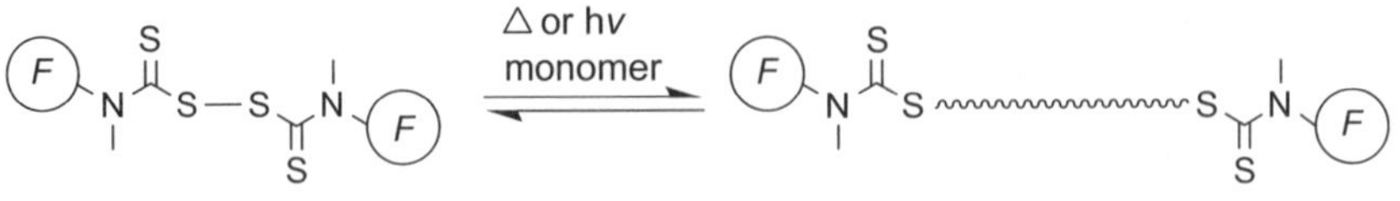

Scheme 3.6 Synthesis of telechelic polymers by the iniferter method.

Table 3.4 Functional iniferters.

Functionality	Iniferter	Polymers
–OH		PMMA[110]
–OH		PMMA[42]
–COOH		PMMA[82]
–COOH		PSt[111]
–NH$_2$		PIP[112]
–OH and –COOH		PSt,[89] PDMAm[89]
–OH and –COOH		PSt,[89] PDMAm[89]
–NH$_2$ and –COOH		PSt,[89] PDMAm[89]
–Chloride		PSt,[36,113] PMMA[36,113]
–Phosphorylamide		PSt,[114] PMMA[114]
–Furanyl	R: -phenyl, -tolyl, -mestyl	PSt,[115] PMMA,[115] PnBA,[115] PSTA,[115] PPCPHA[116]

copolymers can be synthesized by the sequential addition polymerization of different monomer units using the same chemistry. In the iniferter polymerization, the resulting polymers are normally end-capped with an iniferter group, thus making it possible to prepare block polymers. Various block copolymers containing either hydrophobic or hydrophilic blocks can be synthesized by both thermally and photochemically activated iniferter polymerizations. Moreover, the photoiniferter method can be applied to prepare a pure block copolymer without contamination by homopolymers. For example, dithiocarbamate groups were attached to polystyrene beads through a hydrolyzable ester spacer and the resulting polymer was used as a photoiniferter (Scheme 3.7).[117] After the polymerization of styrene, the whole polymer was isolated and extracted with benzene to separate the homopolystyrene. The grafted polystyrene containing dithiocarbamate end groups was further used for the polymerization of methyl methacrylate. After the extraction and hydrolysis of the attached block copolymer attached on to beads, a pure block copolymer, polystyrene-*b*-poly(methyl methacrylate), was obtained successfully. In a similar way, several triblock and tetrablock copolymers have also been prepared in good yields.

As stated previously, the iniferter method is not suitable for all monomers. Therefore, the application of a single mode polymerization technique for the synthesis of block copolymers is severely limited, and suffers from the requirement for care in the order of monomer addition. In order to extend the range of monomers for the synthesis of block copolymers, a mechanistic transformation approach was proposed, by which the polymerization

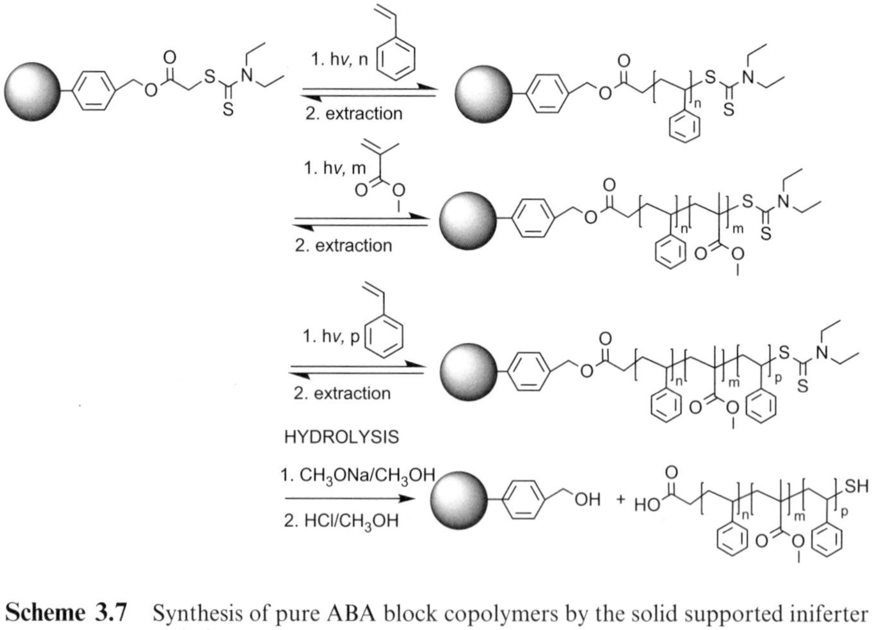

Scheme 3.7 Synthesis of pure ABA block copolymers by the solid supported iniferter technique.

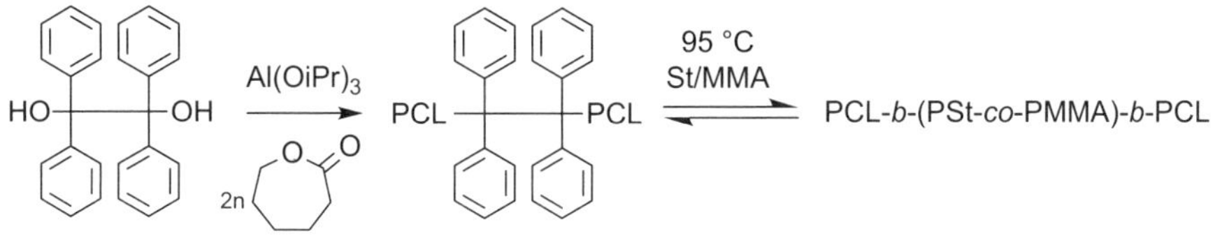

Scheme 3.8 Synthesis of block copolymers by combination of ring-opening polymerization and iniferter polymerization.

mechanism could be changed from one to another which is suitable for the respective monomers.[118] For example, a transformation reaction was applied for the preparation of poly(ε-caprolactone)-*b*-poly(styrene-*co*-methyl methacrylate)-*b*-poly(ε-caprolactone) block copolymers by using the iniferter technique in the controlled radical polymerization step. Substituted tetraphenylethanes represent a class of thermal iniferters applicable to the radical polymerization of many monomers in a controlled manner. The initiation of ring-opening polymerization of ε-caprolactone by aluminum triisopropoxide in the presence of benzopinalcohol leads to the formation of polymers possessing the iniferter structure in the middle of the chain.[119,120] The benzopinacolate groups incorporated into the polymer chain then initiate the polymerization of St and MMA *via* a controlled radical mechanism at 95 °C to yield the desired block copolymers (Scheme 3.8)

Graft copolymers are another class of segmented copolymers and generally consist of a linear backbone of one composition and randomly distributed branches of a different composition. Graft copolymers can be obtained with three general methods: (i) grafting-onto, in which side chains are preformed, and then attached to the backbone; (ii) grafting-from, in which the monomer is grafted from the backbone; and (iii) grafting-through, in which the macromonomers are copolymerized. In the case of iniferter polymerization, the grafting-from method has been mostly applied for the synthesis of graft copolymers. Here, the backbone acts as a side-functional macroiniferter in the polymerization of a certain monomer and graft chains are grown out of a polymer backbone.[121] Table 3.5 lists a variety of photoiniferters used for the synthesis of block and graft copolymers.

Polymers with a dithiocarbamate photoiniferter group in a side-chain end were also synthesized by a chemical reaction. For example, halogenated linear polymers, such as poly(vinyl chloride)[122,123], partially chloromethylated polystyrene[124] and partially chlorinated polydimethylsiloxane[125] were reacted with sodium dialkyldithiocarbamate to obtain the corresponding polymeric macroiniferters. UV irradiation of these polymers in the presence of radically polymerizable monomers triggered graft copolymers. In Scheme 3.9, the reaction mechanism is illustrated for the case of chlorinated polydimethylsiloxane which was first reacted with sodium diethyldithiocarbamate and subsequently with either 2-hydroxyethyl methacrylate or acrylamide or 2-methylaminoethyl methacrylate or methacrylic acid (Scheme 3.9).[125]

Table 3.5 Synthesis of block and graft copolymers *via* iniferter polymerization.

Iniferter	Conditions	Polymers
	at 70 and 120 °C, in DMF, under inert atmosphere	PSt-*co*-PNMI[126,127]
	at 60 or 80 °C, in bulk or toluene, under nitrogen atmosphere	PSt,[128] PSt-*co*-PAN,[128] PSt-*co*-PMAnh[129]
	at 75 °C, in various solvents, under inert atmosphere	PAN,[130] PHFBA,[131] PMAc,[132] PMMA,[133,134] PSt,[133,134] PVP,[135] PVBCl,[136] PSMA,[137] PSt-*b*-PMMA[133,134]
	at 75 °C, in water, under nitrogen atmosphere	PMMA[138]
	at 70 and 120 °C, in DMF, under inert atmosphere	PSt-*co*-PNMI[126,127]
	at 60 or 80 °C, in bulk or toluene, under nitrogen atmosphere	PSt,[128] PSt-*co*-PAN,[128] PSt-*co*-PMAnh[129]
	at 75 °C, in water, under nitrogen atmosphere	PMMA[138]
	UV light, at room temperature, in bulk, under nitrogen atmosphere	P*n*BA,[139] PMA[140,141]
	UV light, at r.t., in benzene, under inert atmosphere	PSt[142]
	UV light, at r.t., in methanol, under argon atmosphere	PNIPAM[143]

Table 3.5 (*Continued*)

Iniferter	Conditions	Polymers
(structure: N–C(=S)–S~(PB)~S–C(=S)–N)	UV light, at r.t., in bulk, under nitrogen atmosphere	PSt-*co*-PAN[144]
(structure: (PU)~N–C(=S)–S—S–C(=S)–N~(PU))	UV light, at r.t., in DMF, under nitrogen atmosphere	PAAm,[145] PMMA,[20], PVP[145]
(structure: N–C(=S)–S~(PTHF)~S–C(=S)–N)	UV light, at r.t., in dicholoro methane, inert atmosphere	PMMA[146]
(structure: (polyester)~N–C(=S)–S—S–C(=S)–N~(polyester))	UV light, at r.t., in bulk, under nitrogen atmosphere	PMMA,[147] PMMA-*b*-PSt[147]
(structure: (PSt)–S–C(=S)–N)	UV light, at r.t., in THF, under nitrogen atmosphere	P*t*BMA[61,62]
(structure: (PIP)–S–C(=S)–N)	UV light, at r.t., in bulk, under nitrogen atmosphere	PDMAMP,[148] PDMMEP,[148] PMMA[149]
(structure: (polyphenylene)–S–C(=S)–N)	UV light, at r.t., in THF, under inert atmosphere	PMA,[150] P*t*BA,[150] PSt-*co*-PVBCl[151]
(structure: (PU)–S–C(=S)–N)	UV light, at r.t., in bulk, under nitrogen atmosphere	PDMAm,[152] PEGMA[152]
(structure: (PECH)–S–C(=S)–N)	UV light, at r.t., in bulk, under nitrogen atmosphere	PSt,[153] PMMA[70,71]
(structure: (PIB)–S–C(=S)–N)	UV light, at r.t., in benzene, under inert atmosphere	PMMA[154]

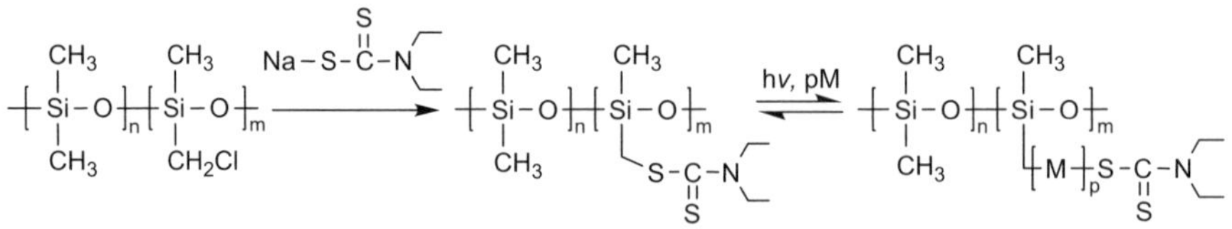

Scheme 3.9 Synthesis of graft copolymers *via* side chain dithiocarbamate functionalized polydimethylsiloxane as macrophotoiniferter.

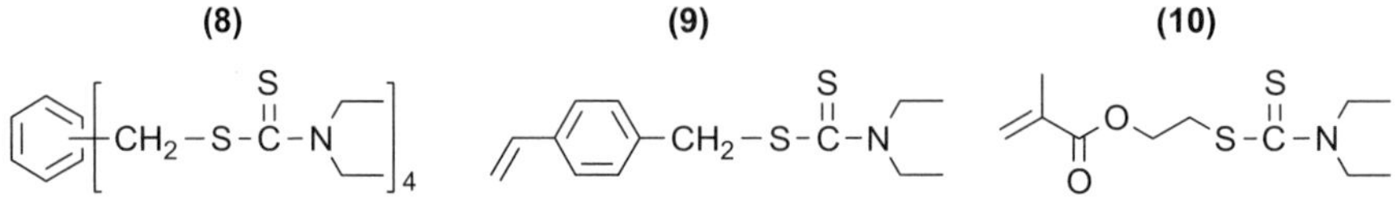

Chart 3.3 Examples of polyfunctional iniferters.

Polyfunctional iniferters have also been employed to prepare branched, network and star polymers (**8-10**, Chart 3.3). For example, Ishizu *et al.*[155] showed that functionalized iniferters ((*N*,*N*-diethyldithiocarbamyl) methyl styrene or 2-(*N*,*N*-diethyldithiocarbamyl)ethyl methacrylate) can be utilized for the synthesis of hyperbranched polymers (UV irradiation), star polymers (copolymerization of vinyl head with crosslinking agent in dark condition), and rigid polymer brushes (vinyl homopolymerization of macroiniferters and subsequent treatment by internal domain locking with diamine compounds).

Photolysis of (*N*,*N*-diethyldithiocarbamyl) methyl styrene leads to the initiating benzyl radical with the inactive diethyldithiocarbamate radical. This benzyl radical can add to the vinyl group of a second molecule of iniferter to produce a dimer. The dimer corresponds to an AB_2 monomer with one polymerizable vinyl group and two initiating/propagating sites. By repeating these elementary reactions, this self-condensing vinyl polymerization[156] system proceeds to form hyper-branched polymers (Scheme 3.10).

3.4.4 Surface Modification by Iniferters

The photoiniferter technique is presented as a simple microlithographic technique that affords flexibility in fabricating and subsequently modifying polymer substrates with various chemistries. Spatial patterns can be easily achieved by simply controlling the location of light exposure across the surface. In fact, the photoiniferter technique allows grafting with the same spatial micropatterning resolution as is achievable with traditional photolithographic techniques. It is also useful for the creation of gradients with nanometre scale resolution in polymer layer thickness, simply by varying the light intensity or exposure time across the surface. Grafting *via* the photoiniferter technique is much faster, usually providing thicknesses of hundreds of nanometres in a few

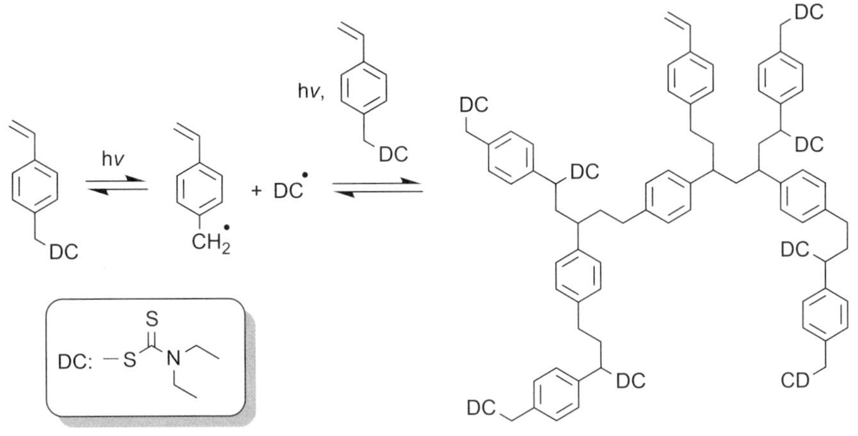

Scheme 3.10 Synthesis of hyper-branched polymers using polyfunctional iniferters.

minutes. Additionally, it can be performed at room temperature with or without solvent (pure monomer), requires no catalyst/ligand system, and is compatible with a wide range of vinyl monomers. In the case of photoiniferter based surface initiated polymerization, the iniferter molecules can be easily immobilized on surfaces *via* self-assembly or by chemically modifying the precursor surfaces through the introduction of dithiocarbamate moieties.[157]

More recently, several groups extensively used the photoiniferter technique to prepare a wide variety of polymer brushes from benzyl-*N*,*N*-diethyldithiocarbamate functionalized substrates.[158–162] For instance, De Boer *et al.* used a trimethoxysilane-modified benzyl-*N*,*N*-diethyldithiocarbamate derivative to modify the surface of silicon substrates and grow PSt brushes (Scheme 3.11).[157] They reported the successful preparation of up to 270 nm-thick PSt-*b*-PMMA brushes within 25 h of irradiation with 365 nm UV light.

To study the living nature of this surface initiated polymerization, several groups have performed kinetic studies.[32,157,163,164] They reported that the nonlinear growth of the polymer brushes as a function of irradiation time was mainly attributed to bimolecular termination reactions, rather than chain transfer to monomer. To avoid irreversible termination reactions, a strategy to increase the amount of deactivating species by adding tetraethylthiuram disulfide to the polymerization mixture, which is mandatory to provide a controlled radical polymerization behavior, was introduced.[165,166]

The photoiniferter polymerization has been used for the modification of various surfaces including polymeric films, gold, silica, glass and silicone surfaces. Although, this technique is limited for microchannels, tubes, or small cavities, which are very difficult to modify and have very limited light accessibility, it provides a versatile route to 2D- and 3D-microstructured polymer brushes with various types of monomers.[167] Table 3.6 summarizes the surfaces, conditions and polymers obtained by surface initiated photoiniferter polymerization.

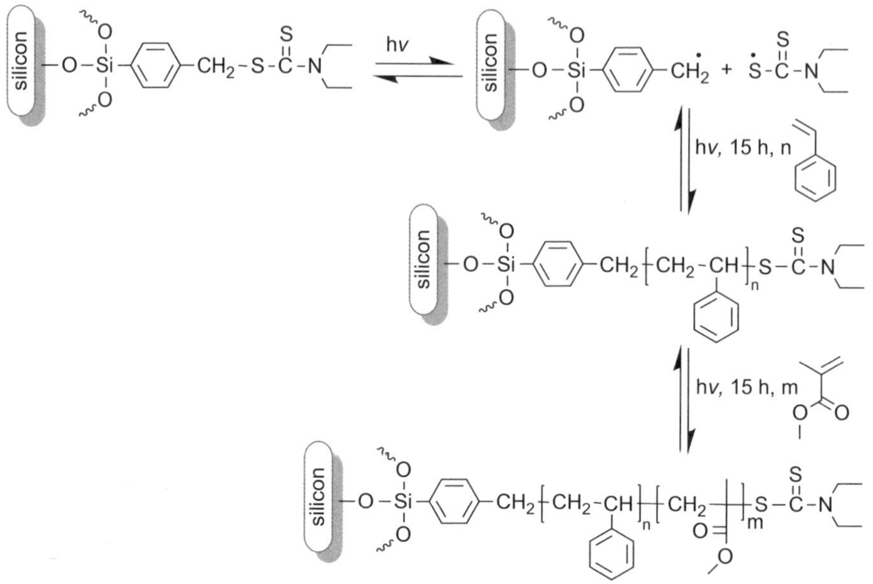

Scheme 3.11   Preparation of PS-*b*-PMMA brushes by surface initiated photoiniferter polymerization from silicon substrates.

Table 3.6   Photoiniferter polymerization in the preparation of surface brushes.

Iniferter	Conditions	Brush polymers
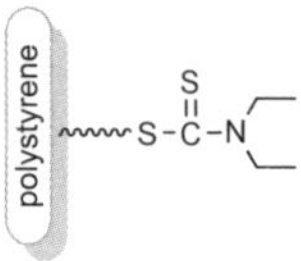 polystyrene ~~~~S–C–N	UV light, at r.t., in various solvents, under inert or nitrogen atmosphere	PAAc,[168,169] PAAm,[158,168–172] PNaAAc,[170] PDMAm,[32,158,163,173,174] PDMAEMA,[163,175] PDMAPAm,[32,170,173], PEA,[163] PEGMA,[92,158,176] PEMA,[163] PHEMA,[163,173,175] PMA,[163] PMAc,[32,158,168,175,176] PMMA,[168] PNaMA,[158] PNaMAc,[173] PNIPAM,[175,177–179] PSt,[32,117,163,174,176] PSMAK,[173] PVP,[163] PDMAPAm-*b*-PDMAm,[180] PSt-*b*-PMMA,[117,175] PSt-*b*-PMMA-*b*-PSt,[117] PSt-*b*-PMMA-*b*-PMA,[117] PSt-*b*-PVBCl-*b*-PMMA,[117] PAAm-*co*-PAAc,[169] PMAc-*co*-PEGDMA,[181] PNIPAM-*co*-PAAc,[178,179] PVBCl-*co*-PDMAm,[182] PVBCl-*co*-DMAEMA[182]
polyacrylate ~~~~S–C–N	UV light, at r.t., in various solvents, under inert or nitrogen atmosphere	PAAm,[183] PAEMA,[184] PDMAEMA,[175] PEGMA,[185–191] PHEMA,[175,186] PMAc,[175] PMMA,[175] PNIPAM,[175] POctA,[186] POctMA,[165,192] P*t*BA,[186] PTFEA,[186,187] PEGMA-*co*-PNaMAc[154]

Table 3.6 (*Continued*)

Iniferter	Conditions	Brush polymers
biopolymers $\sim\!\sim\!S\!-\!\overset{\overset{S}{\|}}{C}\!-\!N$	UV light, at r.t., in water, under nitrogen atmosphere	PNIPAM,[193–197] PMAc-*co*-PEGDMA[198]
polypropylene $\sim\!\sim\!S\!-\!\overset{\overset{S}{\|}}{C}\!-\!N$	UV light, at r.t., in DMF, under nitrogen atmosphere	PAAm[199]
gold $\sim\!\sim\!S\!-\!\overset{\overset{S}{\|}}{C}\!-\!N$	UV light, at r.t., in bulk or THF, under argon or nitrogen atmosphere	PAPBA,[200] PDMAm,[200] PMAc,[201,202], PNIPAM,[203,204] PMEGlc,[202,205] PAPBA-*co*-PDMAPMA[206]
silica $\sim\!\sim\!S\!-\!\overset{\overset{S}{\|}}{C}\!-\!N$	UV light, at r.t., in bulk or toluene, under nitrogen atmosphere	PDMAm,[207] PEGDMA,[208] PMMA,[209] PSt,[209,210] PMAc-*co*-PEGDMA[181,211,212]
silicon $\sim\!\sim\!S\!-\!\overset{\overset{S}{\|}}{C}\!-\!N$	UV light, at r.t., in toluene, under inert atmosphere	PMAc,[213] PMMA,[164,166,214], PSt,[157] PSt-*b*-PMMA,[157] PNIPAM,[215] PMAc-*b*-PNIPAM,[216] PMMA-*b*-PSt[166]
glass $\sim\!\sim\!S\!-\!\overset{\overset{S}{\|}}{C}\!-\!N$	UV light, at r.t., in various solvents, under inert atmosphere	PAAc,[159] PDMAm,[159,161,162] PMAc,[217] PMMA,[218] PNIPAM,[160,162] PMMA-*co*-PSt,[218] PTFMA-*co*-PEGDMAP[217]
CNT or C$_{60}$ $\sim\!\sim\!S\!-\!\overset{\overset{S}{\|}}{C}\!-\!N$	UV light, at r.t., in bulk, under inert atmosphere	PMMA,[219] PMAc-*co*-PEGDMA,[220] PMMA-*co*-PMA,[219] PMMA-*co*-PSt[219]

3.5 Conclusions

In this chapter, the huge potential of iniferter-based polymerization to prepare well-defined polymers and successful surface modification has been discussed. Since its discovery, the iniferter technique has brought about a clear breakthrough in polymer science. It allows the design of polymers from a broad range of monomers with controlled molecular weight, molecular weight distributions and well-defined architectures. Moreover, the iniferter technique is easy to implement and suitable for the modification of different types of surfaces in organic and aqueous solvents. All these important features make the iniferter technique a valuable tool in the development of new materials.

Abbreviations

C/LRP	controlled/living radical polymerization
RAFT	reversible addition fragmentation chain transfer
St	styrene
MA	methyl acrylate
MMA	methyl methacrylate
PSt-MDC	PSt-morpholino dithiocarbamate

Polymers:

PAAc	poly(acrylic acid)
PAAm	polyacrylamide
PAN	polyacrylonitrile
PAOA	poly(acetone oxime acrylate)
PAPBA	poly(3-acrylamidophenyl boronic acid)
PBMA	poly(butyl methacrylate)
PBzMA	poly(benzyl methacrylate)
PDMAEMA	poly(2-dimethylaminoethyl methacrylate)
PDMAm	poly(N,N-dimethylacrylamide)
PDMAMP	poly(dimethyl (acryloyloxymethyl) phosphonate)
PDMAPAm	poly(N,N-dimethylaminopropyl acrylamide)
PDMAPAAm	poly(N-[3-(dimethylamino)propyl] acrylamide)
PDMAPMA	(N,N-dimethylaminopropyl methacrylamide)
PDMMEP	poly(dimethyl(methacryloyloxyethyl) phosphonate)
PEA	poly(ethyl acrylate)
PEGDA	poly(ethylene glycol diacrylate)
PEGDMA	poly(ethylene glycol dimethacrylate)
PEGDMAP	poly(ethylene glycol methacrylate phosphate)
PEMA	poly(ethyl methacrylate)
PFOMA	poly(1,1'-dihydroperfluorooctyl methacrylate)
PHEMA	poly(2-hydroxyethyl methacrylate)
PHFMA	poly(hexafluorobutyl methacrylate)
PIP	polyisoprene
PMA	poly(methyl acrylate)
PMAc	poly(methacrylic acid)
PMAN	polymethacrylonitrile
PMAnh	poly(maleic anhydride)
PMEGlc	poly(2-methacryloyloxyethyl D-glucopyranoside)
PMMA	poly(methyl methacrylate)
PMOEP	poly(2-methacryloyloxyethyl phosphorylcholine)
PNaAAc	poly(sodium acrylic acid)
PNaMA	poly(sodium methacrylate)
PNaMAc	poly(sodium methacrylic acid)
P*n*BA	poly(*n*-butyl acrylate)
PNIPAM	poly(N-isopropylacrylamide)
PNMI	poly(N-methylmaleimide)

POctA	poly(*n*-octyl acrylate)
POctMA	poly(*n*-octyl methacrylate)
PPCPHA	poly(6-[4-(4-propoxyphenoxycarbonyl)phenoxylhexyl acrylate)
PSMA	poly(solketal methacrylate),
PSMAK	poly(3-sulfopropyl methacrylate potassium salt)
PSTA	poly(stearyl acrylate)
PSt	polystyrene
PTAN	(poly(1,1,2,2-tetrahydroperfluorooctyl acrylate)
P*t*BMA	poly(*tert*-butyl methacrylate)
PTFEA	poly(trifluoroethyl acrylate)
PTEGDMA	poly(tetraethylene glycol dimethacrylate)
PTFMA	poly(2-trifluoromethacrylic acid)
PVAc	poly(vinyl acetate)
PVBCl	poly(vinylbenzyl chloride)
PVP	poly(*N*-vinyl-2-pyrrolidon)

References

1. J. Chiefari, Y. K. Chong, F. Ercole, J. Krstina, J. Jeffery, T. P. T. Le, R. T. A. Mayadunne, G. F. Meijs, C. L. Moad, G. Moad, E. Rizzardo and S. H. Thang, *Macromolecules*, 1998, **31**, 5559–5562.
2. C. J. Hawker, A. W. Bosman and E. Harth, *Chem. Rev.*, 2001, **101**, 3661–3688.
3. M. Kamigaito, T. Ando and M. Sawamoto, *Chem. Rev.*, 2001, **101**, 3689–3745.
4. K. Matyjaszewski and J. H. Xia, *Chem. Rev.*, 2001, **101**, 2921–2990.
5. W. A. Braunecker and K. Matyjaszewski, *Prog. Polym. Sci.*, 2007, **32**, 93–146.
6. T. E. Ferington and A. V. Tobolsky, *J. Am. Chem. Soc.*, 1955, **77**, 4510–4512.
7. E. Borsig, M. Lazar, M. Capla and S. Florian, *Angew. Makromol. Chem.*, 1969, **9**, 89–95.
8. D. Braun, *Makromol. Symp*, 1996, **111**, 63–71.
9. T. Otsu, M. Yoshida and T. Tazaki, *Makromol. Chem. Rapid Commun.*, 1982, **3**, 133–140.
10. J. P. Kennedy and R. A. Smith, *J. Polym. Sci., Part A: Polym. Chem.*, 1980, **18**, 1523–1537.
11. T. Otsu and M. Yoshida, *Makromol. Chem. Rapid Commun.*, 1982, **3**, 127–132.
12. A. Goto and T. Fukuda, *Prog. Polym. Sci.*, 2004, **29**, 329–385.
13. S. I. Kuchanov, *Russ. Chem. Rev.*, 1991, **60**, 689.
14. S. I. Kuchanov, *J. Polym. Sci., Part A: Polym. Chem.*, 1994, **32**, 1557–1568.
15. T. Otsu, *J. Polym. Sci., Part A: Polym. Chem.*, 2000, **38**, 2121–2136.

16. T. Otsu and A. Matsumoto, in *Advances in Polymer Science*, Springer, 1998, pp. 75–137.
17. A. R. Kannurpatti, K. J. Anderson, J. W. Anseth and C. N. Bowman, *J. Polym. Sci., Part B: Polym. Phys.*, 1997, **35**, 2297–2307.
18. A. R. Kannurpatti, S. X. Lu, G. M. Bunker and C. N. Bowman, *Macromolecules*, 1996, **29**, 7310–7315.
19. J. H. Ward and N. A. Peppas, *Macromolecules*, 2000, **33**, 5137–5142.
20. A. Patel and K. Mequanint, *Polymer*, 2009, **50**, 4464–4470.
21. K. Ishizu, R. A. Khan, Y. Ohta and M. Furo, *J. Polym. Sci., Part A: Polym. Chem.*, 2004, **42**, 76–82.
22. K. Ishizu, H. Katsuhara and K. Itoya, *J. Polym. Sci., Part A: Polym. Chem.*, 2005, **43**, 230–233.
23. K. Ishizu, H. Katsuhara, S. Kawauchi and M. Furo, *J. Appl. Polym. Sci.*, 2005, **95**, 413–418.
24. T. Otsu, M. Yoshida and A. Kuriyama, *Polym. Bull.*, 1982, 7, 45–50.
25. E. V. Chernikova, Z. A. Pokataeva, E. S. Garina, M. B. Lachinov and V. B. Golubev, *Polym. Sci. Ser. B Polym. Chem.*, 1998, **40**, 221–227.
26. K. Matyjaszewski and T. P. Davis, *Handbook of radical polymerization*, Wiley-Interscience, Hoboken, N.J., 2002.
27. P. Lambrinos, M. Tardi, A. Polton and P. Sigwalt, *Eur. Polym. J.*, 1990, **26**, 1125–1135.
28. R. T. A. Mayadunne, E. Rizzardo, J. Chiefari, Y. K. Chong, G. Moad and S. H. Thang, *Macromolecules*, 1999, **32**, 6977–6980.
29. K. Matyjaszewski, *Macromol. Rapid Commun.*, 2005, **26**, 135–136.
30. S. R. Turner and R. W. Blevins, *Macromolecules*, 1990, **23**, 1856–1859.
31. N. Luo, A. T. Metters, J. B. Hutchison, C. N. Bowman and K. S. Anseth, *Macromolecules*, 2003, **36**, 6739–6745.
32. Y. Nakayama and T. Matsuda, *Macromolecules*, 1996, **29**, 8622–8630.
33. Y. Yagci, A. C. Aydogan and A. E. Sizgek, *J. Polym. Sci., Part C: Polym. Lett.*, 1984, **22**, 103–106.
34. M. H. Acar, Y. Yagci and W. Schnabel, *Polym. Int.*, 1998, **46**, 331–335.
35. A. Bledzki and D. Braun, *Macromol. Chem. Phys.*, 1986, **187**, 2599–2608.
36. D. Braun, T. Skrzek, S. Steinhauerbeisser, H. Tretner and H. J. Lindner, *Macromol. Chem. Phys.*, 1995, **196**, 573–591.
37. H. A. P. Dejongh, H. J. M. Sinnige, W. G. Huysmans, C. R. H. Jonge, H. Jaspers, W. J. Mijs and W. J. D. Klein, *Macromol. Chem.*, 1972, **157**, 279.
38. E. V. Chernikova, Z. A. Pokataeva, E. S. Garina, M. B. Lachinov and V. B. Golubev, *Macromol. Chem. Phys.*, 2001, **202**, 188–193.
39. T. Otsu and T. Tazaki, *Polym. Bull.*, 1986, **16**, 277–284.
40. M. H. Acar and Y. Yagci, *J. Macromol. Sci., Pure Appl. Chem.*, 1991, **A28**, 177–183.
41. T. Otsu, A. Matsumoto and T. Tazaki, *Polym. Bull.*, 1987, **17**, 323–330.
42. A. Bledzki, D. Braun and K. Titzschkau, *Macromol. Chem. Phys.*, 1983, **184**, 745–754.
43. E. Borsig, M. Lazar and M. Capla, *Macromol. Chem.*, 1967, **105**, 212.

44. A. Bledzki and D. Braun, *Macromol. Chem. Phys.*, 1981, **182**, 1047–1056.
45. A. Bledzki, H. Balard and D. Braun, *Macromol. Chem. Phys.*, 1981, **182**, 1057–1062.
46. H. Balard, A. Bledzki and D. Braun, *Macromol. Chem. Phys.*, 1981, **182**, 1063–1071.
47. A. Bledzki, H. Balard and D. Braun, *Macromol. Chem. Phys.*, 1981, **182**, 3195–3206.
48. E. Borsig, M. Lazar, M. Capla and S. Florian, *Angew. Makromol. Chem.*, 1969, **9**, 89–95.
49. M. Kumar and T. Kannan, *Polym. J.*, 2010, **42**, 916–922.
50. X. P. Chen, K. Y. Qiu, G. Swift, D. G. Westmoreland and S. Q. Wu, *Eur. Polym. J.*, 2000, **36**, 1547–1554.
51. H. Zhou, J. G. Jiang and K. D. Zhang, *J. Polym. Sci., Part A: Polym. Chem.*, 2005, **43**, 2567–2573.
52. S. H. Qin, K. Y. Qiu, G. Swift, D. G. Westmoreland and S. G. Wu, *J. Polym. Sci., Part A: Polym. Chem.*, 1999, **37**, 4610–4615.
53. S. H. Qin, D. Q. Qin, K. Y. Qiu, D. G. Westmoreland, W. Lau, S. G. Wu and G. Swift, *J. Macromol. Sci., Pure Appl. Chem.*, 2001, **38**, 57–65.
54. S. H. Qin, K. Y. Qiu, D. G. Westmoreland, W. Lau, S. Wu and G. Swift, *J. Appl. Polym. Sci.*, 2001, **80**, 2566–2572.
55. S. H. Qin, K. Y. Qiu, G. Swift, S. G. Wu and D. G. Westmoreland, *Polym. Int.*, 2001, **50**, 284–289.
56. T. Ozturk and I. Cakmak, *J. Appl. Polym. Sci.*, 2010, **117**, 3277–3281.
57. Y.-Q. Xu, J.-M. Lu, N.-J. Li, F. Yan, X.-W. Xia and Q.-F. Xu, *Eur. Polym. J.*, 2008, **44**, 2404–2411.
58. S. H. Qin and K. Y. Qiu, *J. Polym. Sci., Part A: Polym. Chem.*, 2000, **38**, 2115–2120.
59. S. H. Qin and K. Y. Qiu, *Polymer*, 2001, **42**, 3033–3042.
60. S. H. Qin and K. Y. Qiu, *Polym. Bull.*, 2000, **44**, 123–128.
61. S. H. Qin and K. Y. Qiu, *J. Polym. Sci., Part A: Polym. Chem.*, 2001, **39**, 1450–1455.
62. D. R. Suwier, M. J. Monteiro, A. Vandervelden and E. Koning, *E-Polymers*, 2002, No: 025.
63. K. Endo, K. Murata and T. Otsu, *Macromolecules*, 1992, **25**, 5554–5556.
64. T. S. Kwon, K. Suzuki, K. Takagi, H. Kunisada and Y. Yuki, *J. Macromol. Sci., Pure Appl. Chem.*, 2001, **38**, 591–604.
65. P. Demircioglu, M. H. Acar and Y. Yagci, *J. Appl. Polym. Sci.*, 1992, **46**, 1639–1643.
66. I. W. Cho and J. Kim, *Polymer*, 1999, **40**, 1577–1580.
67. M. A. Tasdelen, Y. Y. Durmaz, B. Karagoz, N. Bicak and Y. Yagci, *J. Polym. Sci., Part A: Polym. Chem.*, 2008, **46**, 3387–3395.
68. J. Watanabe, K. Kano and M. Akashi, *Mater. Sci. Eng., C*, 2009, **29**, 2287–2293.
69. N. Metz and P. Theato, *Eur. Polym. J.*, 2007, **43**, 1202–1209.
70. J. Wootthikanokkhan and B. Tongrubbai, *J. Appl. Polym. Sci.*, 2003, **88**, 921–927.

71. M. E. Arnold, K. Nagai, R. J. Spontak, B. D. Freeman, D. Leroux, D. E. Betts, J. M. DeSimone, F. A. DiGiano, C. K. Stebbins and R. W. Linton, *Macromolecules*, 2002, **35**, 3697–3707.
72. J. Li, B. Zu, Y. Zhang, X. Guo and H. Zhang, *J. Polym. Sci., Part A: Polym. Chem.*, 2010, **48**, 3217–3228.
73. C. Vankerckhoven, H. Vandenbroeck, G. Smets and J. Huybrechts, *Macromol. Chem. Phys.*, 1991, **192**, 101–114.
74. A. Kongkaew and J. Wootthikanokkhan, *J. Appl. Polym. Sci.*, 2000, **75**, 938–944.
75. J. Lalevee, N. Blanchard, M. El-Roz, X. Allonas and J. P. Fouassier, *Macromolecules*, 2008, **41**, 2347–2352.
76. T. S. Kwon, S. Kumazawa, S. Kondo, K. Takagi, H. Kunisada and Y. Yuki, *J. Macromol. Sci., Pure Appl. Chem.*, 1998, **A35**, 1895–1913.
77. K. Rathore, K. R. Reddy, N. S. Tomer, S. M. Desai and R. P. Singh, *J. Appl. Polym. Sci.*, 2004, **93**, 348–355.
78. T. S. Kwon, S. Kumazawa, T. Yokoi, S. Kondo, H. Kunisada and Y. Yuki, *J. Macromol. Sci., Pure Appl. Chem.*, 1997, **A34**, 1553–1567.
79. S. E. Shim, Y. Shin, J. W. Jun, K. Lee, H. Jung and S. Choe, *Macromolecules*, 2003, **36**, 7994–8000.
80. S. E. Shim, Y. Shin, H. Lee, H. J. Jung, Y. H. Chang and S. Choe, *J. Ind. Eng. Chem.*, 2003, **9**, 619–628.
81. J. Kim, J. Kwak and D. Kim, *J. Appl. Polym. Sci.*, 2007, **106**, 3816–3822.
82. J. Kwak, P. Lacroix-Desmazes, J. J. Robin, B. Boutevin and N. Torres, *Polymer*, 2003, **44**, 5119–5130.
83. D. Miyamoto, J. Watanabe and K. Ishihara, *Biomaterials*, 2004, **25**, 71–76.
84. X. M. Yang and K. Y. Qiu, *J. Macromol. Sci., Pure Appl. Chem.*, 1997, **A34**, 315–325.
85. Y. Z. You, C. Y. Hong, R. K. Bai, C. Y. Pan and J. Wang, *Macromol. Chem. Phys.*, 2002, **203**, 477–483.
86. A. Nordborg and K. Irgum, *J. Appl. Polym. Sci.*, 2010, **117**, 2781–2789.
87. J. Lalevee, M. El-Roz, X. Allonas and J. P. Fouassier, *J. Polym. Sci., Part A: Polym. Chem.*, 2007, **45**, 2436–2442.
88. J. Lalevee, X. Allonas and J. P. Fouassier, *Macromolecules*, 2006, **39**, 8216–8218.
89. S. Arimori, S. Ohashi and T. Matsuda, *React. Funct. Polym.*, 2007, **67**, 1346–1360.
90. K. Ishizu, Y. Ohta and S. Kawauchi, *Macromolecules*, 2002, **35**, 3781–3784.
91. K. Ishizu, T. Shibuya and S. Kawauchi, *Macromolecules*, 2003, **36**, 3505–3510.
92. Y. K. Joung, J. H. Choi, J. W. Bae and K. D. Park, *Acta Biomater.*, 2008, **4**, 960–966.
93. K. Ishizu, A. Mori and T. Shibuya, *Polymer*, 2001, **42**, 7911–7914.
94. H. Kitano, M. Chibashi, S. Nakamata and M. Ide, *Langmuir*, 1999, **15**, 2709–2713.

95. K. Ishizu, H. Katsuhara and K. Itoya, *J. Polym. Sci., Part A: Polym. Chem.*, 2006, **44**, 3321–3327.

96. L. G. Lovell, B. J. Elliott, J. R. Brown and C. N. Bowman, *Polymer*, 2001, **42**, 421–429.

97. K. Endo, K. Murata and T. Otsu, *Polymer*, 1992, **33**, 3976–3977.

98. P. K. Bhuyan and D. K. Kakati, *J. Appl. Polym. Sci.*, 2005, **98**, 2320–2328.

99. Y. Nemoto, A. Borovkov, Y.-M. Zhou, Y. Takewa, E. Tatsumi and Y. Nakayama, *Bioconjugate Chem.*, 2009, **20**, 2293–2299.

100. Y. Nemoto and Y. Nakayama, *J. Polym. Sci., Part A: Polym. Chem.*, 2008, **46**, 4505–4512.

101. Y. Nakayama, C. Kakei, A. Lshikawa, Y.-M. Zhou, Y. Nemoto and K. Uchida, *Bioconjugate Chem.*, 2007, **18**, 2037–2044.

102. Y. Nemoto, Y.-M. Zhou, E. Tatsumi and Y. Nakayama, *Bioconjugate Chem.*, 2008, **19**, 2513–2519.

103. T. Otsu, T. Matsunaga, A. Kuriyama and M. Yoshioka, *Eur. Polym. J.*, 1989, **25**, 643–650.

104. T. Otsu, T. Matsunaga, T. Doi and A. Matsumoto, *Eur. Polym. J.*, 1995, **31**, 67–78.

105. J. Lalevee, X. Allonas, S. Jradi and J. P. Fouassier, *Macromolecules*, 2006, **39**, 1872–1879.

106. M. A. Tasdelen, M. U. Kahveci and Y. Yagci, *Prog. Polym. Sci.*, 2011, **36**, 455–567.

107. R. M. Pierson, A. J. Costanza and A. H. Weinstein, *J. Polym. Sci., Part A: Polym. Chem.*, 1955, **17**, 221–246.

108. A. Shefer, A. J. Grodzinsky, K. L. Prime and J. P. Busnel, *Macromolecules*, 1993, **26**, 2240–2245.

109. C. R. Nair, M. C. Richou and G. Clouet, *Macromol. Chem. Phys.*, 1991, **192**, 579–590.

110. C. P. R. Nair, G. Clouet and P. Chaumont, *J. Polym. Sci., Part A: Polym. Chem.*, 1989, **27**, 1795–1809.

111. J. M. Lu, X. W. Xia, X. Guo, Q. F. Xu, F. Yan and L. H. Wang, *J. Appl. Polym. Sci.*, 2008, **108**, 3430–3434.

112. G. Clouet and H. J. Juhl, *Macromol. Chem. Phys.*, 1994, **195**, 243–251.

113. D. Braun and T. Skrzek, *Macromol. Chem. Phys.*, 1995, **196**, 4039–4055.

114. C. P. R. Nair and G. Clouet, *Macromol. Chem. Phys.*, 1989, **190**, 1243–1252.

115. D. Edelmann and H. Ritter, *Macromol. Chem. Phys.*, 1993, **194**, 2375–2384.

116. D. Edelmann and H. Ritter, *Macromol. Rapid Commun.*, 1994, **15**, 791–796.

117. T. Otsu, T. Ogawa and T. Yamamoto, *Macromolecules*, 1986, **19**, 2087–2089.

118. Y. Yagci and M. A. Tasdelen, *Prog. Polym. Sci.*, 2006, **31**, 1133–1170.

119. Z. R. Guo, D. C. Wan and J. L. Huang, *Macromol. Rapid Commun.*, 2001, **22**, 367–371.

120. A. B. Duz, G. Hizal and Y. Yagci, *Eur. Polym. J.*, 2000, **36**, 1373–1378.
121. H. Miyama, N. Harumiya, Y. Mori and H. Tanzawa, *J. Biomed. Mater. Res.*, 1977, **11**, 251–265.
122. P. Liu and J. Guo, *J. Appl. Polym. Sci.*, 2006, **102**, 3385–3390.
123. P. Liu and L. Zhang, *J. Macromol. Sci., Pure Appl. Chem.*, 2008, **45**, 17–21.
124. T. Otsu, K. Yamashita and K. Tsuda, *Macromolecules*, 1986, **19**, 287–290.
125. H. Inoue and S. Kohama, *J. Appl. Polym. Sci.*, 1984, **29**, 877–889.
126. D. R. Suwier, P. A. M. Steeman, M. N. Teerenstra, M. A. J. Schellekens, B. Vanhaecht, M. J. Monteiro and C. E. Koning, *Macromolecules*, 2002, **35**, 6210–6216.
127. D. R. Suwier, M. N. Teerenstra, B. Vanhaecht and C. E. Koning, *J. Polym. Sci., Part A: Polym. Chem.*, 2000, **38**, 3558–3568.
128. E. Kroeze, G. Tenbrinke and G. Hadziioannou, *Macromolecules*, 1995, **28**, 6650 6656.
129. E. Kroeze, G. TenBrinke and G. Hadziioannou, *J. Macromol. Sci., Pure Appl. Chem.*, 1997, **A34**, 439–450.
130. K. Tharanikkarasu and G. Radhakrishnan, *J. Polym. Sci., Part A: Polym. Chem.*, 1996, **34**, 1723–1731.
131. G. Jiang, X. Tuo, D. Wang and Q. Li, *J. Polym. Sci., Part A: Polym. Chem.*, 2009, **47**, 3248–3256.
132. B. K. Kim, K. Tharanikkarasu and J. S. Lee, *Colloid. Polym. Sci.*, 1999, **277**, 285–290.
133. K. Tharanikkarasu and G. Radhakrishnan, *Polym. Int.*, 1997, **43**, 13–21.
134. K. Tharanikkarasu, C. V. Thankam and G. Radhakrishnan, *Eur. Polym. J.*, 1997, **33**, 1771–1777.
135. S. Sundar, K. Tharanikkarasu, A. Dhathathreyan and G. Radhakrishnan, *Colloid. Polym. Sci.*, 2002, **280**, 915–921.
136. G. N. Mahesh, A. Sivaraman, K. Tharanikkarasu and G. Radhakrishnan, *J. Polym. Sci., Part A: Polym. Chem.*, 1997, **35**, 1237–1244.
137. K. Mequanint, A. Patel and D. Bezuidenhout, *Biomacromolecules*, 2006, **7**, 883–891.
138. K. Tharanikkarasu and B. K. Kim, *J. Appl. Polym. Sci.*, 1999, **73**, 2993–3000.
139. B. Yan, J. He, X. Du, K. Zhang, S. Wang, C. Pan and Y. Wang, *Liq. Cryst.*, 2009, **36**, 933–938.
140. B. Yan, J. He, R. Bao, X. Bai, S. Wang, Y. Zeng and Y. Wang, *Eur. Polym. J.*, 2008, **44**, 952–958.
141. Y. Y. Durmaz, B. Karagoz, N. Bicak and Y. Yagci, *Polym. Int.*, 2008, **57**, 1182–1187.
142. Y. Nakayama, M. Miyamura, Y. Hirano, K. Goto and T. Matsuda, *Biomaterials*, 1999, **20**, 963–970.
143. I. K. Kwon and T. Matsuda, *Biomaterials*, 2006, **27**, 986–995.
144. E. Kroeze, B. deBoer, G. tenBrinke and G. Hadziioannou, *Macromolecules*, 1996, **29**, 8599–8605.

145. A. Patel and K. Mequanint, *J. Polym. Sci., Part A: Polym. Chem.*, 2008, **46**, 6272–6284.

146. M. H. Acar, A. Gulkanat, S. Seyren and G. Hizal, *Polymer*, 2000, **41**, 6709–6713.

147. I. Cakmak, *Eur. Polym. J.*, 1998, **34**, 1561–1563.

148. D. Derouet, P. Intharapat, Q. N. Tran, F. Gohier and C. Nakason, *Eur. Polym. J.*, 2009, **45**, 820–836.

149. D. Derouet, Q. N. Tran and J. L. Leblanc, *J. Appl. Polym. Sci.*, 2009, **112**, 788–799.

150. K. Nagesh and S. Ramakrishnan, *Synth. Methods*, 2005, **155**, 320–323.

151. J. Wootthikanokkhan, C. Thanachayanont and N. Seeponkai, *J. Appl. Polym. Sci.*, 2010, **116**, 433–440.

152. H. J. Lee and T. Matsuda, *J. Biomed. Mater. Res.*, 1999, **47**, 564–567.

153. I. Cakmak, H. Baykara and B. Set, *J. Appl. Polym. Sci.*, 2008, **107**, 1604–1608.

154. N. Luo, A. T. Metters, J. B. Hutchison, C. N. Bowman and K. S. Anseth, *Macromolecules*, 2003, **36**, 6739–6745.

155. K. Ishizu, H. Kakinuma and J. Park, *J. Polym. Sci., Part A: Polym. Chem.*, 2004, **42**, 3644–3648.

156. J. M. J. Fréchet, M. Henmi, I. Gitsov, S. Aoshima, M. R. Leduc and R. B. Grubbs, *Science*, 1995, **269**, 1080–1083.

157. B. de Boer, H. K. Simon, M. P. L. Werts, E. W. van der Vegte and G. Hadziioannou, *Macromolecules*, 2000, **33**, 349–356.

158. H. J. Lee, Y. Nakayama and T. Matsuda, *Macromolecules*, 1999, **32**, 6989–6995.

159. S. Kidoaki, Y. Nakayama and T. Matsuda, *Langmuir*, 2001, **17**, 1080–1087.

160. S. Kidoaki, S. Ohya, Y. Nakayama and T. Matsuda, *Langmuir*, 2001, **17**, 2402–2407.

161. T. Matsuda, M. Kaneko and S. R. Ge, *Biomaterials*, 2003, **24**, 4507–4515.

162. T. Matsuda and S. Ohya, *Langmuir*, 2005, **21**, 9660–9665.

163. Y. Nakayama and T. Matsuda, *Macromolecules*, 1999, **32**, 5405–5410.

164. S. B. Rahane, S. M. Kilbey and A. T. Metters, *Macromolecules*, 2005, **38**, 8202–8210.

165. N. Luo, J. B. Hutchison, K. S. Anseth and C. N. Bowman, *Macromolecules*, 2002, **35**, 2487–2493.

166. S. B. Rahane, A. T. Metters and S. M. Kilbey, II, *Macromolecules*, 2006, **39**, 8987–8991.

167. R. Barbey, L. Lavanant, D. Paripovic, N. Schüwer, C. Sugnaux, S. Tugulu and H.-A. Klok, *Chem. Rev.*, 2009, **109**, 5437–5527.

168. A. Ajayaghosh and S. Das, *J. Appl. Polym. Sci.*, 1992, **45**, 1617–1622.

169. G. Duner, H. Anderson, A. Myrskog, M. Hedlund, T. Aastrup and O. Ramstrom, *Langmuir*, 2008, **24**, 7559–7564.

170. Y. Nakayama, J. M. Anderson and T. Matsuda, *J. Biomed. Mater. Res.*, 2000, **53**, 584–591.

171. L. Qin, X. W. He, W. Zhang, W. Y. Li and Y. K. Zhang, *J. Chromatogr. A*, 2009, **1216**, 807–814.

172. F. Rong, X. G. Feng, P. Li, C. W. Yuan and D. G. Fu, *Chin. Sci. Bull.*, 2006, **51**, 2566–2571.

173. J. Higashi, Y. Nakayama, R. E. Marchant and T. Matsuda, *Langmuir*, 1999, **15**, 2080–2088.

174. T. Matsuda, J. Nagase, A. Ghoda, Y. Hirano, S. Kidoaki and Y. Nakayama, *Biomaterials*, 2003, **24**, 4517–4527.

175. A. M. I. Ali and A. G. Mayes, *Macromolecules*, 2010, **43**, 837–844.

176. J. H. Ward, R. Bashir and N. A. Peppas, *J. Biomed. Mater. Res.*, 2001, **56**, 351–360.

177. S. Tsuji and H. Kawaguchi, *Macromolecules*, 2006, **39**, 4338–4344.

178. S. Tsuji and H. Kawaguchi, *Langmuir*, 2004, **20**, 2449–2455.

179. H. Kawaguchi, Y. Isono and S. Tsuji, *Makromol. Symp*, 2002, **179**, 75–87.

180. Y. Nakayama and T. Matsuda, *Langmuir*, 1999, **15**, 5560–5566.

181. B. Ruckert, A. J. Hall and B. Sellergren, *J. Mater. Chem.*, 2002, **12**, 2275–2280.

182. Y. Nakayama, M. Sudo, K. Uchida and T. Matsuda, *Langmuir*, 2002, **18**, 2601–2606.

183. I. Mijangos, A. Guerreiro, E. Piletska, M. J. Whitcombe, K. Karim, I. Chianella and S. Piletsky, *J. Sep. Sci.*, 2009, **32**, 3340–3346.

184. H. M. Simms, C. M. Brotherton, B. T. Good, R. H. Davis, K. S. Anseth and C. N. Bowman, *Lab Chip*, 2005, **5**, 151–157.

185. P. S. Hume and K. S. Anseth, *Biomaterials*, 2010, **31**, 3166–3174.

186. R. P. Sebra, K. S. Anseth and C. N. Bowman, *J. Polym. Sci., Part A: Polym. Chem.*, 2006, **44**, 1404–1413.

187. S. K. Reddy, R. P. Sebra, K. S. Anseth and C. N. Bowman, *J. Polym. Sci., Part A: Polym. Chem.*, 2005, **43**, 2134–2144.

188. D. Lakshmi, A. Bossi, M. J. Whitcombe, I. Chianella, S. A. Fowler, S. Subrahmanyam, E. V. Piletska and S. A. Piletsky, *Anal. Chem.*, 2009, **81**, 3576–3584.

189. R. P. Sebra, A. M. Kasko, K. S. Anseth and C. N. Bowman, *Sens. Actuators, B*, 2006, **119**, 127–134.

190. R. P. Sebra, K. S. Masters, C. Y. Cheung, C. N. Bowman and K. S. Anseth, *Anal. Chem.*, 2006, **78**, 3144–3151.

191. R. P. Sebra, K. S. Masters, C. N. Bowman and K. S. Anseth, *Langmuir*, 2005, **21**, 10907–10911.

192. N. Luo, J. B. Hutchison, K. S. Anseth and C. N. Bowman, *J. Polym. Sci., Part A: Polym. Chem.*, 2002, **40**, 1885–1891.

193. S. Ohya, Y. Nakayama and T. Matsuda, *Biomacromolecules*, 2001, **2**, 856–863.

194. S. Ohya and T. Matsuda, *J. Biomater. Sci., Polym. Ed.*, 2005, **16**, 809–827.

195. N. Morikawa and T. Matsuda, *J. Biomater. Sci., Polym. Ed.*, 2002, **13**, 167–183.

196. S. Ohya, S. Kidoaki and T. Matsuda, *Biomaterials*, 2005, **26**, 3105–3111.

197. T. Magoshi, H. Ziani-Cherif, S. Ohya, Y. Nakayama and T. Matsuda, *Langmuir*, 2002, **18**, 4862–4872.

198. K. Hattori, M. Hiwatari, C. Iiyama, Y. Yoshimi, F. Kohori, K. Sakai and S. A. Piletsky, *J. Membr. Sci.*, 2004, **233**, 169–173.

199. D. M. He and M. Ulbricht, *Macromol. Chem. Phys.*, 2009, **210**, 1149–1158.

200. H. Kitano, S. Morokoshi, K. Ohhori, M. Gemmei-Idea, Y. Yokoyama and K. Ohno, *J. Colloid Interface Sci.*, 2004, **273**, 106–114.

201. E. M. Benetti, E. Reimhult, J. de Bruin, S. Zapotoczny, M. Textor and G. J. Vancso, *Macromolecules*, 2009, **42**, 1640–1647.

202. H. Kitano and K. Ohhori, *Langmuir*, 2001, **17**, 1878–1884.

203. O. Tagit, N. Tomczak, E. M. Benetti, Y. Cesa, C. Blum, V. Subramaniam, J. L. Herek and G. J. Vancso, *Nanotechnology*, 2009, **20**.

204. E. M. Benetti, S. Zapotoczny and J. Vancso, *Adv. Mater.*, 2007, **19**, 268–271.

205. S. Morokoshi, K. Ohhori, K. Mizukami and H. Kitano, *Langmuir*, 2004, **20**, 8897–8902.

206. H. Kitano, Y. Anraku and H. Shinohara, *Biomacromolecules*, 2006, **7**, 1065–1071.

207. O. Moriya, S. Yamamoto, T. Kumon, T. Kageyama, A. Kimura and T. Sugizaki, *Chem. Lett.*, 2004, **33**, 224–225.

208. F. Barahona, E. Turiel, P. A. G. Cormack and A. Martin-Esteban, *J. Polym. Sci., Part A: Polym. Chem.*, 2010, **48**, 1058–1066.

209. J. Bai, K. Y. Qiu and Y. Wei, *Polym. Int.*, 2003, **52**, 853–858.

210. P. Liu and Z. X. Su, *J. Photochem. Photobiol., A*, 2004, **167**, 237–240.

211. B. Sellergren, B. Ruckert and A. J. Hall, *Adv. Mater.*, 2002, **14**, 1204–1208.

212. S. Su, M. Zhang, B. Li, H. Zhang and X. Dong, *Talanta*, 2008, **76**, 1141–1146.

213. B. P. Harris, J. K. Kutty, E. W. Fritz, C. K. Webb, K. J. L. Burg and A. T. Metters, *Langmuir*, 2006, **22**, 4467–4471.

214. B. P. Harris and A. T. Metters, *Macromolecules*, 2006, **39**, 2764–2772.

215. L. Liang, P. C. Rieke, G. E. Fryxell, J. Liu, M. H. Engehard and K. L. Alford, *J. Phys. Chem. B*, 2000, **104**, 11667–11673.

216. S. B. Rahane, J. A. Floyd, A. T. Metters and S. M. Kilbey, *Adv. Funct. Mater.*, 2008, **18**, 1232–1240.

217. A. Bossi, M. J. Whitcombe, Y. Uludag, S. Fowler, I. Chianella, S. Subrahmanyam, I. Sanchez and S. A. Piletsky, *Biosens. Bioelectron.*, 2010, **25**, 2149–2155.

218. M. Yasutake, Y. Andou, S. Hiki, H. Nishida and T. Endo, *Macromol. Chem. Phys.*, 2004, **205**, 492–499.

219. J. D. He, J. Wang, S. D. Li and M. K. Cheung, *J. Appl. Polym. Sci.*, 2001, **81**, 1286–1290.

220. H. Y. Lee and B. S. Kim, *Biosens. Bioelectron.*, 2009, **25**, 587–591.

CHAPTER 4

Controlled/Living Radical Polymerization Mediated by Stable Organic Radicals

PETER NESVADBA

BASF Schweiz AG, Mattenstrasse 22, R-1059.6.05, 4058 Basel, Switzerland
E-mail: peter.nesvadba@basf.com

4.1 Introduction

Life as we know it would not be possible without polymers. About 230 million tons[1] were produced worldwide in 2009. The reader interested in the history of polymers will find a good overview in the book of Morawetz.[2] Synthetic polymers are prepared from low molecular weight monomers by a process called polymerization, which can be divided from a mechanistic point of view[3] into several classes. Amongst the most important polymerization techniques is radical polymerization which belongs to the class of chain-growth polymerizations.

In accordance with the recent IUPAC recommendation[4] the term radical polymerization (RP) will be used in this text instead of *free* radical polymerization (and analogously only radical instead of *free* radical). Polymers produced by RP represent roughly 40–45% of all industrial polymers. The dominant position of RP in industry originates from several unique characteristics, which differentiate it from other polymerization methods. Thus, in contrast to ionic or coordination polymerization, RP is tolerant to protic solvents and trace impurities like oxygen, CO_2, or monomer stabilizers.

RSC Polymer Chemistry Series No. 4
Fundamentals of Controlled/Living Radical Polymerization
Edited by Nicolay V Tsarevsky and Brent S Sumerlin
© The Royal Society of Chemistry 2013
Published by the Royal Society of Chemistry, www.rsc.org

Consequently, it can be conducted in polar solvents such as alcohols or, more importantly, water, with monomers which are not rigorously dried or purified. Furthermore, a large number of inexpensive monomers suitable for this technology are available. Thus, from an industrial and economical point of view, RP is the technique of choice. For example, the cost to polymerize styrene by anionic polymerization is about 50% higher[5] than for polymerization of the same monomer by RP. No details of conventional RP will be discussed in this article; a comprehensive treatment of the chemical and mechanistic aspects of RP is provided, *e.g.*, by the outstanding monographs of Moad and Solomon,[6] Matyjaszewski and Davis[7] and others.[8,9]

Prior to the mid 1980's, classical RP did not allow the synthesis of polymers with precisely designed molecular architectures, such as block, comb, or star copolymers, or polymers with defined terminal functional groups or narrow molecular weight distributions (Scheme 4.1). Many of these polymeric architectures were only accessible by non-radical living polymerizations. The first living polymerization technique, which was discovered[10,11] in 1956 by Michael Szwarz, was living anionic polymerization. Over the years, other controlled and living polymerizations such as group transfer polymerization, living carbo-cationic polymerization, living ring opening metathesis polymerization, or living transition metal catalyzed alkene polymerization were developed. For a recent overview of the state of the art of controlled and living polymerizations see the monograph[12] of Müller and Matyjaszewski. Its preface also contains clear definitions of the sometimes controversial terms "controlled" and "living". Accordingly, and as defined[13] by IUPAC, a living polymerization is a *"chain polymerization* from which *chain termination* and irreversible *chain transfer* are absent"*.

The term controlled polymerization was coined[14] by Matyjaszewski and Müller in 1997 and is defined[12] "as a synthetic method to prepare polymers which are well defined with respect to topology (*e.g.* linear, star-shaped, comb-shaped, dendritic, and cyclic), terminal functionality, composition, and arrangement of co-monomers (*e.g.*, statistical, periodic, block, graft, and gradient, see Scheme 4.1), and which have molecular weights predetermined by

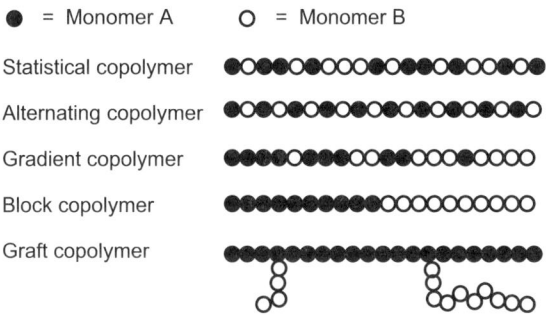

Scheme 4.1   Different types of polymeric architectures.

the ratio of concentrations of reacted monomer to introduced initiator, as well as a designed (not necessarily narrow) molecular weight distribution (MWD). Thus, according to the definition given above, a living polymerization is not always controlled and a controlled polymerization is not always strictly living. In the ideal case, a living polymerization is also controlled; however, in some systems such as in a radical polymerization, termination can never be entirely avoided although its contribution can be sometimes significantly reduced."

Unfortunately, non-radical, in particular ionic, polymerization processes are incompatible with many functional groups and require highly pure monomers and solvents as well as the exclusion of oxygen and water. In contrast, the controlled and living radical polymerization combines the robustness of a classical RP with the power of living polymerization and thus allows the preparation of complex polymeric structures.

Several controlled and living radical polymerization techniques are available today. All are based on the reversible deactivation of growing chains. Consequently, a 2010 IUPAC recommendation[13] proposes the term *controlled reversible-deactivation radical polymerization* (CRDRP) for polymerizations previously referred to as "controlled" radical (CRP) or "living" radical (LRP) polymerization. Nevertheless, due to their widespread acceptance, the terms controlled/living radical polymerization will also be used in this chapter.

The topic of this chapter is one of the most important CRDRP techniques, namely the *controlled/living radical polymerization mediated by stable radicals* or, as recommended by IUPAC,[13] *stable-radical-mediated polymerization* (SRMP).

The other CRDRP techniques are *atom-transfer radical polymerization* (ATRP),[15-19] *reversible-chain-transfer catalyzed polymerization* (RTCP)[20-24] and *degenerate-transfer radical polymerization* (DTRP).[15,25,26] The most prominent variant of the latter is *reversible-addition-fragmentation chain-transfer poly-merization* (RAFT).[26-33] For discussion of these topics, the reader is referred to the cited literature and the corresponding chapters in this book.

4.2 Discovery of Controlled/Living Radical Polymerization Mediated by Stable Radicals

4.2.1 Stable Radicals

Before discussing polymerization mediated by stable radicals, it is important to remember the most fundamental properties of these unique compounds. First of all, the somewhat controversial terms *stable, persistent* and *transient* radical were clearly defined[34] by Griller and Ingold. The chemistry of stable radicals has experienced tremendous growth since the discovery[35,36] of the first "stable" radical, triphenylmethyl (Ph$_3$C$^\bullet$), by Gomberg in 1900. The topic has been reviewed[37,38] several times in the past, most recently by Veciana[39] and

Hicks,[40,41] who gives an overview of the chemistry and applications of different families of stable radicals, including nitroxides.[42] Of the known classes of stable radicals, only a few have found use as mediators of radical polymerization; the most important being the nitroxide radicals. A brief overview of other stable radicals amenable to SRMP is given in Chapter 4.8.

4.2.2 Nitroxide Radicals

Nitroxide radicals **1** (Scheme 4.2) $R^1R^2NO^•$ (R^1, R^2 = alkyl, aryl, or forming together a ring) are also called nitroxides, nitroxyls, *N*-oxyls or, according to the IUPAC recommendation,[13] aminoxyls. Nitroxides belong to the most important persistent organic radicals. Some organic nitroxides were known already at the beginning of the 20th century, but the most prominent nitroxide, 2,2,6,6-tetramethylpiperidine *N*-oxyl (TEMPO) **2** (Scheme 4.2), was only discovered in 1959.[43] Since then, nitroxides have developed from laboratory curiosities to versatile molecules with many applications in organic synthesis, biochemistry, polymer- and materials science. Detailed information on properties of nitroxides, their synthesis and applications is available in several monographs[41,44–47] and review articles.[43,48–55]

Today, the nitroxide-mediated polymerization (NMP) is the most important SRMP method. In this context it is important to illustrate the basic reactions and properties of nitroxides. The unpaired electron resides predominantly in the antibonding π^* singly occupied molecular orbital. The spin density on N and O atoms is comparable[56] (~ 0.5) but is to some extent influenced by the polarity of the solvent.

In polar media, the dipolar resonance structure **1a** is favored and the spin density is shifted towards the N-atom. X-Ray analysis[42,57–61] of several cyclic nitroxides revealed that the $R^1R^2 > NO^•$ moiety is planar for five-membered nitroxides but clearly pyramidal for six- and seven-membered ring nitroxides. The N-O bond length is not much influenced by the ring size. Nitroxides do not form stable, isolable O–O dimers[62–64] because the loss of the delocalization energy[65] ($1 \leftrightarrow 1a \sim 120\,\text{kJ mol}^{-1}$ nitroxide) is greater than the peroxide bond

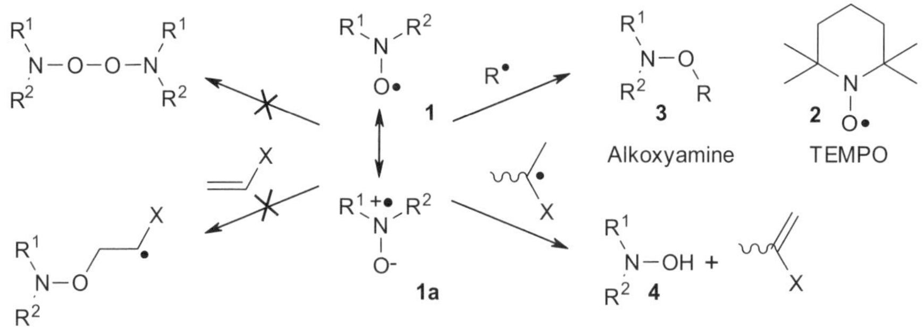

Scheme 4.2   Nitroxide radicals and their basic reactions.

strength. On the other hand, nitroxides react very rapidly with carbon centered radicals $R^\bullet$ forming alkoxyamines R^1R^2NOR **3**, the most important intermediates in NMP.[66–71] Reactions of nitroxides with O-centered radicals are generally not significant in NMP. Oxyaminoethers $R^1R^2N^+(O^-)OR$ may be formed from alkoxy radicals $RO^\bullet$ and trioxides R^1R^2NOOOR from peroxy radicals $ROO^\bullet$. These unstable species are supposed to be involved in the mechanism by which hindered amine light stabilizers (HALS) protect polymers against photooxidative degradation. High level computations[72,73] of this mechanism were recently reported. Reactions of nitroxides with peroxy radicals under acidic conditions take place at nearly diffusion-controlled rates by proton-coupled electron transfer from the protonated nitroxide to afford hydroperoxides ROOH and oxoammonium cations $R^1R^2N^+=O$.[74] The addition of nitroxides to double bonds of monomers ($CH_2=CHX$) does not occur under typical conditions of NMP, *i.e.* nitroxides do not initiate polymerization. However, the double addition product of TEMPO across the vinylic double bond of styrene[75] was obtained in 20–40% yield when an equimolar mixture of TEMPO and styrene was heated at $120\,^\circ C$ for 2 h. Typically, nitroxides suitable for NMP do not abstract hydrogen atoms from unactivated hydrocarbons because the bond dissociation energy[56] of the R^1R^2NO-H bond in the corresponding hydroxylamines **4** is very low ($\sim 293\,kJ\,mol^{-1}$). Only nitroxides $R^1R^2NO^\bullet$ bearing strongly electron withdrawing substituents R^1, R^2 are good hydrogen abstractors.[76–78] Nevertheless, **2** and derivatives thereof slowly abstract hydrogens from allylic[79,80] and benzylic[75,79,80] substrates. However, this reaction is not so important in NMP. On the other hand, H-abstraction from certain C-radicals to yield hydroxylamines **4** and olefins is a serious problem, particularly in the NMP of methacrylates, see 4.3.5.3. The stability of nitroxides under NMP polymerization conditions is also very important, see 4.3.5.4.

4.2.3 Early Developments

The non-living and non-controlled nature of classical RP arises from its mechanism[6] and the inherent reactivity of the intermediate C-centered radicals. The inevitable *irreversible and very fast* termination reactions of propagating C-radicals (*via* combination and disproportionation reactions) limit the chain growth period to only ≈ 1 s and lead to "dead" chains with a more or less time- and conversion-independent degree of polymerization ($\sim 10^3$–10^4). Additionally, various chain transfer reactions also occur. The polymer is not living because the ends of the "dead" chains cannot be reactivated. The mass distribution of the polymeric chains is broad, and the ratio M_w/M_n (known as polydispersity index PDI) is typically $\gg 1.5$, the lowest theoretical limit[81] obtainable in conventional RP.

In 1982, Otsu coined the basic idea[82] of living radical polymerization: "therefore, in order to find a system of living radical polymerization, one must try to form propagating polymer chain ends which may dissociate into

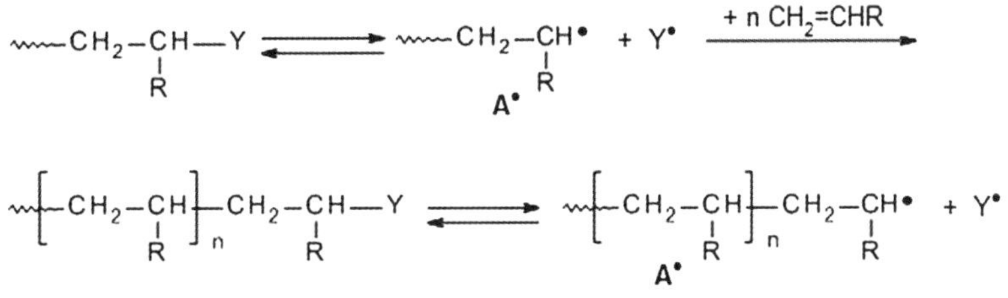

Scheme 4.3   Otsu's proposal of living radical polymerization.

polymers with a radical chain ends $A^{\bullet}$ and small radicals $Y^{\bullet}$ which must be stable enough not to initiate a new polymer chain. Such a radical polymerization would proceed *via* a living mechanism..." (see Scheme 4.3).

Otsu indeed observed[82] some features of living/controlled radical polymerization when polymerizing methyl methacrylate (MMA) with the special azo-initiator **5** (Scheme 4.4). Thus, number average degree of polymerization DP_n increased linearly with monomer conversion, and the molecular weight of the polymer increased when it was heated at $80\,^{\circ}C$ in the presence of fresh monomer.

Thermal decomposition of **5** provides the reactive (transient) phenyl radical **6** which initiates the MMA polymerization and the relatively stable (persistent) triphenylmethyl radical **7** which plays the role of $Y^{\bullet}$.

Otsu used the term *initers* (from *ini*tiator-*ter*minator) for compounds such as **5** or, because some of them were also able to undergo chain transfer to initiator, "*iniferters*" (from *ini*tiator-trans*fer* agent-*ter*minator). Nevertheless, the essential features of a true living polymerization such as accurately controlled molecular weights and low polydispersities were absent with **5** (and other iniferters too) since the trityl radical **7** can also to some extent initiate polymerization.

A good review of the history and development of the iniferter concept is available.[83]

4.2.4   Discovery of Nitroxide-mediated Polymerization

The first true stable radical-mediated polymerization method was the nitroxide-mediated radical polymerization (NMP) discovered by Solomon, Rizzardo and Cacioli at the Commonwealth Scientific and Industrial Research Organization (CSIRO) in 1985. The discovery is recounted in the article[84] of Solomon. This pioneering work, first published as a patent[85] only, attracted broad attention in 1993 when Georges *et al.* reported[86] the synthesis of polystyrene with low polydispersity *via* NMP of styrene mediated by TEMPO **2**. Today, NMP is, by far, the most important and versatile SRMP method. It is an attractive, simple and straightforward technique, because typically only monomer and an appropriate alkoxyamine initiator R^1R^2NOR are needed to conduct the polymerization.

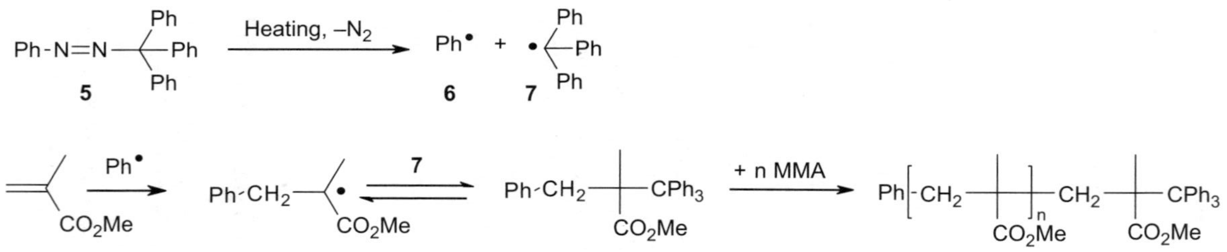

Scheme 4.4 Example of Otsu's iniferter-living radical polymerization.

4.3 Mechanism of NMP

Several reviews[87–92] of the NMP mechanism are available. In view of the restricted size of this chapter, only the most salient mechanistic aspects and the latest findings will be discussed. The mathematical treatment of NMP kinetics goes beyond the scope of this text. Hence, only the most important kinetic aspects of NMP are presented without derivations. A more thorough treatment of this subject can be found in the cited literature.

4.3.1 General Mechanism of NMP

Similar to classic RP, the general mechanism of NMP or any SRMP is composed of initiation, propagation and termination stages, as shown in Scheme 4.5.

In most cases, NMP is started with an alkoxyamine initiator R^1R^2NOR. A solution of this initiator in monomer is heated to the appropriate polymerization temperature at which the alkoxyamine undergoes reversible

Initiation   Reaction

$R^1R^2NOR \xrightarrow{k_d} R^1R^2NO^\bullet + R^\bullet$ (1)

$R^1R^2NO^\bullet + R^\bullet \xrightarrow{k_c} R^1R^2NOR$ (2)

$R^\bullet + M \xrightarrow{k_{add}} RM_1^\bullet$ (3)

Propagation

$RM_1^\bullet + M \xrightarrow{k_p} RM_2^\bullet \xrightarrow{k_p} RM_n^\bullet$ (4)

$R^1R^2NO^\bullet + RM_n^\bullet \xrightarrow{k_{c,ac}} R^1R^2NOM_nR$ (5)

$R^1R^2NOM_nR \xrightarrow{k_{d,dc}} R^1R^2NO^\bullet + {}^\bullet M_nR$ (6)

Termination

$RM_n^\bullet + RM_m^\bullet \xrightarrow{k_{t,ac}} \text{Dead chains}$ (7)

Side reactions

$R^1R^2NO^\bullet + RM_n^\bullet \xrightarrow{k_{Di}} R^1R^2NOH + \text{Alkene}$ (8)

$R^1R^2NOM_nR \xrightarrow{k_{CE}} R^1R^2NOH + \text{Alkene}$ (9)

$R^1R^2NO^\bullet \xrightarrow{k_{ND}} \text{Decomposition products}$ (10)

Scheme 4.5   General mechanism of NMP.

homolysis (rate constant k_d) to yield a persistent nitroxide $R^1R^2NO^\bullet$ and a transient radical $R^\bullet$. The highly reactive $R^\bullet$ can either recombine (k_c) with the nitroxide, terminate with itself or add to a monomer molecule (k_{add}) to afford a new C-centered radical $RM_1^\bullet$. This addition is well understood and has been comprehensively reviewed by Fischer and Radom.[93,94] $RM_1^\bullet$ can either reversibly recombine with the nitroxide or propagate by adding to additional monomer molecules (k_p) to provide the polymeric radicals $RM_n^\bullet$ (active chains). The latter in turn can either further react with monomers or reversibly couple ($k_{c,ac}$ $k_{d,dc}$) with $R^1R^2NO^\bullet$ to form a new alkoxyamine $R^1R^2NOM_nR$ (dormant chains). Hawker *et al.*[95] have demonstrated by radical crossover experiments that nitroxide radicals are free to diffuse out of the reaction cage during NMP and that statistical exchange of the chain ends occurs during the polymerization.

The final result of many iterations of this process is a polymer $R^1R^2NOM_pR$ consisting of long polymeric macro-alkoxyamine chains. The molecular weight distribution of this polymer is narrow because all chains essentially grow in parallel, in other words the polymerization is controlled. Its number average degree of polymerization DP_n after sufficiently long time[96] is given by the ratio [Converted Monomer]/$[R^1R^2NOR]_0$. Moreover, as the chains of the polymer bear reactivable nitroxide end groups, the polymer is living and polymerization and chain growth will resume upon addition of fresh monomer. Similarly, addition of a different monomer will result in a block copolymer. Note that the polymer is telechelic, bearing two different functionalities (R^1R^2NO and R) at the extremities of its chains, and could be used to prepare more complex polymeric structures (see Section 7.4).

It is important to note that irreversible termination ($k_{t,ac}$) of the C-radicals ($R^\bullet$, $RM_1^\bullet$,...$RM_n^\bullet$) also occurs in NMP. However, the extent of such termination reactions is small in comparison to reactions between C-radicals and $R^1R^2NO^\bullet$ which produce the desired dormant chains and eventually the living and controlled polymer $R^1R^2NOM_pR$. This surprising selectivity can be explained by the persistent radical effect (PRE), which is briefly treated in the next section of this chapter.

In reality, NMP is not as simple as portrayed in Scheme 4.5 (reactions 1–7). Side reactions of the nitroxide or alkoxyamine (reactions 8–10) can occur, and there may be additional sources of alkyl radicals, *e.g.*, from deliberately added conventional radical initiator (in order to speed up the polymerization) or from autoinitiation[97] of the monomer. Furthermore, free nitroxide may be present or added at the start of the polymerization. The effects of these side reactions are discussed in Section 3.5.

4.3.2 Persistent Radical Effect

The persistent radical effect (PRE) is a kinetic effect observed when transient and persistent radicals are formed with equal or similar rates. The cross-coupling of these radicals proceeds with high selectivity and dominates the homo-coupling of the transient radicals. The first example of PRE was most likely observed[98] in 1936 by Bachmann and Wiselogle during experiments on thermolysis of pentaarylethanes.

The first kinetic analysis of this phenomenon was made in 1986 by Fischer,[99] who also highlighted[100] its importance in living radical polymerization. The first numerical simulation of PRE in the context of NMP was performed by Johnson *et al.*[101] The term *persistent radical effect* was proposed in 1992 by Daikh and Finke.[14] Several excellent reviews covering the applications of PRE in organic and polymer chemistry are available.[89,102–104]

PRE explains the low level of termination observed in NMP and other SRMP's as well as in ATRP. Reversible homolysis of alkoxyamines, the key step of NMP, is a typical example of a reaction governed by the PRE. The kinetic equations describing this process were experimentally verified in 1998 by Fischer[105] and coworkers with the model alkoxyamine **8** (Scheme 4.6).

The reversible thermal homolysis (k_d) of cumyl-TEMPO **8** (R^1R^2NOR) affords TEMPO **2** ($R^1R^2NO^\bullet$) and the cumyl radical **9** ($R^\bullet$) in equal amounts. The main reaction, which the persistent nitroxide **2** undergoes, is the coupling (k_c) with the transient cumyl radical **9**. Radical **9**, however, can either couple with **2** or undergo self-termination predominantly by dimerization but also disproportionation. Note that hydrogen transfer (also called disproportionation) between **2** and **9** also occurs, affording the hydroxylamine **10** and α-methylstyrene **11**. This leads to consumption of both **2** and **9**. The impact of this side reaction on NMP is discussed in Section 3.5.3.

The kinetic behaviour of the system, disregarding disproportionation, is described by the coupled non-linear equations (1) and (2). These equations are simple but unfortunately do not have a closed analytical solution. The term $2k_t$ instead of k_t is used in eqn (2), in accordance with the IUPAC's recommendation,[106] because a single termination step consumes two radicals.

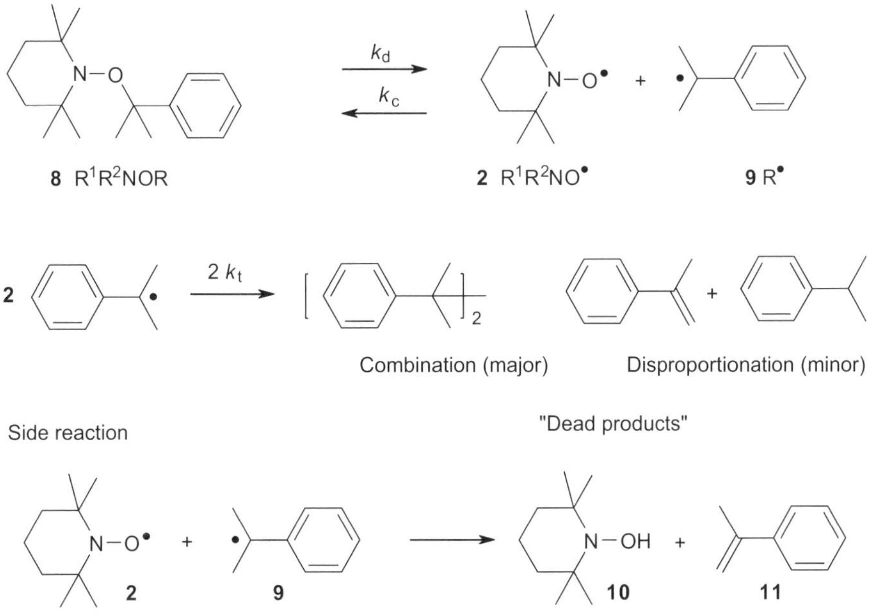

Scheme 4.6   Thermolysis of cumyl-TEMPO.

However, one should note that Fischer used k_t instead of $2k_t$ in his fundamental kinetic equations for NMP (see Section 3.3) to avoid overcrowding of the theoretical equations by the numerical factor 2 (see ref. 20 in ref. 107).

$$\frac{d\left[R^1R^2NO^\bullet\right]}{dt} = k_d\left[R^1R^2NOR\right] - k_c\left[R^1R^2NO^\bullet\right]\left[R^\bullet\right] \tag{4.1}$$

$$\frac{d\left[R^\bullet\right]}{dt} = k_d\left[R^1R^2NOR\right] - k_c\left[R^1R^2NO^\bullet\right]\left[R^\bullet\right] - 2k_t\left[R^\bullet\right]^2 \tag{4.2}$$

The numerical integration of eqn (4.1) and (4.2) with the time dependent concentrations of **2** and **9** is shown in Figure 4.1.

Figure 4.1 can be broken down into three different time ranges: (i) the initial ($t < 0.1$ s), (ii) intermediate (0.1 s $< t < 10^7$ s) and (iii) final ($t > 10^7$ s).

(i) The initial range lasts for less than 100 ms from the onset of the homolysis. The dissociation of **8** predominates since [**2**] and [**9**] are too small for the backward reaction to be of significance and therefore both [**2**] and [**9**] can be assumed to increase linearly with time according to eqn (4.3)

$$\left[R^1R^2NO^\bullet\right] = \left[R^\bullet\right] = k_d\left[R^1R^2NOR\right]_0 \times t \tag{4.3}$$

However, the irreversible bimolecular termination of **9** begins very soon after the onset of the homolysis, and the concentration of the persistent radical **2** starts to builds up. A consequence of the build up of **2** is an increase of rate of the cross-coupling between **2** and **9** to regenerate **8**. Concomitantly, the rate of the homo-coupling between **9** diminishes as the reaction proceeds.

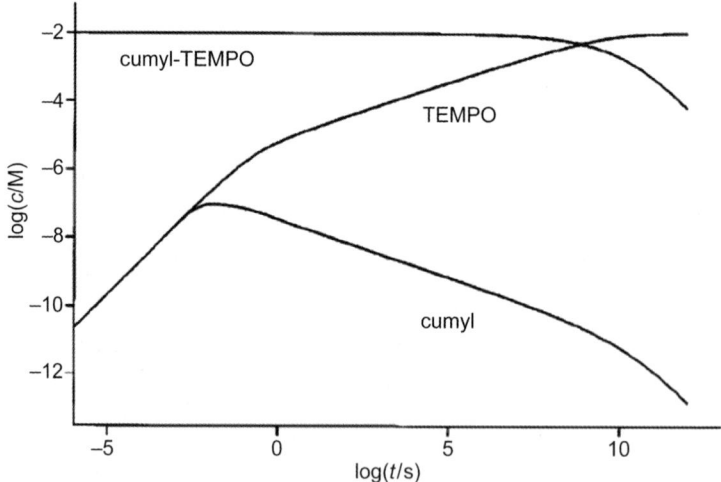

Figure 4.1   Numerically calculated time dependences of concentrations of **2** (TEMPO) and **9** (cumyl) for thermolysis of **8** (cumyl-TEMPO) in double logarithmic presentation. Temperature 83 °C, $[8]_0 = 10^{-2}$ M, $k_d = 2.3 \times 10^{-3}$ s^{-1}, $k_c = 9.1 \times 10^7$ M^{-1} s^{-1}, $2k_t = 2.3 \times 10^9$ M^{-1} s^{-1}.
[Reproduced from ref. 105, Copyright Royal Society of Chemistry, 1998].

The concentration of the transient radicals **9** reaches a maximum value of about 10^{-7} M after ~ 10 ms and then starts to decrease. At this point, only $\sim 0.035\%$ of cumyl-TEMPO **8** has decomposed.

(ii) The long intermediate period lasts for about 1.5×10^7 s $= 4200$ h. The evolution of concentrations of **2** and **9** obeys the unusual $t^{1/3}$ and $t^{-1/3}$ time dependences and can be very well described by the approximative eqn (4.4) and (4.5).

$$[R^1R^2NO^\bullet] = [R^1R^2NOR]_0^{\frac{2}{3}} \times \left(\frac{3k_d^2 \times 2k_t}{k_c^2}\right)^{\frac{1}{3}} \times t^{\frac{1}{3}} \qquad (4.4)$$

$$[R^\bullet] = [R^1R^2NOR]_0^{\frac{1}{3}} \times \left(\frac{k_d}{3k_c2k_t}\right)^{\frac{1}{3}} \times t^{-\frac{1}{3}} \qquad (4.5)$$

Approximately 19% of **8** has decomposed at the end of the intermediate period. The decomposition products are TEMPO in an amount corresponding to the decomposed **8**, and the "dead products" of the cumyl radical. Without the back-reaction between **2** and **9**, the decomposition of **8** would obey classical first order kinetics described by eqn (4.6), and **8** would behave as a classical radical initiator, such as azo-isobutyronitrile (AIBN), *i.e.*, its half life time ($t_{1/2} = \ln2/k_d$) would be only 301 s at 83 °C.

$$\frac{-d[R^1R^2NOR]}{dt} = k_d \times [R^1R^2NOR] \quad \text{or} \quad [R^1R^2NOR] = [R^1R^2NOR]_0 e^{-k_dt}$$
$$(4.6)$$

Thus, the PRE increases the lifetime of **8** by many orders of magnitude.

Decomposition of **8** under PRE-free conditions can be achieved by adding a large excess of a suitable scavenger of C-radicals (*e.g.* oxygen or galvinoxyl[108]) which does not react with **2**. This is demonstrated in Figure 4.2 which shows the evolution of the concentration of **2**, as determined by EPR spectroscopy, in the thermolysis of **8** in deoxygenated *tert*-butylbenzene (a) in the absence of and (b) in the presence of 25-fold excess of galvinoxyl scavenger.

Under PRE conditions (Figure 4.2a), $\sim 2\%$ of **8** are decomposed in 8 h whereas 100% decomposition of **8** is observed in a short time in the presence of the scavenger. This kinetic isolation of the dissociation reaction is the method of choice to determine the rate constant k_d for **8** and other alkoxyamines as well, see Section 6.1.6.

Multiplication of eqn (4.4) and (4.5) gives eqn (4.7) which reveals that there is an equilibrium between the alkoxyamine **8** (R^1R^2NOR) and radicals **2** and **9** in the intermediate range, characterized by the equilibrium constant K.

$$[R^\bullet][R^1R^2NO^\bullet] = [R^1R^2NOR]_0 \times \frac{k_d}{k_c} \quad \text{or}$$

$$\frac{k_d}{k_c} = \frac{[R^\bullet][R^1R^2NO^\bullet]}{[R^1R^2NOR]_0} = K$$
$$(4.7)$$

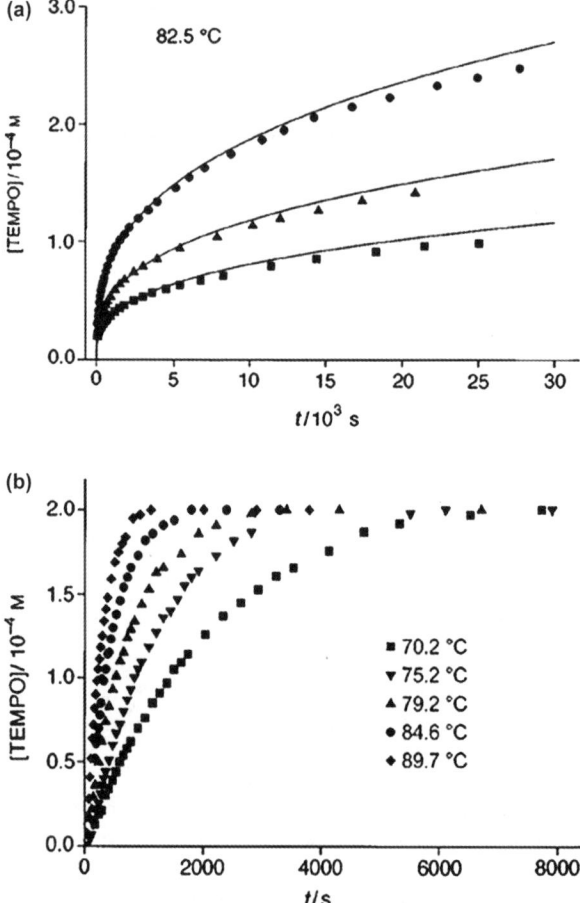

Figure 4.2 Time dependence of the TEMPO concentration during thermolysis of cumyl–
TEMPO **8**. (a) in *tert*-butylbenzene, $[8]_0$: ● $= 1.16 \times 10^{-2}$ M; ▲ $= 5.10 \times 10^{-3}$ M;
■ $= 2.52 \times 10^{-3}$ M; (b) in *tert*-butylbenzene, $[8]_0 = 2 \times 10^{-4}$ M, [galvinoxyl] $=$
5×10^{-3} M.
[Reproduced from ref. 105, Copyright Royal Society of Chemistry, 1998].

However, eqn (4.7) differs from the law of mass action in the usual sense since
it contains the initial concentration $[R^1R^2NOR]_0$ instead of the momentary
concentration $[R^1R^2NOR]$. Moreover, the radical concentrations $[R^1R^2NO^\bullet]$
and $[R^\bullet]$ are time-dependent, even though the time dependence cancels out. For
these reasons, the equilibrium (4.7) is often called a *quasi-equilibrium*. Fukuda
et al.[87,90] independently derived the eqn (4.4) and (4.5) using the *ad hoc*
assumption[109] of the equilibrium (4.7).
  (iii) The final time domain until 99% decay of **8** spans thousands of years
        which perfectly demonstrates the PRE. During this time, all **8** will be
        converted into **2** and the inactive, "dead", products of cumyl
        termination.

In 2006, Tang, Fukuda and Matyjaszewski developed a set of new equations describing the PRE in NMP even more precisely than eqn (4.4) and (4.5), in particular at a very early stage (less than 100s) and after a very long time.[110]

The reversible homolysis of **8** (Scheme 4.6) and the reversible reactions between the dormant and active chains in the NMP (Scheme 4.5) are similar reactions obeying the same kinetic laws. The only difference is that the starting alkoxyamine R^1R^2NOR is rapidly transformed into the macroalkoxyamines $R^1R^2NOM_nR$ (dormant chains) and at the end of the process into the final polymer $R^1R^2NOM_pR$. If the conversion of monomer occurs within the intermediate time range (Figure 4.1), the result will be the living and controlled polymer $R^1R^2NOM_pR$ contaminated with only small amounts of free nitroxide $R^1R^2NO^\bullet$ and dead chains resulting from significantly diminished termination reactions.

4.3.3 Fundamental Kinetic Equations of NMP

Using the kinetic description of PRE discussed above, Fischer *et al.*[89,96,111–113] developed a set of equations (see below) characterizing the ideal NMP in which only the reactions 1–7 (Scheme 4.5) occur. It must be stressed that these equations were derived under the following assumptions: (1) chain length independent rate constants (k_p, $k_{d,dc}$, $k_{c,ac}$, $k_{t,ac}$); (2) termination by disproportionation only; and (3) the initiation stage can be neglected (*i.e.*, assuming $k_d \geq k_{d,dc}$ which guarantees practically simultaneous initiation of all growing chains). Furthermore, the $t^{1/3}$ and $t^{-1/3}$ time dependences of the persistent and transient radicals (see eqn (4.4) and (4.5)) and the intermediate time regime (Figure 4.1) characterized by the quasi-equilibrium (4.7) exist only if the rate constants fulfill the conditions given in eqn (4.8)

$$k_{d,dc} \leq k_{t,ac}[I]_0 \quad k_{d,dc} \ll k_{c,ac}[I]_0 \quad \frac{k_{d,dc}}{k_{c,ac}} = K \ll k_{c,ac}[I]_0/k_{t,ac} \quad (4.8)$$

where $[I]_0$ is the concentration of the alkoxyamine initiator R^1R^2NOR at time $t = 0$.

Moreover, living and controlled polymer is only obtained if the monomer conversion occurs in the intermediate time regime where the quasi-equilibrium (4.7) holds. This requires that the conditions (4.9) are also satisfied.

$$k_p \ll 3k_{t,ac} \quad k_p \ll 3k_{c,ac} \quad (4.9)$$

Fortunately, both conditions (4.8) and (4.9) are fulfilled for most NMP's. If the insignificant amount of polymerization occurring during the few milliseconds before establishment of the quasi-equilibrium is neglected, the kinetics of NMP can be summarized in the following equations.

(i) The logarithm of the quotient of monomer conversion $\ln[M]_0/[M]$ in ideal NMP obeys a $t^{2/3}$ time dependence given by eqn (4.10)

$$\ln \frac{[M]_0}{[M]} = \frac{3}{2} k_p \left(\frac{k_{d,dc}[I]_0}{3k_{c,ac}k_{t,ac}} \right)^{1/3} t^{2/3} \quad (4.10)$$

This contrasts with the linear dependence (4.11) observed in classical RP with constant rate of initiation and a constant (steady state) radical concentration $[R^\bullet]_{st}$.

$$\ln\frac{[M]_0}{[M]} = k_p[R^\bullet]_{st}t \tag{4.11}$$

The difference between eqn (4.10) and (4.11) originates from the time dependence of radical concentration in NMP described by eqn (4.5).

However, in practice the time dependence of $\ln[M]_0/[M]$ is often linear. This may result from the inaccuracy of the measurement or side effects such as varying termination constants, extra radical generation, initial excess of the nitroxide, or its decay (see Section 3.5).

(ii) The time t_c to reach a preset monomer conversion $C = ([M]_0 - [M])/[M]_0$ is given by eqn (4.12).

$$t_c = \frac{\left(2\ln\dfrac{1}{1-C}\right)^{3/2}}{3k_p^{3/2}}\left(\frac{k_{c,ac}k_{t,ac}}{k_{d,dc}[I]_0}\right)^{1/2} \quad\text{or}$$

$$\frac{k_{d,dc}}{k_{c,ac}} = K = \frac{\left(2\ln\dfrac{1}{1-C}\right)^3 k_{t,ac}}{9[I]_0 k_p^3 t_c^2} \tag{4.12}$$

where $[I]_0$ is the original concentration of the alkoxyamine initiator R^1R^2NOR.

(iii) Based on Scheme 4.5 and the stoichiometry, the concentration of living polymer chains $R^1R^2NOM_pR$ is equal to $[I]_0 - [\text{Dead chains}]$. The fraction of dead chains Φ can be predicted with eqn (4.13).

$$\Phi = \frac{[\text{Dead chains}]}{[I]_0} = \left(\frac{2k_{d,dc}k_{t,ac}\ln\dfrac{1}{1-C}}{k_{c,ac}k_p[I]_0}\right)^{1/2} \quad\text{or}$$

$$\frac{k_{d,dc}}{k_{c,ac}} = K = \frac{k_p[I]_0}{2k_{t,ac}\ln\dfrac{1}{1-C}}\Phi^2 \tag{4.13}$$

(iv) The rather complicated full expression[96] for the polydispersity index $PDI = M_w/M_n$ can be approximated for long polymerization times by eqn (4.14)

$$PDI_\infty = 1 + \frac{[I]_0}{[M]_0} + \left(\frac{\pi k_p^3[I]_0}{k_{d,dc}k_{c,ac}k_{t,ac}}\right)^{1/2} \quad\text{or}$$

$$k_{d,dc}k_{c,ac} = \frac{\pi k_p^3[I]_0}{k_{t,ac}}\frac{1}{\left(PDI_\infty - 1 - \dfrac{[I]_0}{[M]_0}\right)^2} \tag{4.14}$$

(v) The number average degree of polymerization DP_n is derived from eqn (4.15) and reaches the final value $[M]_0/[I]_0$ which is characteristic for a controlled polymerization. Deviations from the linear increase at short times are caused by the finite time for initiator decomposition.

$$DP_n = \frac{[M]_0 - [M]}{[I]_0(1 - e^{-k_d t})} \qquad (4.15)$$

One should note that the polymerization rate (4.10) and livingness (4.13) depend on the ratio $k_{d,dc}/k_{c,ac}$, *i.e.*, on the equilibrium constant K. On the other hand, the polydispersity (control) of the polymer (4.14) depends on the product of $k_{d,dc}$ and $k_{c,ac}$. Thus, livingness and control are not necessarily mutually inclusive.

Reliable rate constants k_p are available for styrene,[114] *n*-butyl acrylate,[115] methyl[116] and other alkyl methacrylates.[117,118] Data for other monomers can be found in Polymer Handbook[119] or in the recent review by Beuermann and Buback.[120]

The termination rate constants $k_{t,ac}$ of polymeric radicals are often significantly smaller[106,119,121,122] than those[123] of simple alkyl radicals and should be used in eqn (4.12)–(4.14). Similarly, $k_{d,dc}$ and $k_{c,ac}$ for the macro-alkoxyamines instead of the k_d and k_c of the low molecular alkoxyamine initiator R^1R^2NOR are recommended. Unfortunately, only a little data for *macro*alkoxyamines is available in the literature. The chain length effect on $k_{d,dc}$ is weak in polystyrene based alkoxyamines,[124,125] stronger in polyacrylate[125] and very strong in polymethacrylate alkoxyamines.[126] Even less information is available at this time on chain length effect on $k_{c,ac}$.

Numerical modeling[127] of NMP kinetics described by the general Scheme 4.5 using the PREDICI[128] software has been performed by Guillaneuf and coworkers. In particular, the roles of $k_{d,dc}$ and $k_{c,ac}$, initiation step, targeted degree of polymerization $DP_n = [M]_0/[I]_0$, side reactions, chain transfer to solvent or initial excess of nitroxide were investigated. Very detailed kinetic modeling[129] of NMP of styrene mediated by TEMPO **2** or the so-called SG-1 nitroxide **35** (Scheme 4.12) have been carried out.

The mean lifetime τ of radical $R^\bullet$ in conventional RP is the time from its initiation to its termination, typically ≈ 1 s. In CRDRP, τ is the time of activity of growing chains between two dormant states. For NMP, τ is given by eqn (4.16) and lies in the range[130,131] of $\approx 10^{-4}$ to 10^{-2} s.

$$\tau = \frac{[R^\bullet]}{k_{c,ac}[R^\bullet][R^1R^2NO^\bullet]} = \frac{1}{k_{c,ac}[R^1R^2NO^\bullet]} \qquad (4.16)$$

The average number[130] of monomer units added per cycle of radical activity ν can be determined by eqn (4.17).

$$\nu = k_p[M]\tau \qquad (4.17)$$

4.3.4   Fischer Phase Diagram

For the polymerization of a given monomer at a specific temperature with known k_p and $k_{t,ac}$ and a targeted degree of polymerization, there exists a range

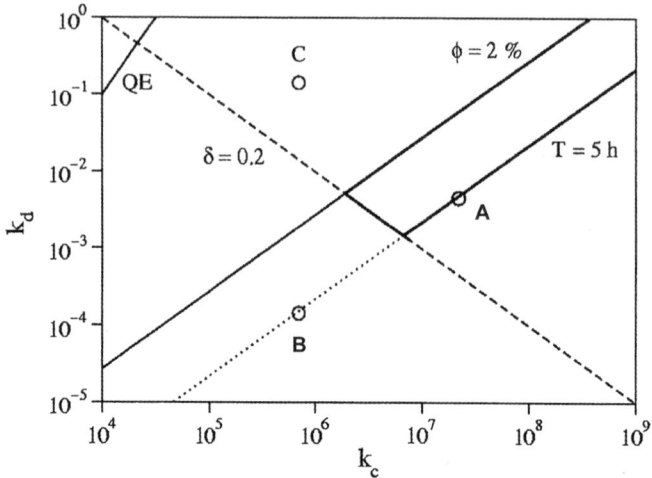

Figure 4.3   Range of rate constants $k_{d,dc}$ and $k_{c,ac}$ for optimal NMP in 5 hours at 90%
conversion with less than 2% of dead polymer and polydispersity δ smaller
than 0.2 ($\delta = \text{PDI} - 1 - [I]_0/[M]_0 \approx \text{PDI} - 1$). $[I]_0 = 0.1$ M, $k_p = 5000$ M^{-1}
s^{-1}, $k_{t,ac} = 10^8$ M^{-1}s^{-1}.
[Reproduced from ref. 111, Copyright Wiley-VCH Verlag GmbH, 2001.]

of rate constants $k_{d,dc}$ and $k_{c,ac}$ for which living and controlled polymerization
should be observed. This range is easily identified in the so called phase diagram
developed by Fischer and Souaille.[89,111,112] The diagram is built for the desired
values of t_c, Φ and PDI using the eqn (4.12)–(4.14) on the assumption that only
$k_{d,dc}$ and $k_{c,ac}$ are unknown (see Figure 4.3). For simplicity, $k_{d,dc}$ and $k_{c,ac}$ were
assumed to be identical to k_d and k_c for the low molecular weight alkoxyamine
initiator R^1R^2NOR. The region where the preset values of t_c, Φ and PDI will be
obtained is outlined in bold. Above line QE the quasi-equilibrium conditions
(4.8) are not met, and therefore, no living and controlled radical polymerization
can occur. The equilibrium constants K are identical at points A and B as are
the time of polymerization (5 h) and conversion (90%). However, as the
product $k_{d,dc} \times k_{c,ac}$ is smaller for B than for A, the polymer obtained at B is
mainly living but not well controlled ($\text{PDI}_A = 1.1$ *vs.* $\text{PDI}_B = 3$).On the other
hand, the polymer obtained at C is mainly controlled but less living. In fact, at
C the product of $k_{d,dc}$ and $k_{c,ac}$ is the same as at A but the equilibrium constant
K is larger leading to a larger fraction Φ ($\approx 50\%$) of dead polymer and to faster
polymerization.

The Fischer phase diagram is a useful tool which aids in the selection of an
appropriate nitroxide for a particular monomer and the ideal polymerization
temperature for NMP. However, it is important to use the rate constants for
the polymeric species $k_{d,dc}$, $k_{c,ac}$, $k_{t,ac}$ whenever available.

4.3.5   Side Reactions in NMP

Several undesired side reactions of NMP are briefly discussed below.

4.3.5.1 Additional Radical Generation

In the course of NMP, the persistent radical effect (PRE) leads to a steady increase in excess nitroxide. This slows the polymerization rate down and leads to longer polymerization times. As shown by Matyjaszewski,[132] Fukuda [133,134] and Miura,[135] introduction of a conventional radical initiator which slowly decomposes under the reaction conditions considerably enhances the conversion rate. Even a low rate of external initiation ($\approx 1\%$ of the initial internal initiation, reaction 1, Scheme 4.5) leads to a considerable reduction in the polymerization time while the livingness, polydispersity and controlled degree of polymerization remain virtually unchanged. The extra radicals reduce the concentration of persistent nitroxides rather than initiating new chains. The kinetic aspects of additional initiation were studied by Fukuda *et al.*[87,90] and Fischer *et al.*[89,111,112,136] In summary, if the rate R_i of generation of additional radicals is much smaller than the rate of radical generation from R^1R^2NOR, the time dependent terms $t^{1/3}$ and $t^{-1/3}$ for the persistent and transient radicals are replaced at specific time by steady state concentrations. The concentration of transient radicals $[R^{\bullet}]_{st} = (R_i/k_{t,ac})^{1/2}$ is then dependent on the initiation and termination rates, as in a conventional RP, and concentration of persistent radicals is determined by the equilibrium described by eqn (4.7). The rate of polymerization R_p is given by eqn (4.18) and the monomer conversion obeys the relation (4.11).

$$R_p = -\frac{d[M]}{dt} = k_p[R^{\bullet}]_{st}[M] = k_p\left(\frac{R_i}{k_{t,ac}}\right)^{1/2}[M] \qquad (4.18)$$

Additional initiation may arise not only from deliberately added initiators but also from radical generating impurities or autoinitiation[97] of the monomers. Most important is the autoinitiation of styrene.[137,138] In agreement with the experiment,[139] eqn (4.18) predicts that R_p of NMP of styrene, which is inevitably accompanied by autoinitiation, will not depend on the concentration of the R^1R^2NOR initiator. Of course, large external radical generation rates (*e.g.*, 20% of radical generation from R^1R^2NOR) lead to faster monomer conversion but uncontrolled polymerization.

4.3.5.2 Initial Excess of Nitroxide Radicals

The effect of initial presence of $R^1R^2NO^{\bullet}$ radicals has been analyzed theoretically by Fischer and Souaille[107] and by Fukuda *et al.*[87] Different amounts of added nitroxide with respect to the initiator R^1R^2NOR lead to different kinetic behaviors, characterized by either $t^{2/3}$ (for very small added amounts) or the linear time dependence of $\ln([M]_0/[M])$. In general, addition of free nitroxide slows the polymerization down but may significantly improve livingness and control. Addition of a few percent of free nitroxide is an effective strategy to achieve controlled NMP of monomers with very high propagation rate constants such as acrylates[140,141] or acrylic acid.[142]

4.3.5.3   Hydrogen Transfer Reactions

An important side reaction which complicates the thermolysis of alkoxyamines and particularly the NMP of methacrylates,[143–147] is the intermolecular β-hydrogen transfer (k_{Di}) from the propagating radicals to the nitroxide, also called disproportionation (path (a), Scheme 4.7 or reaction (4.8), Scheme 4.5).

The products are the corresponding hydroxylamine **4** and the alkene **12** which is a ω-unsaturated polymer in case of a polymeric alkoxyamine. In principle, **4** and **12** may also be formed by a concerted intramolecular nonradical Cope-type elimination of the alkoxymine (k_{CE}) (path (b) or reaction (9), Scheme 4.5).

Both reactions were studied theoretically[107,148] and experimentally[105,149,150] by Fischer *et al.* Their main consequence is lower monomer conversion, increase of polydispersity and loss of livingness in the final stages of polymerization. As long as the conditions (4.8) and (4.9) are fulfilled, the system will reach the quasi-equilibrium regime characterized by eqn (4.19),

$$[R^{\bullet}] \, [R^1 R^2 NO^{\bullet}] = \left([R^1 R^2 NOR]_0 - [R^1 R^2 NOH]\right) \times \frac{k_{d,dc}}{k_{c,ac}} \qquad (4.19)$$

where $R^{\bullet} = HCH_2C^{\bullet}R^3R^4$ and $[R^1 R^2 NOR]_0 - [R^1 R^2 NOH]$ is the concentration of dormant chains at time t.

The concentration of hydroxylamine $R^1 R^2 NOH$ increases according to eqn (4.20)

$$[R^1 R^2 NOH] = [R^1 R^2 NOR]_0 \left(1 - \exp(-f_{Di} k_{d,dc} t)\right) \qquad (4.20)$$

where $f_{Di} = k_{Di}/(k_{Di} + k_{c,ac})$ is the fraction of disproportionation in the cross-coupling between the nitroxide and the propagating radicals. As $[R^1 R^2 NOH]$ approaches $[R^1 R^2 NOR]_0$, the dormant chains are mainly converted to hydroxylamine and the ω-unsaturated polymer.

Equations predicting the maximum attainable conversion in the presence of disproportionation and defining the upper limit for f_{Di} for a set monomer conversion with the desired fraction of living chains were developed by Souaille and Fischer.[148] The tolerable amount of disproportionation depends strongly on other rate parameters as well. Generally, $f_{Di} < 0.5\%$ is generally insignificant and even 2% may be acceptable with monomers possessing a large propagation constant k_p. In fact, f_{Di} can be larger if a low degree of polymerization is desired.

It was shown[148] that both pathways (a) and (b) in Scheme 4.7 influence the final outcome of NMP to nearly the same extent. Therefore (4.20) and the other relations are also valid for path (b) if f_{Di} is replaced by $f_{CE} = k_{CE}/k_{d,dc}$, the fraction of Cope-type elimination in the cleavage of dormant chains. The disproportionation path (a) seems to be preferred because of its low activation energy barrier E_a ($<15 \, kJ \, mol^{-1}$) and high values of $k_{Di} > 10^6 \, M^{-1} \, s^{-1}$.[151] However, density functional theory (DFT) calculations indicated significantly higher values of E_a for NMP of acrylonitrile[152] and styrene[153] mediated by TEMPO. On the other hand, the high E_a ($\approx 140 \, kJ \, mol^{-1}$) and restricted transition state geometry make the Cope-type elimination path (b) less

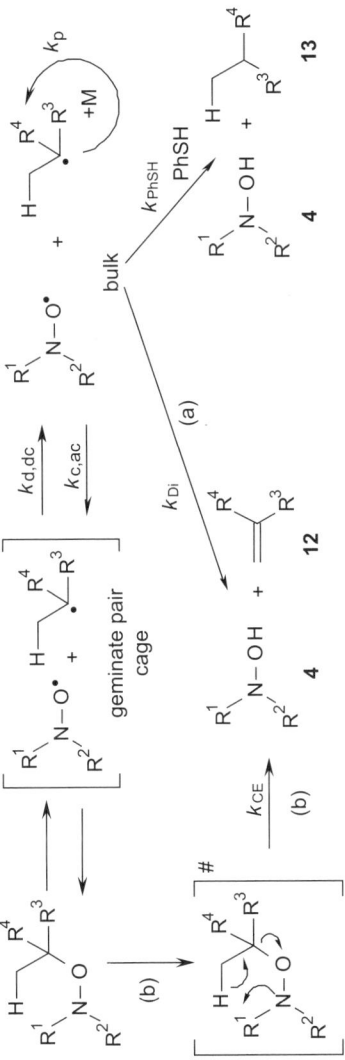

Scheme 4.7   H-Transfer reactions of nitroxides and alkoxyamines.

favorable.[151] Nevertheless, several examples of its demonstrated[149,151] or probable[154,155] occurrence are known.

A simple method developed by Bagryanskaya *et al.*[151,156] allows discrimination between paths (a) and (b). The presence of an excess of a radical scavenger (thiophenol, PhSH) during thermolysis of alkoxyamines efficiently suppresses the disproportionation and cross-coupling reaction ($k_{c,ac}$) through rapid reduction of alkyl radicals[157] into alkanes **13** and nitroxides (slower)[158] into hydroxylamines **4**.

Consequently, all the alkene **12** detected under such conditions was formed by the Cope-type elimination (b). On the other hand, alkene formed in the absence of thiophenol was generated by both paths. Recently, Marque *et al.*[159] showed that thermolysis of alkoxyamines in the presence of the scavenger PhSH induces chemically induced dynamic nuclear polarization (CIDNP). The CIDNP effect, detected by ^{1}H-NMR, indicates that the intermolecular H-transfer occurs, in contrast to earlier reports,[148] already in the spin correlated geminate (cage) radical pair, formed immediately after the cleavage of the NO–C bond.

All the factors influencing the extent of H-transfer reactions which depend on both nitroxide and radical are not yet elucidated. In general, H-transfer is negligible with polystyryl radicals but can be significant between polyacrylate radicals and certain nitroxides. The significant H-transfer occurring between polymethacrylyl radicals and the majority of nitroxides known today makes NMP of methacrylates[143–147] quite challenging, despite recent[156,160] progress (see Section 4.7).

4.3.5.4 Decomposition of Nitroxide Radicals

The concentration of persistent species has a decisive influence on the fate of every SRMP. Consequently, the intrinsic stability of persistent radicals during this process is an important issue. Their decomposition leads to an increased rate of polymerization since the retarding effect of the persistent species is diminished and the transient radicals which are often (but not always) produced through their decomposition can initiate new chains or form new dormant species.[107,161] This subject has been reviewed by Nielsen and Braslau.[162] One has to differentiate between the intrinsic stability of the nitroxide and its stability to reactants, which depends both on the nature of the reactant and structure of the nitroxide. In the absence of other reagents, nitroxides **1** (Scheme 4.2) in which the substituents R^1 and R^2 are aliphatic (see formula **1b**, Scheme 4.8) decompose *via* the three main pathways (i)–(iii) (Scheme 4.8):

(i) α-fragmentation,[163] which occurs predominantly with acyclic nitroxides to yield a new radical $R^{2a}R^{2b}R^{2c}C^{\bullet}$ and the nitroso compound **14**;

(ii) β-fragmentation,[151] to yield a new radical species $R^{2a\bullet}$ and a nitrone **15**. Both α and β fragmentations are the reverse reactions of the well-known[164] trapping of C-radicals with nitrones and nitroso compounds;

(iii) disproportionation, which occurs with nitroxides **1c** bearing at least one H-atom in the α-position to the N-atom. According to Ingold, the

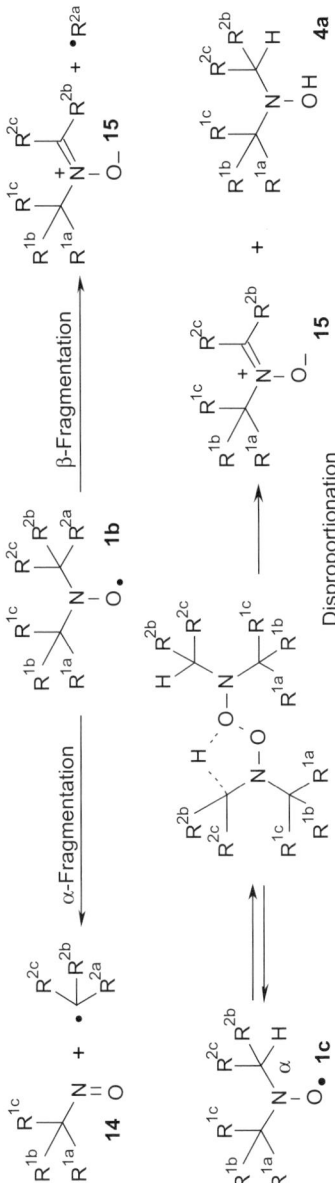

Scheme 4.8 Decomposition pathways of nitroxide radicals.

mechanism[62] of disproportionation consists of reversible dimerization followed by syn-elimination affording the nitrone **15** and the hydroxylamine **4a**. Consequently, only nitroxides **1c** having the α-hydrogen in a syn-orientation with respect to the nitroxide O-atom readily disproportionate.

The α-H atom is located anti to the nitroxide O-atom in the preferred conformation of the important nitroxides **34** and similar compounds[57,59,60] and **35**,[58] (Scheme 4.12) both in the solid state and in solution. Consequently, these nitroxides do not disproportionate. A different mechanism for disproportionation of α-hydrido-nitroxides involving head to tail nitroxide dimers has been recently proposed.[162] The half-life times of several nitroxides at 120 °C in *t*-butylbenzene solution were measured by Marque.[165] As a general rule, cyclic nitroxides are more stable then acyclic ones.

Nitroxides in which the one or both of the substituents R^1 and R^2 are aromatic decompose by a different mechanism[50] starting with the intermolecular coupling of the nitroxide oxygen atom with the (generally) *para* position of the aromatic substituent of another nitroxide molecule. Some nitroxides, such as **36** (Scheme 4.12) are stable[166] even at 200 °C.

The decrease of concentration of nitroxides in NMP can be caused not only by their intrinsic instability but also through deliberate addition of suitable additives, which will transform them into non-radical products. One possibility is the reduction of nitroxides **1** into the corresponding hydroxylamines **4** through reaction with good H-donors. Georges *et al.* used ascorbic acid[152] or diverse α-hydroxycarbonyl compounds[153] for the reduction of the excess of nitroxide during the NMP of *n*-butyl acrylate with TEMPO. Through use of such a strategy the TEMPO mediated NMP of *n*-butyl acrylate becomes possible and proceeds in a moderately controlled manner. This was attributed to diminished retardation due to the reduction of the concentration of TEMPO.

Georges *et al.*[167] showed that addition of camphorsulfonic acid significantly increases the rate of TEMPO mediated styrene polymerization. However, the mechanism of action was not unequivocally elucidated. One possible explanation[168] is the reduction of the TEMPO concentration through its reaction with acid. It is well known that nitroxides disproportionate under acidic conditions into hydroxylamine **4** and oxoammonium salt **16** (Scheme 4.9). The mechanism of this reaction was recently investigated.[169]

Acid inhibition of styrene autoinitiation[170] may also play a role. A review by Jianying[171] summarizes various acidic additives which have since been

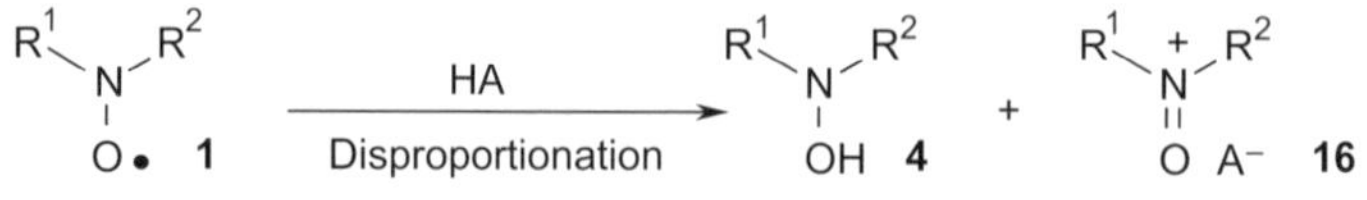

Scheme 4.9   Acid catalyzed disproportionation of nitroxides.

developed to enhance the rate of NMP. However, their mechanism of action remains speculative. Hawker *et al.*[172] showed that acylating agents such as acetic anhydride (Ac$_2$O) significantly increase the rate of TEMPO-mediated styrene polymerization. As for the effect of acids, the detailed mode of action remains unclear. Fukuda *et al.*[173] demonstrated that TEMPO is stable in the presence of Ac$_2$O at 110 °C over a period of 15 h and, therefore, proposed formation of a strong complex between TEMPO and Ac$_2$O. As only free TEMPO efficiently couples with persistent radicals, the effect is the same as a decreased concentration of free TEMPO which leads to an increased rate of polymerization. Recent experimental and computational work of Lansalot *et al.*[174] supports this hypothesis.

4.3.5.5 Chain Transfer Reactions

Chain transfer reactions (CT) almost always accompany conventional RP. Lovell *et al.*[131] showed that CT to polymer and the resulting branching is significantly reduced in NMP of *n*-butyl acrylate (BA) (the same effect is observed in ATRP and RAFT as well). This phenomenon, also expected with other monomers, was explained by the lower concentration of highly reactive short-chain radicals in CRDRP. Very recently, the existence of mid-chain radicals in NMP of BA mediated by **35** was demonstrated.[175]

4.4 Initiating Systems for NMP

4.4.1 Unimolecular Alkoxyamines

Today, NMP is most often initiated (reaction 1, Scheme 4.5) with a unimolecular alkoxyamine initiator R^{1}R^{2}NOR (see Section 4.6). The concentration of the initiator is selected according to the targeted number average degree of polymerization DP$_n$ = conversion × [M]$_0$/[R^{1}R^{2}NOR]$_0$. The solution of the alkoxyamine in the monomer M is rapidly heated to the necessary reaction temperature, typically 100–130 °C, and then stirred under inert atmosphere until the desired conversion is reached. The advantage of this approach is that the alkoxyamine can be prepared in a highly pure form and that its cleavage directly affords the two radicals in the right stoichiometry.

4.4.1.1 Peculiarities of Initiation

The kinetic equations of NMP (see Section 4.3.3) were derived under the assumption that all growing chains start to propagate simultaneously so that the initiation stage can be neglected, *i.e.* $k_d \geq k_{d,dc}$. Bertin *et al.*[176] demonstrated experimentally that k_d of the initiating alkoxyamine indeed plays a pivotal role. Thus, the control of polymerization of styrene at 90 °C initiated with the alkoxyamines (Scheme 4.12) **35A2** ($k_d = 5 \times 10^{-5}$ s^{-1}), **35A1** ($k_d = 2.3 \times 10^{-4}$ s^{-1}) and **35A4** ($k_d = 0.28$ s^{-1}) increased from **35A2** to **35A4** – the faster the initiation, the better the control. In fact, successful NMP of *n*-butyl acrylate (BA) was achieved at 90 and 120 °C with **35A4** without addition[176] of free nitroxide **35**. The rapidly cleaving alkoxyamine **35A4** immediately produces a large amount of nitroxide and alkyl radicals. As some of the latter disappear by self-termination, an excess of nitroxide arises which permits good control of polymerization of rapidly propagating monomers such as BA. For successful NMP, therefore, the k_d of the alkoxyamine initiator and the propagation constant k_p of the monomer should fulfill[176] the condition $k_p/k_d \leq 6.0 \times 10^5$ L mol^{-1}.

However, alkoxyamines R^1R^2NOR with high k_d can only be used for initiation if the H-transfer reaction between $R^1R^2NO^\bullet$ and $R^\bullet$ (Scheme 4.5, reaction (8)) is insignificant ($< 0.5\%$) and cannot compete with the addition of $R^\bullet$ to the monomer. Bagryanskaya *et al.*[177] showed that NMP of styrene at 120 °C was controlled with **28A1** but failed with **28A5**, which undergoes faster homolysis. Around 3% of H-transfer occurs[151] between **28** and **A5** compared to $\approx 0\%$ with **A1**. On the other hand, **A5** adds to styrene more slowly than **A1** but couples faster with **28**. Consequently, the H-transfer reaction is preferred over the addition of **A5** to monomer and disrupts NMP.

4.4.2 Bimolecular Systems

In the early days of NMP, a system[86,178–180] consisting of a free nitroxide and a conventional thermal radical initiator (TRI) such as AIBN, azocyclohexyl-carbonitrile (ACCN) or dibenzoylperoxide was used. In this case, the primary radicals from the cleavage of the TRI either combine with the nitroxide forming the alkoxyamine initiator *in-situ* or start growing chains which rapidly engage in the NMP process (reactions 4–6, Scheme 4.5). However, due to cage reactions, the yield of primary radicals is never 100%. Consequently, this requires that the optimal TRI/nitroxide ratio has to be determined experimentally for each given system.

4.4.3 *In-Situ* NMP

Alternatively, the nitroxide radicals (and the related alkoxyamines) can be formed in the polymerizing mixture by the reaction of conventional radical initiators with nitric oxide, nitroso compounds, or nitrones. Formation of nitroxides and alkoxyamines from nitric oxide, nitroso compounds and nitrones is exemplified in Scheme 4.10 for compounds **17**[181,182] and **18–19**.[164] This approach attracted considerable interest due to its potential for industrial applications because the nitroxide precursors are inexpensive. However, these systems are much more complex than those using a pure unimolecular initiator, and the underlying mechanisms must be well understood before the systems can be optimized and applied industrially. The state of the art of so-called *in-situ* NMP as of 2008 was reviewed by Detrembleur *et al.*[183]

Recently, a new variant of *in-situ* NMP was reported by Barner-Kowollik *et al.*[184,185] First, a *macromolecular mid-chain* alkoxyamine **21** is obtained *via* addition of two macroradicals to a nitrone, *e.g.* phenyl-*t*-butyl nitrone **20**

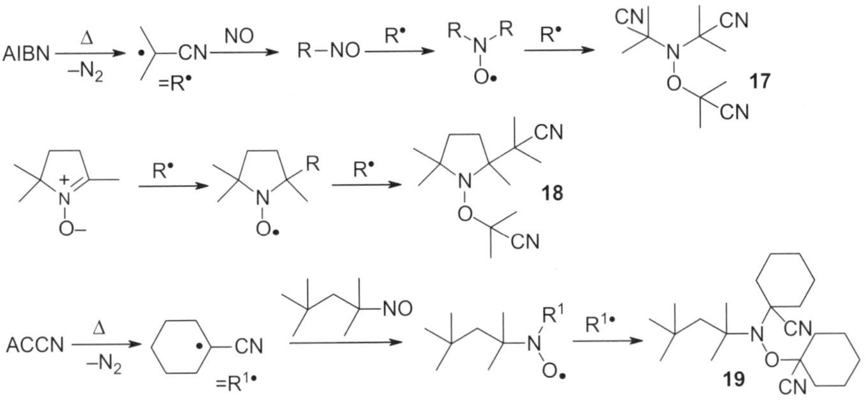

Scheme 4.10   Synthesis of alkoxyamine initiators from NO, nitroso compounds or nitrones.

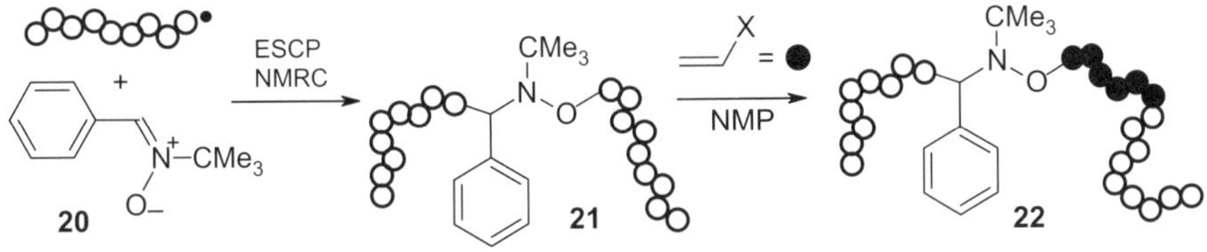

Scheme 4.11   Mid chain macroalkoxyamines for NMP.

(Scheme 4.11). The macroradicals are generated either through conventional polymerization initiated by TRI or from dormant chains pre-made by ATRP. The process is called ESCP[184] (enhanced spin capturing polymerization) in the first case and NMRC[185] (nitrone-mediated radical coupling) in the second. The formation of **21** is irreversible at the reaction temperature (<80 °C), however, **21** will behave as NMP-macroinitiator when heated to appropriate temperature (>100 °C) in the presence of a monomer. The result is an ABA block copolymer **22** or even more complex polymeric architecture if a polyfunctional nitrone is used. A review[186] of the state of the art of these techniques as of 2011 is available.

4.5   Main Classes of Nitroxides for NMP

Since the advent of NMP, at least 80 different nitroxides have been reported as mediators. The majority of these is listed in the outstanding review by Bagryanskaya and Marque[70] on alkyl radical scavenging by nitroxides. The most relevant and efficient nitroxides are shown in Scheme 4.12, along with the most important alkyl radicals derived from the related alkoxyamines (see Section 4.6).

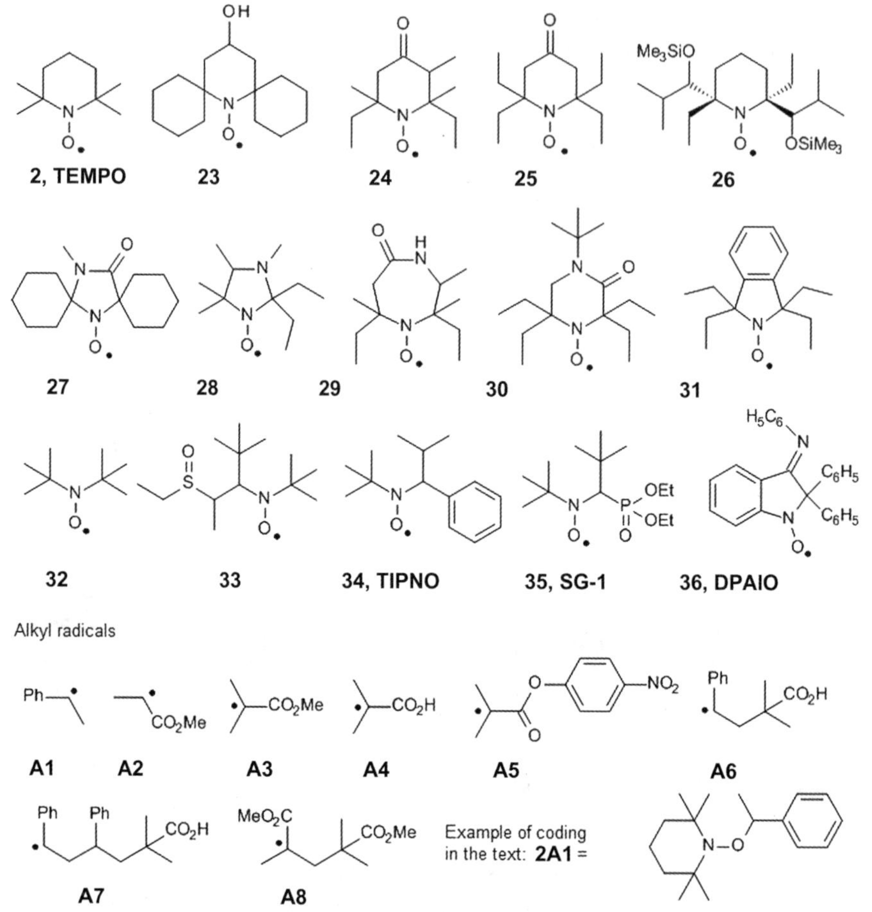

Scheme 4.12   Important nitroxides and alkoxyamines for NMP.

TEMPO **2** is suitable for NMP of styrene[86] (ST), 4-vinylpyridine[187] (VP), butadiene[188] (BD) and isoprene[189] (IP), but not for acrylates or methacrylates.[145] However, Georges *et al.* showed that addition of nitroxide reducing additives[152,153] (see Section 4.3.5.4) allows living polymerization of *n*-butyl acrylate (BA) with PDI < 1.4 in up to 50% yield.

2,6-Bis-cycloalkylidenepiperidine nitroxides[135] such as **23** are suitable for NMP of ST and for random copolymerizations of St with BA or methyl methacrylate (MMA).

The readily available sterically-hindered piperidine nitroxide **24** is efficient for NMP of ST and BA.[190,191] The even more hindered tetraethyl piperidone nitroxide **25** is amongst the best choices for BA[192] and also for *N*-isopropyl-acrylamide.[193] An improved synthesis of **25** was recently published.[194] The extremely hindered nitroxide **26** of Studer *et al.*[195] is of interest from a theoretical point of view. The substituents in the 2- and 6-position are already so bulky that the coupling reaction of **26** with C-radicals is too slow (too low $k_{c,ac}$). Therefore, the alkoxyamine **26A1** behaves like a classical radical initiator and **26** as spectator without any mediating capacity. Moad *et al.*[179] developed the 2,2,5,5-tetraalkyl imidazolidinone nitroxides, such as **27**. They perform better than **2** in the NMP of ST and fairly well for acrylates. Imidazolidine nitroxides, *e.g.* **28** are also efficient mediators for NMP of ST.[177] The seven membered-ring nitroxide **29** introduced by Nesvadba *et al.*[191,196] is a very efficient mediator for both BA and ST. A more sterically hindered derivative[197] of **29** and other 7- and 8-membered ring nitroxides[198] were prepared by Studer *et al.* The highly hindered piperazinone nitroxide[196,199–201] **30** along with **24**, **25** and **29** are some of the most efficient cyclic nitroxides for NMP of ST and acrylates. The 1,1,3,3-tetraalkylisoindolin nitroxides, such as **31**, investigated by Moad and Solomon[84] showed some efficiency for ST but only modest activity for acrylates and methacrylates.

Di-*t*-butylnitroxide **32** is efficient for NMP of ST[139] and acrylates.[134] The β-sulfinyl nitroxide **33** allows low temperature (<90 °C) NMP of St[202] and acrylates.[203] The nitroxides **34** (nicknamed TIPNO) and **35** (SG-1), both discovered by Tordo and coworkers,[204] play a prominent role in NMP. Hawker *et al.*[140,205] showed that **34** or its "universal" alkoxyamine **34A1** is efficient for NMP of ST, acrylates, acrylamide and acrylonitrile allowing accurate control of molecular weights and to achieve polydispersities as low as 1.06. Efficient formation of random and block copolymers from a wide selection of monomer units containing reactive functional groups, such as amino, carboxylic acid, and glycidyl was also possible. A real breakthrough is the readily available nitroxide **35** and its related alkoxyamine[206–212] **35A4**, which has been commercialized by Arkema[213] under the tradename BlockBuilder®. SG-1 (or the frequently used related alkoxyamine **35A4**) is very efficient for NMP of ST,[58,214] BA,[58,130] acrylic acid,[142,215] *t*-butylacrylamide[216] or *N,N*-diethylacrylamide.[217,218] Successful copolymerization of MMA was possible with small amounts (<10%) of St,[219,220] acrylonitrile[221] or 9-(4-vinylbenzyl)-9*H*-carbazole[222] comonomers. Similarly, copolymerization[223] of methacrylic acid with sodium styrene sulfonate was possible using **35**. Several analogues of SG-1 have been prepared and evaluated.[224] The aromatic nitroxide **36** (nicknamed DPAIO) of Greci *et al.*[225] is the first nitroxide which allowed, in the form of its alkoxyamine **36A5**, living and controlled homopolymerization[160] of MMA at 100 °C affording poly-MMA in 60% yield and PDI < 1.4.

4.6 Alkoxyamines for NMP

The various methods for the synthesis of alkoxyamines R^1R^2NOR have been reviewed by Nesvadba,[226] Marque *et al.*,[227] Detrembleur *et al.*[183] and Greene and Grubbs.[228] Most important is the coupling of nitroxides with C-centered radicals R• (Schemes 4.13 and Scheme 4.10, 4.11). At least 10 variants[183,226–228] of this technique exist, depending on the source of C-radicals. By far most versatile is the atom transfer radical addition (ATRA) developed

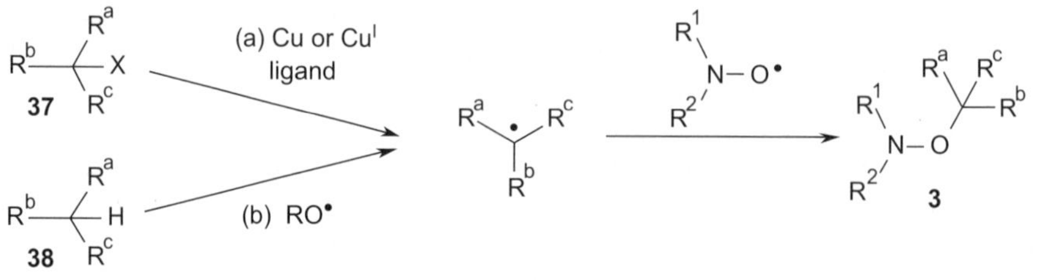

Scheme 4.13   Synthesis of alkoxyamines by trapping of C-radicals.

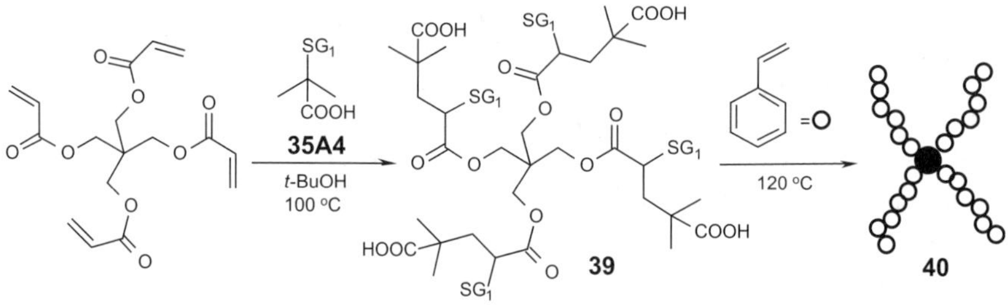

Scheme 4.14   Intermolecular 1,2-radical addition, IRA.

by Matyjaszewski *et al.*[229,230] It involves a reaction of an activated organic halide $R^a R^b R^c CX$, X = Br, Cl) **37** with elemental Cu or Cu^I salts in the presence of an amine ligand to produce alkyl radicals which are then trapped by the nitroxide to give the desired alkoxyamine (Scheme 4.13, path (a)). The C-centered radicals can also be obtained by abstraction of H-atoms (path (b)) from activated hydrocarbons **38** (*e.g.* PhEt) with O-centered radicals that do not readily react with nitroxides (see Section 4.2.2). The latter can be generated by thermal (120–130 °C)[178] or photochemical[231] decomposition of dialkyl-peroxides, or low temperature (<50 °C) decomposition of dialkylperoxalates[232] or dialkylhyponitrites.[233] A method for the large scale, industrial synthesis of alkoxyamines uses the generation of O-centered radicals from *t*-butylhydro-peroxide[234] under catalysis with copper ions or quarternary ammonium salts.

Alkoxyamines with low NO–C bond dissociation energy (BDE_{NO-C}) can be transformed into new alkoxyamines with higher BDE_{NO-C} by intermolecular 1,2-radical addition (IRA).[235] Particularly suitable for IRA is BlockBuilder® **35A4** (Scheme 4.12) which is a crystalline material with an extremely weak NO–C bond ($BDE_{NO-C} \approx 112$ kJ mol^{-1}).[208]

IRA is a powerful method for synthesis of highly functionalized alkoxy-amines which can be used for preparation of complex macromolecular archi-tectures. One example[236] out of many,[235] is the transformation of **35A4** (partly dissociated at the reaction temperature) into the tetra-alkoxyamine **39** (estimated[208] $BDE_{NO-C} \approx 127$ kJ mol^{-1}, virtually undissociated at the reaction temperature) and its use for synthesis of the 4-arms star polymer **40** (Scheme 4.14).

4.6.1 Factors Influencing k_d

The rate constant for cleavage of alkoxyamines R^1R^2NOR k_d (reaction 1, Scheme 4.5) depends on several effects. These have been extensively studied and are reviewed by Marque *et al.*[91,227]

The numerical values of k_d span ten orders of magnitude $(10^{-10}\,s^{-1} < k_d < 5\,s^{-1})$.[208,227]

The strong temperature dependency of k_d obeys the Arrhenius relation (4.21)

$$k_d = A_d\, e^{-\frac{E_{ad}}{RT}} \tag{4.21}$$

The reported frequency factors A_d for alkoxyamines range from $10^6\,s^{-1}$ to $10^{17}\,s^{-1}$, although some data may be erroneous. The values for most NMP-relevant alkoxyamines center[165,168,227,237] around $2.4 \times 10^{14}\,s^{-1}$. Activation energies E_{ad} range[208,237,238] from ≈ 105 to $140\,kJ\,mol^{-1}$, *e.g.* $111.7\,kJ\,mol^{-1}$ (**35A4**),[208] $129.6\,kJ\,mol^{-1}$ (**34A1**),[168] $133\,kJ\,mol^{-1}$ (**2A1**).[168] The E_{ad} for simple primary alkyl alkoxyamines is much higher (*O*-methyl-TEMPO: $E_{ad} \approx 200\,kJ\,mol^{-1}$).[79] E_{ad}'s are close to NO–C bond dissociation energies of the alkoxyamines.[165,239]

4.6.1.1 Dependence of k_d on the Alkyl Fragment

Mulder *et al.*[79] and later Marque *et al.*[165,168] have shown that k_d correlates with the bond dissociation energy of the C–H bond (BDE_{C-H}) in the parent alkane of the released radical $R^\bullet$. In other words, the more stabilized the radical, the faster the homolysis (Figure 4.4).

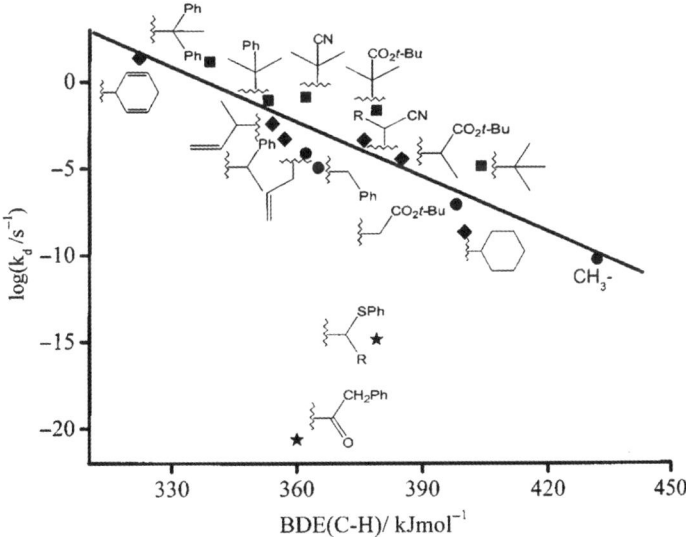

Figure 4.4 Rate constants $\log(k_d)$ for the NO–C bond cleavage of TEMPO derivatives of trialkylhydroxylamines *vs.* BDE_{C-H} of the corresponding hydrocarbon: (■) tertiary alkoxyamines, (♦) secondary alkoxyamines, (●) primary alkoxyamines, (*) other alkoxyamines.
[Reproduced from ref. 168. Copyright American Chemical Society, 2001).]

However, as shown in Figure 4.4, the correlation is not very good, and for some alkoxyamines it does not exist at all. In fact, k_d depends in a more complex way on both the structure of the nitroxide and the released alkyl radical. The latter influences k_d through several effects (E; arrows indicate the respective influence of an increasing E on the rate constant k_d): stabilization of the alkyl radical (E↑k_d↑),[240] steric effect (E↑k_d↑),[240] polar effect (E↑k_d↑),[240,241] remote polar effect (E↑k_d↑),[177,242,243] remote steric effect (E↑k_d↓),[244] anomeric effect (E↑k_d↓),[79,239] anchimeric effect (E↑k_d↑),[168,245] and penultimate unit effect (E↑k_d↑).[125]

For TEMPO **2** and SG-1 **35** based alkoxyamines Marque *et al.*[240] developed equations allowing prediction of k_d for a given alkyl radical $R^\bullet$ characterised by a set of steric, stabilisation and polar parameters.

4.6.1.2 Effect of Penultimate Unit on k_d

The use of the Fischer's phase diagram to predict the outcome of NMP experiments requires knowledge of $k_{d,dc}$ for the macro-alkoxyamine or at least for the alkoxyamine carrying the penultimate unit.[125] Unfortunately, only a little data is available, mostly for alkoxyamines based on the SG-1 nitroxide **35**. For example,[125] an 18-fold increase of k_d is observed from **35A1** to **35A6** but only 1.5-fold increase from **35A1** to **35A7**. On the other hand, the change from **35A4** to **35A6** leads to a 3-fold decrease of k_d and from **35A6** to **35A7** to a 13-fold decrease. The magnitude of the penultimate effect (PUE) influences dramatically the fate of NMP.[126,219] For example, NMP of MMA failed with **35A3** even though the Fischer's diagram predicted a successful experiment. On the other hand, when k_d and k_c of **35A8**, which more closely resembles the poly-MMA macro radical, were used, the MMA polymerization was predicted to be uncontrolled with **35**. A multiparameter structure–activity relatioship[125] developed for SG-1 alkoxyamines efficiently models the effects of penultimate and antepenultimate units and allows to predict their k_d. The PUE also explains[92] the chain length dependence of $k_{d,dc}$ of macroalkoxyamines.

4.6.1.3 Dependence of k_d on the Nitroxide Fragment

Similar effects operate in the nitroxide fragment. Moad and Rizzardo[246] showed that k_d increases with increasing steric congestion around the NO–R bond of the alkoxyamine. This effect was later nicely demonstrated[200] on homologues of the alkoxyamine **30A1** with systematically varied substituents. Figure 4.5 shows a logarithmic plot of k_d at 393 K *vs.* the steric substituent constant ΔE_s,[200] which quantifies the bulkiness of the α,α' substituents. In the same work the recoupling constants k_c at 373 K were measured. As expected, the dependence of k_d and k_c is inversely related to the steric hindrance. As a consequence, the equilibrium constants $K = k_d/k_c$ for the 3,3,5,5-tetramethyl- ($K_{393} = 4.2 \times 10^{-13}$ M) and the 3,3,5,5-tetraethyl- ($K_{393} = 1.4 \times 10^{-10}$ M) derivatives differ by more than 2 orders of magnitude.

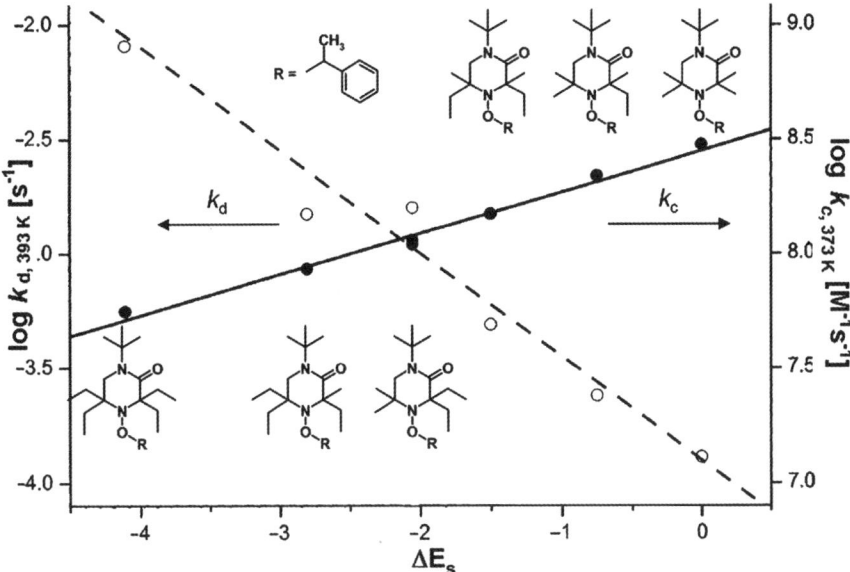

Figure 4.5   Rate constants for the cleavage (k_d, 393 K, open symbols) and the reformation (k_c, 373 K, closed symbols) of various piperazinone derived alkoxyamines in a semilogarithmic plot *vs.* the steric substituent constant ΔE_s.

[Modified from ref. 200. Copyright American Chemical Society, 2003).]

More examples for positive correlation of k_d with the bulkiness of groups surrounding the NOR group are known.[195,197,246] Other effects which influence k_d are ring size[191,198,237,238,246] (as a rule of thumb, k_d increases from 5- to 7-membered rings), hybridization[198] of one of the ring atoms, substituents on the ring and their electronic nature,[237,238] as well as intramolecular hydrogen bonding[168,205,247] which lowers the energy of the transition state through stabilization of the generated nitroxide. The structure–activity relationship equation of Marque, Nesvadba and coworkers[238] allows the prediction of k_d values for alkoxyamines releasing the phenethyl radical **A1** (PhCH•CH$_3$) using the steric and polar/stabilization descriptors of the nitroxide. This equation and those mentioned in Sections 4.6.1.1 and 4.6.1.2 may help to reduce the amount of synthetic work for design of novel efficient alkoxyamines.

4.6.1.4   Diastereoisomeric Effect

The values of k_d of the individual diastereoisomers of alkoxyamines containing two and more asymmetric centers can differ by a factor of ≈ 5. Most work has been done on diastereomeric alkoxyamines based on SG-1 **35**,[150,245,248–251] sulfinyl nitroxide **33**[203] or TEMPO-styrene dimer alkoxyamines.[252] Matyjaszewski and Annachenko showed[248] that polymerization of *tert*-butyl acrylate using a diastereomerically enriched analogue (*t*-Bu instead of Me ester) of **35A2** proceeded without control of polymer tacticity.

4.6.1.5   Effect of Solvents, Salts and Acids

k_d depends only slightly (by a factor of ≈ 2) on solvent.[168,246] Slightly faster homolysis is observed in polar solvents through the stabilization of the dipolar resonance structure **1a** of the nitroxide. Only a minor effect on k_d of one diastereoisomer was detected by Marque *et al.*[251] in a series of salts of **35A4**. Acids can influence NMP kinetics *via* destruction of the nitroxide (see 3.5.4) but addition of camphorsulfonic acid had no effect on the value of k_d for TEMPO-based alkoxyamines in organic media.[168] However, dramatic dependence of k_d on pH is observed with alkoxyamines containing protonable groups. Thus, Marque and Brémont[253] reported that protonation of **41** (Scheme 4.15) on the pyridine ring in organic solvent caused a 15-fold increase in k_d whereas a 64-fold increase[254] was observed during protonation in aqueous solution. Similarly, an increase of k_d values was observed upon quarternization[255] of the pyridine nitrogen in **41**. In contrast, a 4-fold decrease of k_d was observed upon protonation of **42**.[256] As suggested already by Marx *et al.*,[257] who prepared the pyridyl modified TIPNO **43** and some related alkoxyamines, this pH-induced switching of k_d (and to a lesser extent k_c) may open new possibilities for NMP.

4.6.1.6   Determination of k_d

Measurement of k_d requires the suppression of the cross-coupling reaction (see Section 3.2).

The method of choice is scavenging of the alkyl radical, *e.g.* with O$_2$,[165,252] galvinoxyl,[105] or thiophenol.[156] In the absence of an efficient scavenger, the experimentally determined value of k_d will be underestimated. k_d is obtained by measuring the increase of the nitroxide concentration by ESR[105,165] or through monitoring of the alkoxyamine decay with ^{1}H-NMR[201,252] or ^{31}P-NMR[258] (for alkoxyamines derived from **35**). Alternatively, the transient radicals can be trapped with an excess of a profluorescent nitroxide,[259] so that the buildup of the fluorescence mirrors the decay of the original alkoxyamine.[260]

The scope and limitations of each method as well as additional methods are discussed in the review of Marque *et al.*[227]

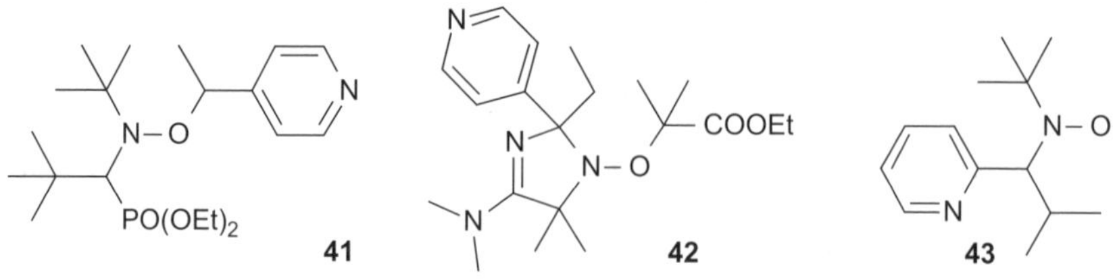

Scheme 4.15   Alkoxymines and nitroxides with protonable functions.

4.6.2   Factors Influencing k_c

In contrast to the dissociation rate constant, significantly less work has been devoted to the recombination rate constant k_c. The latter depends, similarly to k_d, on different effects. Detailed discussions can be found in the articles of Marque *et al.*,[91,227,261] Gigmes and Marque[92] and especially the very recent and comprehensive review[70] by Bagryanskaya and Marque. Here, only the most important features will be discussed.

Values[70,227] of k_c are in the range of $10^5\,M^{-1}\,s^{-1} < k_c < 10^9\,M^{-1}\,s^{-1}$. The coupling reaction between nitroxides and C-centered radicals is controlled[262] by a large negative activation entropy, and the temperature dependency of k_c is very weak and often does not obey the Arrhenius relationship (4.21). In some cases the frequency factors A_c and activation energies E_{ac} are rather low, *e.g.* for **8**[105] $A_c = 2.5 \times 10^8\,M^{-1}s^{-1}$ and $E_{ac} = 3.7\,kJ^{-1}\,mol$. In general, k_d and k_c exhibit opposite substituent effects; that is, a faster cleavage of an alkoxyamine goes in hand with a slower radical cross-reaction. Increasing steric congestion around both the nitroxide[195,200,261,263,264] (see Figure 4.5) and the alkyl radical[263] decreases k_c. Increasing stabilisation[68] of the alkyl radical has the same effect. Electron withdrawing substituents on the nitroxide desta-bilize[261,264] its dipolar structure **1a** and increase the unpaired electron density on the oxygen atom and, in turn, lead to an increase in k_c. Macroradicals usually exhibit lower k_c values than small alkyl radicals.[70] For example, coupling of **35** with polystyryl and polyacrylyl radicals is about 10-times slower, and poly-MMA radicals 100-times slower compared to coupling with the respective (analogous) small radicals **A1**, **A2**, **A3**.[126] Ingold *et al.*[69] showed that k_c slowly decreases with increasing polarity and viscosity of the solvent. Multiparameter correlations of k_c with descriptors characterizing the radical stabilization, bulkiness and polarity were developed by Marque *et al.*[70,261,264]

Recently, Coote *et al.*[265] developed correlations of k_c and k_d with ionization potentials (IP) and radical stabilization energies (RSE) of the alkyl radicals and RSE of the nitroxide using high-level *ab initio* methods. These correlations allow computational prediction of k_c and k_d for a given alkoxyamine.

4.6.2.1   Determination of k_c

A detailed overview of experimental techniques for measurement of k_c can be found in the review[70] of Bagryanskaya and Marque. The most popular method is laser flash photolysis-kinetic absorption spectroscopy.[262] A pulsed lamp polymerization-size exclusion chromatography method developed by Guillaneuf *et al.*[266] allows measurement of k_c for polymeric radicals.

4.6.3   NO–C *vs.* N–OC Homolysis

Cleavage of the NO–C bond in alkoxyamines is preferred over the N–OC cleavage if the bond dissociation energy BDE_{NO-C} is lower than BDE_{N-OC} (Scheme 4.16). The BDE_{NO-C} values are close to the Arrhenius activation

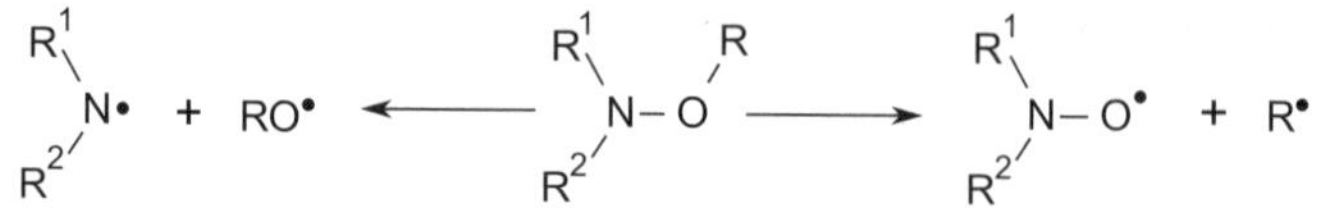

Scheme 4.16 NO–C *vs.* N–OC homolysis of alkoxyamines.

energies E_{ad}.[165,239] Increasingly reliable quantum chemical methods have been used for their calculation.[239,246,267–271] Coote *et al.*[269] modeled the cleavage mode (NO–C *vs.* N–OC) of a large set of alkoxyamines. Accordingly, N–OC homolysis competes or dominates over NO–C homolysis only when the released alkyl radical is not sufficiently stabilized or destabilized by an α-heteroatom, or alternatively if the aminyl radical is stabilized by delocalization. In fact, **36A1** or an analogue of **36A3** (Et instead Me ester) undergoes predominantly NO–C homolysis, whereas the O-hexyl- or O-isopropyl alkoxamines decomposed exclusively into alkoxy and aminyl radicals.[272] Calculations[273] suggest that *N*-acyloxyamines (Scheme 4.16, R=COR′) undergo N–OC homolysis affording transient aminyl and acyloxy radicals R′COO•. Indeed, *N*-acyloxyamines do not lead to controlled/living polymerization but have found applications as conventional radical initiators.[274]

4.7 Polymerizations

Since the first article[86] (in the open literature) on NMP of styrene mediated by TEMPO, the range of monomers amenable to NMP has increased significantly; a detailed list is available in the review by Gigmes and Marque.[92] Nevertheless, most of the polymers produced by NMP still have molecular weights significantly lower than 10^5 g mol^{-1}. Polystyrene with M_n close to 2×10^5 g mol^{-1} and a fraction of dead chains $\Phi < 0.4$ and PDI = 1.6 was recently prepared using **35A4**.[174] Efficient NMP of fast propagating monomers such as acrylic esters became possible with the acyclic nitroxides, in particular **34** and **35**. Originally, small amounts (2–5%) of the free nitroxide **34**[140] or **35**[130] were used in combination with the alkoxyamine initiator. In this way, the growing radicals were provided with enough stable radicals to form dormant species and thus avoid significant irreversible termination. The addition of extra nitroxide is not needed with fast homolysing alkoxyamines such as **35A4** or **24A1**, **25A1**, **29A1**, **30A1**. Nevertheless, polymerization of acrylates remains more delicate than that of styrene and is somewhat dependent on experimental conditions (monomer purity, heating ramp, type of polymerization vessel).[176]

The challenging polymerization of MMA deserves a special mention. H-transfer reactions (see Section 3.5.3) and large equilibrium constants *K* are the most important complications of NMP of MMA. The first prevents reaching large conversions and producing nitroxide terminated polymers when TEMPO

2 is used as mediating nitroxide.[145,148,149] Unsuccessful homopolymerization of MMA with SG-1 **35** is mainly due to the large activation–deactivation equilibrium constant K. In fact, the bulky penultimate tertiary unit strongly accelerates the homolysis rate and strongly decreases the recombination rate.[126] This leads to a high concentration of propagating macroradicals and, hence, favors their irreversible self-termination. Consequently, active chains disappear within a short time, and large excess of free SG-1 builds up. H-transfer is significantly less important with SG-1, but it still causes[146,147] decomposition of macromolecular alkoxymines, thus preventing living polymerization. Charleux *et al.*[275] showed that addition of a small amount of a comonomer with a low K leads to a greatly reduced *average* activation-deactivation equilibrium constant $<K>$. Indeed, addition of only 8.8 mol% of styrene allowed the living and controlled polymerization of MMA with **35A4** and little free **35** in high conversion (75%, PDI = 1.3).

Later it was shown[219] that the alkoxyamine end group was connected to a single styrene terminal unit (Scheme 4.17) and that the MMA penultimate unit caused a significantly lower cleavage temperature. Consequently, the copolymerization of MMA with a small amount of styrene could be performed at 78 °C, an unprecedented low temperature for NMP.

Using this concept, successful NMP of MMA was achieved with small amount of styrene in ionic liquids,[220] acrylonitrile[221] or 9-(4-vinylbenzyl)-9*H*-carbazole.[222] Methacrylic acid was copolymerized with sodium styrene sulfonate.[223,276]

The best homopolymerization of MMA so far was obtained[160] with the alkoxyamine **36A5**. It is believed that the delocalization[277] of the spin density from the O-atom of **36** into the aromatic ring decreases its propensity for H-transfer reactions. This and the right value of K would explain the successful NMP. The delocalization of the unpaired electron may explain the relatively good results reported by Grubbs and Greene for alkoxyamines obtained by the addition of C-radicals to nitrosobenzene.[278,279]

As far as the kinetics and thermodynamics of controlled/living statistical copolymerization is concerned, similar values of reactivity ratios (also called copolymerization parameters) are expected since the macroradicals and monomers involved possess the same intrinsic reactivity.[275]

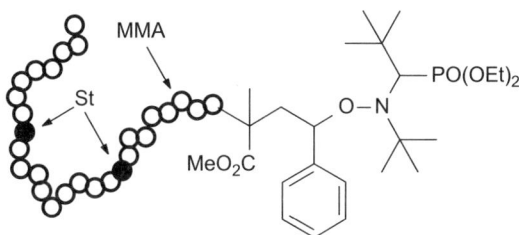

Scheme 4.17 Schematic representation of poly-MMA initiated by the BlocBuilder® alkoxyamine **35A4** with a small percentage of styrene as a comonomer.

4.7.1 Polymerizations in Dispersed Systems and from Surfaces

Several reviews[280–284] cover the considerable progress of NMP and other CRP techniques in aqueous miniemulsion or emulsion polymerizations. Preparation of hybrid latex particles and core-shell particles in aqueous media using NMP was summarized by Charleux *et al.*[285]

Surface initiated NMP (SI-NMP) is a powerful method for surface modification of a variety of substrates or the synthesis of organic-inorganic hybrid materials. Several reviews of the topic have been published.[286–290]

4.7.2 Complex Polymeric Architectures by NMP

Synthesis of complex polymeric molecules such as random and block copolymers, star and graft polymers, hyperbranched and dendritic structures by NMP and other CRP techniques has been reviewed several times.[88,291] Good overviews of telechelic polymers[292] or the coupling of NMP with other polymerization techniques[293–297] are available. Synthesis of bioconjugates through CRDRP, including NMP, is treated by Nicolas *et al.*[298]

4.7.3 Photochemically Triggered NMP

Photochemically triggered RP is a rapidly growing technology used to make coatings (*e.g.* on furniture, flooring, metal, plastics, automotive parts, paper, optical fibers), printing inks and adhesives, or in the production of microelectronic devices.[299–302] In contrast, controlled photopolymerizations are much less developed. Some progress has been achieved with photoiniferters[303] introduced by Otsu[83] (see Section 2.3). The principle feasibility of nitroxide-mediated photopolymerization (NMPP) through photochemical homolysis of TEMPO-based alkoxyamines carrying a photosensitizing group has been demonstrated during the last decade.[304,305] Significant progress and a deeper understanding of NMPP was attained through investigation of new alkoxyamines based on TEMPO and **35** as well as on some new nitroxides.[306–309] The chromophore must be attached close to the nitroxide moiety to ensure an efficient intramolecular energy transfer from the chromophore to the NO–C bond which should also undergo homolysis in the macroalkoxyamine. However, direct attachment of the aromatic chromophore to the alkoxyamine nitrogen, as in **44**, leads to competitive cleavage of the N–OC bond because the resulting aminyl radical is resonance stabilized (see Section 6.3). On the other hand, alkoxyamine **45** undergoes a selective NO-C photolysis.[306,310] Linear increase of M_n with conversion was observed during NMPP of *n*-butyl acrylate with **45,** as expected for controlled polymerization (Scheme 4.18).[310]

Photosensitization of TEMPO and SG-1 based alkoxyamines with photosensitizers (*e.g.* benzophenone or 2-isopropylthioxanthone) also leads to the desired NO-C photolysis. However, side reactions were also observed.[308,311] The photo-NMP with 4-methoxy-TEMPO in conjunction with a conventional

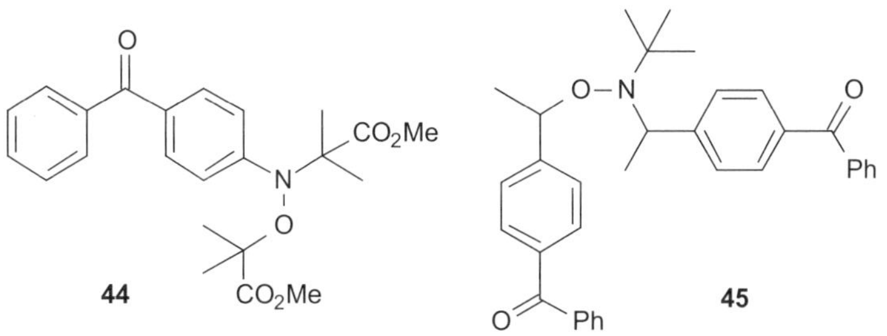

Scheme 4.18 Photosensitive alkoxyamines for NMPP.

radical initiator *and* a photoacid generator has been reported several times,[312–314] however the mechanism remains unclear.

4.7.4 Functionalization of NMP-Polymers

4.7.4.1 *Functionalization of the α-Terminus*

Functionalization of the α-terminus is readily achieved using properly substituted alkoxyamines. A well-known example is the BlocBuilder® alkoxyamine **35A5**, which introduces a COOH group at the α-end of every chain. Access to other functionalized alkoxyamines is rather straightforward and only limited by the compatibility of the desired functional group with the method of the alkoxyamine synthesis. A review of Matyjaszewski *et al.*[315] summarizes the synthesis of functionalized alkylhalides suitable for ATRA synthesis of alkoxyamines.

4.7.4.2 *Functionalization of the ω-Terminus*

ω-Functionalization is more difficult because it must be performed on the polymer. In contrast to ATRP[315,316] and RAFT[317,318] polymers, the removal or transformation of the nitroxide moiety in NMP polymers has received less attention. Nevertheless, the ω-macroalkoxyamine can be elaborated into a variety of functional groups. The nitroxide moiety can be easily exchanged with another nitroxide[95,319–322] (path a, Scheme 4.19). Higaki *et al.* showed that heating a mixture of two homopolymers, one terminated with a nitroxide moiety (polymer-NO•) and the other with an alkoxyamine (polymer-ONR¹R²), led to a block copolymer with M_n equal to the sum of the two original homopolymers.[323] Nanogels were prepared by a thermal radical crossover reaction between two diblock copolymers carrying complementarily reactive alkoxyamine units.[324] Zinc/acetic acid,[85,321] LiAlH₄[325] and oxidation with cerium ammonium nitrate[326] or oxygen[321] (followed by the reduction of the intermediate hydroperoxide) have been used to introduce an OH group (path b). Reaction with H-donors, such as 2,6-di-tert-butyl-4-methylphenol, affords H-terminated polymers[325] (path c) and oxidation with *m*-chloroperbenzoic

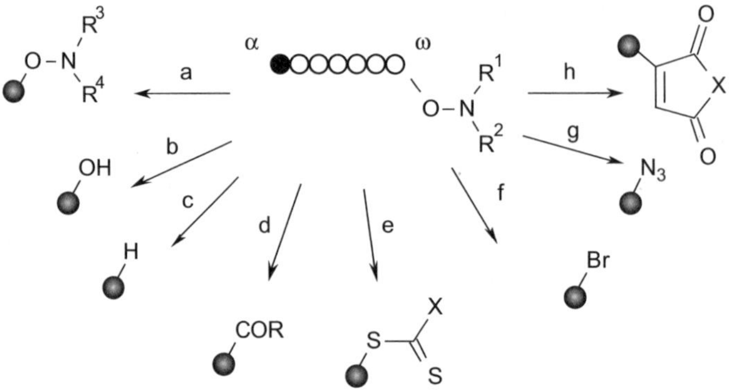

Scheme 4.19   Functionalization of an NMP polymer.

acid[325] (path d) gives the corresponding ketone (*e.g.* in the case of styrene the group R will by C_6H_5). Heating with a RAFT agent, *e.g.* benzyl dithiobenzoate, leads to the formation of the corresponding macro-RAFT agent[297] (path e). Treatment with ethyl-2-bromoisobutyrate (path f) or ethanesulfonylazide (path g) provides access to the bromo and azido terminated polymers.[321]

Reaction with maleic anhydride and maleimides leads to efficient introduction of just one of these molecules at end of the polymeric chain[155] (path h).

4.7.5   Industrial Applications of NMP

A detailed review of the industrial obstacles and achievements of CRDRP is available.[327] CRDRP opens completely new possibilities for the design and synthesis of novel materials. As is often the case when revolutionary technologies emerge, very optimistic forecasts of their commercial potential appeared. In 2000, the potential market for CRDRP products was anticipated to exceed 20 billion US$ year.[327] Today, 12 years later, the known industrial applications of CRDRP are few. It is now clear that CRDRP will not replace classical RP for production of large volume commodity polymers. However, its potential for high-value polymeric specialties remains intact. In 2003, Ciba (now BASF) announced the first industrial use of NMP for synthesis of acrylate block copolymer pigment dispersants.[328] Another major step towards acceptance of NMP in industrial chemistry was the commercialization of the BlocBuilder® alkoxyamine **35A5** by Arkema.[213] Commercial block copolymers (Nanostrength®) made with BlocBuilder® are now available.[212]

4.8   Other Stable Radicals as Polymerization Mediators

NMP is today arguably the most important SRMP method. However, other techniques using purely organic stable radicals or organometallic radicalloid species exist and are briefly mentioned below.

4.8.1 Non-nitroxide Stable Organic Radicals

Moderately successful controlled/living polymerization of styrene[329,330] and BA[330] was achieved with unimolecular initiators based on verdazyl radicals such as **46** (Scheme 4.20). Braun and coworkers showed that the relatively persistent diphenyl cyanomethyl radicals **47**, readily generated by thermolysis of tetraphenyl succinimide, can be used for the preparation α,ω-difunctionalized poly-MMA oligomers. These oligomers can be further reacted with styrene to afford A-B-A block copolymers.[331,332] Muellen *et al.* reported controlled polymerization of styrene, acrylates, methacrylates, and vinyl acetate with the combination of triazolinyl radicals such as **48**[333–335] and conventional radical initiators and demonstrated formation of block copolymers. Radical **48** slowly decomposes[161] into triphenyltriazol and an initiating phenyl radical. This process has the same rate enhancement effect as the partial removal of nitroxides by suitable additives (see Section 3.5.4). The relatively stable borinate radicals **49** formed by thermolysis of *in situ* generated alkyl boryl peroxides exerted moderate control during polymerization of MMA.[336–338] Druliner[339] reported a linear increase of molecular weight with conversion during polymerization of BA and MMA mediated with arylazooxy radical **50**, but the resulting polymers exhibited broad molecular weight distributions (PDI>3). The thio-analogue of TEMPO **51** did not control polymerization of styrene but rather behaved as S-centered iniferter radical.[340]

A mechanistic "inversion" of stable free-radical polymerization is the thioketone-mediated polymerization (TKMP) invented by Barner-Kowollik *et al.*[341–343] TKMP functions on the same principle of reversible activation-deactivation as NMP except that a non-radical thioketone such as **52** instead of a nitroxide couples with the propagating radicals to afford the dormant persistent radical **53** (Scheme 4.21).

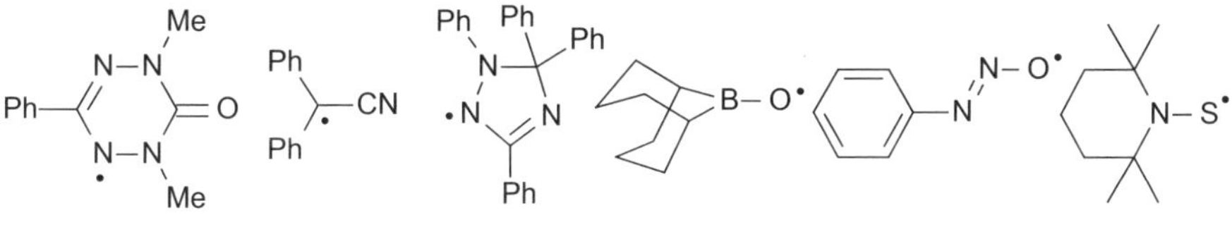

Scheme 4.20 Additional stable radicals for SRMP.

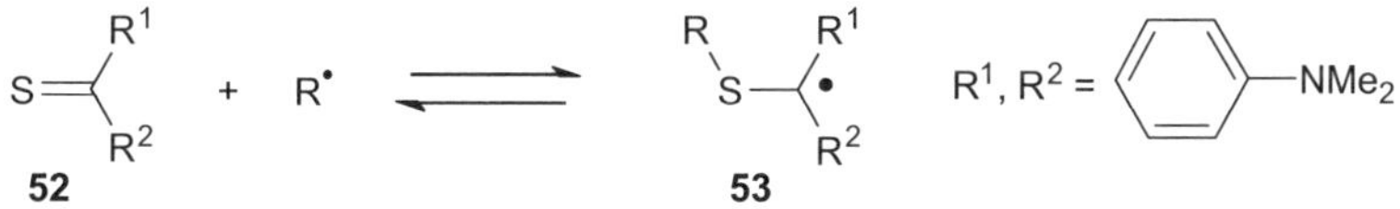

Scheme 4.21 Thioketone-mediated radical polymerization.

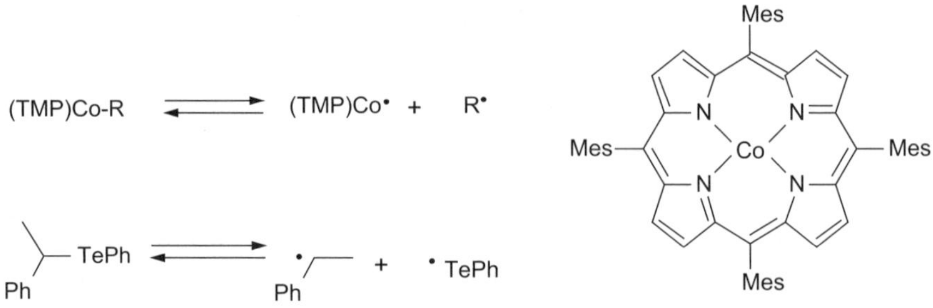

Mes = 2,4,6-trimethylphenyl

Scheme 4.22 Examples of reagents for OMRP.

4.8.2 Organometallic Radicals

Organometallic-mediated radical polymerization (OMRP) is based on the fast and reversible homolytic cleavage (Scheme 4.22) of a metal–carbon bond (metal: Co, Rh, Mo, Os, Fe, Pd, Ni, Ti, Cr, V, Te, Bi, Sb). In some cases it can also proceed *via* degenerative transfer which does not exhibit a persistent radical effect. OMRP has been recently comprehensively reviewed by Shaver *et al.*[344] Wayland *et al.*[345] reported in 1992 the first example of OMRP based on a porphyrin complex of rhodium. One of the most successful OMRP techniques is the cobalt-mediated radical polymerization[346] (CMRP). It allows the CRP of acrylic esters, acrylic acid, vinyl esters, acrylonitrile and vinylpyrrolidone and affords well-defined polymers, in some cases with high molecular weights ($>200\,000\,\mathrm{g\,mol^{-1}}$). The polymerization can be performed at rather low temperatures; *e.g.*, acrylates[347,348] and vinyl acetate[349] can be polymerized at 60 °C with cobalt(ii) tetramesitylporphyrin Co(TMP) **54** (Scheme 4.22).

Yamago *et al.*[350] discovered the reversible radical generation upon thermolysis or photolysis of organotellurium compounds. Indeed, organotellurium compounds such as **55** promote highly controlled CRP. Similar reactivity has been observed with organostibines[351] and organobismuthines.[352] A recent review of these techniques is available.[353] Despite the progress achieved by OMRP, problems related to the recovery and the recycling of the metal catalysts must be resolved before an industrial application of OMRP is conceivable.

List of Abbreviations

AIBN	Azoisobutyronitrile
ATRP	Atom-transfer radical polymerization
BA	*n*-Butyl acrylate
CMRP	Cobalt-mediated radical polymerization
CRDRP	Controlled reversible-deactivation radical polymerization
CRP	Controlled radical polymerization
DP$_n$	Number average degree of polymerization

DTRP	Degenerate-transfer radical polymerization
ESCP	Enhanced spin capturing polymerization
IUPAC	International Union of Pure and Applied Chemistry
LRP	Living radical polymerization
MMA	Methyl methacrylate
MWD	Molecular weight distribution
NMP	Nitroxide-mediated polymerization
NMPP	Nitroxide-mediated photopolymerization
NMRC	Nitrone-mediated radical coupling
OMRP	Organometallic-mediated radical mediated polymerization
PDI	Polydispersity index, PDI = M_w/M_n
PRE	Persistent radical effect
RAFT	Reversible-addition-fragmentation chain-transfer polymerization
RP	Radical polymerization
RTCP	Reversible-chain-transfer catalyzed polymerization
SRMP	Stable-radical mediated polymerization
ST	Styrene
TEMPO	2,2,6,6-Tetramethylpiperidine *N*-oxyl
TRI	Thermal radical initiator

References

1. PlacticsEurope, *Plastics – the Facts 2010*, <http://www.plasticseurope.org/plastics-industry/market-data.aspx>.
2. H.Morawetz, *Polymers: The Origin and Growth of a Science*, John Wiley & Sons, New York, 1985.
3. H.-G. Elias, *Macromolecules*, Wiley-VCH, Weinheim, 2005, vol. 1, p. 153.
4. S. Penczek and G. Moad, *Pure Appl. Chem.*, 2008, **80**, 2163–2193.
5. J. Scheirs and D. Priddy (eds.), *Modern Styrenic Polymers*, Wiley, UK, 2003, p. 557.
6. G. Moad and D. Solomon, *The Chemistry of Radical Polymerization*, 2nd edn, Elsevier, Oxford, 2006.
7. K. Matyjaszewski and T. P. Davis (eds.), *Handbook of Radical Polymerization*, John Wiley & Sons, Hoboken, 2002.
8. K. Matyjaszewski and W. A. Braunecker in *Macromolecular Engineering*, (eds.) K. Matyjaszewski, Y. Gnanou and L. Leibler, Wiley-VCH, Weinheim 2007, vol. 1, pp. 161–215.
9. M. Buback and A. M. v. Herk (eds.), *Radical Polymerization: Kinetics and Mechanism*, Wiley-VCH, Weinheim, 2007.
10. M. Szwarz, *Nature*, 1956, **176**, 1168–1169.
11. K. Matyjaszewski and A. H. E. Mueller, *Prog. Polym. Sci.*, 2006, **31**, 1039–1040.
12. A. H. E. Müller and K. Matyjaszewski (eds.), *Controlled and Living Polymerizations*, Wiley-VCH, Weinheim, 2010.
13. A. D. Jenkins, R. G. Jones and G. Moad, *Pure Appl. Chem.*, 2010, **82**, 483–491.

14. B. E. Daikh and R. G. Finke, *J. Am. Chem. Soc.*, 1992, **114**, 2938–2943.
15. W. A. Braunecker and K. Matyjaszewski, *Prog. Polym. Sci.*, 2007, **32**, 93–146.
16. N. V. Tsarevsky and K. Matyjaszewski, *Chem. Rev.*, 2007, **107**, 2270–2299.
17. B. M. Rosen and V. Percec, *Chem. Rev.*, 2009, **109**, 5069–5119.
18. K. Matyjaszewski (ed.), *ACS Symp. Ser.*, 2009, 1023.
19. F. di Lena and K. Matyjaszewski, *Prog. Polym. Sci.*, 2010, **35**, 959–1021.
20. A. Goto, Y. Tsujii and T. Fukuda, *Polymer*, 2008, **49**, 5177–5185.
21. A. Goto, T. Wakada, T. Fukuda and Y. Tsujii, *Macromol. Chem. Phys.*, 2010, **211**, 594–600.
22. A. Wolpers, L. Ackermann and P. Vana, *Macromol. Chem. Phys.*, 2011, **212**, 259–265.
23. Y. Kitayama, M. Yorizane, H. Minami and M. Okubo, *Polym. Chem.*, 2012, **3**, 1394–1398.
24. Y. Kitayama, M. Yorizane, H. Minami and M. Okubo, *Macromolecules*, 2012, **45**, 2286–2291.
25. G. David, C. Boyer, J. Tonnar, B. Ameduri, P. Lacroix-Desmazes and B. Boutevin, *Chem. Rev.*, 2006, **106**, 3936–3962.
26. K. Matyjaszewski, ed., *ACS Symp. Ser.*, 2009, **1024**.
27. G. Moad, Y. K. Chong, A. Postma, E. Rizzardo and S. H. Thang, *Polymer*, 2005, **46**, 8458–8468.
28. G. Moad, E. Rizzardo and S. H. Thang, *Aust. J. Chem.*, 2005, **58**, 379–410.
29. G. Moad, *Aust. J. Chem.*, 2006, **59**, 661–662.
30. S. Perrier and C. Barner-Kowollik, *J. Polym. Sci., Part A: Polym. Chem.*, 2008, **46**, 5715–5723.
31. G. Moad, E. Rizzardo and S. H. Thang, *Acc. Chem. Res.*, 2008, **41**, 1133–1142.
32. C. Barner-Kowollik, ed., *Handbook of RAFT Polymerization*, Wiley-VCH, Weinheim, 2008.
33. G. Moad, E. Rizzardo and S. H. Thang, *Aust. J. Chem.*, 2009, **62**, 1402–1472.
34. D. Griller and K. U. Ingold, *Acc. Chem. Res.*, 1976, **9**, 13–19.
35. M. Gomberg, *J. Am. Chem. Soc.*, 1900, **22**, 757–771.
36. L. Eberson, *Chem. Intell.*, 2000, **6**, 44–49.
37. A. L. Buchachenko, *Stable Radicals*, Consultants Bureau Enterprises, New York, 1965.
38. A. R. Forrester, J. M. Hay and R. H. Thomson, *Organic Chemistry of Stable Free Radicals*, Academic Press, London & New York, 1968.
39. I. Ratera and J. Veciana, *Chem. Soc. Rev.*, 2012, **41**, 303–349.
40. R. G. Hicks, *Org. Biomol. Chem.*, 2007, **5**, 1321–1338.
41. R. G. Hicks, ed., *Stable Radicals: Fundamentals and Applied Aspects of Odd-Electron Compounds*, John Wiley & Sons Ltd, Chichester, United Kingdom, 2010.

42. H. Karoui, F. Le Moigne, O. Ouari and P. Tordo, in *Stable Radicals*, ed. R. G. Hicks, John Wiley & Sons Ltd., Chichester, 2010, pp. 173–229.
43. T. Vogler and A. Studer, *Synthesis*, 2008, 1979–1993.
44. E. G. Rozantsev, *Free Nitroxyl Radicals*, Plenum, New York, 1970.
45. E. Breuer, H. G. Aurich and A. Nielsen, *Nitrones, Nitronates, and Nitroxides*, John Wiley and Sons, Chichester, UK, 1989.
46. L. B. Volodarsky, V. A. Reznikov and V. I. Ovcharenko, *Synthetic Chemistry of Stable Nitroxides*, CRC, Boca Raton, Florida, USA, 1994.
47. G. I. Likhtenshtein, J. Yamauchi, S. Nakatsuji, A. I. Smirnov and R. Tamura, *Nitroxides. Aplications in Chemistry, Biomedicine, and Materials Science*, Wiley-VCH, Weinheim, 2008.
48. E. G. Rozantsev and V. D. Sholle, *Synthesis*, 1971, 190–202.
49. E. G. Rozantsev and V. D. Sholle, *Synthesis*, 1971, 401–414.
50. J. F. W. Keana, *Chem. Rev.*, 1978, **78**, 37–64.
51. K. Hideg, T. Kalai and C. P. Sar, *J. Heterocyclic. Chem.*, 2005, **42**, 437–450.
52. B. P. Soule, F. Hyodo, K.-i. Matsumoto, N. L. Simone, J. A. Cook, M. C. Krishna and J. B. Mitchell, *Antioxid. Redox Signaling*, 2007, **9**, 1731–1744.
53. A. Studer and T. Vogler, *Science of Synthesis*, 2009, **40b**, 845–853.
54. L. Tebben and A. Studer, *Angew. Chem., Int. Ed.*, 2011, **50**, 5034–5068.
55. E. Zottler and G. Gescheidt, *J. Chem. Res.*, 2011, **35**, 257–267.
56. P. S. Billone, P. A. Johnson, S. Lin, J. C. Scaiano, G. A. DiLabio and K. U. Ingold, *J. Org. Chem.*, 2011, **76**, 631–636.
57. V. A. Reznikov, I. A. Gutorov, Y. V. Gatilov, T. V. Rybalova and L. B. Volodarsky, *Russ. Chem. Bull.*, 1996, **45**, 384–392.
58. C. Le Mercier, S. Acerbis, D. Bertin, F. Chauvin, D. Gigmes, O. Guerret, M. Lansalot, S. Marque, F. Le Moigne, H. Fischer and P. Tordo, *Macromol. Symp.*, 2002, **182**, 225–247.
59. A. Studer, K. Harms, C. Knoop, C. Mueller and T. Schulte, *Macromolecules*, 2004, **37**, 27–34.
60. O. Lagrille, N. R. Cameron, P. A. Lovell, R. Blanchard, A. E. Goeta and R. Koch, *J. Polym. Sci., Part A: Polym. Chem.*, 2006, **44**, 1926–1940.
61. G. D. Frey and E. Herdtweck, *Z. Naturforsch., B: J. Chem. Sci.*, 2010, **65**, 475–478.
62. D. F. Bowman, T. Gillan and K. U. Ingold, *J. Am. Chem. Soc.*, 1971, **93**, 6555–6561.
63. G. D. Mendenhall and K. U. Ingold, *J. Am. Chem. Soc.*, 1973, **95**, 6390–6394.
64. G. D. Mendenhall and K. U. Ingold, *J. Am. Chem. Soc.*, 1973, **95**, 6395–6400.
65. I. Novak, L. J. Harrison, B. Kovac and L. M. Pratt, *J. Org. Chem.*, 2004, **69**, 7628–7634.
66. J. Chateauneuf, J. Lusztyk and K. U. Ingold, *J. Org. Chem.*, 1988, **53**, 1629–1632.

67. A. L. J. Beckwith, V. W. Bowry and G. Moad, *J. Org. Chem.*, 1988, **53**, 1632–1641.
68. V. W. Bowry and K. U. Ingold, *J. Am. Chem. Soc.*, 1992, **114**, 4992–4996.
69. A. L. J. Beckwith, V. W. Bowry and K. U. Ingold, *J. Am. Chem. Soc.*, 1992, **114**, 4983–4992.
70. E. G. Bagryanskaya and S. R. A. Marque, *Chem. Rev.*, 2013, submitted.
71. I. W. C. E. Arends, P. Mulder, K. B. Clark and D. D. M. Wayner, *J. Phys. Chem.*, 1995, **99**, 8182–8189.
72. J. L. Hodgson and M. L. Coote, *Macromolecules*, 2010, **43**, 4573–4583.
73. D. L. Marshall, M. L. Christian, G. Gryn'ova, M. L. Coote, P. J. Barker and S. J. Blanksby, *Org. Biomol. Chem.*, 2011, **9**, 4936–4947.
74. R. Amorati, G. F. Pedulli, D. A. Pratt and L. Valgimigli, *Chem. Commun.*, 2010, **46**, 5139–5141.
75. T. J. Connolly and J. C. Scaiano, *Tetrahedron Lett.*, 1997, **38**, 1133–1136.
76. C. Galli, P. Gentili and O. Lanzalunga, *Angew. Chem., Int. Ed.*, 2008, **47**, 4790–4796.
77. B. Patel, J. Carlisle, S. E. Bottle, G. R. Hanson, B. M. Kariuki, L. Male, J. C. McMurtrie, N. Spencer and R. S. Grainger, *Org. Biomol. Chem.*, 2011, **9**, 2336–2344.
78. A. J. P. Cardenas, B. J. Culotta, T. H. Warren, S. Grimme, A. Stute, R. Frohlich, G. Kehr and G. Erker, *Angew. Chem., Int. Ed.*, 2011, **50**, 7567–7571.
79. M. V. Ciriano, H.-G. Korth, W. B. van Scheppingen and P. Mulder, *J. Am. Chem. Soc.*, 1999, **121**, 6375–6381.
80. J. E. Babiarz, G. T. Cunkle, A. D. DeBellis, D. Eveland, S. D. Pastor and S. P. Shum, *J. Org. Chem.*, 2002, **67**, 6831–6834.
81. G. G. Odian, *Principles of Polymerization*, 2nd edn, John Wiley & Sons, New York, 1981, p. 280.
82. T. Otsu, M. Yoshida and T. Tazaki, *Makromol. Chem., Rapid Commun.*, 1982, **3**, 133–140.
83. T. Otsu, *J. Polym. Sci., Part A: Polym. Chem.*, 2000, **38**, 2121–2136.
84. D. H. Solomon, *J. Polym. Sci., Part A: Polym. Chem.*, 2005, **43**, 5748–5764.
85. D. H. Solomon, G. Wawerly, E. Rizzardo, W. Hill and P. Cacioli, US Patent 4 581 429, 1986.
86. M. K. Georges, R. P. N. Veregin, P. M. Kazmaier and G. K. Hamer, *Macromolecules*, 1993, **26**, 2987–2988.
87. T. Fukuda, A. Goto and K. Ohno, *Macromol. Rapid Commun.*, 2000, **21**, 151–165.
88. C. J. Hawker, A. W. Bosman and E. Harth, *Chem. Rev.*, 2001, **101**, 3661–3688.
89. H. Fischer, *Chem. Rev.*, 2001, **101**, 3581–3610.
90. A. Goto and T. Fukuda, *Prog. Polym. Sci.*, 2004, **29**, 329–385.
91. D. Bertin, D. Gigmes, S. R. A. Marque and P. Tordo, *Chem. Soc. Rev.*, 2011.

92. D. Gigmes and S. R. A. Marque, in *Encyclopedia of Radicals in Chemistry, Biology and Materials*, (eds.) C. Chatgilialoglu and A. Studer, John Wiley & Sons, Chichester, UK, 2012, vol. 4, pp. 1813–1850.
93. H. Fischer and L. Radom, *Angew. Chem. Int. Ed.*, 2001, **40**, 1340–1371.
94. H. Fischer and L. Radom, *Macromol. Symp.*, 2002, **182**, 1–14.
95. C. J. Hawker, G. G. Barclay and J. Dao, *J. Am. Chem. Soc.*, 1996, **118**, 11467–11471.
96. M. Souaille and H. Fischer, *Macromolecules*, 2000, **33**, 7378–7394.
97. H. K. Hall, *Angew. Chem., Int. Ed.*, 1983, **22**, 440–455.
98. W. E. Bachmann and F. Y. Wiselogle, *J. Org. Chem.*, 1936, **01**, 354–382.
99. H. Fischer, *J. Am. Chem. Soc.*, 1986, **108**, 3925–3927.
100. H. Fischer, *Macromolecules*, 1997, **30**, 5666–5672.
101. C. H. J. Johnson, G. Moad, D. H. Solomon, T. H. Spurling and D. J. Vearing, *Aust. J. Chem.*, 1990, **43**, 1215–1230.
102. A. Studer, *Chem.–Eur. J.*, 2001, **7**, 1159–1164.
103. A. Studer, *Chem. Soc. Rev.*, 2004, **33**, 267–273.
104. K.-S. Focsaneanu and J. C. Scaiano, *Helv. Chim. Acta*, 2006, **89**, 2473–2482.
105. T. Kothe, S. Marque, R. Martschke, M. Popov and H. Fischer, *J. Chem. Soc., Perkin Trans. 2*, 1998, **2**, 1553–1560.
106. M. Buback, M. Egorov, R. G. Gilbert, V. Kaminsky, O. F. Olaj, G. T. Russell, P. Vana and G. Zifferer, *Macromol. Chem. Phys.*, 2002, **203**, 2570–2582.
107. M. Souaille and H. Fischer, *Macromolecules*, 2002, **35**, 248–261.
108. P. D. Bartlett and T. Funahashi, *J. Am. Chem. Soc.*, 1962, **84**, 2596–2601.
109. T. Fukuda, T. Terauchi, A. Goto, K. Ohno, Y. Tsujii, T. Miyamoto, S. Kobatake and B. Yamada, *Macromolecules*, 1996, **29**, 6393–6398.
110. W. Tang, T. Fukuda and K. Matyjaszewski, *Macromolecules*, 2006, **39**, 4332–4337.
111. H. Fischer and M. Souaille, *Macromol. Symp.*, 2001, **174**, 231–240.
112. H. Fischer and M. Souaille, *Chimia*, 2001, **55**, 109–113.
113. H. Fischer, *ACS Symp. Ser.*, 2003, **854**, 10–23.
114. M. Buback, R. G. Gilbert, R. A. Hutchinson, B. Klumperman, F.-D. Kuchta, B. G. Manders, K. F. O'Driscoll, G. T. Russell and J. Schweer, *Macromol. Chem. Phys.*, 1995, **196**, 3267–3280.
115. J. M. Asua, S. Beuermann, M. Buback, P. Castignolles, B. Charleux, R. G. Gilbert, R. A. Hutchinson, J. R. Leiza, A. N. Nikitin, J.-P. Vairon and A. M. van Herk, *Macromol. Chem. Phys.*, 2004, **205**, 2151–2160.
116. S. Beuermann, M. Buback, T. P. Davis, R. G. Gilbert, R. A. Hutchinson, O. F. Olaj, G. T. Russell, J. Schweer and A. M. van Herk, *Macromol. Chem. Phys.*, 1997, **198**, 1545–1560.
117. S. Beuermann, M. Buback, T. P. Davis, R. G. Gilbert, R. A. Hutchinson, A. Kajiwara, B. Klumperman and G. T. Russell, *Macromol. Chem. Phys.*, 2000, **201**, 1355–1364.
118. S. Beuermann, M. Buback, T. P. Davis, N. García, R. G. Gilbert, R. A. Hutchinson, A. Kajiwara, M. Kamachi, I. Lacík and G. T. Russell, *Macromol. Chem. Phys.*, 2003, **204**, 1338–1350.

119. J.Brandup, E. H.Immergut, E. A.Grulke, *Polymer Handbook, Vol. II, 4th edn,* John Wiley & Sons, New York, 1999, pp. 77–95.

120. S. Beuermann and M. Buback, *Prog. Polym. Sci.*, 2002, **27**, 191–254.

121. J. Barth, M. Buback, P. Hesse and T. Sergeeva, *Macromolecules*, 2009, **42**, 481–488.

122. J. Barth, M. Buback, G. T. Russell and S. Smolne, *Macromol. Chem. Phys.*, 2011, **212**, 1366–1378.

123. H. Fischer and H. Paul, *Acc. Chem. Res.*, 1987, **20**, 200–206.

124. D. Bertin, F. Chauvin, S. Marque and P. Tordo, *Macromolecules*, 2002, **35**, 3790–3791.

125. D. Bertin, P.-E. Dufils, I. Durand, D. Gigmes, B. Giovanetti, Y. Guillaneuf, S. R. A. Marque, T. Phan and P. Tordo, *Macromol. Chem. Phys.*, 2008, **209**, 220–224.

126. Y. Guillaneuf, D. Gigmes, S. R. A. Marque, P. Tordo and D. Bertin, *Macromol. Chem. Phys.*, 2006, **207**, 1278–1288.

127. D. Gigmes, D. Bertin, C. Lefay and Y. Guillaneuf, *Macromol. Theory Simul.*, 2009, **18**, 402–419.

128. M. Wulkow, *Macromol. React. Eng.*, 2008, **2**, 461–494.

129. L. Bentein, D. R. D'Hooge, M.-F. Reyniers and G. B. Marin, *Macromol. Theory Simul.*, 2011, **20**, 238–265.

130. D. Benoit, S. Grimaldi, S. Robin, J.-P. Finet, P. Tordo and Y. Gnanou, *J. Am. Chem. Soc.*, 2000, **122**, 5929–5939.

131. N. M. Ahmad, B. Charleux, C. Farcet, C. J. Ferguson, S. G. Gaynor, B. S. Hawkett, F. Heatley, B. Klumperman, D. Konkolewicz, P. A. Lovell, K. Matyjaszewski and R. Venkatesh, *Macromol. Rapid Commun.*, 2009, **30**, 2002–2021.

132. D. Greszta and K. Matyjaszewski, *J. Polym. Sci., Part A: Polym. Chem.*, 1997, **35**, 1857–1861.

133. A. Goto and T. Fukuda, *Macromolecules*, 1997, **30**, 4272–4277.

134. A. Goto and T. Fukuda, *Macromolecules*, 1999, **32**, 618–623.

135. Y. Miura, N. Nakamura, I. Taniguchi and A. Ichikawa, *Polymer*, 2003, **44**, 3461–3467.

136. M. Souaille and H. Fischer, *Macromolecules*, 2002, **35**, 248–261.

137. F. R. Mayo, *J. Am. Chem. Soc.*, 1968, **90**, 1289–1295.

138. A. W. Hui and A. E. Hamielec, *J. Appl. Polym. Sci.*, 1972, **16**, 749–769.

139. J. M. Catala, F. Bubel and S. O. Hammouch, *Macromolecules*, 1995, **28**, 8441–8443.

140. D. Benoit, V. Chaplinski, R. Braslau and C. J. Hawker, *J. Am. Chem. Soc.*, 1999, **121**, 3904–3920.

141. P. Lacroix-Desmazes, J.-F. Lutz, F. Chauvin, R. Severac and B. Boutevin, *Macromolecules*, 2001, **34**, 8866–8871.

142. L. Couvreur, C. Lefay, J. Belleney, B. Charleux, O. Guerret and S. Magnet, *Macromolecules*, 2003, **36**, 8260–8267.

143. G. Moad, F. Ercole, J. Krstina, C. L. Moad, E. Rizzardo and S. H. Thang, *Polym. Prepr. (Am. Chem. Soc., Div. Polym. Chem.)*, 1997, **38**, 744–745.

144. G. Moad, A. G. Anderson, F. Ercole, C. H. J. Johnson, J. Krstina, C. L. Moad, E. Rizzardo, T. H. Spurling and S. H. Thang, *ACS Symp. Ser.*, 1998, **685**, 332–360.
145. C. Burguiere, M.-A. Dourges, B. Charleux and J.-P. Vairon, *Macromolecules*, 1999, **32**, 3883–3890.
146. R. McHale, F. Aldabbagh and P. B. Zetterlund, *J. Polym. Sci., Part A: Polym. Chem.*, 2007, **45**, 2194–2203.
147. C. Dire, J. Belleney, J. Nicolas, D. Bertin and S. Magnet, *J. Polym. Sci., Part A: Polym. Chem.*, 2008, **46**, 6333.
148. M. Souaille and H. Fischer, *Macromolecules*, 2001, **34**, 2830–2838.
149. G. S. Ananchenko and H. Fischer, *J. Polym. Sci., Part A: Polym. Chem.*, 2001, **39**, 3604–3621.
150. G. S. Ananchenko, M. Souaille, H. Fischer, C. Le Mercier and P. Tordo, *J. Polym. Sci., Part A: Polym. Chem.*, 2002, **40**, 3264–3283.
151. M. V. Edeleva, I. A. Kirilyuk, D. P. Zubenko, I. F. Zhurko, S. R. A. Marque, D. Gigmes, Y. Guillaneuf and E. G. Bagryanskaya, *J. Polym. Sci., Part A: Polym. Chem.*, 2009, **47**, 6579–6595.
152. M. K. Georges, J. L. Lukkarila and A. R. Szkurhan, *Macromolecules*, 2004, **37**, 1297–1303.
153. A. Debuigne, T. Radhakrishnan and M. K. Georges, *Macromolecules*, 2006, **39**, 5359–5363.
154. A. Studer, *Angew. Chem. Int. Ed.*, 2000, **39**, 1108–1111.
155. E. Harth, C. J. Hawker, W. Fan and R. M. Waymouth, *Macromolecules*, 2001, **34**, 3856–3862.
156. M. Edeleva, S. R. A. Marque, D. Bertin, D. Gigmes, Y. Guillaneuf, S. V. Morozov and E. G. Bagryanskaya, *J. Polym. Sci., Part A: Polym. Chem.*, 2008, **46**, 6828–6842.
157. J. A. Franz, B. A. Bushaw and M. S. Alnajjar, *J. Am. Chem. Soc.*, 1989, **111**, 268–275.
158. T. Doba and K. U. Ingold, *J. Am. Chem. Soc.*, 1984, **106**, 3958–3963.
159. M. Edeleva, S. R. A. Marque, D. Bertin, D. Gigmes, Y. Guillaneuf and E. Bagryanskaya, *Polymers*, 2010, **2**, 364–377.
160. Y. Guillaneuf, D. Gigmes, S. R. A. Marque, P. Astolfi, L. Greci, P. Tordo and D. Bertin, *Macromolecules*, 2007, **40**, 3108–3114.
161. M. Steenbock, M. Klapper, K. Muellen, C. Bauer and M. Hubrich, *Macromolecules*, 1998, **31**, 5223–5228.
162. A. Nilsen and R. Braslau, *J. Polym. Sci., Part A: Polym. Chem.*, 2006, **44**, 697–717.
163. S. Jousset and J. M. Catala, *Macromolecules*, 2000, **33**, 4705–4710.
164. M.-O. Zink, A. Kramer and P. Nesvadba, *Macromolecules*, 2000, **33**, 8106–8108.
165. S. Marque, C. Le Mercier, P. Tordo and H. Fischer, *Macromolecules*, 2000, **33**, 4403–4410.
166. A. Alberti, P. Carloni, L. Greci, P. Stipa and C. Neri, *Polym. Degrad. Stab.*, 1993, **39**, 215–224.

167. M. K. Georges, R. P. N. Veregin, P. M. Kazmaier, G. K. Hamer and M. Saban, *Macromolecules*, 1994, **27**, 7228–7229.
168. S. Marque, H. Fischer, E. Baier and A. Studer, *J. Org. Chem.*, 2001, **66**, 1146–1156.
169. V. D. Sen and V. A. Golubev, *J. Phys. Org. Chem.*, 2009, **22**, 138–143.
170. W. C. Buzanowski, J. D. Graham, D. B. Priddy and E. Shero, *Polymer*, 1992, **33**, 3055–3059.
171. H. Jianying, L. Jian, L. Minghua, L. Qiang, D. Lizong and Z. Yousi, *J. Polym. Sci., Part A: Polym. Chem.*, 2005, **43**, 5246–5256.
172. E. E. Malmström, R. D. Miller and C. J. Hawker, *Tetrahedron*, 1997, **53**, 15225–15236.
173. A. Goto, Y. Tsujii and T. Fukuda, *Chem. Lett.*, 2000, 788–789.
174. M. Lansalot, Y. Guillaneuf, B. Luneau, S. Acerbis, P.-E. Dufils, A. Gaudel-Siri, D. Gigmes, S. R. A. Marque, P. Tordo and D. Bertin, *Macromol. React. Eng.*, 2010, **4**, 403–414.
175. L. Hlalele and B. Klumperman, *Macromolecules*, 2011, **44**, 5554–5557.
176. F. Chauvin, P.-E. Dufils, D. Gigmes, Y. Guillaneuf, S. R. A. Marque, P. Tordo and D. Bertin, *Macromolecules*, 2006, **39**, 5238–5250.
177. E. Bagryanskaya, D. Bertin, D. Gigmes, I. Kirilyuk, S. R. A. Marque, V. Reznikov, G. Roshchupkina, I. Zhurko and D. Zubenko, *Macromol. Chem. Phys.*, 2008, **209**, 1345–1357.
178. C. J. Hawker, G. G. Barclay, A. Orellana, J. Dao and W. Devonport, *Macromolecules*, 1996, **29**, 5245–5254.
179. Y. K. Chong, F. Ercole, G. Moad, E. Rizzardo and S. H. Thang, *Macromolecules*, 1999, **32**, 6895–6903.
180. J. R. Lizotte, B. M. Erwin, R. H. Colby and T. E. Long, *J. Polym. Sci., Part A: Polym. Chem.*, 2002, **40**, 583–590.
181. L. L. T. Vertommen, H. J. W. Van Den Haak, P. Hope, C. P. M. Lacroix, J. Meijer and A. Talma, *WO 98/13392*, 1998.
182. P. Nesvadba, A. Kramer, A. Steinmann and W. Stauffer, *EP 891 986*, 1999.
183. V. Sciannamea, R. Jerome and C. Detrembleur, *Chem. Rev.*, 2008, **108**, 1104–1126.
184. E. H. H. Wong, T. Junkers and C. Barner-Kowollik, *J. Polym. Sci., Part A: Polym. Chem.*, 2008, **46**, 7273–7279.
185. E. H. H. Wong, C. Boyer, M. H. Stenzel, C. Barner-Kowollik and T. Junkers, *Chem. Commun.*, 2010, **46**, 1959–1961.
186. E. H. H. Wong, T. Junkers and C. Barner-Kowollik, *Polym. Chem.*, 2011, **2**, 1008–1017.
187. A. Fischer, A. Brembilla and P. Lochon, *Macromolecules*, 1999, **32**, 6069–6072.
188. J. L. Pradel, B. Boutevin and B. Ameduri, *J. Polym. Sci., Part A: Polym. Chem.*, 2000, **38**, 3293–3302.
189. B. Keoshkerian, M. Georges, M. Quinlan, R. Veregin and B. Goodbrand, *Macromolecules*, 1998, **31**, 7559–7561.
190. A. Kramer and P. Nesvadba, US Patent 6 353 107, 2002.

191. P. Nesvadba, L. Bugnon and R. Sift, *J. Polym. Sci., Part A: Polym. Chem.*, 2004, **42**, 3332–3341.

192. C. Wetter, J. Gierlich, C. A. Knoop, C. Mueller, T. Schulte and A. Studer, *Chem.–Eur. J.*, 2004, **10**, 1156–1166.

193. T. Schulte, K. O. Siegenthaler, H. Luftmann, M. Letzel and A. Studer, *Macromolecules*, 2005, **38**, 6833–6840.

194. Y. Kinoshita, K.-i. Yamada, T. Yamasaki, H. Sadasue, K. Sakai and H. Utsumi, *Free Radical Res.*, 2009, **43**, 565–571.

195. K. O. Siegenthaler and A. Studer, *Macromolecules*, 2006, **39**, 1347–1352.

196. P. Nesvadba, A. Kramer and M.-O. Zink, *US 6 479 608*,2002.

197. C.-C. Chang, K. O. Siegenthaler and A. Studer, *Helv. Chim. Acta*, 2006, **89**, 2200–2210.

198. T. Schulte and A. Studer, *Macromolecules*, 2003, **36**, 3078–3084.

199. P. Nesvadba, A. Kramer and M.-O. Zink, US 6 479 608, 2002.

200. S. Marque, J. Sobek, H. Fischer, A. Kramer, P. Nesvadba and W. Wunderlich, *Macromolecules*, 2003, **36**, 3440–3442.

201. S. Miele, P. Nesvadba and A. Studer, *Macromolecules*, 2009, **42**, 2419–2427.

202. E. Drockenmuller and J.–M. Catala, *Macromolecules*, 2002, **35**, 2461–2466.

203. E. Drockenmuller, J.-P. Lamps and J.-M. Catala, *Macromolecules*, 2004, **37**, 2076–2083.

204. S. Grimaldi, F. Lemoigne, J.-P. Finet, P. Tordo, P. Nicol and M. Plechot, WO Patent 96/24620, 1996.

205. E. Harth, B. Van Horn and C. J. Hawker, *J. Chem. Soc., Chem. Commun.*, 2001, 823–824.

206. J. Nicolas, B. Charleux, O. Guerret and S. Magnet, *Angew. Chem. Int. Ed.*, 2004, **43**, 6186–6189.

207. J. Nicolas, B. Charleux, O. Guerret and S. Magnet, *Macromolecules*, 2004, **37**, 4453–4463.

208. F. Chauvin, J.-L. Couturier, P.-E. Dufils, P. Gerard, D. Gigmes, O. Guerret, Y. Guillaneuf, S. R. A. Marque, D. Bertin and P. Tordo, *ACS Symp. Ser.*, 2006, **944**, 326–341.

209. J. Vinas, N. Chagneux, D. Gigmes, T. Trimaille, A. Favier and D. Bertin, *Polymer*, 2008, **49**, 3639–3647.

210. Y. Guillaneuf, J.-L. Couturier, D. Gigmes, S. R. A. Marque, P. Tordo and D. Bertin, *J. Org. Chem.*, 2008, **73**, 4728–4731.

211. D. Gigmes, J. Vinas, N. Chagneux, C. Lefay, T. N. T. Phan, T. Trimaille, P.-E. Dufils, Y. Guillaneuf, G. Carrot, F. Boue and D. Bertin, *ACS Symp. Ser.*, 2009, **1024**, 245–262.

212. P. Gerard, L. Couvreur, S. Magnet, J. Ness and S. Schmidt, *ACS Symp. Ser.*, 2009, **1024**, 361–373.

213. Arkema. *BlocBuilder*, <http://www.arkema-inc.com/index.cfm?pag = 1257>.

214. S. Grimaldi, J.-P. Finet, F. L. Moigne, A. Zeghdaoui, P. Tordo, D. Benoit, M. Fontanille and Y. Gnanou, *Macromolecules*, 2000, **33**, 1141–1147.

215. C. Lefay, J. Belleney, B. Charleux, O. Guerret and S. Magnet, *Macromol. Rapid Commun.*, 2004, **25**, 1215–1220.

216. O. Gibbons, W. M. Carroll, F. Aldabbagh, P. B. Zetterlund and B. Yamada, *Macromol. Chem. Phys.*, 2008, **209**, 2434–2444.

217. G. Delaittre, J. Rieger and B. Charleux, *Macromolecules*, 2011, **44**, 462–470.

218. G. Delaittre, M. Save and B. Charleux, *Macromol. Rapid Commun.*, 2007, **28**, 1528–1533.

219. J. Nicolas, C. Dire, L. Mueller, J. l. Belleney, B. Charleux, S. R. A. Marque, D. Bertin, S. p. Magnet and L. Couvreur, *Macromolecules*, 2006, **39**, 8274–8282.

220. S. Brusseau, O. Boyron, C. Schikaneder, C. C. Santini and B. Charleux, *Macromolecules*, 2011, **44**, 215–220.

221. J. Nicolas, S. Brusseau and B. Charleux, *J. Polym. Sci., Part A: Polym. Chem.*, 2010, **48**, 34–47.

222. B. Lessard, E. J. Y. Ling, M. S. T. Morin and M. Maric, *J. Polym. Sci., Part A: Polym. Chem*, 2011, **49**, 1033–1045.

223. S. Brusseau, F. D'Agosto, S. Magnet, L. Couvreur, C. Chamignon and B. Charleux, *Macromolecules*, 2011, **44**, 5590–5598.

224. J. Marchand, L. Autissier, Y. Guillaneuf, J.-L. Couturier, D. Gigmes and D. Bertin, *Aust. J. Chem.*, 2010, **63**, 1237–1244.

225. C. Berti, M. Colonna, L. Greci and L. Marchetti, *Tetrahedron*, 1975, **31**, 1745–1753.

226. P. Nesvadba, *Chimia*, 2006, **60**, 832–840.

227. D. Bertin, D. Gigmes and S. R. A. Marque, *Recent Res. Dev. Org. Chem.*, 2006, **10**, 63–121.

228. A. C. Greene and R. B. Grubbs, *ACS Symp. Ser.*, 2009, **1024**, 81–93.

229. K. Matyjaszewski, B. E. Woodworth, X. Zhang, S. G. Gaynor and Z. Metzner, *Macromolecules*, 1998, **31**, 5955–5957.

230. T. Pintauer and K. Matyjaszewski, *Chem. Soc. Rev.*, 2008, **37**, 1087–1097.

231. T. J. Connolly, M. V. Baldovi, N. Mohtat and J. C. Scaiano, *Tetrahedron Lett.*, 1996, **37**, 4919–4922.

232. Y. Miura, K. Hirota, H. Moto and B. Yamada, *Macromolecules*, 1998, **31**, 4659–4661.

233. R. Braslau, L. C. Burrill, II, M. Siano, N. Naik, R. K. Howden and L. K. Mahal, *Macromolecules*, 1997, **30**, 6445–6450.

234. H.-J. Kirner, F. Schwarzenbach, P. A. Van Der Schaaf, A. Hafner, V. Rast, M. Frey, P. Nesvadba and G. Rist, *Adv. Synth. Catal.*, 2004, **346**, 554–560.

235. D. Gigmes, P.-E. Dufils, D. Gle, D. Bertin, C. Lefay and Y. Guillaneuf, *Polym. Chem.*, 2011, **2**, 1624–1631.

236. P.-E. Dufils, N. Chagneux, D. Gigmes, T. Trimaille, S. R. A. Marque, D. Bertin and P. Tordo, *Polymer*, 2007, **48**, 5219–5225.

237. S. Marque, *J. Org. Chem.*, 2003, **68**, 7582–7590.

238. H. Fischer, A. Kramer, S. R. A. Marque and P. Nesvadba, *Macromolecules*, 2005, **38**, 9974–9984.

239. A. Gaudel-Siri, D. Siri and P. Tordo, *ChemPhysChem*, 2006, **7**, 430–438.
240. D. Bertin, D. Gigmes, S. R. A. Marque and P. Tordo, *Macromolecules*, 2005, **38**, 2638–2650.
241. D. Bertin, D. Gigmes, C. Le Mercier, S. R. A. Marque and P. Tordo, *J. Org. Chem.*, 2004, **69**, 4925–4930.
242. D. Bertin, D. Gigmes, S. R. A. Marque, S. Milardo, J. Peri and P. Tordo, *Collect. Czech. Chem. Commun.*, 2004, **69**, 2223–2238.
243. G. Ananchenko, E. Beaudoin, D. Bertin, D. Gigmes, P. Lagarde, S. R. A. Marque, E. Revalor and P. Tordo, *J. Phys. Org. Chem.*, 2006, **19**, 269–275.
244. D. Bertin, D. Gigmes, S. Marque, R. Maurin and P. Tordo, *J. Polym. Sci., Part A: Polym. Chem.*, 2004, **42**, 3504–3515.
245. E. Beaudoin, D. Bertin, D. Gigmes, S. R. A. Marque, D. Siri and P. Tordo, *Eur. J. Org. Chem.*, 2006, 1755–1768.
246. G. Moad and E. Rizzardo, *Macromolecules*, 1995, **28**, 8722–8728.
247. C. A. Knoop and A. Studer, *J. Am. Chem. Soc.*, 2003, **125**, 16327–16333.
248. G. Ananchenko and K. Matyjaszewski, *Macromolecules*, 2002, **35**, 8323–8329.
249. G. Ananchenko, S. Marque, D. Gigmes, D. Bertin and P. Tordo, *Org. Biomol. Chem.*, 2004, **2**, 709–715.
250. A. Blachon, S. R. A. Marque, V. Roubaud and D. Siri, *Polymers (Basel, Switz.)*, 2010, **2**, 353–363.
251. D. Bertin, D. Gigmes, S. R. A. Marque, D. Siri, P. Tordo and G. Trappo, *Chem Phys Chem.*, 2008, **9**, 272–281.
252. L. Li, G. K. Hamer and M. K. Georges, *Macromolecules*, 2006, **39**, 9201–9207.
253. P. Bremond and S. R. A. Marque, *Chem. Commun.*, 2011, **47**, 4291–4293.
254. E. Bagryanskaya, P. Brémond, M. Edeleva, S. R. A. Marque, D. Parkhomenko, V. Roubaud and D. Siri, *Macromol. Rapid Commun.*, 2012, **33**, 152–157.
255. P. Bremond, A. Koïta, S. R. A. Marque, V. Pesce, V. Roubaud and D. Siri, *Org. Lett.*, 2012, **14**, 358–361.
256. M. V. Edeleva, I. A. Kirilyuk, I. F. Zhurko, D. A. Parkhomenko, Y. P. Tsentalovich and E. G. Bagryanskaya, *J. Org. Chem.*, 2011, **76**, 5558–5573.
257. L. Marx and P. Hemery, *Polymer*, 2009, **50**, 2752–2761.
258. D. Bertin, D. Gigmes, S. Marque and P. Tordo, *e-Polym.*, 2003, **2**, 1–9.
259. J. P. Blinco, K. E. Fairfull-Smith, B. J. Morrow and S. E. Bottle, *Aust. J. Chem.*, 2011, **64**, 373–389.
260. O. G. Ballesteros, L. Maretti, R. Sastre and J. C. Scaiano, *Macromolecules*, 2001, **34**, 6184–6187.
261. E. G. Bagryanskaya, S. R. A. Marque and Y. P. Tsentalovich, *J. Org. Chem.*, 2012, **77**, 4996–5005.
262. J. Sobek, R. Martschke and H. Fischer, *J. Am. Chem. Soc.*, 2001, **123**, 2849–2857.

263. K. Iwao, K. Sakakibara and M. Hirota, *J. Comput. Chem.*, 1998, **19**, 215–221.

264. H. Fischer, S. R. A. Marque and P. Nesvadba, *Helv. Chim. Acta*, 2006, **89**, 2330–2340.

265. J. L. Hodgson, C. Y. Lin, M. L. Coote, S. R. A. Marque and K. Matyjaszewski, *Macromolecules*, 2010, **43**, 3728–3743.

266. Y. Guillaneuf, D. Bertin, P. Castignolles and B. Charleux, *Macromolecules*, 2005, **38**, 4638–4646.

267. P. Marsal, M. Roche, P. Tordo and P. de Claire, *J. Phys. Chem. A*, 1999, **103**, 2899–2905.

268. C.-Y. Lin, S. R. A. Marque, K. Matyjaszewski and M. L. Coote, *Macromolecules*, 2011, **44**, 7568–7583.

269. J. L. Hodgson, L. B. Roskop, M. S. Gordon, C. Y. Lin and M. L. Coote, *J. Phys. Chem. A*, 2010, **114**, 10458–10466.

270. P. M. Kazmaier, K. A. Moffat, M. K. Georges, R. P. N. Veregin and G. K. Hamer, *Macromolecules*, 1995, **28**, 1841–1846.

271. E. Megiel, A. Kaim and M. K. Cyranski, *J. Phys. Org. Chem.*, 2010, **23**, 1146–1154.

272. D. Gigmes, A. Gaudel-Siri, S. R. A. Marque, D. Bertin, P. Tordo, P. Astolfi, L. Greci and C. Rizzoli, *Helv. Chim. Acta*, 2006, **89**, 2312–2326.

273. P. Nesvadba, L. Bugnon and T. Wagner, *Chimia*, 2010, **64**, 56–58.

274. M. Roth, R. Pfaendner, P. Nesvadba and M.-O. Zink, WO Patent 2001/090113, 2001.

275. B. Charleux, J. Nicolas and O. Guerret, *Macromolecules*, 2005, **38**, 5485–5492.

276. S. Brusseau, J. Belleney, S. Magnet, L. Couvreur and B. Charleux, *Polym. Chem.*, 2010, **1**, 720–729.

277. P. Stipa, *Chem. Phys.*, 2006, **323**, 501–510.

278. A. C. Greene and R. B. Grubbs, *Macromolecules*, 2010, **43**, 10320–10325.

279. A. C. Greene and R. B. Grubbs, *Macromolecules*, 2009, **42**, 4388–4390.

280. M. Cunningham, M. Lin, J.-A. Smith, J. Ma, K. McAuley, B. Keoshkerian and M. Georges, *Prog. Colloid Polym. Sci.*, 2004, **124**, 88–93.

281. M. Save, Y. Guillaneuf and R. G. Gilbert, *Aust. J. Chem.*, 2006, **59**, 693–711.

282. P. B. Zetterlund, Y. Kagawa and M. Okubo, *Chem. Rev.*, 2008, **108**, 3747–3794.

283. J. K. Oh, *J. Polym. Sci., Part A: Polym. Chem.*, 2008, **46**, 6983–7001.

284. M. F. Cunningham, *Prog. Polym.Sci.*, 2008, **33**, 365–398.

285. B. Charleux, F. D'Agosto and G. Delaittre, *Adv. Polym. Sci.*, 2011, **233**, 125–183.

286. J. Pyun and K. Matyjaszewski, *Chem. Mater.*, 2001, **13**, 3436–3448.

287. P. Liu, *e-Polym.*, 2007, Nr. 62.

288. M. K. Brinks and A. Studer, *Macromol. Rapid Commun.*, 2009, **30**, 1043–1057.

289. R. Barbey, L. Lavanant, D. Paripovic, N. Schuwer, C. Sugnaux, S. Tugulu and H.-A. Klok, *Chem. Rev.*, 2009, **109**, 5437–5527.

290. A. Olivier, F. Meyer, J.-M. Raquez, P. Damman and P. Dubois, *Prog. Polym. Sci.*, 2012, **37**, 157–181.

291. N. Hadjichristidis, H. Iatrou, M. Pitsikalis and J. Mays, *Prog. Polym. Sci.*, 2006, **31**, 1068–1132.

292. M. A. Tasdelen, M. U. Kahveci and Y. Yagci, *Prog. Polym. Sci.*, 2011, **36**, 455–567.

293. Y. Yagci and M. Atilla Tasdelen, *Prog. Polym. Sci.*, 2006, **31**, 1133–1170.

294. K. V. Bernaerts and F. E. Du Prez, *Prog. Polym. Sci.*, 2006, **31**, 671–722.

295. A. P. Dove, *Chem. Commun.*, 2008, 6446–6470.

296. D. Zehm, A. Laschewsky, H. Liang and J. P. Rabe, *Macromolecules*, 2011, **44**, 9635–9641.

297. A. Favier, B. Luneau, J. Vinas, N. Laissaoui, D. Gigmes and D. Bertin, *Macromolecules*, 2009, **42**, 5953–5964.

298. B. Le Droumaguet and J. Nicolas, *Polym. Chem.*, 2010, **1**, 563–598.

299. H. F. Gruber, *Prog. Polym. Sci.*, 1992, **17**, 953–1044.

300. K. Dietliker, R. Hüsler, J. L. Birbaum, S. Ilg, S. Villeneuve, K. Studer, T. Jung, J. Benkhoff, H. Kura, A. Matsumoto and H. Oka, *Prog. Org. Coat.*, 2007, **58**, 146–157.

301. J. P. Fouassier, X. Allonas (eds.), *Basics and Applications of Photopolymerization Reactions Vol. 1–3*, Research Signpost, Trivandrum, 2010.

302. Y. Yagci, S. Jockusch and N. J. Turro, *Macromolecules*, 2010, **43**, 6245–6260.

303. J. Lalevée, X. Allonas and J. P. Fouassier, *Macromolecules*, 2006, **39**, 8216–8218.

304. S. Hu, J. H. Malpert, X. Yang and D. C. Neckers, *Polymer*, 2000, **41**, 445–452.

305. A. Goto, J. C. Scaiano and L. Marettib, *Photochem. Photobiol. Sci.*, 2007, **6**, 833–835.

306. D.-L. Versace, Y. Guillaneuf, D. Bertin, J. P. Fouassier, J. Lalevee and D. Gigmes, *Org. Biomol. Chem.*, 2011, **9**, 2892–2898.

307. D.-L. Versace, J. Lalevee, J.-P. Fouassier, Y. Guillaneuf, D. Bertin and D. Gigmes, *Macromol. Rapid Commun.*, 2010, **31**, 1383–1388.

308. D.-L. Versace, J. Lalevee, J.-P. Fouassier, D. Gigmes, Y. Guillaneuf and D. Bertin, *J. Polym. Sci., Part A: Polym. Chem.*, 2010, **48**, 2910–2915.

309. Y. Guillaneuf, D. Bertin, D. Gigmes, D.-L. Versace, J. Lalevee and J.-P. Fouassier, *Macromolecules*, 2010, **43**, 2204–2212.

310. Y. Guillaneuf, D.-L. Versace, D. Bertin, J. Lalevee, D. Gigmes and J.-P. Fouassier, *Macromol. Rapid Commun.*, 2010, **31**, 1909–1913.

311. J. C. Scaiano, T. J. Connolly, N. Mohtat and C. N. Pliva, *Can. J. Chem.*, 1997, **75**, 92–97.

312. E. Yoshida, *Colloid Polym. Sci.*, 2011, **289**, 837–841.

313. E. Yoshida, *Colloid Polym. Sci.*, 2011, **289**, 1127–1132.

314. E. Yoshida, *Colloid Polym. Sci.*, 2011, **289**, 1625–1630.

315. V. Coessens, T. Pintauer and K. Matyjaszewski, *Prog. Polym. Sci.*, 2001, **26**, 337–377.

316. K. Nakatani, M. Ouchi and M. Sawamoto, *Macromolecules*, 2008, **41**, 4579–4581.

317. H. Willcock and R. K. O'Reilly, *Polym. Chem.*, 2010, **1**, 149–157.
318. G. Moad, E. Rizzardo and S. H. Thang, *Polym. Int.*, 2011, **60**, 9–25.
319. N. J. Turro, G. Lem and I. S. Zavarine, *Macromolecules*, 2000, **33**, 9782–9785.
320. J. P. Blinco, K. E. Fairfull-Smith, A. S. Micallef and S. E. Bottle, *Polym. Chem.*, 2010, **1**, 1009–1012.
321. Y. Guillaneuf, P.-E. Dufils, L. Autissier, M. Rollet, D. Gigmes and D. Bertin, *Macromolecules*, 2010, **43**, 91–100.
322. A. Debuigne, D. Chan-Seng, L. Li, G. K. Hamer and M. K. Georges, *Macromolecules*, 2007, **40**, 6224–6232.
323. Y. Higaki, H. Otsuka and A. Takahara, *Macromolecules*, 2006, **39**, 2121–2125.
324. Y. Amamoto, Y. Higaki, Y. Matsuda, H. Otsuka and A. Takahara, *Chem. Lett.*, 2007, **36**, 774–775.
325. H. Malz, H. Komber, D. Voigt and J. Pionteck, *Macromol. Chem. Phys.*, 1998, **199**, 583–588.
326. G. O'Brya and R. Braslau, *Macromolecules*, 2006, **39**, 9010–9017.
327. M. Destarac, *Macromol. React. Eng.*, 2010, **4**, 165–179.
328. C. Auschra, E. Eckstein, R. Knischka and P. Nesvadba, *Asia Pacific Coatings Journal*, 2003, **16**, 20–23.
329. S. J. Teertstra, E. Chen, D. Chan-Seng, P. O. Otieno, R. G. Hicks and M. K. Georges, *Macromol. Symp.*, 2007, **248**, 117–125.
330. E. K. Y. Chen, S. J. Teertstra, D. Chan-Seng, P. O. Otieno, R. G. Hicks and M. K. Georges, *Macromolecules*, 2007, **40**, 8609–8616.
331. D. Braun, T. Skrzek, S. Steinhauer-Beiåer, H. Tretner and H. J. Lindner, *Macromol. Chem. Phys.*, 1995, **196**, 573–591.
332. D. Braun, *Macromol. Symp.*, 1996, **111**, 63–71.
333. M. Steenbock, M. Klapper and K. Muellen, *Macromol. Chem. Phys.*, 1998, **199**, 763–769.
334. N. S. Khelfallah, M. Peretolchin, M. Klapper and K. Muellen, *Polym. Bull. (Heidelberg, Ger.)*, 2005, **53**, 295–304.
335. A. Dasgupta, M. Klapper and K. Muellen, *Polym. Bull. (Heidelberg, Ger.)*, 2008, **60**, 199–210.
336. T. C. Chung, W. Janvikul and H. L. Lu, *J. Am. Chem. Soc.*, 1996, **118**, 705–706.
337. T. C. Chung and H. Hong, *ACS Symp. Ser.*, 2003, **854**, 481–495.
338. T. C. M. Chung, *Polym. News*, 2003, **28**, 238–244.
339. J. D. Druliner, *Macromolecules*, 1991, **24**, 6079–6082.
340. N. Bricklebank and A. Pryke, *J. Chem. Soc., Perkin Trans. 1*, 2002, **1**, 2048–2051.
341. A. A. Toy, H. Chaffey-Millar, T. P. Davis, M. H. Stenzel, E. I. Izgorodina, M. L. Coote and C. Barner-Kowollik, *Chem. Commun.*, 2006, 835–837.
342. T. Junkers, M. H. Stenzel, T. P. Davis and C. Barner-Kowollik, *Macromol. Rapid Commun.*, 2007, **28**, 746–753.

343. F. Guenzler, T. Junkers and C. Barner-Kowollik, *J. Polym. Sci., Part A: Polym. Chem.*, 2009, **47**, 1864–1876.
344. L. E. N. Allan, M. R. Perry and M. P. Shaver, *Prog. Polym. Sci.*, 2012, **37**, 127–156.
345. B. B. Wayland, G. Poszmik and M. Fryd, *Organometallics*, 1992, **11**, 3534–3542.
346. A. Debuigne, R. Poli, C. Jerome, R. Jerome and C. Detrembleur, *Prog. Polym. Sci.*, 2009, **34**, 211–239.
347. B. B. Wayland, L. Basickes, S. Mukerjee, M. Wei and M. Fryd, *Macromolecules*, 1997, **30**, 8109–8112.
348. B. B. Wayland, G. Poszmik, S. L. Mukerjee and M. Fryd, *J. Am. Chem. Soc.*, 1994, **116**, 7943–7944.
349. C.-H. Peng, J. Scricco, S. Li, M. Fryd and B. B. Wayland, *Macromolecules*, 2008, **41**, 2368–2373.
350. S. Yamago, K. Iida and J.-i. Yoshida, *J. Am. Chem. Soc.*, 2002, **124**, 2874–2875.
351. S. Yamago, B. Ray, K. Iida, J.-i. Yoshida, T. Tada, K. Yoshizawa, Y. Kwak, A. Goto and T. Fukuda, *J. Am. Chem. Soc.*, 2004, **126**, 13908–13909.
352. S. Yamago, E. Kayahara, M. Kotani, B. Ray, Y. Kwak, A. Goto and T. Fukuda, *Angew. Chem., Int. Ed.*, 2007, **46**, 1304–1306.
353. S. Yamago, *Chem. Rev.*, 2009, **109**, 5051–5068.

CHAPTER 5

Mechanistic Aspects of Living Radical Polymerization Mediated by Organometallic Complexes

ZICHUAN YE AND BRADFORD B. WAYLAND*

Department of Chemistry, Temple University, Philadelphia, PA 19122, USA
*Email: bwayland@temple.edu

5.1 Introduction

Organo-transition metal complexes which are by definition held together by metal to carbon bonds are particularly well recognized for producing living polymerization by sequential monomer insertions[1-4] at the metal center and more recently have been shown to mediate living radical polymerization (LRP) where freely diffusing radicals propagate polymer chain growth in the solution where they are outside of the influence of the metal site.[5-9] Monomer insertion reactions that occur at a metal center provide elegant control of polymer architecture including stereoselective living polymerizations, but are at present largely limited to a relatively small set of α-olefin monomers.[1,2] Organometallic mediated living radical polymerization does not yet provide the control of polymer stereochemistry, but the functional group tolerance of radicals allows LRP methods to be applied to a much wider variety of monomers.[10] The potential to obtain sequential living radical polymerization and insertion processes promises to provide an array of polymer materials that cannot be

RSC Polymer Chemistry Series No. 4
Fundamentals of Controlled/Living Radical Polymerization
Edited by Nicolay V Tsarevsky and Brent S Sumerlin
© The Royal Society of Chemistry 2013
Published by the Royal Society of Chemistry, www.rsc.org

obtained by a single type of polymerization process.[11] This chapter exclusively addresses processes where organometallic transition metal complexes function as mediators to achieve living characteristics for authentic radical polymerization where freely diffusing radicals propagate with monomers in solution. The general abbreviation of M-R will be used throughout this chapter to denote a metal–carbon bonded organometallic complex where the organo radical (R$^{\bullet}$) is a carbon-centered radical and M-P stands for an organometallic complex where the organo unit is a polymeric radical (P$^{\bullet}$).

Effective approaches to obtain living radical polymerization (LRP) are separated by reaction mechanism into two broad categories called reversible termination (RT) and degenerative transfer (DT).[6,8,9,12,13] Both reversible termination (RT) and degenerative transfer (DT) pathways to obtain living radical polymerization (LRP) utilize a dormant species (X-P) as a latent reversible source of the propagating polymeric radical chains (P$^{\bullet}$). When the X group in the dormant species (X-P) for a LRP process is a transition metal complex, then this type of LRP process was named by Poli[14] as an organometallic mediated radical polymerization (OMRP), and this definition will be consistently used in this chapter. Carbon-centered polymeric radical chains (P$^{\bullet}$) are reversibly released into solution from the dormant complex (X-P) by either homolytic dissociation or associative radical interchange processes (Scheme 5.1).

Reversible termination (RT) and degenerative transfer (DT) mechanisms for LRP are clearly distinguished by the source and concentration of radicals in solution as well as through the origin of the living character manifested by the process. The dormant complex (X-P) is the exclusive source of radicals (P$^{\bullet}$) for all RT processes which encompass the subcategories of homolytic X-P dissociation (5.1) and atom transfer (5.2). The ideal radicals X$^{\bullet}$ and Y$^{\bullet}$ depicted in eqn (1) and (2) are unable either to dimerize or initiate polymerization.

$$X\text{-}P \rightleftharpoons X^{\bullet} + {}^{\bullet}P \tag{5.1}$$

$$Y^{\bullet} + X\text{-}P \rightleftharpoons X\text{-}Y + {}^{\bullet}P \tag{5.2}$$

The freely diffusing radical and the dormant species establish a quasi-equilibrium which produces a nearly constant low concentration of radicals that slowly decrease in concentration through radical termination. Control of the radical concentration ([P$^{\bullet}$]) by cross coupling with a stable persistent radical (X$^{\bullet}$) to form a dormant complex (X-P) contributes to the living character of an

Dissociative radical interchange

$$X\text{-}P \rightleftharpoons X^{\bullet} + {}^{\bullet}P$$
$$X^{\bullet} + {}^{\bullet}P' \rightleftharpoons X\text{-}P'$$

Associative radical interchange

$$P'^{\bullet} + X\text{-}P \rightleftharpoons X\text{-}P' + {}^{\bullet}P$$

Scheme 5.1 Interchange of carbon-centered radicals (P$^{\bullet}$) between a dormant complex (X-P) and freely diffusing radicals (P'$^{\bullet}$) in solution.

RT process through the persistent radical effect (PRE).[5,15–17] The persistent radical effect is an inherent kinetic phenomenon that suppresses the rate of bimolecular termination through the reversible binding of the propagating radicals ($P^\bullet$) with a stable radical ($X^\bullet$). Radical polymerizations have a second mechanism that gives living character through the decrease in the bimolecular radical termination rate constant (k_t) as the molecular weight of the polymer radical and solution viscosity increase.[6,9,18] Decrease in the termination rate constant (k_t) results in an increase in the number of propagation events per termination and thus increases the living character for radical polymerization.

Living radical polymerization (LRP) processes that occur by a degenerative transfer (DT) mechanism, such as organo-main group element polymerization[19–22] and RAFT,[23] use a continual influx of small radicals ($R^\bullet$) from an external source such as dialkyl azo compounds (RN_2R) like AIBN and V-70 to initiate monomer polymerization and to exchange with larger polymeric radicals ($P^\bullet$) in the dormant complex (X-P). The concentration of radicals in solution for a DT process is primarily determined by the concentration of the external radical source ([I]) such as AIBN and V-70, and the rate constants for radicals to enter solution (k_i) and terminate (k_t) ($[R^\bullet] = (k_i[I]/2k_t)^{1/2}$) just as in regular radical polymerization.[6] The concentration of radicals in solution and rates of polymerization for DT processes thus approach the values for regular uncontrolled radical polymerization. The dormant complex (X-P) in a DT process does not control the radical concentration and thus the persistent radical effect cannot contribute to the living character for DT processes.[6] The living radical character for radical polymerization controlled by a DT process results only from the inherent decrease in the radical termination rate constant (k_t) as the polymer molecular weight and solution viscosity increase.[6,9,18]

5.2 Organometallic Mediated Radical Polymerization (OMRP)

Organo cobalt porphyrin ($(por)Co-R$)[5] and related low spin organo-cobalt complexes[24] were the first organometallic compounds intentionally designed for the control of radical polymerization. Subsequently, high spin cobalt complexes of acetylacetonate[25,26] and related ligands[27–29] and a variety of d-transition metal complexes of titanium,[30–33] vanadium,[34] iron,[35–39] chromium,[40,41] molybdenum,[42–44] nickel[11] and palladium[45–48] have been evaluated for control of living radical polymerization. There are several excellent reviews of OMRP[7–9,49] and an outstanding comprehensive review[39] that covers OMRP and organo-main group LRP mediators. From the polymer synthesis and materials perspectives, $Co(acac)_2$ and Cp_2TiCl are particularly important OMRP reagents because they mediate LRP of monomers that are difficult to control by other methods.[23,49–61] Cp_2TiCl has the unusual capability of controlling both styrenic and isoprene monomers[62–64] that have applications[62–66] in materials. $Co(acac)_2$ is most notable for the remarkable control observed for vinyl acetate (VAc)[8] and this system has produced new classes of polymeric materials that have wide ranging technological applications[7–9,67–86].

The M-R bond dissociation free energy for the M-R unit is the central factor in reversible termination (RT) processes and also a significant contributing term in associative radical exchange required for effective degenerative transfer (DT).[6] The virtually limitless variability of the metal complexes for OMRP provides a broader range of opportunities for tuning the behavior than other LRP methods through choice of both the metal centers and ligand arrays. Organometallic complexes also provide additional dimensions for the tuning of M-R bond dissociation energetics through structural and electronic rearrangements and spin state changes that accompany organo-metal bond homolysis and radical exchange. Cobalt porphyrins have structural and electronic properties that result from the strong tetrapyrrole donors and aromaticity of porphyrins that facilitate and simplify evaluation of the kinetic and thermodynamic factors relevant to the mediation of living radical polymerization by OMRP.[5,6,87–92] Organo-cobalt porphyrins are used in this chapter to illustrate the criteria for RT and DT mechanisms for LRP and as prototype complexes for the general organo-metal mediated living radical polymerization.

5.3 Organo-cobalt Porphyrin Mediated Radical Polymerization

Cobalt(II) porphyrins are planar four-coordinate low spin (d^7, $S = 1/2$) metal-centered radicals ((por)Co$^{II•}$) and organo-cobalt porphyrins ((por)Co-R) are diamagnetic (d^6, $S = 0$) five-coordinate sixteen-electron cobalt(III) complexes. Organo-cobalt porphyrin bond dissociation energies fall in the weak to medium range (15–35 kcal mol^{-1})[93–95] which can provide thermal and photolytic sources of organic and metal-centered radicals for reversible termination (RT) processes. Coordinate and electronic unsaturation of five coordinate sixteen-electron organo-cobalt complexes are important properties for facile exchange of radicals by an associative (S_H2) pathway,[6,87,90] which is a necessary requirement for effective degenerative transfer (DT). Diamagnetism of organo-Co(por) complexes is convenient for NMR observations and the magnetic anisotropy or ring current of the aromatic porphyrin macrocycle shifts the ^{1}H NMR positions in the organic groups bonded to cobalt to the high field region ($\delta < 0$) where they can be easily observed and assigned.[5,87,90] Facile chemical substitution on the periphery of the porphyrin ligands is another useful property both to tune steric and electronic properties, and to produce favorable solubility in organic and donor media including water[89] (Figure 5.1).

Cobalt(II) porphyrins ((por)Co$^{II•}$) are stable and relatively unreactive metal-centered radicals except with species that have unpaired electrons such as organic radicals (R$^•$) and dioxygen. The relatively rigid porphyrin macrocycle

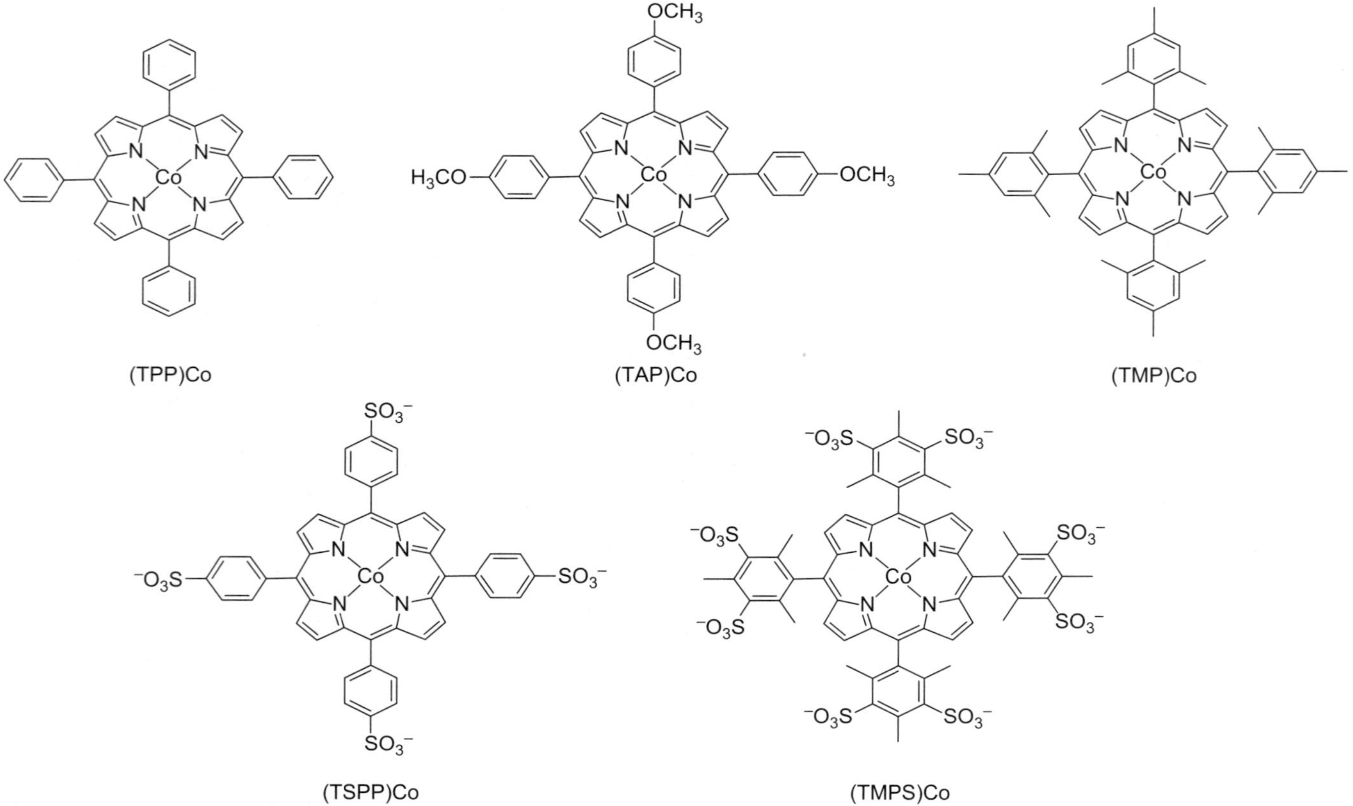

Figure 5.1 Representative cobalt porphyrin complexes for low and high ligand steric demands and for use in organic solvents ((TPP)Co, (TAP)Co, (TMP)Co) and aqueous media ((TSPP)Co, (TMPS)Co).

restricts ligand structural changes and placement of the unpaired electron in the d_{z^2} orbital minimizes electronic reorganization in forming diamagnetic organo-cobalt(III) porphyrin complexes ((por)Co-R). These structural and electronic features make cobalt porphyrins valuable prototype transition metal complexes for mechanistic studies of organo-metal mediated radical polymerization (OMRP) in living radical polymerizations (LRP) that occur by either reversible termination (RT) or degenerative transfer (DT) pathways. Studies of cobalt porphyrins in LRP processes will be used in this chapter as introductory illustrations of the more general area of OMRP. The model studies of living radical polymerization (LRP) generally use methyl acrylate (MA) as the monomer because of the relatively large propagation constant (k_p) ($k_{p(MA)(333K)} = 1.38 \times 10^4$) and small rate constants for chain transfer to monomer (MA) and polymer (PMA) which are properties that are favorable for living character in radical polymerizations.

5.3.1 Low Spin Cobalt(II) ($S = 1/2$) and Organo-cobalt(III) ($S = 0$) in Organometallic Mediated Radical Polymerization of Vinyl Monomers

Organo-cobalt porphyrin complexes have Co-R bond dissociation free energies (BDFE) in a range ($\Delta G \approx 10$–$25\,kcal\,mol^{-1}$)[93–95] that encompass the ideal values ($\Delta G^0 \approx 14$–$16\,kcal\,mol^{-1}$) for thermal production of radicals ($R^{\bullet}$) at concentrations that give useful rates of radical polymerization at mild conditions ($T < 100\,°C$).[5] Organo-cobalt bond homolysis generates a stable and relatively nonreactive cobalt(II) porphyrin metal-centered radical ((por)Co$^{II\bullet}$) and a highly reactive carbon-centered organic radical ($R^{\bullet}$) capable of initiating olefin polymerizations.[5] Organo-cobalt porphyrin complexes thus function as latent thermal precursors for both organic radicals ($R^{\bullet}$) to initiate radical polymerization of monomers and a persistent cobalt(II) ($S = 1/2$) metal-centered radical for reversible termination (RT) of the growing polymer radical chain ($P^{\bullet}$).[5] Metallo radicals of low spin cobalt(II) porphyrins are not capable of initiating radical polymerization, but reactions with dioxygen and hydrogen abstraction from organic radicals can have unfavorable influences on the radical polymerization process. Cobalt(II) porphyrins and other low spin cobalt(II) complexes react with oxygen to form paramagnetic superoxide and diamagnetic peroxide complexes (eqn (5.7), (5.8)) which remove cobalt(II) from the system.[96] Hydrogen atom abstraction by a cobalt(II) metallo-radical ((por)Co$^{II\bullet}$) from a growing polymer radical produces a transient cobalt hydride ((por)Co-H) (5.9) that either terminates by dihydrogen formation (5.10) or gives catalytic chain transfer by reaction with a monomer to form an organometallic ((por)Co-R) that dissociates and reinitiates polymerization to yield a new polymer radical chain (eqn (5.11), (5.12)).[87,90,97–101] In the case of acrylate monomers this pathway for chain transfer can be suppressed by introducing sterically demanding substituents on the porphyrin such as mesityl groups that sterically prohibit achieving the transition state for hydrogen abstraction.[5]

$$(por)Co\text{-}R \rightleftharpoons (por)Co^{II\bullet} + {}^{\bullet}R \tag{5.3}$$

$$R^{\bullet} + CH_2{=}CH(X) \xrightarrow{k_p} RCH_2\text{-}CHX^{\bullet} \tag{5.4}$$

$$RCH_2\text{-}CHX^{\bullet} + CH_2{=}CH(X) \rightleftharpoons P^{\bullet} \tag{5.5}$$

$$(por)Co^{II\bullet} + {}^{\bullet}P \rightleftharpoons (por)Co\text{-}P \tag{5.6}$$

$$(por)Co^{\bullet} + O_2 \rightleftharpoons (por)Co\text{-}O_2 \tag{5.7}$$

$$2(por)Co^{\bullet} + O_2 \rightleftharpoons (por)Co\text{-}O\text{-}O\text{-}Co(por) \tag{5.8}$$

$$(por)Co^{\bullet} + {}^{\bullet}CH(X)CH_3 \rightarrow (por)Co\text{-}H + CH_2{=}CH(X) \tag{5.9}$$

$$2(por)Co\text{-}H \rightarrow 2(por)Co^{II\bullet} + H_2 \tag{5.10}$$

$$(por)Co\text{-}H + CH_2{=}CHX \rightleftharpoons (por)Co\text{-}CH(CH_3)X \tag{5.11}$$

$$(por)Co\text{-}CH(CH_3)X \rightleftharpoons (por)Co^{II\bullet} + {}^{\bullet}CH(CH_3)X \tag{5.12}$$

5.3.2 Organo-cobalt Porphyrins in LRP of Methyl Acrylate by the Reversible Termination Mechanism

Polymerization of methyl acrylate (MA) is initiated by heating neopentyl-cobalt porphyrins including the tetraphenyl (TPP) and tetramesityl (TMP) derivatives at 60 °C in the presence of methyl acrylate (MA) monomer[5] (Figure 5.2). When the relatively low steric demand $(TPP)Co\text{-}CH_2C(CH_3)_3$ is used as a neopentyl radical initiator and $(por)Co^{II\bullet}$ functions as the mediator, the number average molecular weight (M_n) increases to a maximum of about 20 000 at 20% conversion and then slightly decreases as the conversion continues (Figure 5.2A). The PDI is small at low monomer conversion (PDI = 1.10, $M_n = 5500$), but increases to values approaching uncontrolled radical polymerization at moderate conversion ($M_n = 20\,000$, PDI = 1.8–2.0). This behavior for (TPP)Co-P is characteristic of a relatively slow catalytic chain transfer process where the monomer propagates an average of about 200 times per chain transfer event. When the more sterically demanding tetramesityl porphyrin derivative $((TMP)Co\text{-}CH_2C(CH_3)_3)$ is used to initiate and mediate a radical polymerization of methyl acrylate, the living character of the polymerization increases substantially from that obtained by $(TPP)Co\text{-}CH_2C(CH_3)_3$.[5] The linear increase in the number average molecular

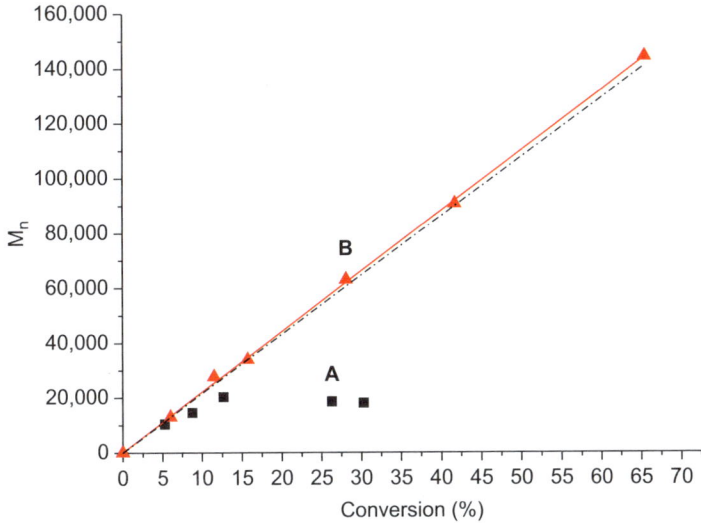

Figure 5.2 Number average molecular weight of poly(methyl acrylate) (PMA) as a function of conversion in radical polymerization initiated and mediated by (por)Co-CH$_2$C(CH$_3$)$_3$ ([(por)Co-CH$_2$C(CH$_3$)$_3$] = 1.0 × 10^{-3} M) complexes in benzene at 333 K when [MA]$_i$ = 2.5 M. (A) (TPP)Co-CH$_2$C(CH$_3$)$_3$ and (B) (TMP)Co-CH$_2$C(CH$_3$)$_3$. The dashed line represents the theoretical number average molecular weight for a living process based on [(por)Co$^{II•}$].

weight (M_n) with conversion and retention of relatively small PDI clearly indicates that the (TMP)Co system produces a living radical polymerization that has thousands of polymer propagation events with only a small fraction of the polymer chains truncated by radical termination and chain transfer (Figure 5.2B). Sequential additions of MA and butyl acrylate (BA) produce block copolymers that have low PDI and linear increase in molecular weight with conversion.[5]

(TMP)Co-CH(CO$_2$CH$_3$)CH$_3$ (**1**) was also used in the polymerization of MA in order to initiate the process with an organic radical (•CH(CO$_2$CH$_3$)CH$_3$) that more nearly emulates the radicals in the PMA oligomer complex of Co(TMP).[5] Reaction of (TMP)Co-CH(CO$_2$CH$_3$)CH$_3$ with MA at 60 °C in benzene results in formation of PMA with relatively small PDI (1.1–1.2) and linear increase in M_n (1 × 10^4 to 1.7 × 10^5) with MA conversion (5–80%; DP = 125–2000).

A related similar thermal reaction of (TMP)Co-CH(CO$_2$CH$_3$)CH$_3$ with a much smaller ratio of 50 molecules of MA per cobalt slows the rate of polymer growth sufficiently to permit the direct ^{1}H-NMR observation of the sequential transformation of (TMP)Co-CH(CO$_2$CH$_3$)CH$_3$ to (TMP)Co-CH(CO$_2$CH$_3$)CH$_2$CH(CO$_2$CH$_3$)CH$_3$ which corresponds to one propagation event and then to an oligomer complex (TMP)Co-CH(CO$_2$CH$_3$)CH$_2$CH(CO$_2$CH$_3$)CH$_2$-PMA (Figure 5.3). At higher monomer concentrations ([MA]/[Co] = 2500), only the complex of the polymer radical ((TMP)Co-P) is observed.

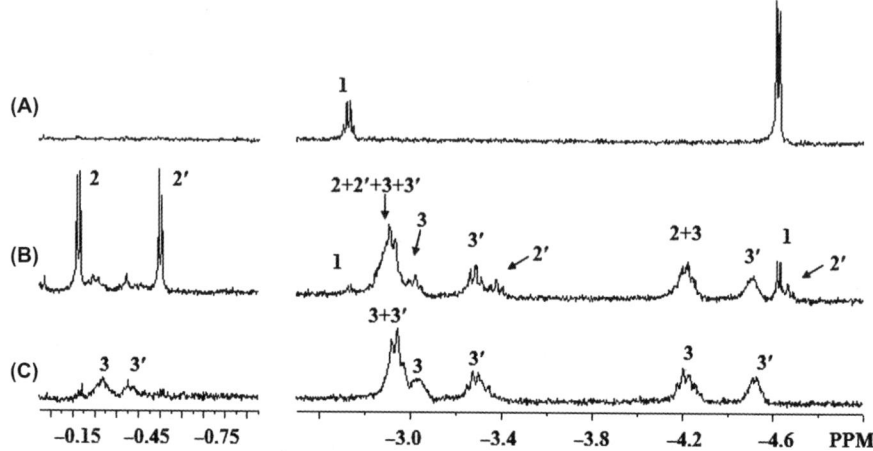

Figure 5.3 Changes in ^{1}H NMR (500 MHz, C_6D_6) high-field region for organo-cobalt complexes during the polymerization of methyl acrylate at 313 K. Initial concentrations: $[(TMP)Co^{II•}]_i = 1.48 \times 10^{-3}$ M, $[V-70]_i = 2.00 \times 10^{-3}$ M, $[MA]_i = 0.14$ M. (A)100 min; (B) 200 min; (C) 300 min. Assignments of the ^{1}H NMR for organo-cobalt species: (**1**) $(TMP)Co\text{-}CH(CO_2CH_3)CH_3$, (**2**) $(TMP)Co\text{-}CH(CO_2CH_3)CH_2CH(CO_2CH_3)CH_3$, and (**3**) $(TMP)Co\text{-}PMA$. Reprinted with permission from ref. 90. Copyright 2011 American Chemical Society.

Polymerization of methyl acrylate initiated by organo-Co(TMP) complexes is envisioned as occurring by the reaction sequence given by 5.13–5.17, in which $X = CO_2CH_3$ and $R = CH(CO_2CH_3)CH_3$. Bond homolysis of (TMP)Co-R produces a carbon-centered radical ($R^•$) (5.13) that initiates polymerization by reacting with an acrylate monomer to form $RCH_2CHX^•$ (5.14), which either combines reversibly with $(TMP)Co^{II•}$ (5.15) or reacts with additional acrylate monomers to form an oligomer radical (eqn 5.16) that subsequently combines reversibly with $(TMP)Co^{II•}$ (eqn (5.17)).

$$(TMP)Co\text{-}R \rightleftharpoons (TMP)Co^{II•} + {}^•R \tag{5.13}$$

$$R^• + CH_2{=}CHX \rightarrow RCH_2CHX^• \tag{5.14}$$

$$RCH_2CHX^• + (TMP)Co^{II•} \rightleftharpoons RCH_2CHX\text{-}Co(TMP) \tag{5.15}$$

$$RCH_2CHX^• + (n+1)CH_2{=}CHX \rightarrow RCH_2CHX(CH_2CHX)_nCH_2CHX^• \tag{5.16}$$

$$RCH_2CHX(CH_2CHX)_nCH_2CHX^• + (TMP)Co^{II•}$$
$$\rightleftharpoons RCH_2CHX(CH_2CHX)_nCH_2CHX\text{-}Co(TMP) \tag{5.17}$$

The equilibrium constant at 333 K to form the (TMP)Co-PMA oligomer complex was determined to be 8.6×10^9 ($K_{17}(333 \text{ K}) = 8.6(.2) \times 10^9$; ΔG^0 (333 K) = $-15.1 \text{ kcal mol}^{-1}$).[90] The estimate of 23 kcal mol^{-1} for the (TMP)Co-PMA BDE using $\Delta S^0 \sim 27 \text{ cal K}^{-1} \text{mol}^{-1}$ compares favorably with the BDE of 25 kcal mol^{-1} directly evaluated for the less sterically demanding tetra p-methoxy phenyl porphyrin (TAP) derivative (TAP)Co-CH(CO$_2$CH$_3$)CH$_3$.[95] Repetition of reactions 5.14–5.17 without radical termination or chain transfer would result in a fully living radical polymerization process. The real polymerization process cannot be fully living because of inherent bimolecular radical termination (eqn (5.18), (5.19)) and H$^\bullet$ transfer reactions with monomer, polymer, solvent and (TMP)Co$^{II\bullet}$ (eqn (5.20), (5.21)) which produce non-living polymer chains.

$$2 \sim\!\!\text{CH}_2\text{CHX}^\bullet \rightarrow \sim\!\!\text{CH}_2\text{CHX} - \text{CHXCH}_2 \sim\!\!\!\quad\quad\quad (5.18)$$

$$2 \sim\!\!\text{CH}_2\text{CHX}^\bullet \rightarrow \sim\!\!\text{CH}_2\text{CH}_2\text{X} + \sim\!\!\text{CH}=\text{CHX} \quad\quad\quad (5.19)$$

$$\sim\!\!\text{CH}_2\text{CHX}^\bullet + \text{HT} \rightarrow \sim\!\!\text{CH}_2\text{CH}_2\text{X} + {}^\bullet\text{T} \quad\quad\quad (5.20)$$

$$\sim\!\!\text{CH}_2\text{CHX}^\bullet + (\text{TMP})\text{Co}^{II\bullet} \rightarrow \text{CH}=\text{CHX} + (\text{TMP})\text{Co-H} \quad\quad\quad (5.21)$$

HT = radical transfer reagent

In spite of the processes that can limit polymer growth (eqn (5.18)–(5.21)), observation of linear increase in M_n with conversion, formation of block copolymers, and relatively small PDI demonstrate that organo-Co(TMP) complexes thermally initiate and mediate an effective living radical polymerization of acrylates. ^{1}H NMR and GPC observations of the polymerization process indicate that the vast majority (>95%) of the polymer chains are living even at relatively high conversions. The small fraction of radicals that irreversibly terminate produce an increase in the (TMP)Co$^{II\bullet}$ concentration which suppresses the radical (R$^\bullet$, P$^\bullet$) concentration through the equilibrium with the organo-cobalt complex ((TMP)Co-P) (eqn (5.13), (5.15) and (5.17)) until further irreversible radical termination events (eqn (5.18) and (5.19)) become improbable relative to polymer propagation (5.16). This self-regulating suppression of radical termination by the presence of the stable (TMP)Co$^{II\bullet}$ metallo radical is an example of the persistent radical effect.[6] The capping or reversible termination agent in a living radical polymerization must additionally not initiate polymerization or produce any irreversible reactions with the polymer radical such as β-H abstraction which terminate polymer chain growth. These criteria are fulfilled by (TMP)Co$^{II\bullet}$ because cobalt-carbon bonds are too weak for the metal-centered radical to initiate methyl acrylate polymerization and the steric requirements of the TMP ligand effectively prohibit β-H abstraction from the propagating oligomer radicals by (TMP)Co$^{II\bullet}$.[5]

5.3.3 Organo-cobalt Porphyrins in LRP of Methyl Acrylate by a Degenerative Transfer (DT) Mechanism

Organo-cobalt porphyrin complexes also function as prototype organo-metal complexes for the mediation of living radical polymerization (LRP) by a degenerative transfer (DT) mechanism.[6] Effective LRP by a DT pathway requires fast exchange of organic radicals (R'$^\bullet$) in solution with radicals (P$^\bullet$) in the dormant organometallic complex (M-P). An external source of radicals is required for polymer propagation to occur in a degenerative transfer process.[6] Azo compound (RN=NR) such as AIBN (R$^\bullet$=(C(CN)(CH$_3$)$_2$)$_2$) and V-70 (R$^\bullet$=(C(CN)(CH$_3$)CH$_2$C(OCH$_3$)(CH$_3$)$_2$)$_2$) that have half-lives at 333 K of 18 hours and 11 minutes respectively are used as fast and slow external radical sources in probing mechanistic features of the DT controlled LRP.

$$R'^\bullet + M\text{-}P \rightleftharpoons R'\text{-}M + P^\bullet$$

Radical polymerization data for methyl acrylate ([MA]=2.5 M) initiated by V-70 ([V-70]=0.52×10^{-4} M) as the source of radicals entering solution in the absence and presence of (TMP)Co-CH(CO$_2$CH$_3$)CH$_3$ (1.0×10^{-3} M) as the organometallic mediator are shown in Figure 5.4. Radical polymerization of MA initiated by radicals from V-70 in the presence of the organo-Co(TMP) complex produced poly(methyl acrylate) (PMA) with relatively low PDI (1.05–1.11) and molecular weight close to the theoretical values for one living polymer chain per (TMP)Co unit (65% conversion, $M_{n\,exp}$=142 000, $M_{n\,theory}$=139 900, PDI=1.08). In the absence of the organo-cobalt mediator

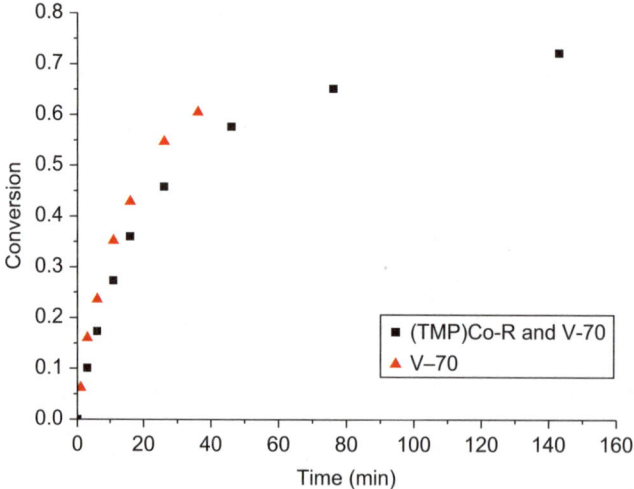

Figure 5.4 Radical polymerization of methyl acrylate ([MA]$_i$=2.50 M, T=333 K, C$_6$D$_6$ initiated by (A) [V-70]$_i$=5.2×10^{-4} M, (B) [V-70]$_i$=5.2×10^{-4} M, and [(TMP)Co-CH(CH$_3$)CO$_2$CH$_3$]$_i$=1.0×10^{-3} M; $M_{n(PMA)}$=142 000 and PDI=1.08 at 65% conversion (theory for LRP M_n=140 000).
Reprinted with permission from ref. 6. Copyright 2011 American Chemical Society.

uncontrolled radical polymerization occurred (45% conversion, $M_n = 230\,000$, PDI $= 2.4$). Continuous injection of radicals from V-70 into solution keeps the rate of monomer conversion relatively large both in the presence and absence of the organo-Co(TMP) mediator (Figure 5.4). The rate of conversion is slightly larger in the absence of the (TMP)Co-R mediator which is tentatively ascribed to having larger molecular weight polymeric radicals and thus smaller radical termination constants (k_t) and larger radical concentration $([P^\bullet] = (k_i[I]/2)k_t)^{1/2})$ in the uncontrolled radical polymerization.[6] The most significant result is that the azo initiated radical polymerization process in the presence of an organo-cobalt mediator remains controlled with high living character even at the large polymerization rates that result from continuous injection of radicals from the external radical source. These results illustrate that the dormant organometallic complex ((TMP)Co-P) mediates a living radical polymerization of MA by a degenerative transfer (DT) pathway as well as the earlier recognized reversible termination (RT) route. Observation of the DT process infers that there is rapid exchange of radicals in solution $(R^\bullet)$ with the latent radical $(P^\bullet)$ in the dormant complex (TMP)Co-P in order to account for obtaining narrow polydispersity. Detailed kinetic studies for the exchange process experimentally confirmed that the rate constants for associative radical exchange for (TMP)Co-P with $R^\bullet$ in benzene $(k_{(333\,K)} = 7 \times 10^5\,M^{-1}\,s^{-1})$ are large enough to account for the observed narrow PDI.[87,90] Evaluating the PDI for the MA polymerization at 50% conversion, $k_p = 1.3 \times 10^4\,s^{-1}$ and $k_{ex} = 7 \times 10^5$ $M^{-1}s^{-1}$ gives $M_w/M_n = 1.06$ ($M_w/M_n = 1 + ((2/c)-1)/C_e$; $C_{ex} = k_{ex}/k_p$, $c = $ fractional conversion)[61] which is close to values observed in the living radical polymerization of MA by the DT mechanism.[6]

5.3.3.1 In Situ *Generation of Organo-cobalt Porphyrin Mediators*

Organo-cobalt porphyrin complexes are generated in the polymerization medium from reactions of (TMP)Co$^{II\bullet}$, V-70 and MA which circumvents the necessity for isolation and purification of the organo-cobalt complexes.[88] Solutions of methyl acrylate (2.5 M), V-70 ((0.7–2.0) $\times 10^{-3}$ M), and (TMP)Co$^{II\bullet}$ (1.2×10^{-3} M) in benzene were heated to 333 K, and the conversion to polymer was followed by ^1H NMR (Figure 5.5). The induction time periods prior to observing polymerization decrease with increasing concentration of V-70, and correspond to the time required to inject sufficient radicals into solution to convert effectively all of the (TMP)Co$^{II\bullet}$ into an organometallic complex (TMP)Co-CH(CO$_2$CH$_3$)CH$_3$. Injection of modest excesses of total organic radicals into solution from V-70 compared to the initial concentration of (TMP)Co$^{II\bullet}$ results in large increases in the rate of polymerization without substantial increase in the polydispersity or deviation of the molecular mass (M_n) from the theoretical value of one living chain per cobalt (Figure 5.5). The polymers formed when using this more convenient procedure showed a linear increase in number-average molecular weight (M_n) with conversion, unusually low polydispersity (1.05), and molecular weights close to the theoretical values corresponding to one polymer chain per cobalt porphyrin (Figure 5.6).

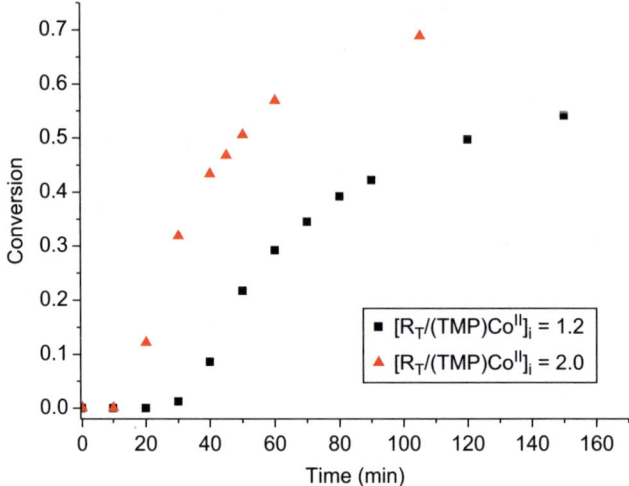

Figure 5.5   Kinetic plot for the radical polymerization of methyl acrylate in benzene at 333 K with $[MA]_i = 2.5\,M$ and $[(TMP)Co^{II\bullet}]_i = 1.2 \times 10^{-3}\,M$. V-70 was added to produce different ratios for the total moles of external radicals that can enter solution to the total moles of cobalt porphyrin. (▲) $[R^{\bullet}_T/(TMP)Co^{II\bullet}]_i = 1.2$, conversion $= 54\%$, $t = 150\,min$, $M_n = 9.5 \times 10^4$, PDI $= 1.06$; (■) $[R^{\bullet}_T/(TMP)Co^{II\bullet}]_i = 2.0$, conversion $= 69\%$, $t = 105\,min$, $M_n = 11.6 \times 10^4$, PDI $= 1.06$.
Reprinted with permission from ref. 88. Copyright 2011 American Chemical Society.

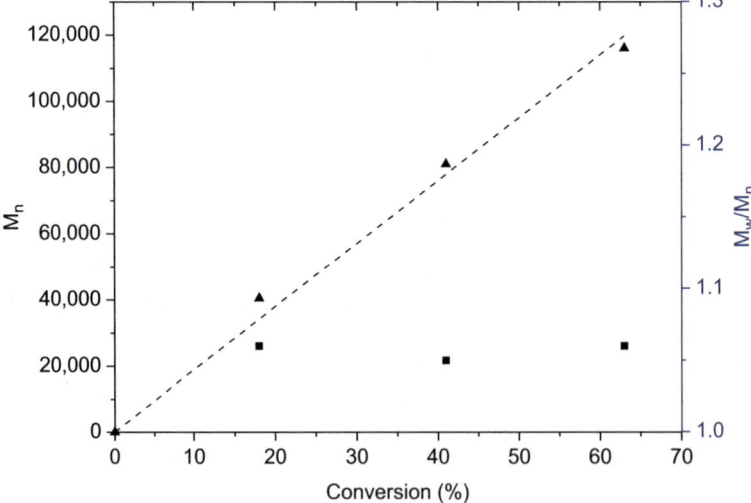

Figure 5.6   Change in the number-average molecular weight and polydispersity with methyl acrylate conversion to PMA initiated by V-70 when $[MA]_i = 2.5\,M$ and $[(TMP)Co^{II\bullet}]_i = 1.2 \times 10^{-3}\,M$, and $[R^{\bullet}_T/(TMP)Co^{II\bullet}]_i = 1.2$ at 333 K in benzene.
Reprinted with permission from ref. 88. Copyright 2011 American Chemical Society.

2.1) V-70 $\longrightarrow$ 2 $^\bullet$CH(CH$_3$)(R)(CN) (=R$^\bullet$)

2.2) (TMP)Co$^{II}\bullet$ + $^\bullet$CH(CH$_3$)(R)(CN) $\longrightarrow$ (TMP)Co-H + (CN)(R)C=CH$_2$

2.3) (TMP)Co-H + CH$_2$=CH(CO$_2$CH$_3$) $\rightleftharpoons$ (TMP)Co-CH(CH$_3$)CO$_2$CH$_3$

2.4) (TMP)Co-CH(CH$_3$)CO$_2$CH$_3$ $\rightleftharpoons$ (TMP)Co$^{II}\bullet$ + $^\bullet$CH(CH$_3$)CO$_2$CH$_3$

2.5) CH$_3$CO$_2$(CH$_3$)HC$^\bullet$ + (n+1) CH$_2$=CH(CO$_2$CH$_3$) $\xrightarrow{\;k_p\;}$

$\qquad$ CH$_3$(CO$_2$CH$_3$)HC [CH$_2$CH(CO$_2$CH$_3$)]$_n$ CH$_2$(CO$_2$CH$_3$)(H)C$^\bullet$ (=P$^\bullet$)

2.6) (TMP)Co$^{II}\bullet$ + $^\bullet$P $\rightleftharpoons$ (TMP)Co-P

Scheme 5.2 Dominant reactions during the induction period

3.1) R$^\bullet$ + CH$_2$=CH(CO$_2$CH$_3$) $\xrightarrow{\;k_p\;}$ RCH$_2$(CO$_2$CH$_3$)HC$^\bullet$ (=R'$^\bullet$)

3.2) R'$^\bullet$ + (TMP)Co-P $\rightleftharpoons$ (TMP)Co-R' + $^\bullet$P

3.3) P$^\bullet$ + m $\xrightarrow{\;k_p\;}$ P'$^\bullet$

3.4) P'$^\bullet$ + (TMP)Co-P $\rightleftharpoons$ (TMP)Co-P' + P$^\bullet$

3.5) 2 R$^\bullet$ $\longrightarrow$ R-R

Scheme 5.3 Dominant reactions during the post induction period.

The dominant reactions that occur during the early portion of the induction period when significant amounts of (TMP)Co$^{II}\bullet$ are present are shown in Scheme 5.2, eqn (5.2.1)–(5.2.6). Hydrogen abstraction from a methyl group on the tertiary carbon radical produced by V-70 forms a transient cobalt hydride intermediate (TMP)Co-H which adds with the methyl acrylate monomer to produce an organometallic complex, (TMP)Co-CH(CH$_3$)CO$_2$CH$_3$ (eqn (5.2.2), (5.2.3)). Thermal homolytic dissociation of (TMP)Co-CH(CH$_3$)CO$_2$CH$_3$ produces an organic radical ($^\bullet$CH(CH$_3$)CO$_2$CH$_3$) that initiates methyl acrylate (MA) polymerization (eqn (5.2.4)) to form an oligomer radical that binds with (TMP)Co$^{II}\bullet$. During the last stage of the induction period (eqn (5.2.4)–(5.2.6)) growth of the polymeric radical has living character from the persistent radical effect.

The reactions that predominate during the post-induction period where rapid polymerization occurs are given in Scheme 5.3 by eqn (5.3.1)–(5.3.4). Radicals (R$^\bullet$) entering solution from V-70 initiate polymerization of MA (eqn (5.3.1)), and the radicals in solution exchange with radicals in the dormant (TMP)Co-P complex by dissociative and associative pathways (eqn (5.3.2), (5.3.4)). In the specific case of the methyl acrylate oligomer organometallic complex ((TMP)Co-PMA) homolytic dissociation at 333 K is sufficiently fast to account for the observed low polydispersity.

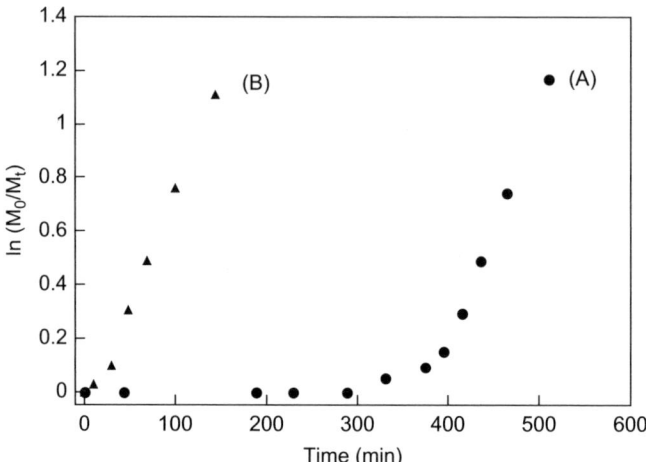

Figure 5.7   First order rate plot for polymerization of methyl acrylate in benzene at 333 K
([MA]$_i$ = 2.5 M; [(TMP)Co^{II}]$_i$ = 1.0 × 10^{-3} M). (A) [AIBN]$_i$ = 4.8 × 10^{-3} M;
(B) [V-70]$_i$ = 4.1 × 10^{-3} M.

First order rate plots for the polymerization of methyl acrylate initiated
by V-70 and AIBN at 333 K in the presence of (TMP)Co^{II} and
(TMP)Co-CH(CO$_2$CH$_3$)CH$_3$ are illustrated in Figure 5.7. AIBN has a half life
of about 18 hours at 333 K in benzene and the much slower entry of radicals
into solution compared to V-70 ($t_{1/2}$(333 K) = 11 minutes) permits attaining a
near constant steady state radical concentration which simplifies the kinetic
analysis. The rate of MA polymerization after the induction period follows first
order rate behavior where the slope of ln([M]$_0$/[M]$_t$) *versus* time is proportional
to the square root of the initiator concentration ([AIBN]$^{1/2}$). This demonstrates
that the rate of MA polymerization is controlled by the AIBN concentration
and not by the organo-Co(TMP) mediator complex which is a signature
criterion for a degenerative transfer (DT) process.[6,12]

Using AIBN as the external radical source lengthens the induction period for
convenient observation of the changes in the solution species by ^{1}H NMR.
The induction period observed for the polymerization in the presence of
the cobalt(II) porphyrin corresponds to the time required to form slightly
more than one radical from AIBN per cobalt(II) in solution. Conversion
of the (TMP)Co$^{II•}$ metallo radical to the organometallic complex
((TMP)Co-CH(CH$_3$)CO$_2$CH$_3$) is directly observed by ^{1}H NMR during the
induction period and demonstrates the formation and reaction of (TMP)Co-H.
The pyrrole hydrogens of ((TMP)Co-CH(CO$_2$CH$_3$)CH$_3$) and (TMP)Co-PMA
are diastereotopic and appear as an AB pattern in the ^{1}H NMR because of the
chiral centers in the organo ligand bonded to cobalt (Figure 5.8).

Rapid interchange of radicals in solution (R′$^•$) with the latent radicals (R$^•$) in
the dormant complex is a necessary condition for degenerative transfer to
provide control.

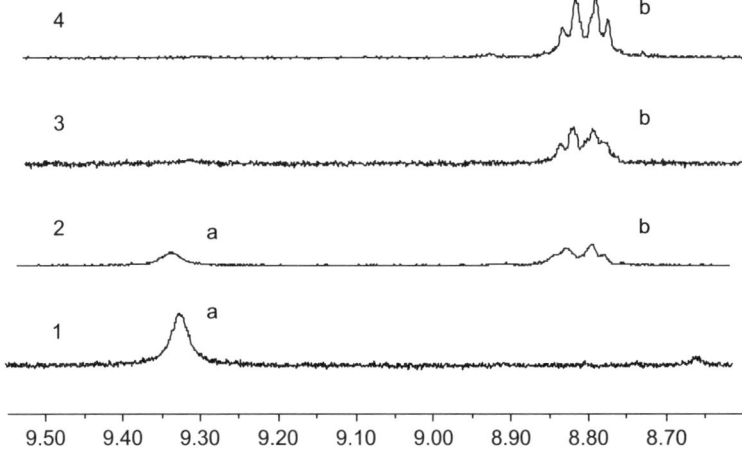

Figure 5.8 Porphyrin ^{1}H NMR (300 MHz) spectra in C_6D_6 illustrate that the formation of an organo-cobalt complex during the induction period for the organo-Co(TMP) mediated LRP of MA. (a) (TMP)Co^{II} *m*-phenyl H, (b) (TMP)Co-CH(CH$_3$)CO$_2$CH$_3$ pyrrole AB pattern. (1) (TMP)Co^{II}; (2) (TMP)Co^{II} and (TMP)Co-CH(CH$_3$)CO$_2$CH$_3$; (3) 90% (TMP)Co-R, 1–3% polymer formed; and (4) (TMP)Co-R, rapid polymer propagation starts.

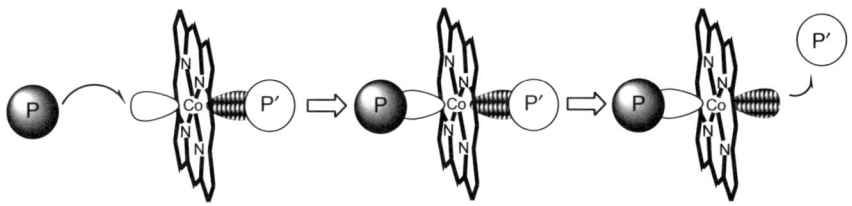

Figure 5.9 Radical interchange for organo-cobalt porphyrin complexes with solution polymeric radicals from solution.

$$R'^{\bullet} + M\text{-}R \rightleftharpoons R'\text{-}M + R^{\bullet}$$

Comparable numbers of radical exchange and polymer propagation events are needed to achieve limiting small PDI. Organo-cobalt porphyrins are five coordinate sixteen-electron complexes and this coordinate and electronic unsaturation provides a highly favorable situation for fast associative radical interchange (Figure 5.9).

Detailed kinetic studies and kinetic simulations for the formation and decay of complexes that can only occur by radical interchange were analyzed to give a radical exchange constant (k_{ex}) for the vinyl acetate[87] and methyl acrylate[90] polymerizations in the presence of (TMP)Co-P. The rate constants deduced for radical exchange ($k_{ex}(333\,K) = 0.5\text{--}1.0 \times 10^6 \, M^{-1}s^{-1}$) ($P_m{}^{\bullet} + (TMP)Co\text{-}P_n \rightleftharpoons P_m\text{-}Co(TMP) + P_n{}^{\bullet}$) in the (TMP)Co system are comparable to or larger than the values reported for exchange rates of radicals with dithioesters $(RAFT)^{23,102,103}$ and a series of organo main group species.[13,104,105] The observed rate constants for the radical exchange processes in the organo-cobalt mediated LRP of VAc $(k_{ex}(333K) = 7 \times 10^5 \, M^{-1} \, s^{-1})^{87}$ are large enough to account for the relatively low polydispersity. DFT calculations provide further support for a low activation energy pathway for radical exchange with 5-coordinate (por)Co-R complex.[87] The organo-cobalt porphyrin systems were the first transition metal complexes to provide direct measurements for exchange of radicals in solution with an organometallic derivative,[89,91] but observations of LRP controlled by DT mechanisms give indirect evidence for fast radical exchange with several other organo-transition metal complexes.[25,31,106]

Methyl acrylate (MA) polymerizations are used to illustrate the organo-cobalt mediated RT and DT processes (Figures 5.2, 5.5 and 5.7) that simultaneously operate in the (TMP)Co system. When the ratio of total moles of radicals injected into solution from the radical source to the initial moles of $(TMP)Co^{II\bullet}$ is less than unity, an RT process controls the LRP. Polymerizations controlled by an RT mechanism are relatively slow because the radical is maintained at a low concentration by a quasi-equilibrium between $(TMP)Co^{II\bullet}$ and the dormant complex (TMP)Co-P which fulfills Fukuda's criteria for a dissociation combination type of RT process.[13] The living character of this RT process benefits from the persistent radical effect.[6,15]

Radical polymerization data of MA for the condition where the total moles of radicals that enter solution from external radical sources like V-70 and AIBN exceeds the initial moles of $(TMP)Co^{II\bullet}$ are illustrated in Figures 5.5 and 5.7. An induction period where a few percent conversion of the monomer occurs is followed by the onset of rapid polymerization at the time when effectively all of the $(TMP)Co^{II\bullet}$ has been converted to an organo-cobalt complex ((TMP)Co-P). The rate of polymerization at this condition approaches the rate for an uncontrolled radical polymerization at the same concentration of V-70 or AIBN in the absence of $(TMP)Co^{II}$ (Figure 5.4) which fulfills the criteria for a degenerative transfer (DT) process.[6] The continued influx of radicals from V-70 or AIBN changes the polymerization process from a cobalt(II)-mediated RT pathway to an organo-cobalt-mediated DT mechanism.[6] The persistent radical effect ensures the living character for the polymer formed during the RT portion of the processes. The most important feature of the polymerization is that during the period of fast polymerization the process remains controlled and produces low polydispersity polymers where M_n increases linearly with monomer conversion. After a period of 90 minutes the influx of new radicals from V-70 has ended, and the polymerization reaction reverts to a slower RT pathway (Figure 5.4).

Low-polydispersity poly(methyl acrylate)–poly(butyl acrylate) block copolymers (PMA-b-PBA) are formed by initially building one block of PMA, replacing MA with BA under vacuum, and heating to 333 K to reinitiate polymerization. The linear increase in the number-average molecular mass (M_n) with conversion and low polydispersity $(M_w/M_n = 1.07)$ demonstrates the formation of a PMA-b-PBA block copolymer (Figure 5.10).

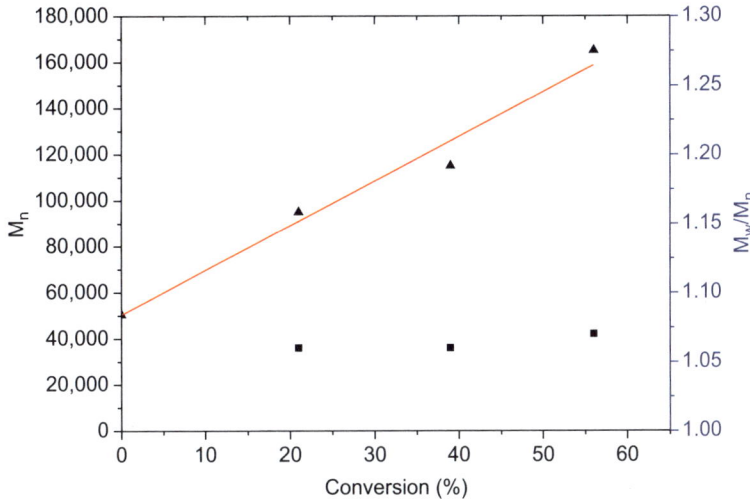

Figure 5.10 Change in number-average molecular weight and polydispersity with conversion in forming a block copolymer of PMA with PBA using a (TMP)Co-PMA macroinitiator ($M_n = 51\ 000$) and V-70 with BA (2.5 M) at 333 K in C_6D_6.
Reprinted with permission from ref. 88. Copyright 2011 American Chemical Society.

5.3.3.2 Reduction of Ligand Steric Demands for LRP by the DT Pathway

When the smaller steric requirement tetra(p-methoxyphenyl)-porphyrin derivative (TAP)Co^{II} is substituted for (TMP)Co^{II} as a mediator at the conditions for the RT polymerization of acrylates $(([R^{\bullet}{}_T]/[(TMP)Co^{II\bullet}{}_T]) = 0.9)$, the resulting polymers only grow to a maximum degree of polymerization of around 200 before there is unit probability for cobalt(II) catalytic chain transfer to occur through β-H abstraction and readdition to monomer (eqn (5.22) and (5.23)) (Figure 5.11A). Under the DT conditions of continual influx of radicals $(([R^{\bullet}{}_T]/[(TMP)Co^{II\bullet}{}_T]) = 1.1)$ the cobalt(II) is maintained at a sufficiently low concentration such that β-H abstraction is effectively quenched. The polymers grow much larger and have relatively small polydispersities (1.10–1.15) at the condition for a DT process (Figure 5.11 B,C). The effective absence of cobalt(II) metalloradicals during the organo-cobalt-mediated DT process removes the requirement for having sterically demanding complexes as an approach to suppress β-H abstraction from the growing oligomer/polymer radicals.

$$
\begin{aligned}
(TAP)Co^{II\bullet} + {}^{\bullet}CH(CH_2\text{-}P)Co_2CH_3 &\rightharpoonup (TAP)Co\text{-H} \\
+ (Co_2CH_3)(H)C=CH(P) &
\end{aligned}
\tag{5.22}
$$

$$
(TAP)Co\text{-H} + CH_2=CH(Co_2CH_3) \rightharpoonup (TAP)Co\text{-}CH(CH_3)Co_2CH_3 \tag{5.23}
$$

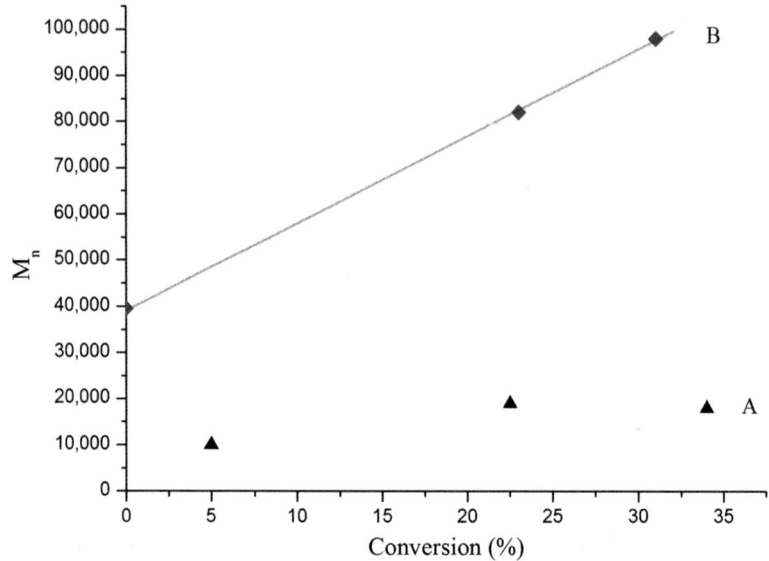

Figure 5.11   Changes in number-average molecular weight with conversion of MA to
PMA in an RT process using V-70 as the external radical source. Poly-
merization mediated by (A) (TAP)Co^{II}; and block copolymerization
mediated by (B) (TAP)Co-PtBA (PDI = 1.09, M_n = 3.86 × 10^4).
Reprinted with permission from ref. 6. Copyright 2011 American
Chemical Society.

5.3.4   Organo-cobalt Porphyrin Mediated Polymerization of Acrylic Acid in Water

Sulfonated tetraphenyl porphyrin derivatives provide a series of water-soluble
cobalt(II) metalloradical and organocobalt complexes (Figure 5.1) to
evaluate the efficacy of cobalt porphyrins in mediating the LRP of acrylates
in aqueous media.[89] The polymerization in water mediated by
(TMPS)Co-CH(CO$_2$H)CH$_2$P that is given in Figure 5.12 shows the same
general features as those observed for methyl acrylate (MA) in organic solvents
except that the polymerization is much faster in water. Kinetic plots for radical
polymerizations of acrylic acid (AA) in water (333 K) using V-70 and AIBN as
radical sources in the presence of cobalt tetra(3,5-disulfonatomesityl)porphyrin
((TMPS)Co^{II}) are shown in Figures 5.12 and 5.13.[89] The polymerization
process involves an induction period followed by a stage of relatively fast
AA polymerization during which radicals from V-70 continue to enter
solution. The measured M_n values are close to the theoretical values based
on one living chain per cobalt porphyrin. The observed conversion *versus*
time plots (Figure 5.12) and first-order kinetic plots (Figure 5.13) are
simulated by using experimental rate constants for radicals entering solution
from V-70 (k_i (333 K) = 1.54 × 10^{-3} s^{-1}), the propagation constant for AA

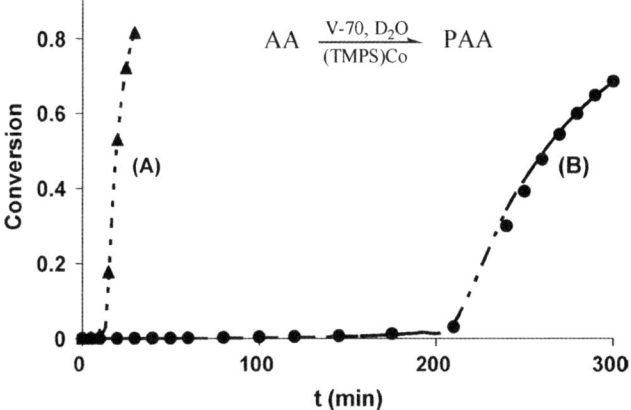

Figure 5.12 Monomer conversion as a function of time in the radical polymerization of acrylic acid (4.72 M) initiated by V-70 (1.68×10^{-3} M) in D_2O at (A) 333 K with $[(TMPS)Co^{II}]_i = 1.30 \times 10^{-3}$ M; conversion = 81.6%; $M_n = 232\,000$ (theory 212 000); $M_w/M_n = 1.30$ and (B) 313 K with $[(TMPS)Co^{II}]_i = 1.24 \times 10^{-3}$ M; conversion = 68.6%; $M_n = 195\,000$ (theory 187 000); $M_w/M_n = 1.33$.
Reprinted with permission from ref. 89. Copyright 2011 American Chemical Society.

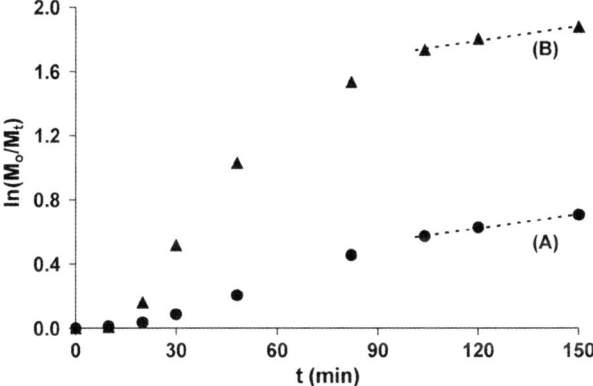

Figure 5.13 Kinetic plots for polymerization of acrylic acid (AA) in D_2O at 333 K with $[(TMPS)Co^{II}]_i = 1.82 \times 10^{-3}$ M, $[V-70]_i/[(TMPS)Co^{II}]_i = 1.30$, and two concentrations of $[AA]_i$: (A) $[AA]_i = 1.00$ M, conversion = 50.7%; $M_n = 18\,000$ (theory 20 000); $M_w/M_n = 1.41$; (B) $[AA]_i = 2.50$ M, conversion = 84.7%; $M_n = 77\,000$ (theory 84 000); $M_w/M_n = 1.38$.
Reprinted with permission from ref 89. Copyright 2011 American Chemical Society.

(k_p(333 K) = $1.28 \times 10^5 \, M^{-1} \, s^{-1}$)[89] and equilibrium constant for dissociation of (TMPS)Co-PAA (k_d(333 K) = 1×10^{-10}).[89] The induction periods shown in Figure 5.12 correspond closely with the time required for V-70 to inject one radical into solution for each $(por)Co^{II}$ molecule. A major contribution to the large observed polydispersities for PAA in water compared to PMA is the larger propagation constant ($M_w/M_n = 1 + (2/c - 1)/C_{ex}$, $C_{ex} = k_{ex}/k_p$).[39]

The AA polymerization using V-70 and (TMPS)Co-CH(CO$_2$CH$_3$)CH$_3$ at 333 K through the time where V-70 has fully decayed (90 minutes) is shown in Figure 5.13. At the end of the induction period effectively all of the (TMPS)Co$^{II\bullet}$ is converted to organocobalt complexes of the poly(acrylic acid) radical (TMPS)Co-PAA). Radicals continue to enter solution from V-70 which in the absence of cobalt(II) trapping reacts with the AA monomer to initiate radical polymerization (Figure 5.13). A period of fast radical polymerization then occurs in the presence of the cobalt-PAA complex ((TMPS)Co-PAA). The rate of polymerization from the end of the induction period through the stage of rapid polymerization is found to be proportional to the square root of the V-70 concentration ($[R^\bullet] = [(k_i/2k_t)[V-70]]^{1/2}$) which demonstrates that the external radical source (V-70) determines the radical concentration.

During the period of rapid polymerization the process must be controlled by the organocobalt complexes through a transition metal degenerative transfer (DT) that involves exchange of propagating radicals in solution and the dormant radicals in the organocobalt complexes. After the external radical source is fully reacted ($t > 90$ min, $T = 60\,^\circ$C) the system contains a dormant polymer organocobalt complex ((TMPS)Co-PAA). Reversible dissociation of the dormant complex provides a source of polymeric radicals (PAA$^\bullet$) that are maintained at a relatively low concentration by a quasi-equilibrium. Acrylic acid polymerization continues at a lower rate under control by reversible termination (RT) of the dormant organo-cobalt complex ((TMPS)Co-PAA). Observation of equal slopes for the two plots in Figure 5.13 at the same concentration of organometallic complex is definitive evidence for an RT mechanism. The AA polymerization controlled by the reversible termination mechanism attains living character through the persistent radical effect.

A representative plot for the number-average molecular weight and polydispersity as a function of monomer conversion for acrylic acid polymerization in the presence of (TMPS)Co complexes in water is given in Figure 5.14. The observed linear increase in number-average molecular weight (M_n) for PAA with conversion of acrylic acid indicates that there is a nearly constant number of chains that are growing at effectively the same rate which is a signature for high living character. The PAA polymers formed are highly linear as evidenced by the inability to detect the ^{13}C NMR resonances in the range ($\delta(^{13}$C)) 36–40 and 47–50 ppm) that are characteristic of acrylate branching. Chain transfer to solvent which could limit polymer growth is also quenched by using aqueous reaction media.

^{1}H NMR permits following the sequence of events in the (TMPS)Co mediated radical polymerization of AA in water (Figure 5.15).[89] The characteristic high field doublet methyl ($\delta = -4.91$ ppm) and C–H quartet ($\delta = -3.33$ ppm) resonances for the first organometallic that forms in the AA polymerization ((TMPS)Co-CH(CO$_2$H)CH$_3$) declines in intensity as insertion

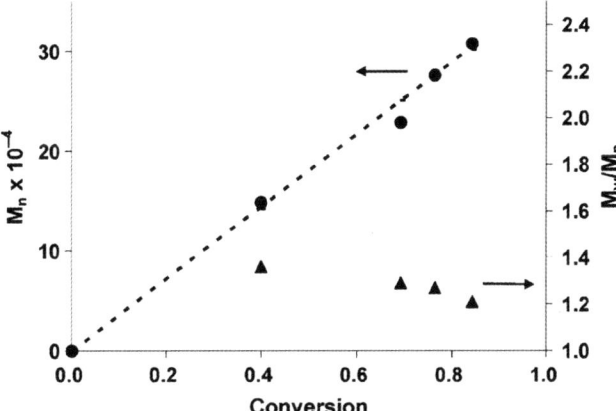

Figure 5.14 Change in the number-average molecular weight (M_n) and polydispersity (M_w/M_n) with acrylic acid (AA) with conversion of AA to PAA at 333 K. $[AA]_i = 4.72$ M; $[(TMPS)Co^{II}]_i = 1.14 \times 10^{-3}$ M; $[V\text{-}70]_i = 1.88 \times 10^{-3}$ M. The dotted line is the theoretical line for one polymer chain per organocobalt complex.
Reprinted with permission from ref. 89. Copyright 2011 American Chemical Society.

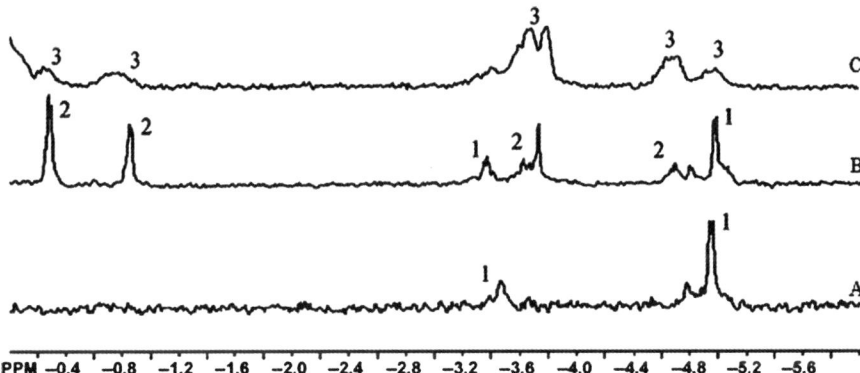

Figure 5.15 ^{1}H NMR showing the time evolution of diamagnetic organocobalt complexes in the aqueous polymerization of AA initiated by V-70 in water at 333 K: (1) (TMPS)Co-CH(CO$_2$H)CH$_3$; (2) (TMPS)Co-CH(CO$_2$H)CH$_2$-CH(CO$_2$H)CH$_3$; (3) (TMPS)Co-PAA.
Reprinted with permission from ref. 89. Copyright 2011 American Chemical Society.

of a second AA occurs ((TMPS)Co-CH-(CO$_2$H)CH$_2$-CH(CO$_2$H)CH$_3$) occurs (Figure 5.15). The pair of doublet resonances at $\delta = -0.28$, -0.84 ppm are assigned as the terminal methyl group for two diastereomers that arise from the two chiral centers in the AA dimer unit. Near the end of the induction period the AA dimer complex is replaced by organocobalt complexes of AA oligomers ((TMPS)Co-CH(CO$_2$H)CH$_2$-CH(CO$_2$H)-CH$_2$-PAA). ^{1}H NMR from -3 to

-5 ppm have six resonances that are characteristic of the first AA unit ((TMPS)Co-CH(CO$_2$H)CH$_2$-) that occur in two diastereomers. Chirality of the organo groups in (TMPS)Co-CH(CO$_2$H)CH$_3$ and (TMPS)Co-CH-(CO$_2$H)CH$_2$-CH(CO$_2$H)CH$_2$-PAA also makes the porphyrin pyrrole hydrogens diastereotopic and the magnetic anisotropy of the carboxylate groups results in observing a characteristic pyrrole AB pattern.[89]

High steric requirement water soluble (TMPS)Co complexes mediate LRP of AA by both DT and RT mechanisms in water. The surprising feature is that radical exchange between solution and organo-cobalt complex in aqueous solution is fast enough to have a DT mechanism be effective in the mediation of a LRP of AA. Coordination of water produces six-coordinate organo-cobalt complexes which should provide a barrier for radical exchange. The larger PDI for forming the PAA in water compared to PMA in hydrocarbon media may reflect both a reduced radical exchange rate and the observed larger propagation constant in water.

5.3.5 Organo-cobalt Porphyrin Mediated Radical Polymerization of Vinyl Acetate

Radicals derived from vinyl acetate ($^\bullet$C(OC(O)CH$_3$)(CH$_2$P)) have higher energy odd electron orbitals and smaller radical stabilization energy by approximately 5–7 kcal mol^{-1} than the values for radicals derived from methyl acrylate.[87] The π electron donating property of acetate (-OC(O)CH$_3$) in contrast to the π electron accepting property of carboxylate (-C(O)OCH$_3$) accounts for the substantially smaller propagation constant for VAc ($k_p(333\,\mathrm{K}) = 800$), and larger radical chain transfer constants compared to acrylate system. The C-H bond dissociation energy (BDE) for H-CH(OC(O)CH$_3$)CH$_3$ is also about 5–7 kcal mol^{-1} larger than the value for H-CH(CO$_2$CH$_3$)CH$_3$.

Substantial effort has been directed toward the control of vinyl acetate (VAc) radical polymerization using living radical polymerization (LRP) methods, including atom transfer radical polymerization,[107,108] degenerative transfer through alkyl iodide,[109] dialkyl tellurium,[13] trithiocarbonate,[110,111] xanthate,[112–114] and cobalt acetylacetonate OMRP.[26,115,116] The focus of interest on polyvinyl acetate (PVAc) stems from the facile conversion to polyvinyl alcohol which is a water soluble biocompatible polymer material. Radical polymerization of vinyl acetate mediated by organo-cobalt porphyrins provides an introduction to the main features of the vinyl acetate polymerization system.[91]

Kinetic plots for polymerization of vinyl acetate (VAc) initiated by azo radical sources (AIBN and V-70) and mediated by cobalt(II) tetramesityl porphyrin ((TMP)Co^{II}) in C$_6$D$_6$[91] (Figure 5.16) appear very similar to the corresponding plots for MA polymerization (Figure 5.7). During the induction period radicals that enter solution from the azo radical sources react and are trapped by (TMP)Co^{II} as the initially formed organo-cobalt complexes ((TMP)Co-CH(OC(O)CH$_3$)CH$_3$ in a manner that parallel the acrylate

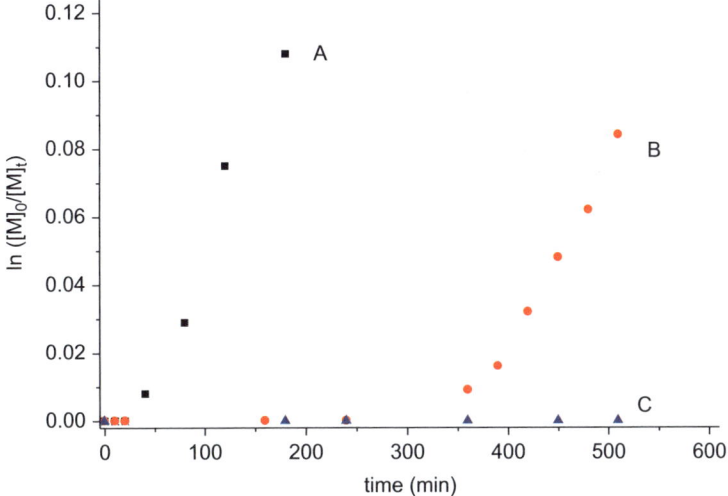

Figure 5.16 First order kinetics plots for the radical polymerization of vinyl acetate (VAc) mediated by cobalt(II) tetramesitylporphyrin ((TMP)Co^{II}) in C$_6$D$_6$ at 333 K: (A) [(TMP)Co^{II}]$_i$ = 6.9 × 10^{-4} M, [V-70]$_i$ = 8.0 × 10^{-4} M, and [VAc]$_i$ = 5.42 M, R$_{(total)}$/Co^{II}$_i$ = 1.4, conversion = 10.11%, PDI (M_w/M_n) = 1.13, M_n = 64 000 ($M_{n(theory)}$ = 68 000); (B) [(TMP)Co^{II}]$_i$ = 5.9 × 10^{-4} M, [AIBN]$_i$ = 7.6 × 10^{-3} M, and [VAc]$_i$ = 1.74 M, R$_{(total)}$/Co^{II}$_i$ = 3.8, conversion = 8.00%, PDI (M_w/M_n) = 1.22, M_n = 14 000 ($M_{n(theory)}$ = 20 000). (C) [(TMP)Co^{II}]$_i$ = 5.9 × 10^{-4} M, [V-70]$_i$ = 3.9 × 10^{-4} M, and [VAc]$_i$ = 1.54 M. Reprinted with permission from ref. 91. Copyright 2011 American Chemical Society.

reactions. At end of the induction period radicals continue to enter solution from the azo radical sources and then relatively rapid radical polymerization of VAc occurs (333 K) (Figure 5.17A, B). During the period of fast VAc polymerization, control of radical polymerization requires a degenerative transfer mechanism which depends on the exchange of radicals between the solution and the organo-cobalt transfer agent. When the number of radicals from the external radical source is insufficient to convert all the (TMP)Co^{II} to organo-cobalt complexes, as illustrated in Figure 5.16C, then this is the condition where the control of radical polymerization uses a reversible termination mechanism and the concentration of radicals in solution is derived exclusively from dissociation of organo-cobalt complex. The vanishingly small rate of polymerization from radicals that result from (TMP)Co-CH(CO$_2$CH$_3$)CH$_3$ bond homolysis (Figure 5.16C) places an upper limit of ~5 × 10^{-13} on the homolytic dissociation constant (K_{dis}(333K) < 5 × 10^{-13}; ΔG^0 > 18.7 kcal mol^{-1} for (TMP)Co-CH(OC(O)CH$_3$)CH$_2$R) for organo-cobalt complexes formed in the VAc polymerization system.

Both the energy for C-H abstraction and the organo-cobalt ((L)Co-CH(OC(O)CH$_3$)CH$_3$) BDE associated with the vinyl acetate system are expected to be about 5–7 kcal mol^{-1} larger than the corresponding values for methyl acrylate system. The BDE of the Co-C bond in (TAP)Co-CH(CO$_2$CH$_3$)CH$_3$ has been evaluated at 25 ± 1 kcal mol^{-1} and the (TMP)Co-CH(CO$_2$CH$_3$)CH$_3$ estimated at 23 kcal mol^{-1} which places the (TMP)Co-CH(OC(O)CH$_3$)CH$_3$ BDE at approximately 28–30 kcal mol^{-1}.

The VAc polymerization process mediated by organo-Co(TMP) complexes is observed to have living character during the period of relatively fast radical polymerization after the induction period (Figure 5.16A, B). At low vinyl acetate conversion, the number average molecular weight increases linearly with conversion and relatively small polydispersities are observed (Figure 5.17). The observed M_n values approach the theoretical values for one living chain per organo-Co(TMP) unit, but deviations toward lower molecular weight regularly increase as the conversion increases (Figure 5.17).

When the conversion is larger than 25% a shoulder appears on the GPC molecular weight distribution at approximately twice that of the most probable M_n which is interpreted as resulting from hydrogen abstraction from the VAc polymer chain and radical coupling. The vinyl acetate radical polymerization at 333 K thus has inherent non-ideal behavior that limit the precision of the controlled radical polymerization.

Block copolymers of vinyl acetate with acrylates are obtained by sequentially growing the acrylate block on (TMP)Co using AIBN as a radical source, removing the acrylate monomer and then adding vinyl acetate to grow the PVAc block. GPC traces for the polymethyl acrylate block attached with (TMP)Co ((TMP)Co-PMA) and the subsequent block copolymer with vinyl acetate ((TMP)Co-PVAc-*b*-PMA) are shown in Figure 5.18. Tailing of the

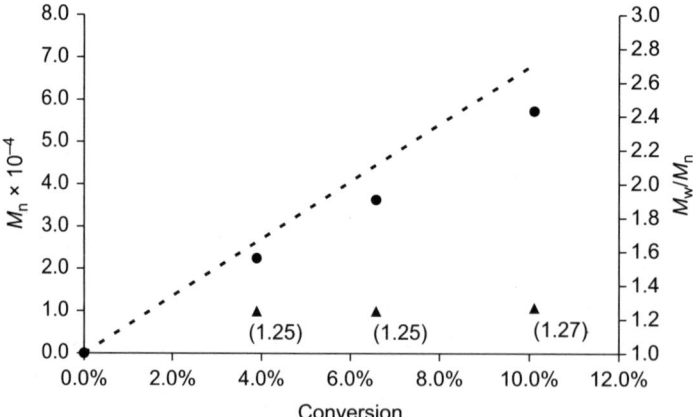

Figure 5.17 Change in the number average molecular weight (M_n) (●) and poly-dispersity (M_w/M_n) (▲) with conversion of vinyl acetate (VAc) to polyvinyl acetate (PVAc) at 333 K in C$_6$D$_6$. [VAc]$_i$ = 5.42 M; [(TMP)Co^{II}]$_i$ = 6.90 × 10^{-4} M; [V-70]$_i$ = 8.00 × 10^{-4} M. The dotted line is the theoretical line for one polymer chain per organo-cobalt complex. Reprinted with permission from ref. 91. Copyright 2011 American Chemical Society.

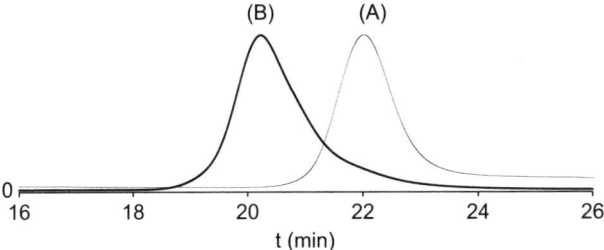

Figure 5.18 GPC traces showing formation of a polymethyl acrylate polyvinyl acetate block copolymer (PMA-*b*-PVAc) mediated by an organo-cobalt complex ((TMP)Co-P). (A) (TMP)Co-PMA block, $M_n = 30\,000$, PDI $(M_w/M_n) = 1.11$; (B) (PMA-*b*-PVAc), $M_n = 85\,000$, PDI $(M_w/M_n) = 1.21$. Reprinted with permission from ref. 91. Copyright 2011 American Chemical Society.

molecular weight distribution of the block copolymer reflects the declining living character with conversion.

Radical polymerization of vinyl acetate initiated by radicals from an azo source can be controlled by forming a dormant organometallic complex ((por)Co-VAc), and used in producing VAc homopolymers and block copolymers (PMA-b-PVAc) . However, the high energy odd electron and relatively small radical stabilization energy for radicals from VAc produce inherent hydrogen abstractions and chain transfer events that place limitations on the living character.

5.4 Cobalt(acac)$_2$ and related cobalt(II) complexes in mediating LRP

Bis acetylacetonate cobalt(II) (Co(acac)$_2$) and cobalt(II) porphyrins have an apparent similarity as four coordinate cobalt(II) complexes ,but (acac)$_2$Co^{II} and (por)Co^{II} are actually distinctly different classes of cobalt(II) complexes. Acetylacetonate (acac) is a mono anionic bidentate oxygen donor ligand compared to porphyrin as a tetrapyrrole nitrogen donor macrocycle. Structural flexibility and weaker cobalt–ligand bonding of the bidentate acetylacetonate ligand compared to the nitrogen donor macrocycles result in substantial structural, electronic and reactivity differences. Co(acac)$_2$ is high spin ($S = 3/2$) and has a structure (D_{2d}) based on a near tetrahedral array of oxygen donor sites[117] in contrast to the low spin planar (D_{4h}) nature of (por)Co^{II} complexes. The bidentate acetylacetonate ligands in Co(acac)$_2$ have facile structural rearrangements that are not available to complexes of the rigid planar aromatic porphyrin macrocycle. An important feature that Co(acac)$_2$ and (por)Co systems have in common is the formation of diamagnetic organo-cobalt(III) complexes that function as the dormant complex of the polymeric radical in the control living radical polymerization.[7,8] The d^7 high spin tetrahedral type structure for (acac)$_2$Co^{II} undergo large electronic and structural changes in the reaction with organic radicals (R•) to form diamagnetic organo-cobalt complexes ((acac)$_2$Co-R).[9] The reorganization energy for the reverse bond homolysis process has contributions from both the Co(II) metal-centered and organic carbon-centered radicals that reduce the magnitude of the Co-X bond dissociation energetics for (acac)$_2$Co-X complexes relative to (por)Co-X.

5.4.1 Organo-Co(acac)$_2$ Controlled LRP of Vinyl Acetate

The observation that Co(acac)$_2$ mediates a living free radical polymerization of vinyl acetate[25,26] constituted a major advance in OMRP that has stimulated an intense effort to obtain fundamental mechanistic understanding and to develop applications in polymer synthesis. A survey of several common monomers showed that Co(acac)$_2$ was capable of giving good control for VAc poly-merization initiated by V-70 at 30 °C but was not able to control the poly-merization of acrylate or styrene monomers.[25,26] Larger radical stabilization energies for MA and Sty compared to VAc resulted in weaker bonded organo-cobalt complexes ((acac)$_2$Co-R) that were too highly dissociated to provide useful control of the radical polymerization.

End group analysis of the PVAc showed that the VAc polymerization was initiated by reaction of the monomer with the radical from V-70 (•C(CN)(CH$_3$)CH$_2$C(CH$_3$)$_2$OCH$_3$)) and not by (acac)$_2$Co-H.[118,119] This obser-vation demonstrates that the small steric demand (acac)$_2$Co^{II} complex does not abstract a hydrogen atom from the radical (•C(CN)(CH$_3$)CH$_2$C(CH$_3$)$_2$OCH$_3$)) to produce (acac)$_2$Co-H. The inability of (acac)$_2$Co^{II} to abstract a H atom from •C(CN)(CH$_3$)R radicals most probably results from (acac)$_2$Co-H having a substantially smaller Co-H BDE compared to (por)Co-H as a consequence of the reorganization energy attributed to the intra-molecular change of configuration involved by two acac ligands.

Isolation and purification of a diamagnetic vinyl acetate oligomer organo-cobalt complex (acac)$_2$Co-PVAc) ((acac)$_2$Co-CH(OC(O)CH$_3$)CH$_2$-(CH(OC(O)CH$_2$)$_n$-C(CN)(CH$_3$)R) from the polymerization system opened the capability for more precise reproducible control of the VAc polymeriza-tion and detailed mechanistic studies.[118] Computational studies for (acac)$_2$Co-CH(OC(O)CH$_3$)CH$_3$ indicate that intra-molecular binding of the acetate carbonyl oxygen with the cobalt(III) center is a probable structural feature that contributes $\sim 6 \, \text{kcal} \, \text{mol}^{-1}$ to the effective Co-VAc bond energy.[7,9] The open coordination site adjacent to the Co-C bond that permits the intra-molecular carbonyl oxygen bonding to give a 5-membered ring is a conse-quence of the structural flexibility of the bidentate acetylacetonate oxygen donor ligands. In contrast with the flexible (acac)$_2$Co system, the organo-cobalt porphyrin vinyl acetate complex ((TMP)Co-CH(OC(O)CH$_3$)CH$_3$) is strictly 5-coordinate with only Co-C bonding because all of the sites *cis* to the Co-C bond are occupied by porphyrin pyrrole donors. The secondary intra-molecular binding of (acac)$_2$Co with the acetate carbonyl has a profound impact on the effective organo-cobalt dissociation energetics and also probably influences the reaction pathway for radical exchange.

The VAc oligomer organo-cobalt complex ((acac)$_2$Co-PVAc) when rigorously purified did not dissociate into radicals to a sufficient extent at 30 °C to reinitiate VAc polymerization at a measurable rate.[7] When the organo-Co bonding in (acac)$_2$Co-PVAc is reinforced by acetate carbonyl oxygen binding of $\sim$ 6 kcal mol^{-1} the effective homolytic dissociation energetics are too large to generate enough radicals to reinitiate VAc polymerization. The use of (acac)$_2$Co-PVAc in conjunction with V-70 at 303 K produced a controlled radical polymerization of VAc that most likely occurs by a degenerative transfer (DT) mechanism.[7] The purified VAc oligomer organo-cobalt complex ((acac)$_2$Co-PVAc) is activated by addition of water as evidenced by the observed re-initiation of a remarkably precise VAc living polymerization ($M_n = 150$ K; PDI $= 1.15$).[91] Displacement of the carbonyl oxygen by water entering the sixth coordination site produces a water bound derivative ((acac)$_2$Co-PVAc(H$_2$O)) that dissociates sufficiently into radicals to initiate polymerization of VAc at 303 K. When water or another donor molecule replaces the acetate carbonyl oxygen, only the organo-cobalt component of the Co-PVAc bond remains. The relatively weakly bonded organo-cobalt unit dissociates homolytically at 303 K to give sufficient radicals to reinitiate polymerization of VAc by a reversible termination (RT) mechanism.[7,9] A multiplicity of reaction pathways are possible for the Co(acac)$_2$ system by structural variability that are not accessible to cobalt porphyrins. Access to an array of reaction pathways is generally advantageous for producing reactivity including polymerization, but it makes acquiring precise mechanistic information more challenging.

Co(acac)$_2$ neither abstracts hydrogen from the radical produced by V-70 (•C(CN)(CH$_3$)CH$_2$C(OCH$_3$)(CH$_3$)$_2$)$_2$) to form (acac)$_2$Co-H nor forms the organometallics ((acac)$_2$Co-CH(CH$_2$CH$_3$)CH$_2$R) with a large enough Co-R BDE to mediate a LRP for methyl acrylate which are prominent reactions observed for (por)Co^{II} complexes. The radical derived by adding a hydrogen atom to vinyl acetate (•CH(OC(O)CH$_3$)CH$_3$) is about 5–7 kcal mol^{-1} less stabilized than the corresponding radical from methyl acrylate (•CH(CO$_2$CH$_3$)CH$_3$). This translates into Co-C bond dissociation energies for cobalt-organo vinyl acetate complexes ((L)Co-CH(OC(O)CH$_3$)CH$_3$) that are substantially larger than those for the corresponding organo-cobalt methyl acrylate complexes ((L)Co-CH(CO$_2$CH$_3$)CH$_3$).

Several factors that are specific to the (acac)$_2$Co system, including intramolecular acetate carbonyl bonding along with spin state and structural reorganization energies, contribute to the distinct features manifested in controlling the LRP of VAc. bis-Acetylacetonate cobalt(II) and organo-cobalt derivatives ((acac)$_2$Co-VAc) are particularly effective in obtaining relatively high molecular weight low PDI vinyl acetate which has been a long standing objective in polymer materials. Mechanistic knowledge for OMRP processes has guided the efforts to expand the controlled polymerization to other classes of monomers including acrylates,[106] acrylonitrile[120] and vinyl pyrrolidone homo and block copolymers.[121] The beta-diketonate class of bidentate ligands provides substantial capability for tuning of the ligand steric demands and donor properties through substitution of organic groups for the methyls of acac which may be used in further expanding applications in OMRP.[27,29,122]

5.4.2 Organo-Co(acac)$_2$ Controlled LRP of Acrylate Monomers

Application of Co(acac)$_2$ in the successful mediation of LRP for acrylate monomers and preparation of block copolymers with vinyl acetate, substantially broadens the capability for polymeric materials synthesis.[123] Initial efforts to use stoichiometric quantities of (acac)$_2$Co to control acrylate radical polymerization were unsuccessful because the extent of the organo-cobalt homolytic dissociation was too large and (acac)$_2$Co^{II} inefficient in trapping the polymeric radicals. Knowledge of the controlled radical polymerization mechanisms resulted in using large excesses of (acac)$_2$Co^{II} as a successful strategy to maintain the polymeric radicals in the dormant organo-cobalt complex form. Success of this strategy to use excess (acac)$_2$Co^{II} is also dependent on the inability for (acac)$_2$Co^{II} to abstract hydrogen from the initiator and polymer radicals which contrasts with the propensity of (por)Co^{II} for hydrogen abstraction and catalytic chain transfer.

5.5 Scope of Cobalt(II) and Organo-cobalt Complexes with Nitrogen and Oxygen Donors in OMRP

Six different series of cobalt(II) complexes that have been evaluated as LRP mediators are shown in Scheme 5.4. Several of the important features for a specific OMRP system include electronic effects that influence M-C and M-H bond energetics, steric effects on access to the metal center for H-abstraction from radicals and M-C bonding, and the mechanistic capability to have reversible termination (RT) and degenerative transfer (DT) operate in mediating an LRP process.

   The types of cobalt(II) complexes shown in Scheme 5.4 provide opportunities to probe the influence of geometric structure, coordination number, donor atom types, electron configuration and spin state on the LRP mediation process. Planar cobalt(II) complexes that involve coordination of two or four nitrogen donor sites occur as low spin ($S = 1/2$) cobalt(II) metallo radicals,[6,106] while the all oxygen (O$_4$) and N$_2$O$_2$ complexes that have structures based on a tetrahedral arrays are high spin ($S = 3/2$).[25,118] The organo-cobalt bond dissociation energies reflect the structures and electronic reorganization energies associated with forming low spin ($S = 0$) organo-Co^{III} derivatives. The porphyrin tetrapyrrole and planar tetra-imine donor ligands (I,II) produce low spin ($S = 1/2$) cobalt(II) complexes, which havethe unpaired electron in dz^2,and manifest the smallest reorganization energies in forming diamagnetic Co-R and Co-H complexes and thus the largest Co-X bond energies.[6] Four coordinate salen type di-imine ligands (III) give low spin cobalt(II) complexes, but the unpaired electron is in the *dyz* orbital. The reorganization energy to transfer an electron from the dyz to the dz^2 configuration produces smaller M-C and M-H

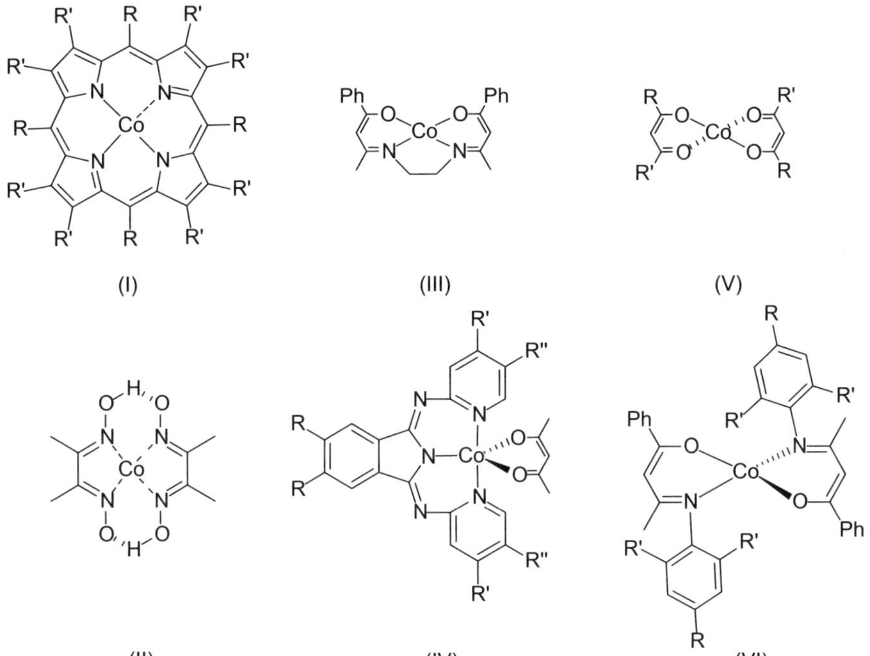

Scheme 5.4 Representative cobalt complexes used in OMRP.

bond dissociation energies. High spin tetrahedral cobalt(II) complexes, like Co(acac)$_2$,[25,26,123] experience the largest reorganization energies and have the smallest M-C and M-H bond dissociation energies. Steric effects are probably responsible for the five coordinate complex (IV) not giving hydrogen abstraction from the radicals formed by AIBN and V-70 and in contrast it is the unfavorable thermodynamics from a small Co-H bond energy that inhibits hydrogen abstraction from radicals by Co(acac)$_2$. Five coordinate organo-cobalt (Co-R) derivatives are capable of mediating LRP by both RT and DT pathways, but six coordinate donor adducts of organo-cobalt(III) complexes can only use the dissociative RT mechanism for LRP because the exchange with radicals is slow.

5.6 Prospects for OMRP

Experimental studies for explicit reactions of organic radicals with transition metal complexes is an under-investigated area of inorganic/organometallic/catalysis chemistry. Early studies by Meyerstein,[124–126] Halpern,[127,128] Espensen and Bakac[129] provided some reference points for the range of expected behavior, and this area is currently attracting more attention from both polymer and transition metal inorganic chemists because of their central role in controlled radical polymerization (CRP).[8,9,39,49] A recent review by Poli describes the first effort to provide a comprehensive systematic classification for reactions of metal complexes with organic radicals and related one-electron processes.[14] Availability of this thoughtfully constructed general review should contribute to expanding investigations in this area by identifying classes of complexes where opportunities for discovery abound. Identification of probable candidates for experimental studies of a particular type of reactivity from among the vast array of metal–ligand combinations, organic radicals, and reaction media will no doubt greatly benefit from computational studies that can survey trends in behavior much more quickly than experiments. The efficacy of computations in OMRP studies is illustrated by recent applications in processes that involve cobalt and chromium complexes.[40,118]

The relative significance for a synthetic method such as OMRP is based on the scope and importance of the materials that are best prepared using that approach. At present, specific types of OMRP are clearly best for homo and block copolymers of vinyl acetate, and thus vinyl alcohol[72,73] and OMRP may also be preferred for isoprene[64] and acrylic acid.[89] In the longer range perspective, the construction of copolymers from monomers with and without functional groups will require catalyst materials that can mediate both living radical polymerization and another type of living polymerization such as ionic or insertion mechanisms. The virtually limitless capability for tuning behavior by choice of metal–ligand combinations could eventually make OMRP the preferred method for LRP in obtaining catalysts that have versatility in controlling several forms of living polymerizations for both polar and nonpolar olefin monomers. Degenerative transfer (DT) LRP provides the greatest versatility for applications to a wide range of monomers and these processes involve the virtually unexplored area of one-electron (S_H2) exchange reactions at transition metal centers. A major challenge in organometallic chemistry is to develop new classes of organo-metal complexes that are coordinately and electronically unsaturated to promote fast exchange with organic radicals. Organo-metal mediated living radical polymerization has expanded to encompass metal complexes of titanium,[30–33] vanadium,[34] chromium,[40,41] iron[35–39] and molybdenum,[42–44] nickel[11] and palladium[45–48] where closely related metal complex derivatives are also known to mediate living alkene polymerizations. Imminent development of catalysts that can mediate living radical and one or more other forms of living polymerization will provide a new dimension for design and synthesis of currently attractive polymer architectures that can be expected to have important applications in technology and medicine.

Acknowledgements

This work was supported by NSF on grant NSFCHE-0501198.

References

1. G. W. Coates, *Chem. Rev.*, 2000, **100**, 1223–1252.
2. G. J. Domski, J. M. Rose, G. W. Coates, A. D. Bolig and M. Brookhart, *Progr. Polym. Sci.*, 2007, **32**, 30–92.
3. B. J. Burger, M. E. Thompson, W. D. Cotter and J. E. Bercaw, *J. Am. Chem. Soc.*, 1990, **112**, 1566–1577.
4. M. Brookhart, F. C. Rix, J. M. DeSimone and J. C. Barborak, *J. Am. Chem. Soc.*, 1992, **114**, 5894–5895.
5. B. B. Wayland, G. Poszmik, S. L. Mukerjee and M. Fryd, *J. Am. Chem. Soc.*, 1994, **116**, 7943–7944.
6. B. B. Wayland, C.-H. Peng, X. Fu, Z. Lu and M. Fryd, *Macromolecules*, 2006, **39**, 8219–8222.
7. A. Debuigne, R. Poli, C. Jerome, R. Jerome and C. Detrembleur, *Progr. Polym. Sci.*, 2009, **34**, 211–239.
8. M. Hurtgen, C. Detrembleur, C. Jerome and A. Debuigne, *Polym. Rev.*, 2011, **51**, 188–213.
9. R. Poli, *Eur. J. Inorg. Chem.*, 2011, **2011**, 1513–1530.
10. K. Matyjaszewski, *ACS Symp. Ser.*, 2009, **1023**, 3–13.
11. A. Leblanc, E. Grau, J.-P. Broyer, C. Boisson, R. Spitz and V. Monteil, *Macromolecules*, 2011, **44**, 3293–3301.
12. A. Goto, Y. Kwak, T. Fukuda, S. Yamago, K. Iida, M. Nakajima and J.-i. Yoshida, *J. Am. Chem. Soc.*, 2003, **125**, 8720–8721.
13. Y. Kwak, A. Goto, T. Fukuda, Y. Kobayashi and S. Yamago, *Macromolecules*, 2006, **39**, 4671–4679.
14. R. Poli, *Angew. Chem. Int. Ed.*, 2006, **45**, 5058–5070.
15. H. Fischer, *Chem. Rev.*, 2001, **101**, 3581–3610.
16. B. E. Daikh and R. G. Finke, *J. Am. Chem. Soc.*, 1992, **114**, 2938–2943.
17. K.-S. Focsaneanu and J. C. Scaiano, *Helv. Chim. Acta*, 2006, **89**, 2473–2482.
18. A. Theis, T. P. Davis, M. H. Stenzel and C. Barner-Kowollik, *Macromolecules*, 2005, **38**, 10323–10327.
19. S. Yamago, K. Iida and J.-i. Yoshida, *J. Am. Chem. Soc.*, 2002, **124**, 2874–2875.
20. S. Yamago, Y. Ukai, A. Matsumoto and Y. Nakamura, *J. Am. Chem. Soc.*, 2009, **131**, 2100–2101.
21. P. Lacroix-Desmazes, R. Severac and B. Boutevin, *Macromolecules*, 2005, **38**, 6299–6309.
22. J. Tonnar, P. Lacroix-Desmazes and B. Boutevin, *Macromolecules*, 2006, **40**, 186–190.
23. G. Moad, E. Rizzardo and S. H. Thang, *Ac. Chem. Res.*, 2008, **41**, 1133–1142.
24. D. Arvanitopoulos Labros, P. Greuel Michael, M. King Brian, K. Shim Anne and H. J. Harwood, in *Controlled Radical Polymerization*, American Chemical Society, 1998, vol. 685 , pp. 316–331.

25. A. Debuigne, J.-R. Caille, C. Detrembleur and R. Jérôme, *Angew. Chem., Int. Ed.*, 2005, **44**, 3439–3442.
26. A. Debuigne, J.-R. Caille and R. Jérôme, *Angew. Chem., Int. Ed.*, 2005, **44**, 1101–1104.
27. H. Kaneyoshi and K. Matyjaszewski, *Macromolecules*, 2005, **38**, 8163–8169.
28. H. Kaneyoshi and K. Matyjaszewski, *Macromolecules*, 2006, **39**, 2757–2763.
29. K. S. S. Kumar, Y. Gnanou, Y. Champouret, J.-C. Daran and R. Poli, *Chem–Eur. J.*, 2009, **15**, 4874–4885.
30. A. D. Asandei and I. W. Moran, *J. Am. Chem. Soc.*, 2004, **126**, 15932–15933.
31. A. D. Asandei, I. W. Moran, G. Saha and Y. Chen, *J. Polym. Sci. Part A: Polym. Chem.y*, 2006, **44**, 2156–2165.
32. D. Asandei Alexandru, P. Simpson Christopher, S. Yu Hyun, I. Adebolu Olumide, G. Saha and Y. Chen, in *Controlled/Living Radical Polymerization: Progress in RAFT, DT, NMP & OMRP*, American Chemical Society, 2009, vol. 1024, pp. 149–163.
33. D. F. Grishin, S. K. Ignatov, A. A. Shchepalov and A. G. Razuvaev, *Appl. Organomet. Chem.*, 2004, **18**, 271–276.
34. M. P. Shaver, M. E. Hanhan and M. R. Jones, *Chem. Commun.*, 2010, **46**, 2127–2129.
35. L. E. N. Allan, M. P. Shaver, A. J. P. White and V. C. Gibson, *Inorg. Chem.*, 2007, **46**, 8963–8970.
36. M. P. Shaver, L. E. N. Allan and V. C. Gibson, *Organometallics*, 2007, **26**, 4725–4730.
37. M. P. Shaver, L. E. N. Allan, H. S. Rzepa and V. C. Gibson, *Angew. Chem., Int. Ed.*, 2006, **45**, 1241–1244.
38. R. Poli and Z. Xue, *Polym. Prepr. (Am. Chem. Soc., Div. Polym. Chem.)*, 2011, **52**, 572–573.
39. L. E. N. Allan, M. R. Perry and M. P. Shaver, *Progr. Polym. Sci.*, 2012, **37**, 127–156.
40. Y. Champouret, U. Baisch, R. Poli, L. Tang, J. L. Conway and K. M. Smith, *Angew. Chem. Int. Ed.*, 2008, **47**, 6069–6072.
41. Y. Champouret, K. C. MacLeod, U. Baisch, B. O. Patrick, K. M. Smith and R. Poli, *Organometallics*, 2009, **29**, 167–176.
42. E. Le Grognec, J. Claverie and R. Poli, *J. Am. Chem. Soc.*, 2001, **123**, 9513–9524.
43. S. Maria, F. Stoffelbach, J. Mata, J.-C. Daran, P. Richard and R. Poli, *J. Am. Chem. Soc.*, 2005, **127**, 5946–5956.
44. F. Stoffelbach, R. Poli and P. Richard, *J. Organomet. Chem.*, 2002, **663**, 269–276.
45. A. C. Albéniz, P. Espinet and R. López-Fernández, *Organometallics*, 2003, **22**, 4206–4212.
46. C. Elia, S. Elyashiv-Barad, A. Sen, R. López-Fernández, A. C. Albéniz and P. Espinet, *Organometallics*, 2002, **21**, 4249–4256.

47. E. Szuromi, H. Shen, B. L. Goodall and R. F. Jordan, *Organometallics*, 2008, **27**, 402–409.
48. G. Tian, H. W. Boone and B. M. Novak, *Macromolecules*, 2001, **34**, 7656–7663.
49. F. di Lena and K. Matyjaszewski, *Progr. Polym. Sci.*, 2010, **35**, 959–1021.
50. W. A. Braunecker and K. Matyjaszewski, *Progr. Polym. Sci.*, 2007, **32**, 93–146.
51. C. J. Hawker, A. W. Bosman and E. Harth, *Chem. Rev.*, 2001, **101**, 3661–3688.
52. M. Kamigaito, T. Ando and M. Sawamoto, *Chem. Rev.*, 2001, **101**, 3689–3746.
53. E. Kayahara, S. Yamago, Y. Kwak and T. Fukuda, *Macromolecules*, 2008, **41**, 527–529.
54. A. B. Lowe and C. L. McCormick, *Progr. Polym. Sci.*, 2007, **32**, 283–351.
55. K. Matyjaszewski and J. Xia, *Chem. Rev.*, 2001, **101**, 2921–2990.
56. G. Moad, E. Rizzardo and S. H. Thang, *Polymer*, 2008, **49**, 1079–1131.
57. M. Ouchi, T. Terashima and M. Sawamoto, *Acc. Chem. Res.*, 2008, **41**, 1120–1132.
58. Y. Sugihara, Y. Kagawa, S. Yamago and M. Okubo, *Macromolecules*, 2007, **40**, 9208–9211.
59. S. Yamago, K. Iida and J.-i. Yoshida, *J. Am. Chem. Soc.*, 2002, **124**, 13666–13667.
60. S. Yamago, E. Kayahara, M. Kotani, B. Ray, Y. Kwak, A. Goto and T. Fukuda, *Angew. Chem., Int. Ed.*, 2007, **46**, 1304–1306.
61. S. Yamago, *Chem. Rev.*, 2009, **109**, 5051–5068.
62. A. D. Asandei and Y. Chen, *Macromolecules*, 2006, **39**, 7549–7554.
63. A. D. Asandei, Y. Chen, C. Simpson, M. Gilbert and I. W. Moran, *Polym. Prepr. (Am. Chem. Soc., Div. Polym. Chem.)*, 2008, **49**, 489–490.
64. A. D. Asandei, C. P. Simpson, H. S. Yu, O. I. Adebolu, G. Saha and Y. Chen, *ACS Symp. Ser.*, 2009, **1024**, 149–163.
65. A. D. Asandei, Y. Chen, G. Saha and I. W. Moran, *Tetrahedron*, 2008, **64**, 11831–11838.
66. A. D. Asandei, G. Saha and A. Ranade, *Polym. Prepr. (Am. Chem. Soc., Div. Polym. Chem.)*, 2006, **47**, 501–502.
67. R. Bryaskova, D. Pencheva, M. Kyulavska, D. Bozukova, A. Debuigne and C. Detrembleur, *J. Colloid Interface Sci.*, 2010, **344**, 424–428.
68. R. Bryaskova, N. Willet, P. Degee, P. Dubois, R. Jerome and C. Detrembleur, *J. Polym. Sci., Part A: Polym. Chem.*, 2007, **45**, 2532–2542.
69. R. Bryaskova, N. Willet, A.-S. Duwez, A. Debuigne, L. Lepot, B. Gilbert, C. Jerome, R. Jerome and C. Detrembleur, *Chem.–Asian J.*, 2009, **4**, 1338–1345.
70. A. M. Brzozowska and Q. Zhang, K. A. de, W. Norde and M. A. C. Stuart, *Colloids Surf., A*, 2010, **368**, 96–104.
71. E. Chiellini, A. Corti, S. D'Antone and R. Solaro, *Prog. Polym. Sci.*, 2003, **28**, 963–1014.

72. A. Debuigne, J.-R. Caille and R. Jerome, *Macromolecules*, 2005, **38**, 5452–5458.
73. A. Debuigne, J.-R. Caille, N. Willet and R. Jerome, *Macromolecules*, 2005, **38**, 9488–9496.
74. A. Debuigne, J. Warnant, R. Jerome, I. Voets, K. A. de, S. M. A. Cohen and C. Detrembleur, *Macromolecules*, 2008, **41**, 2353–2360.
75. C. Detrembleur, A. Debuigne, C. Jerome, T. N. T. Phan, D. Bertin and D. Gigmes, *Macromolecules*, 2009, **42**, 8604–8607.
76. C. Detrembleur, O. Stoilova, R. Bryaskova, A. Debuigne, A. Mouithys-Mickalad and R. Jerome, *Macromol. Rapid Commun.*, 2006, **27**, 498–504.
77. N. Georgieva, R. Bryaskova, A. Debuigne and C. Detrembleur, *J. Appl. Polym. Sci.*, 2010, **116**, 2970–2975.
78. M. Hurtgen, A. Debuigne, A. Mouithys-Mickalad, R. Jerome, C. Jerome and C. Detrembleur, *Chem.–Asian J.*, 2010, **5**, 859–868.
79. H. J. Jeon, Y. C. You and J. H. Youk, *J. Polym. Sci., Part A: Polym. Chem.*, 2009, **47**, 3078–3085.
80. H. J. Jeon and J. H. Youk, *Macromolecules*, 2010, **43**, 2184–2189.
81. J. R. Morton, K. F. Preston, P. J. Krusic, S. A. Hill and E. Wasserman, *J. Am. Chem. Soc.*, 1992, **114**, 5454–5455.
82. M. Mumtaz, E. Ibarboure, C. Labrugere, E. Cloutet and H. Cramail, *Macromolecules*, 2008, **41**, 8964–8970.
83. M. Mumtaz, C. Labrugere, E. Cloutet and H. Cramail, *J. Polym. Sci., Part A: Polym. Chem.*, 2010, **48**, 3841–3855.
84. J.-M. Thomassin, A. Debuigne, C. Jerome and C. Detrembleur, *Polymer*, 2010, **51**, 2965–2971.
85. J.-M. Thomassin, I. Molenberg, I. Huynen, A. Debuigne, M. Alexandre, C. Jerome and C. Detrembleur, *Chem. Commun.*, 2010, **46**, 3330–3332.
86. J.-M. Thomassin, C. Pagnoulle, L. Bednarz, I. Huynen, R. Jerome and C. Detrembleur, *J. Mater. Chem.*, 2008, **18**, 792–796.
87. S. Li, B. d. Bruin, C.-H. Peng, M. Fryd and B. B. Wayland, *J. Am. Chem. Soc.*, 2008, **130**, 13373–13381.
88. Z. Lu, M. Fryd and B. B. Wayland, *Macromolecules*, 2004, **37**, 2686–2687.
89. C.-H. Peng, M. Fryd and B. B. Wayland, *Macromolecules*, 2007, **40**, 6814–6819.
90. C.-H. Peng, S. Li and B. B. Wayland, *Inorg. Chem.*, 2009, **48**, 5039–5046.
91. C.-H. Peng, J. Scricco, S. Li, M. Fryd and B. B. Wayland, *Macromolecules*, 2008, **41**, 2368–2373.
92. B. B. Wayland, L. Basickes, S. Mukerjee, M. Wei and M. Fryd, *Macromolecules*, 1997, **30**, 8109–8112.
93. F. T. T. Ng and G. L. Rempel, *J. Am. Chem. Soc.*, 1982, **104**, 621–623.
94. F. T. T. Ng, G. L. Rempel and J. Halpern, *Inorg. Chim. Acta*, 1983, **77**, L165–L166.
95. D. C. Woska, Z. D. Xie, A. A. Gridnev, S. D. Ittel, M. Fryd and B. B. Wayland, *J. Am. Chem. Soc.*, 1996, **118**, 9102–9109.

96. J. P. Collman, K. E. Berg, C. J. Sunderland, A. Aukauloo, M. A. Vance and E. I. Solomon, *Inorg. Chem.*, 2002, **41**, 6583–6596.
97. A. A. Gridnev and S. D. Ittel, *Chem. Rev.*, 2001, **101**, 3611–3660.
98. A. A. Gridnev, S. D. Ittel, B. B. Wayland and M. Fryd, *Organometallics*, 1996, **15**, 5116–5126.
99. A. A. Gridnev, S. D. Ittel, M. Fryd and B. B. Wayland, *Organometallics*, 1993, **12**, 4871–4880.
100. A. A. Gridnev, S. D. Ittel, M. Fryd and B. B. Wayland, *Organometallics*, 1996, **15**, 222–235.
101. B. de Bruin, W. I. Dzik, S. Li and B. B. Wayland, *Chem.–Eur. J.*, 2009, **15**, 4312–4320.
102. J. Chiefari, Y. K. Chong, F. Ercole, J. Krstina, J. Jeffery, T. P. T. Le, R. T. A. Mayadunne, G. F. Meijs, C. L. Moad, G. Moad, E. Rizzardo and S. H. Thang, *Macromolecules*, 1998, **31**, 5559–5562.
103. A. Goto, K. Sato, Y. Tsujii, T. Fukuda, G. Moad, E. Rizzardo and S. H. Thang, *Macromolecules*, 2001, **34**, 402–408.
104. A. Goto, K. Ohno and T. Fukuda, *Macromolecules*, 1998, **31**, 2809–2814.
105. Y. Kwak, A. Goto, T. Fukuda, S. Yamago and B. Ray, *Z. Phys. Chem.*, 2005, **219**, 283–293.
106. R. K. Sherwood, C. L. Kent, B. O. Patrick and W. S. McNeil, *Chem.l Commun.*, 2010, **46**, 2456–2458.
107. H. Kaneyoshi and K. Matyjaszewski, *J. Polym. Sci. Part A: Polym. Chemi.*, 2007, **45**, 447–459.
108. M. Wakioka, K.-Y. Baek, T. Ando, M. Kamigaito and M. Sawamoto, *Macromolecules*, 2001, **35**, 330–333.
109. M. C. Iovu and K. Matyjaszewski, *Macromolecules*, 2003, **36**, 9346–9354.
110. A. Postma, T. P. Davis, G. Li, G. Moad and M. S. O'Shea, *Macromolecules*, 2006, **39**, 5307–5318.
111. S. R. S. Ting, A. M. Granville, D. Quémener, T. P. Davis, M. H. Stenzel and C. Barner-Kowollik, *Aus. J. Chem.*, 2007, **60**, 405–409.
112. M. Destarac, D. Charmot, X. Franck and S. Z. Zard, *Macromol. Rapid Commun.*, 2000, **21**, 1035–1039.
113. D. Charmot, P. Corpart, H. Adam, S. Z. Zard, T. Biadatti and G. Bouhadir, *Macromol. Symp.*, 2000, **150**, 23–32.
114. A. Theis, T. P. Davis, M. H. Stenzel and C. Barner-Kowollik, *Polymer*, 2006, **47**, 999–1010.
115. A. Debuigne, N. Willet, R. Jérôme and C. Detrembleur, *Macromolecules*, 2007, **40**, 7111–7118.
116. S. Maria, H. Kaneyoshi, K. Matyjaszewski and R. Poli, *Chem.–Eur. J.*, 2007, **13**, 2480–2492.
117. F. A. Cotton and R. H. Soderberg, *Inorg. Chem.*, 1964, **3**, 1–5.
118. A. Debuigne, Y. Champouret, R. Jerome, R. Poli and C. Detrembleur, *Chem.–Eur. J*, 2008, **14**, 4046–4059.
119. S. Maria, H. Kaneyoshi, K. Matyjaszewski and R. Poli, *Chem.–Eur. J.*, 2007, **13**, 2480–2492.

120. A. Debuigne, C. Michaux, C. Jerome, R. Jerome, R. Poli and C. Detrembleur, *Chem.–Eur. J.*, 2008, **14**, 7623–7637.
121. C. Detrembleur, M. Hurtgen, Y. Piette, C. Jerome and A. Debuigne, *Polym. Prepr. (Am. Chem. Soc., Div. Polym. Chem.)*, 2011, **52**, 442–443.
122. H. Kaneyoshi and K. Matyjaszewski, *Macromolecules*, 2006, **39**, 2757–2763.
123. M. Hurtgen, A. Debuigne, C. Jerome and C. Detrembleur, *Macromolecules (Washington, DC, U. S.)*, 2010, **43**, 886–894.
124. H. Cohen and D. Meyerstein, *Inorg. Chem.*, 1986, **25**, 1505–1506.
125. M. Masarwa, H. Cohen and D. Meyerstein, *Inorg. Chem.*, 1986, **25**, 4897–4900.
126. N. Navon, G. Golub, H. Cohen and D. Meyerstein, *Organometallics*, 1995, **14**, 5670–5676.
127. F. T. T. Ng and G. L. Rempel, *J. Am. Chem. Soc.*, 1982, **104**, 621–623.
128. F. T. T. Ng, G. L. Rempel and J. Halpern, *Inorg. Chim. Acta*, 1983, **77**, L165–L166.
129. D. G. Kelley, A. Marchaj, A. Bakac and J. H. Espenson, *J. Am. Chem. Soc.*, 1991, **113**, 7583–7587.

CHAPTER 6

Fundamentals of RAFT Polymerization

GRAEME MOAD,* EZIO RIZZARDO AND
SAN H. THANG

CSIRO Materials Science and Engineering, Bayview Ave, Clayton,
Victoria 3168, Australia
*Email: graeme.moad@csiro.au

6.1   Introduction

Polymerization with reversible addition–fragmentation chain transfer (RAFT) is a reversible deactivation radical polymerization (RDRP);[1] a process which, with appropriate attention to reagents and reaction conditions, can possess the attributes usually associated with living polymerization. These processes are also often called living radical polymerizations or controlled radical poly-merizations. However, the use of these terms in this context is now discouraged.[1]

   In an ideal living polymerization, all chains are initiated, grow at the same rate and survive polymerization; there is no termination or other side reactions. In radical polymerization, all chains cannot be simultaneously active because of the propensity of the propagating radicals to undergo self-reaction or reaction with other radicals by combination or disproportionation. In RDRP living characteristics are provided through reagents that react with the propagating radicals ($P_n^\bullet$) by reversible coupling or reversible chain transfer so as to maintain the majority of living chains in a dormant form (P_n-X).[2]

RSC Polymer Chemistry Series No. 4
Fundamentals of Controlled/Living Radical Polymerization
Edited by Nicolay V Tsarevsky and Brent S Sumerlin
© The Royal Society of Chemistry 2013
Published by the Royal Society of Chemistry, www.rsc.org

The instantaneous concentration of active propagating species in RDRP can be similar to that for the conventional process and under these conditions termination events will occur with similar frequency. However, for a well-controlled polymerization, the target molar mass must be very much less than that for the similar conventional process. Rapid equilibration between the active and dormant polymer ensures that all chains possess an equal chance for growth and that essentially all chains continue to grow, albeit intermittently. This provides for a well-defined molar mass that increases linearly with conversion and a low dispersity (*e.g.*, $Đ = M_w/M_n < 1.1$).

The first RDRP technique to receive widespread attention, nitroxide-mediated polymerization (NMP),[3,4] was developed in our laboratories in the mid 1980's,[5] though the method did not gain popularity until 1993, when Georges *et al.*[6] demonstrated the application of NMP to synthesize narrow dispersity polystyrene. The process involves activation by unimolecular dissociation and deactivation by coupling as shown in Scheme 6.1. A variety of stable radical mediated polymerizations (SRMP) have now been reported.[7]

Atom transfer radical polymerization (ATRP),[8–11] first reported in 1995,[12–14] differs from these processes, in that activation is a bimolecular process in which an atom (or group) is transferred from the dormant species to a catalyst to form a propagating radical, as indicated in Scheme 6.2. Deactivation is the reverse process in which the propagating species is trapped by atom or group transfer.

RAFT (reversible addition fragmentation chain transfer) is a process that provides reversible deactivation by degenerate chain transfer (Scheme 6.3).[2,15–20] The chain transfer step has been termed degenerate because the process involves only an exchange of functionality and the only distinction between the species on the two sides of the equilibrium is their degree of polymerization (*n* and *m*) and, in an effective process, even these will be similar.

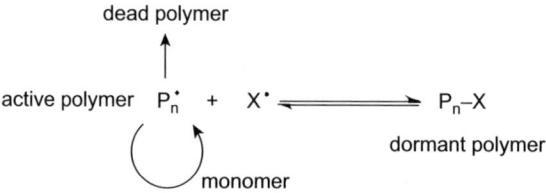

Scheme 6.1   Polymerization with reversible deactivation by coupling.

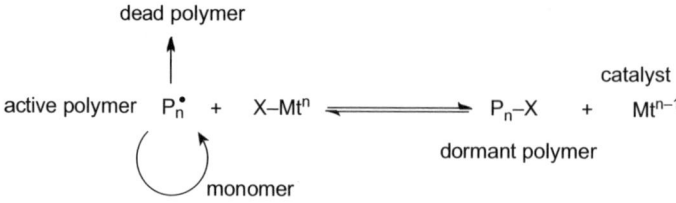

Scheme 6.2   Polymerization with reversible deactivation by atom or group transfer.

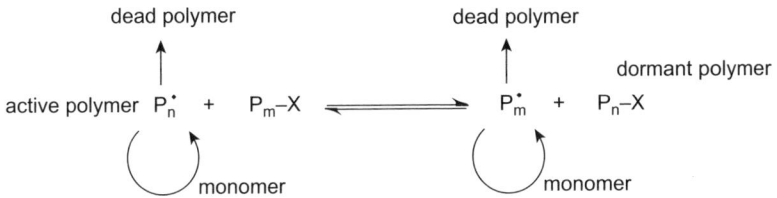

Scheme 6.3 Polymerization with reversible activation/deactivation by degenerate chain transfer.

In 1982, Otsu *et al.*[21,22] proposed that living characteristics, observed for polymerization in the presence of dithiocarbamate photoiniferters, might be attributable to either or both SRMP (Scheme 6.1) and degenerate chain transfer (Scheme 6.3) mechanisms. It is now known that the transfer constants of dithiocarbamates in the polymerization of the more activated monomers examined in those pioneering experiments are extremely low such that degenerate chain transfer is, at best, very minor pathway under the reaction conditions used.[23]

The first reports of radical addition-fragmentation processes appeared in the synthetic organic chemistry literature in the early 1970s.[24,25] Well-known examples of reactions that involve a S_H2' mechanism include allyl transfer reactions with allyl sulfides[26] and stannanes (the Keck reaction)[27] and the Barton–McCombie deoxygenation process with xanthates.[28] The first reports of the direct use of addition–fragmentation transfer agents to control molecular weight and end group functionality of polymer in radical polymerization appeared in the 1980s. The process has been the subject of a number of reviews.[29–36] RAFT to provide living characteristics to polymerization was first disclosed in 1995 with macro-monomer RAFT agents,[37,38] and in 1998, with thiocarbonylthio RAFT agents (dithioesters or trithiocarbonates).[39,40] MADIX (macromolecular design by interchange of xanthate – *i.e.*, polymerization with xanthate RAFT agents) was also first reported in a 1998 patent.[41] RAFT polymerization is now arguably the most convenient and versatile of the RDRP methods.[2,15–20]

6.2 RAFT Polymerization with Macromonomer RAFT Agents

Macromonomer RAFT polymerization has been reviewed within larger reviews on RAFT polymerization[36] or on catalytic chain transfer.[42,43] Macromonomers of general structure **1** act as addition-fragmentation transfer agents as shown in Scheme 6.4,[37,38,44] in which a propagating radical ($P_n{}^•$) adds to the macro-monomer **1** to form an intermediate **2** that fragments either to reform starting materials or provide a new macromonomer (**3**) and a radical ($R^•$) to reinitiate polymerization. The overall consequence is that the macromonomer ($-CH_2-C(Z)=CH_2$) functionality is transferred from R to P_n. The product **3** is also a macromonomer and is immediately subject to the same chemistry.

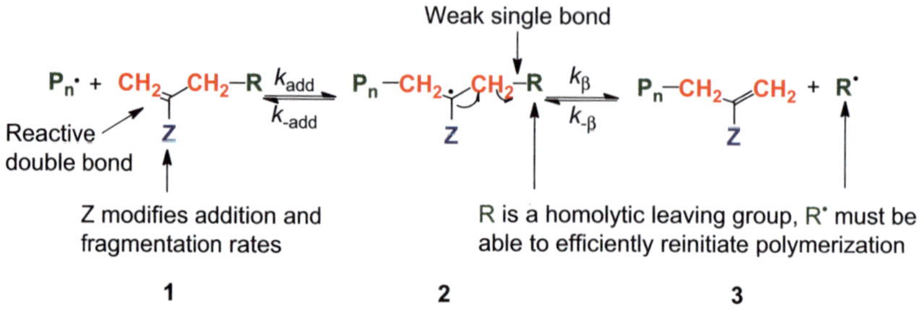

Weak single bond

Reactive double bond — $P_n^\bullet + CH_2{=}CH_2{-}R \xrightarrow[k_{-add}]{k_{add}} P_n{-}CH_2{-}\dot{C}H_2{-}R \xrightarrow[k_{-\beta}]{k_\beta} P_n{-}CH_2{-}CH_2 + R^\bullet$

Z modifies addition and fragmentation rates

R is a homolytic leaving group, R$^\bullet$ must be able to efficiently reinitiate polymerization

1   **2**   **3**

Scheme 6.4   Mechanism for reversible addition-fragmentation chain transfer (RAFT) with macromonomer RAFT agents.

The Z group both modifies the reactivity of the macromonomer transfer agent (**1**) towards radical addition and conveys stability to the intermediate radical (**2**). Z is most often an ester or an aryl group. Rate constants for radical addition to the macromonomer transfer agents are generally similar to that of the common monomers they resemble ($k_{add} \sim 10^2$–$10^3 \, M^{-1} \, s^{-1}$) and transfer constants (C_{tr}) are in the range 0.2–0.3.[44,45] In polymerization with conventional (irreversible) chain transfer, a C_{tr} of unity has been called 'ideal' since it implies that the transfer agent and monomer are consumed at the same rate and, as a consequence, the molar mass remains essentially constant with monomer conversion.[46] For polymerizations with reversible chain transfer it means that the initial transfer agent will not be consumed until complete conversion of monomer. A much higher transfer constant is required to achieve narrow molecular weight distributions in a batch polymerization process (see below; Figure 6.3).[47]

The radicals 'R$^\bullet$' and 'P$_n{}^\bullet$' must both be good homolytic leaving groups under the polymerization conditions. Moreover, the leaving group ability of 'R$^\bullet$' should be similar to or better than that of 'P$_n{}^\bullet$'.

The reactions associated with RAFT equilibria are in addition to those that occur normally during conventional radical polymerization (*i.e.* initiation, propagation and termination). The RAFT agent is a transfer agent. Since radicals are neither formed nor destroyed as a consequence of the RAFT process, termination is not directly suppressed. Retention of the macromonomer end group in the polymeric product is responsible for the living character of RAFT polymerization and renders the process potentially suitable for synthesizing block copolymers and end functional polymers.

Because transfer coefficients of the macromonomers are relatively low it is necessary to use starved feed conditions to achieve low dispersities.[37,38] Best results have been achieved using emulsion polymerization[37,38] where the rate of termination is lowered by compartmentalization effects. Under these conditions, molecular weights that increase linearly with conversion, molecular weight distributions that narrow with conversion, and high block purities >90% have been achieved.[37,38] An example is provided in Figure 6.1.[37]

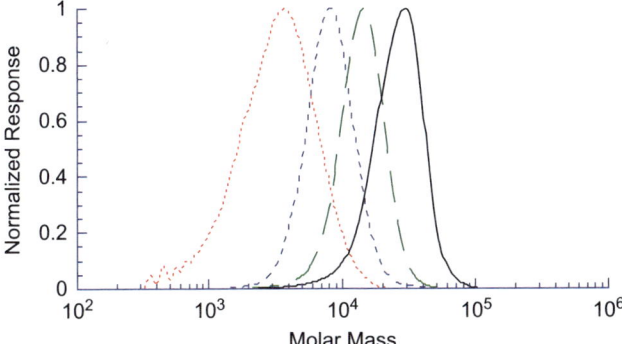

Figure 6.1 Molecular weight distributions for initial poly(methyl methacrylate) macromonomer RAFT agent (8.5 g, M_n 2300, Đ 1.5, ⋯⋯) and for poly(butyl methacrylate)-block-poly(methyl methacrylate) after additions of 33.6 (⋯⋯), 59.8 (– –), and 88.2 g (——) of butyl methacrylate. Emulsion polymerization of methyl methacrylate macromonomer latex (30 g, 28% solids; prepared with 0.3% sodium dodecyl sulfate surfactant) was heated to 80 °C under nitrogen. Potassium persulfate solution (0.4 wt% in water, 19 mL h^{-1}) and butyl methacrylate (10 g h^{-1}) were added by syringe pumps.
Figure reproduced from ref. 37, © American Chemical Society.

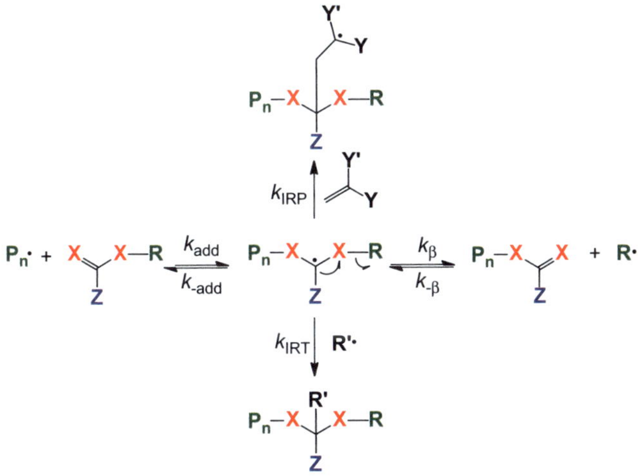

Scheme 6.5 Pathways for formation and disappearance of the intermediate radical in RAFT polymerization.

Rates of fragmentation relative to monomer addition for the intermediate radical are high for 1,1-disubstituted monomers such that intermediate radical propagation or copolymerization of the RAFT agent is usually not a detectable pathway [*i.e.* $(k_\beta + k_{-add}) > k_{IRP}$; refer to Scheme 6.5]. However, for mono-substituted monomers these rates are of similar magnitude such that

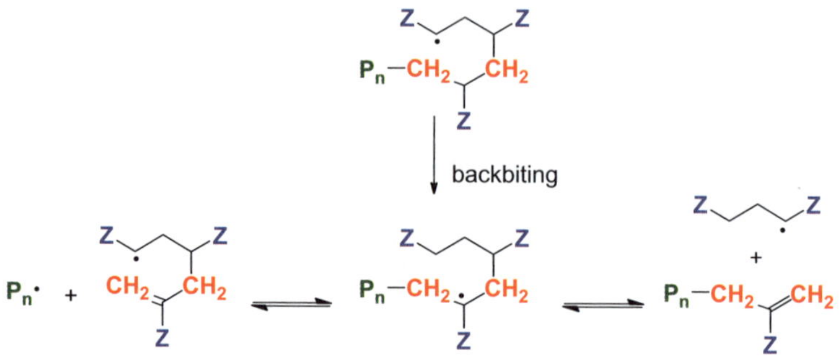

Scheme 6.6 Proposed mechanism for producing macromonomers by mid-chain radical formation by backbiting and β-scission.[48,51] For an acrylate, $Z = CO_2R$.

macromonomer copolymerization is a significant pathway [*i.e.* $(k_\beta + k_{-add}) \sim k_{IRP}$]. Higher reaction temperatures can be used to favour fragmentation over intermediate radical propagation.

The relatively narrow molecular weight distributions of macromonomers formed by high temperature polymerization of acrylates and other mono-substituted monomers can also be attributed to the RAFT mechanism (Scheme 6.5).[48,49] Similarities between the chemistry of RAFT polymerization and that seen in formation and reaction of mid-chain radicals formed during polymerization of acrylates were highlighted in a recent publication[50] (refer to Scheme 6.6).

Macromonomer RAFT polymerization:

- most suited to the polymerization or copolymerizations of methacrylic and similar monomers.
- transfer coefficients of macromonomer RAFT agents are typically in the range 0.1–1.0.
- low dispersities require the use of starved monomer feed conditions.
- shows tolerance of unprotected functionality in monomer, solvent and RAFT agent (*e.g.* OH, COOH). Polymerizations can be carried out in aqueous or protic media.
- is compatible with a wide range of reaction conditions (*e.g.* bulk, organic or aqueous solution, emulsion, miniemulsion, suspension, continuous flow).

6.3 RAFT Polymerization with Thiocarbonylthio RAFT Agents

The chain activation-deactivation mechanism in thiocarbonylthio RAFT polymerization (Scheme 6.7) is directly analogous to that in macromonomer RAFT polymerization (Scheme 4.4). However, rate constants for radical addition to the thiocarbonylthio double bond are typically several orders of magnitude higher than those to analogous carbon–carbon double bonds and

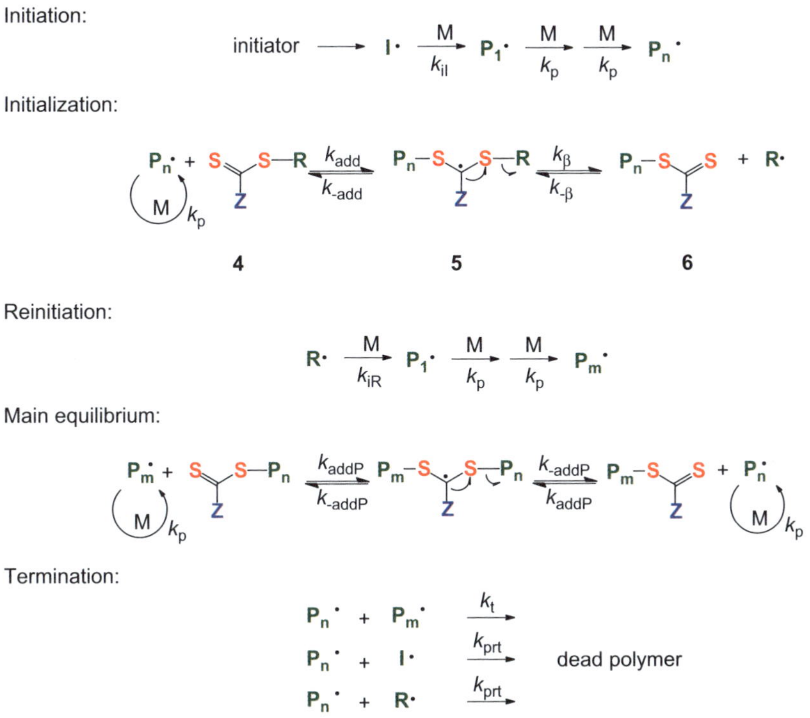

Initiation:

$$\text{initiator} \longrightarrow \text{I} \cdot \xrightarrow[k_{il}]{M} \text{P}_1 \cdot \xrightarrow{M}_{k_p} \xrightarrow{M}_{k_p} \text{P}_n \cdot$$

Initialization:

$$\text{P}_n \cdot + \underset{Z}{\overset{S}{\underset{S}{\bigvee}}}\text{S-R} \underset{k_{-add}}{\overset{k_{add}}{\rightleftharpoons}} \text{P}_n-\underset{Z}{\overset{S}{\underset{S}{\bigvee}}}\text{S-R} \underset{k_{-\beta}}{\overset{k_\beta}{\rightleftharpoons}} \text{P}_n-\underset{Z}{\overset{S}{\underset{S}{\bigvee}}}\text{S} + \text{R} \cdot$$

$$\text{(M)}\,k_p$$

4   **5**   **6**

Reinitiation:

$$\text{R} \cdot \xrightarrow{M}_{k_{iR}} \text{P}_1 \cdot \xrightarrow{M}_{k_p} \xrightarrow{M}_{k_p} \text{P}_m \cdot$$

Main equilibrium:

$$\text{P}_m \cdot + \underset{Z}{\overset{S}{\underset{S}{\bigvee}}}\text{S-P}_n \underset{k_{-addP}}{\overset{k_{addP}}{\rightleftharpoons}} \text{P}_m-\underset{Z}{\overset{S}{\underset{S}{\bigvee}}}\text{S-P}_n \underset{k_{addP}}{\overset{k_{-addP}}{\rightleftharpoons}} \text{P}_m-\underset{Z}{\overset{S}{\underset{S}{\bigvee}}}\text{S} + \text{P}_n \cdot$$

$$\text{(M)}\,k_p \qquad\qquad\qquad\qquad\qquad\qquad\qquad \text{(M)}\,k_p$$

Termination:

$$\text{P}_n \cdot + \text{P}_m \cdot \xrightarrow{k_t}$$
$$\text{P}_n \cdot + \text{I} \cdot \xrightarrow{k_{prt}} \quad \text{dead polymer}$$
$$\text{P}_n \cdot + \text{R} \cdot \xrightarrow{k_{prt}}$$

Scheme 6.7 Mechanism for reversible addition–fragmentation chain transfer (RAFT) with thiocarbonylthio RAFT agents.

transfer constants are correspondingly much higher. Rates of fragmentation relative to monomer addition for the intermediate radical **5** are also high such that copolymerization of the RAFT agent is not a detectable pathway [*i.e.* $k_{IRP}/(k_\beta + k_{-add})$ is zero or undetectably small] (Scheme 6.5). There is evidence for intermediate radical termination under some circumstances (see discussion on side reactions below) but is undetectable with appropriate choice of RAFT agent.

An example of the molar mass distributions that can be produced by RAFT polymerization is shown in Figure 6.2.[39] The molar mass distributions are narrow ($Đ < 1.1$). The mole fraction of dead chains observed is small and consistent with that predicted by kinetic simulation using the usual kinetic parameters.

Thiocarbonylthio RAFT polymerization provides:

- the ability to control polymerization of most monomers polymerizable by radical polymerization. These include (meth)acrylates, (meth)acrylamides, acrylonitrile, styrenes, dienes and vinyl monomers.
- narrow molecular weight distributions, high end-group fidelity and the ability to make blocks and more complex architectures.

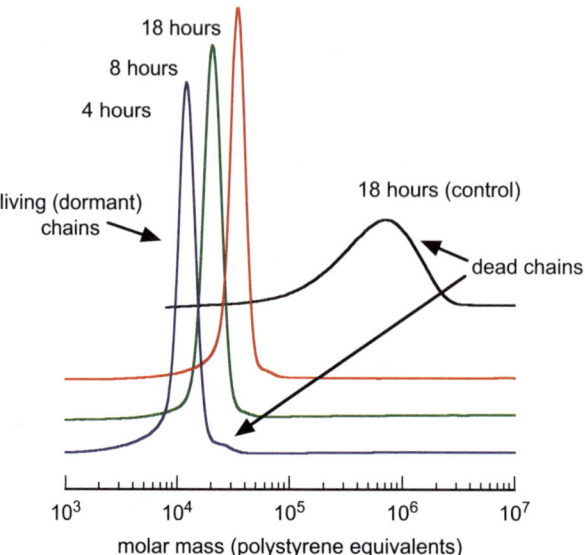

Figure 6.2 Molar mass distributions for poly(St-*co*-AN) polymerized by heating St and AN (62 : 38 mole ratio) at 100 °C in the presence of cumyl dithio-benzoate (0.0123 M) after 4 h ($M_n = 20\,100$; $Đ = 1.04$, ——), 8 h ($M_n = 33\,000$; $Đ = 1.05$, ——), and 18 h ($M_n = 51\,400$; $Đ = 1.07$, ——). The molecular weight distribution for a similar polymerization performed without cumyl dithiobenzoate is also shown (control $M_n = 424\,000$, $Đ = 1.70$, ——).[39]

- transfer constants can be very high (up to 5000) but depend strongly on Z, R and the particular monomer.
- tolerance of unprotected functionality in monomer, solvent and RAFT agent (*e.g.* OH, NR_2, COOH, $CONR_2$, SO_3H). Polymerizations can be carried out in aqueous or protic media.
- compatibility with a wide range of reaction conditions (*e.g.* bulk, organic or aqueous solution, emulsion, miniemulsion, suspension, flow).
- simple in implementation and low cost in relation to competitive technologies. In some cases can involve simply adding a RAFT agent to an otherwise conventional radical polymerization.

6.4   Transfer Coefficients of RAFT Agents

RAFT agents are characterized by two transfer coefficients. A rate coefficient for chain transfer (k_{tr}) can be defined in terms of the rate constant for addition (k_{add}) and a partition coefficient (ϕ) which defines how the adduct is partitioned between products and starting materials (eqn (6.1) and (6.2)).[45]

$$k_{tr} = k_{add} \frac{k_\beta}{k_{-add} + k_\beta} = k_{add}\phi \qquad (6.1)$$

$$\phi = \frac{k_\beta}{k_{-add} + k_\beta} \tag{6.2}$$

The transfer coefficient is then defined in terms of k_{tr} and the propagation rate constant (k_p) in the usual way (eqn (6.3)).

$$C_{tr} = \frac{k_{tr}}{k_p} \tag{6.3}$$

A rate coefficient for the reverse transfer is defined analogously as shown in eqn (6.4):

$$k_{-tr} = k_{-\beta} \frac{k_{-add}}{k_{-add} + k_\beta} \tag{6.4}$$

The reverse transfer constant (C_{-tr}) is then (eqn (6.5)):

$$C_{-tr} = \frac{k_{-tr}}{k_{iR}} \tag{6.5}$$

where k_{iR} is the rate constant for reinitiation by the RAFT agent derived radical, $R^\bullet$ (refer to Scheme 6.7).

The rate of consumption of the initial RAFT agent (T) is then given by eqn (6.6):

$$\begin{aligned}\frac{d[T]}{d[M]} &\approx C_{tr} \frac{[T]}{[M] + C_{tr}[T] + C_{-tr}[P_n^T]} \\ &= C_{tr} \frac{[T]}{[M] + C_{tr}[T] + C_{-tr}([T]_0 - [T])}\end{aligned} \tag{6.6}$$

where [T] and $[P_n^T]$ are the concentrations of RAFT agent (**1** or **4**) and macro-RAFT agent (**3** or **6**) and [M] is the monomer concentration. The derivation of these equations assumes that fragmentation is not rate determining and that the intermediates **2** or **5** undergo no side reactions. The equation (eqn (6.6)) can be solved numerically to give estimates of C_{tr} and C_{-tr}.[52,53] If the reverse reaction can be neglected (very low conversion $P_n^T \sim 0$ and/or $C_{-tr} \sim 0$), eqn (6.6) simplifies to eqn (6.7)

$$\frac{d[M]}{d[T]} \approx C_{tr} \frac{[M]}{[T]} + 1 \tag{6.7}$$

which suggests that a plot of log [M] *vs.* log [T] should be a straight line with the slope providing the transfer constant.[54] This equation has been used to evaluate C_{tr} for a range of macromonomers[45,55,56] and thiocarbonylthio RAFT agents.[47,52,53,57–60] For less active RAFT agents ($C_{tr} \leq 1$), transfer constants for RAFT agents may also be determined using the Mayo and related methods.

For the more active RAFT agents, the values obtained with neglect of C_{-tr} (most values in the literature) should be regarded as apparent transfer constants (C_{tr}^{app}). They are minimum values for C_{tr} and depend on the polymerization conditions and, in particular, the concentration of the RAFT agent.[47,52]

The actual value of C_{tr} may exceed C_{tr}^{app} by several orders of magnitude. Nonetheless, values of C_{tr}^{app} obtained under similar conditions, still provide a valuable method for assessing the relative activity of RAFT agents.

The partition coefficient ϕ (eqn (6.2)) indicates the preference for the intermediate radicals to fragment to products or return to starting materials. For effective RAFT agents, R should be a good homolytic leaving group with respect to the propagating radical (*i.e.*, $\phi > 0.5$).

The equation (6.6) also simplifies for macro-RAFT agents used in RAFT homopolymerization, where for n and m are greater than 2, ϕ is 0.5 and C_{tr} and C_{-tr} are the same.

The calculated dependence of the molecular weight and dispersity on monomer conversion for various values of C_{tr} is shown in Figure 6.3.[47] The predictions are based on equations proposed by Müller *et al.* to describe a hypothetical polymerization controlled by degenerate chain transfer but without termination.[61,62] The degree of polymerization is simply the ratio of [monomer consumed] : [RAFT agent consumed]. The figure shows that a higher degree of polymerization than that predicted for complete utilization of the transfer agent need not indicate some form of 'hybrid behavior' as is suggested in some papers. It can simply indicate a low C_{tr} and that the initial RAFT agent is not fully converted to macro-RAFT agent.[63] Lowered dispersities will generally only be observed when $C_{tr} > 2$, and $C_{tr} > 10$ are usually required to achieve the characteristics most often associated with living polymerization (*i.e.*, significantly narrowed molecular weight distributions, molecular weights that are predictable

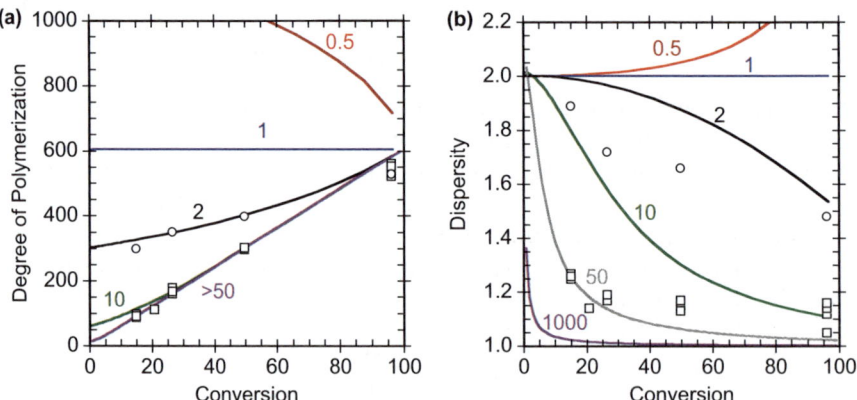

Figure 6.3 Predicted dependence of (a) degree of polymerization and (b) dispersity on conversion in polymerizations involving reversible chain transfer as a function of the chain transfer constant (C_{tr}). Predictions are based on equations proposed by Müller *et al.*[61,62] with $\alpha = 10^{-7}$ (the concentration of active species), β (the transfer constant) as indicated and $\gamma = 605$ (the ratio monomer to transfer agent). Experimental data points are for methyl methacrylate (7.02 M) polymerization in presence of dithiobenzoate esters (0.0116 M) where R = C(Me)$_2$CO$_2$Et ($\bigcirc$) and C(Me)$_2$Ph ($\square$).
(Figures adapted from ref. 47. Copyright 2003 American Chemical Society.)

from reagent concentrations and increase linearly with conversion). The more effective RAFT agents in this context have $C_{tr} \gg 100$.

6.5 RAFT Equilibrium Constants

The properties of RAFT agents **4** are also often discussed in terms of the value of the addition–fragmentation equilibrium constant $K (= k_{add}/k_{-add})$. Rates of addition to RAFT agents **4** are typically high ($k_{add} \sim 10^4$–$10^8\,M^{-1}\,s^{-1}$). Thus, a high equilibrium constant generally implies a low fragmentation rate constant (k_{-add}) and consequently an increased likelihood for retardation and/or side reaction involving the adduct species (**5**). However, values of the equilibrium constants do not, by themselves, provide sufficient information to predict the ability of a RAFT agent to control polymerization.

In a given RAFT polymerization, there are at least four equilibrium constants that need to be considered.

$K_P (= k_{addP}/k_{-addP})$ associated with the chain equilibration process. This step is sometimes called the main equilibrium (Scheme 6.7).

$K (= k_{add}/k_{-add})$ and $K_\beta = (k_{-\beta}/k_\beta)$ associated with the initial reversible chain transfer step, sometimes known as the pre-equilibrium or initialization step (Scheme 6.7).

$K_R (= k_{addR}/k_{-addR})$ associated with the reaction of the expelled radical with the initial RAFT agent.

There may be other equilibrium constants to consider if penultimate group effects are significant (there is theoretical data[64,65] and some experimental evidence[47,66] to indicate that this is the case). There are also a further series of reactions that need to be considered that involve initiator radical-derived RAFT agents. RAFT agents of differing reactivity might be derived from each radical species present.

6.6 Polymer Molar Mass and the Fraction of Living Chains

The molar mass of the polymer by RAFT polymerization formed can be estimated knowing the concentration of the monomer consumed and the initial RAFT agent concentration using the simple relationship (eqn 6.8):

$$M_n(\text{calc}) = \frac{[M]_0 - [M]}{[T]_0} m_M + m_{RAFT} \tag{6.8}$$

where $[M]_0$–$[M]$ is the concentration of monomer consumed and m_M and m_{RAFT} are molar masses of the monomer and the RAFT agent, respectively. Positive deviations from equation (6.8) indicate incomplete usage of RAFT agent. Negative deviations indicate that other sources of polymer chains are significant. These include the initiator-derived chains. It follows from the mechanism of the RAFT process shown in Scheme 6.7 that there must be a fraction of dead chains formed which directly relates to the number of chains initiated by initiator-derived radicals. It is important that this fraction be taken into account when calculating the molecular weights of polymers formed by the RAFT process, particularly when a high ratio of initiator to RAFT agent is used.[47] Thus, a more complete expression is eqn (6.9):

$$M_n(\text{calc}) = \frac{[M]_0 - [M]}{([T]_0 - [T]) + df([I]_0 - [I])} m_M + m_{RAFT} \qquad (6.9)$$

where $[T]_0 - [T]$ is the RAFT agent consumed, d is the number of chains produced from radical–radical termination ($d \sim 1.67$ in MMA and $d \sim 1.0$ in styrene polymerization), $[I]_0 - [I]$ is the concentration of initiator consumed and f is the initiator efficiency. If the initiator decomposition rate constant is known, the initiator consumption can be estimated using eqn (6.10):

$$[I]_0 - [I]_t = [I]_0(1 - e^{-k_d t}). \qquad (6.10)$$

Based on similar considerations it is possible to calculate the fraction of living (dormant) chains (L) in RAFT polymerization (eqn (6.11)).

$$L = \frac{[T]_0}{[T]_0 + df([I]_0 - [I]_0)} \qquad (6.11)$$

It is also possible to calculate L based on the difference between the experimental molecular weight and that predicted by eqn (6.8) using equation (6.12).

$$L = \frac{(\bar{M}_n - m_{RAFT})[T]_0}{([M]_0 - [M])m_M} \qquad (6.12)$$

Figure 6.4 shows the dependence of the fraction of living chains on RAFT agent concentration for a series of polymerizations carried out with the same

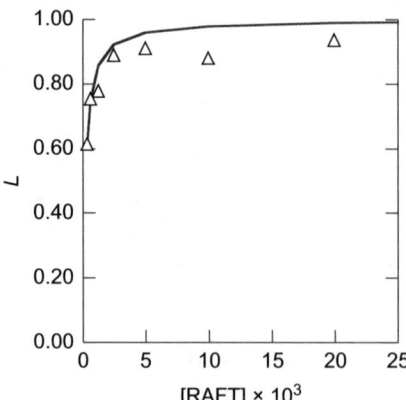

Figure 6.4 Fraction of living chains (L) estimated using eqn (6.12) (Δ) or predicted by eqn (6.9) (—) with cumulative $f = 0.3$, $d = 1.67$ and $k_d = 2.54 \times 10^{-5}\,\text{M}^{-1}\,\text{s}^{-1}$ for poly(MMA) formed by polymerization of MMA (6.55 M in benzene) with 1,1′-azobis(1-cyclohexanenitrile) (0.0018 M) as initiator and various concentrations of RAFT agent **31** for 6 h at 90 °C (corresponds to experiments shown in Figure 6.9 in Section 6.1.12).

reaction times and initiator concentrations. All polymerizations gave narrow dispersity products ($Đ < 1.2$).

6.7 Monomers in RAFT Polymerization

While the reactivity of monomers forms a continuum, it is useful to consider them as belonging to one of two broad classes. The 'less activated' monomers (LAMs, see Figure 6.5) are those where the double bond is adjacent to an oxygen or nitrogen lone pair (*e.g.*, vinyl acetate (VAc), *N*-vinylpyrrolidone (NVP)), the heteroatom of a heteroaromatic ring (*e.g.*, *N*-vinylcarbazole (NVC)) or saturated carbon (*e.g.*, diallyldimethylammonium chloride (DADMAC)).

The 'more-activated' monomers (MAMs, see Figure 6.6) are those where the double bond is conjugated to an aromatic ring (*e.g.*, styrene (St), 4-vinyl pyridine (4VP), acenaphthalene (AcN)), a carbonyl group (*e.g.*, methyl methacrylate (MMA), methyl acrylate (MA), acrylamide (Am), *N*-acryloyl morpholine (NAM), maleic anhydride (MAH)) or a nitrile (*e.g.*, acrylonitrile (AN)).

The choice of an appropriate RAFT agent (ZC(=S)SR) for the particular monomers and reaction conditions is crucial for success. The effectiveness of RAFT agents is determined by the substituents R and Z, and guidelines for their selection have been proposed (see below and Figure 6.7).[15,17] The polymerization of most monomers can be well-controlled so as to provide a high fraction of living chains and minimal retardation by using one of just two RAFT agents. One suited to MAMs and another suited to LAMs. Recently, switchable RAFT agents that can be switched to control both MAMs and LAMs have been described.[67,68] Requirements for specific end-functionality or architecture may dictate the use of other RAFT agents.[69,70]

RAFT polymerization is compatible with a wide range of unprotected functionalities in either the monomer or RAFT agent. Tolerated monomer functionalities include tertiary amino (in 2-dimethylaminoethyl methacrylate (DMAEMA), quaternary amino, carboxylic acid (*e.g.*, in methacrylic acid (MAA)), sulfonate (*e.g.*, in sodium 4-styrenesulfonate (SSNa)) hydroxyl (*e.g.*, in hydroxyethyl methacrylate (HEMA)) and epoxy (*e.g.*, in glycidyl methacrylate (GMA)). RAFT agents are generally not compatible with primary and secondary amino or with thiol functionalities, though it has been shown that RAFT is possible with primary amino functionality as long as it is fully

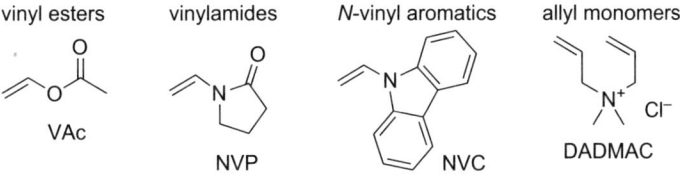

vinyl esters   vinylamides   *N*-vinyl aromatics   allyl monomers

VAc   NVP   NVC   DADMAC

Figure 6.5   Structures of some less-activated monomers (LAMs).

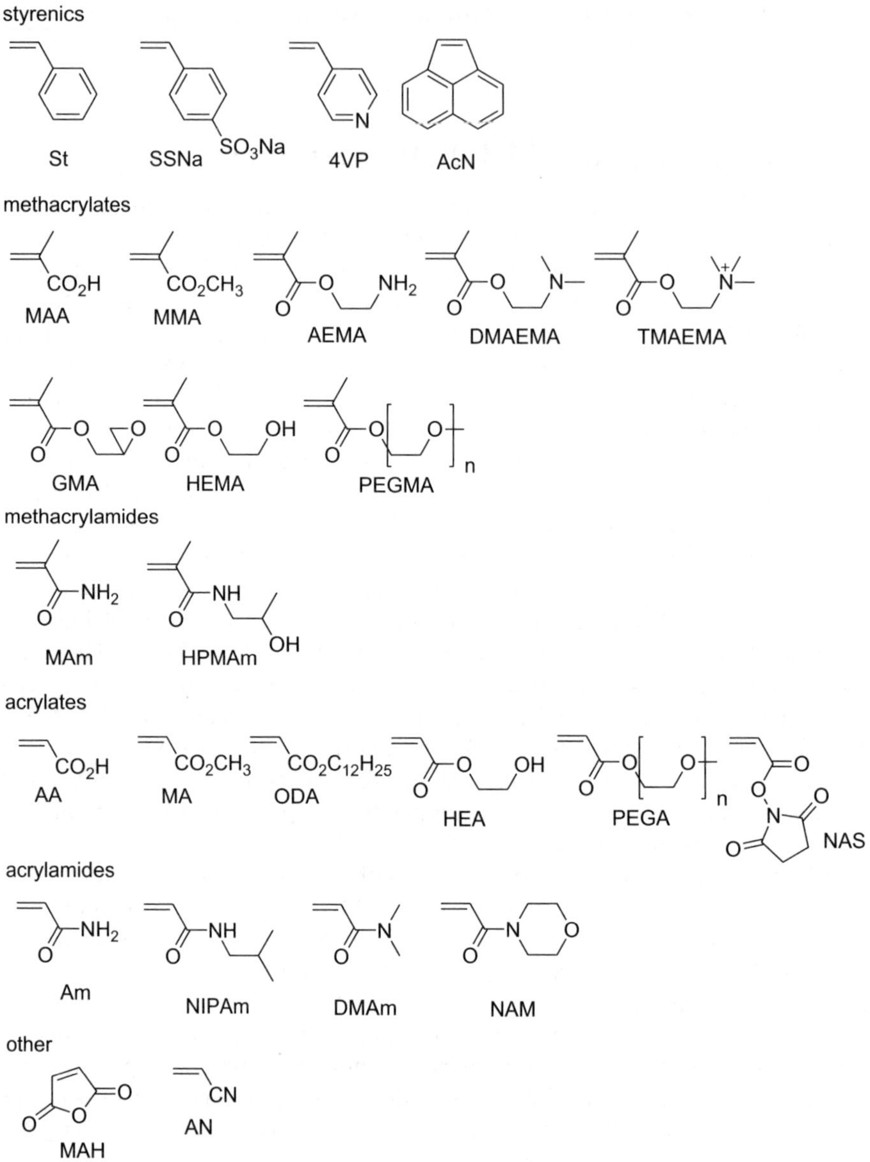

styrenics

methacrylates

methacrylamides

acrylates

acrylamides

other

Figure 6.6 Structures of some more-activated monomers (MAMs).

protonated (*e.g.*, in 2-aminoethyl methacrylate (AEMA)).[71] RAFT poly-merization of functional monomers has been used as a route to structures as diverse as glycopolymers,[5,15,72–78] possible hole or electron transport materials,[79,80] photochromic materials[81,82] and light harvesting polymers (*e.g.*, from AcN[83–85]). RAFT (co)polymerization of activated monomers such as

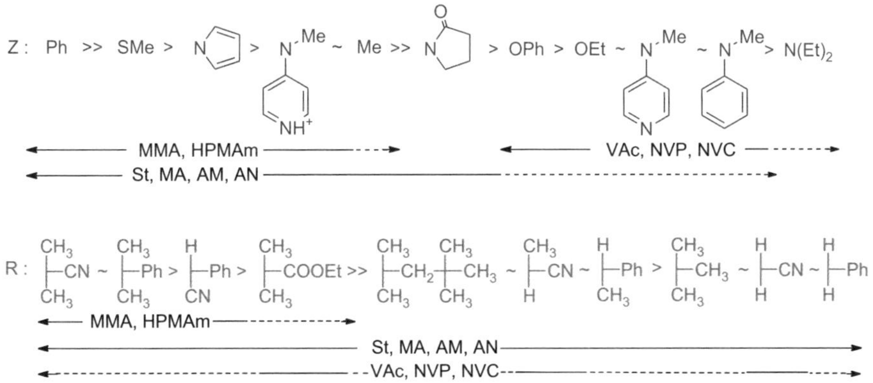

Z : Ph >> SMe > [pyrrole] > [N-Me pyridine] ~ Me >> [N-pyrrolidinone] > OPh > OEt ~ [N-Me pyridine] ~ [N-Me phenyl] > N(Et)$_2$

[NH$^+$ pyridinium]

◄——————— MMA, HPMAm ———————▸ - - - -►   ◄——————— VAc, NVP, NVC ——————▸ - - - - -►
◄——————— St, MA, AM, AN ——————————————— -▸

R : $\overset{CH_3}{\underset{CH_3}{|}}$—CN ~ $\overset{CH_3}{\underset{CH_3}{|}}$—Ph > $\overset{H}{\underset{CN}{|}}$—Ph > $\overset{CH_3}{\underset{CH_3}{|}}$—COOEt >> $\overset{CH_3}{\underset{CH_3}{|}}$—CH$_2$—$\overset{CH_3}{\underset{CH_3}{|}}$ ~ $\overset{CH_3}{\underset{H}{|}}$—CN ~ $\overset{H}{\underset{CH_3}{|}}$—Ph > $\overset{CH_3}{\underset{CH_3}{|}}$—CH$_3$ ~ $\overset{H}{\underset{H}{|}}$—CN ~ $\overset{H}{\underset{H}{|}}$—Ph

◄——— MMA, HPMAm ———————▸
◄———————————————— St, MA, AM, AN ————————————————————————————————▸
◄- - - - - - - - - - - - - - VAc, NVP, NVC- -▸

Figure 6.7 Guidelines for selection of RAFT agents (Z-C(=S)S-R) for various poly-merizations.[15,17] For Z, addition rates and transfer constants decrease and fragmentation rates increase from left to right. For R, fragmentation rates decrease from left to right. A dashed line indicates limited control (*e.g.*, retardation, high dispersity likely).

N-acryloyl succinimide (NAS)[86–88] also provides a means of synthesizing functional (co)polymers.

6.8 The Z Substituent in Thiocarbonylthio RAFT Agents

The Z group modifies both the rate of addition of propagating radicals (P$_n$•) to the thiocarbonyl of **4** and the rate of fragmentation of the intermediate radical **5**. The rate constant k_{add} can be "adjusted" over some 5 orders of magnitude through manipulation of Z.

The most reactive RAFT agents include the dithioesters and trithio-carbonates which have carbon or sulfur adjacent to the thiocarbonylthio group. RAFT agents with a lone pair on nitrogen or oxygen adjacent to the thio-carbonyl, such as the *O*-alkyl xanthates, *N,N*-dialkyldithiocarbamates and *N*-alkyl-*N*-aryldithiocarbamates, have dramatically lower reactivity towards radical addition. Lower rate coefficients for addition are predicted by molecular orbital calculations[52,89,90] and can be qualitatively understood in terms of the importance of the zwitterionic canonical forms **8** and **10** (Figure 6.8). The interaction between the lone pair and the C=S double bond both reduces the double bond character of the thiocarbonyl group and stabilizes the RAFT agent (**4**) relative to the intermediate radical **5**.[19,52,90,91] Dithiocarbamates where the nitrogen lone pair is not as available because it is part of an aromatic ring system (such as a pyrrole in **12**) or where a carbonyl (as in **15**) is α to the nitrogen lone pair (Figure 6.8), have reactivities similar to those of the dithioesters and trithiocarbonates.[52,92,93] In these cases, the importance of

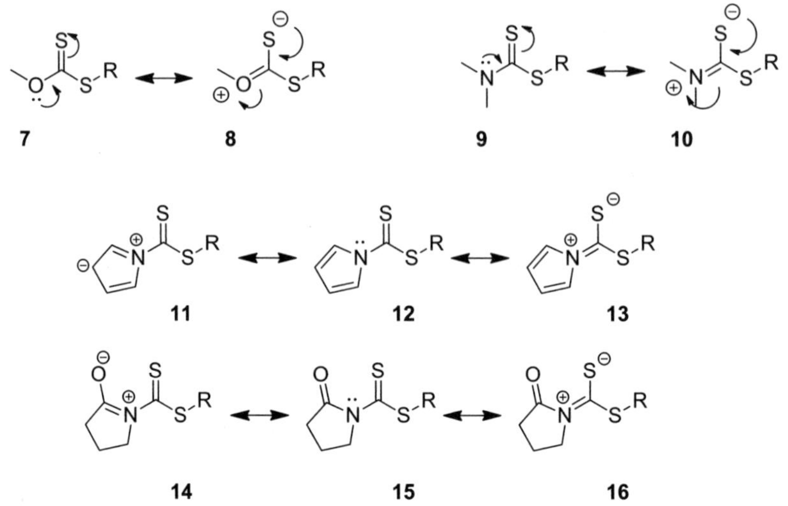

Figure 6.8   Canonical forms of xanthates and dithiocarbamates.

canonical forms such as **11** and **14** effectively reduces the contribution of **13** and **16** respectively. The effectiveness of xanthates is similarly sensitive to the nature of substituents on oxygen.[91]

Propagating radicals with a terminal more-active monomer (MAM) unit are less reactive in radical addition (lower k_p, lower k_{add}) and one of the more active RAFT agents is required for good control. The poly(MAM) propagating radicals are relatively good homolytic leaving groups (higher k_β, k_{-add}); therefore, retardation solely due to slow fragmentation is unlikely. The more active RAFT agents such as the dithioesters, trithiocarbonates and aromatic dithiocarbamates allow the preparation of low dispersity polymers from MAMs whereas the *N*-alkyl-*N*-aryldithiocarbamates and the *O*-alkyl xanthates typically have lower transfer constants and provide poor control.

Propagating radicals with a terminal less-activated monomer (LAM) unit are highly reactive in radical addition (higher k_p, higher k_{add}). Thus, the *N*-alkyl-*N*-aryldithiocarbamates and the *O*-alkyl xanthates have high transfer constants in LAM polymerization. However, the poly(LAM) propagating radicals are relatively poor homolytic leaving groups (lower k_β, k_{-add}) such that inhibition or retardation is likely with the active RAFT agents, such as dithioesters.

General guidelines for selection of Z are shown in Figure 6.7. Irrespective of the class of RAFT agent, the transfer constant is generally enhanced by the presence of electron withdrawing groups on Z and by the capacity of Z to stabilize an adjacent radical centre.[52,89] However, these same factors generally can also increase the likelihood of side reactions. The second important role of Z is to determine the stability of the intermediate radicals **5**. When Z is aryl, the intermediate is stabilized, and the rate of intermediate radical fragmentation is slower than when connecting atom of Z is sp^3 carbon, oxygen or nitrogen or sulfur.

It was proposed[94,95] that RAFT agents where Z is fluorine might perform as a "universal" RAFT agent. In terms of the simple theories expounded above, the high electronegativity of fluorine means that canonical forms analogous to **8** and **10** in which an electron is removed from Z have little importance. The fluorine also provides little stability to the intermediate radical. However, difficulties in synthesis of the so-called F-RAFT agents have meant that such RAFT agents have not been fully tested.

We[67,68,96,97] have designed a new class of stimuli-responsive, switchable, RAFT agents that can be switched to offer good control over polymerization of both MAMs and LAMs. Our motivation has been to provide a direct route to poly(MAM)-*block*-poly(LAM). The *N*-methyl-*N*-(4-pyridinyl)dithiocarbamates (**18**, Scheme 6.8) behave as other *N*-aryl-*N*-alkyldithiocarbamates, and are effective in controlling the polymerization of LAMs but have relatively low transfer constants when used in MAM polymerization. However, in the presence of a strong protic or Lewis acid, the switched form of the of the RAFT agent (**20**) provides excellent control over the polymerization of MAMs.[68] Protonation reduces the importance of canonical forms such as **21** which contain two positive charges in proximity.

The *N*-aryl-*N*-(4-pyridinyl) dithiocarbamates (**22**)[60] appear more effective (dispersities are lower) than the analogous *N*-methyl-*N*-(4-pyridinyl) dithiocarbamates (**18**) with LAMs in the unswitched (neutral) form and more active with MAMs in the switched (protonated) form.

A final consideration is that Z should not cause any side reactions. With xanthates (**4**, Z=OR') it is important that R' be a poor homolytic leaving group. Otherwise fragmentation with loss of R' (irreversible chain transfer) will compete with the desired RAFT process. This requires that R' be primary alkyl or aryl.[98,99] Similar design consideration apply in the case of unsymmetrical trithiocarbonates (**4**, Z=SR').[59]

When Z is strongly electron-withdrawing the thiocarbonyl group may undergo a direct reaction with monomers. Thus, RAFT agents where Z is

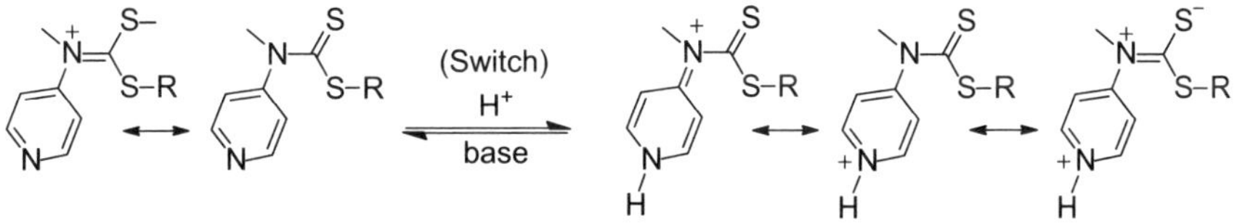

| 17 | 18 | | 19 | 20 | 21 |

Scheme 6.8   *N*-(4-pyridinyl)-*N*-methyldithiocarbamate switchable RAFT agents.

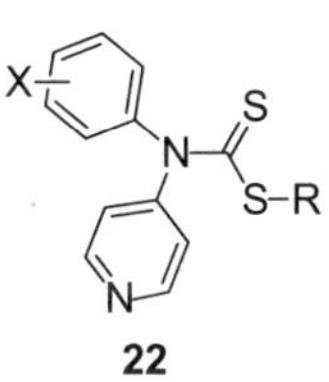

22

alkylsulfonyl or phenylsulfonyl group Z=PhSO$_2$– undergo direct reaction with (meth)acrylate monomers (BA, MA, tBA, and MMA) under polymerization conditions with consumption of the thiocarbonylthio group and ultimately little control over the polymerization.[100] A hetero-Diels Alder mechanism was suggested. Good control was achieved only with isobornyl acrylate where the side reactions with monomer were suppressed by the bulky ester substituent.

The presence of withdrawing groups on Z, which lead to higher transfer coefficients, increases the likelihood of side reactions such as hydrolysis or aminolysis[15,101] and participation in cycloaddition reactions[102] such as the hetero-Diels Alder reaction with diene monomers[100] and 1,3-dipolar cyclo-addition.[103] This is an important consideration in some RAFT agent syntheses, can be critical to the choice of RAFT agent for specific polymerization conditions (*e.g.*, in aqueous media or in emulsion polymerization), and determines the ease of end group transformation processes that may be required post-RAFT polymerization.

6.9   The R Substiuent in Thiocarbonylthio RAFT Agents

The R group of RAFT agents **4** is important in determining the value of the partition coefficient ϕ (eqn (6.3)). For optimal control of a polymerization, the R group of **4** must be a good homolytic leaving group with respect to $P_n{}^\bullet$, such that the intermediate **5**, formed by addition of $P_n{}^\bullet$ to **4**, fragments rapidly and partitions substantially in favor of **6** and R$^\bullet$. The expelled radical (R$^\bullet$) must also be able to reinitiate polymerization efficiently (*i.e.* $k_{i,R} > k_p$), otherwise retardation is likely.[67] Radical stability is important in determining frag-mentation rates. Experimental findings, that the transfer coefficient and the value of ϕ increase in the series primary < secondary < tertiary and with the introduction of substituents which are capable of delocalizing the radical centre, are consistent with this view. However, other factors can be of equal or greater significance.

It is not sufficient for R to be a monomeric analog of the propagating radical because penultimate unit effects are substantial, particularly when R is tertiary. RAFT agents with R = 2-ethoxycarbonyl-2-propyl (**24**), which can be considered as a monomeric model for a methacrylate chain (**23**), provide only poor control over the polymerization of MMA and other methacrylates because R is a poor homolytic leaving group with respect to the poly(MMA) propagating radical.[45,47] For similar reasons, RAFT agent with R = *t*-butyl (**24**) is poor with respect to RAFT agent with R = *t*-octyl (**25**).[47] These differences in RAFT agent activity are attributed to steric factors.

Polar effects are also extremely important in determining the partition coefficient ϕ. lectron-withdrawing groups on R both decrease rates of addition to the thiocarbonyl group and increase rates of fragmentation. The relatively high transfer constants of cyanoalkyl RAFT agents *vs.* similar benzylic RAFT agents is attributed to the influence of polar factors.[47]

Thus, control over the polymerization of methacrylates and methacrylamides (and other 1,1-disubstituted monomers which result in a tertiary $P_n{}^\bullet$) usually requires that R be tertiary (*e.g.*, 2-cyano-2-propyl or cumyl)[47] or secondary aralkyl (*e.g.*, α-cyanobenzyl).[104,105] However, polymerization of monomers with high propagation rate constants (k_p) such as acrylates, acrylamides, vinyl esters (*e.g.*, vinyl acetate) and vinyl amides (*e.g.*, N-vinylpyrrolidone) are best controlled with RAFT agents with primary or secondary R groups (*e.g.*, cyanomethyl or cyanoethyl). Tertiary radicals, such as 2-cyano-2-propyl radical, are inefficient in reinitiating polymerization since k_{iR} is often lower than k_p.[47]

Guidelines to the selection of R are provided in Figure 6.7. The order shown is based on measurements of $C_{tr}{}^{app}$ for dithiobenzoate RAFT agents in St and MMA polymerization[47] and seems to be general based on limited data for other classes of RAFT agent. The benzylic radicals and tertiary alkyl radicals add to most LAMs very slowly (with reference to k_p) and an inhibition period is often observed with these R groups.[106]

6.10 Mechanisms for Retardation and Side Reactions in RAFT Polymerization

Retardation is not intrinsic to the RAFT mechanism. Radicals are neither formed nor destroyed in the RAFT equilibria. RAFT agents can behave as ideal chain transfer agent. Nonetheless, many papers describe retardation or inhibition during RAFT polymerization.

Retardation during RAFT polymerization has been attributed to slow reinitiation by R (poor choice of RAFT agent), poor choice of initiator, impurities in RAFT agent, monomer or the medium, the presence of air or oxygen (poor degassing of the reaction medium), the properties of the inter-mediate radical (slow fragmentation, intermediate radical termination, poor choice of RAFT agent) or some combination of these phenomena. The mechanisms for retardation in RAFT polymerization with dithiobenzoate RAFT agents has been studied extensively and are the subject of an IUPAC report.[107] However, there is still much debate about the exact mechanism for these and other cases.

6.10.1 Slow Reinitiation by R•

Processes which lead to retardation in conventional radical polymerization also affect RAFT polymerization. Thus, slow reinitiation ($k_{iR} < k_p$) will cause retardation as it does in conventional radical polymerization with irreversible chain transfer. The influence of slow reinitiation can be aggravated by the partition coefficient ϕ being unfavourable (*i.e.* $\phi \ll 0.5$ such that reversion to starting materials is favoured). It is important to select R of the initial RAFT agent such that it forms a good initiating radical and is a good homolytic leaving group with reference to the propagating radicals. The 2-cyanoprop-2-yl radical is a good initiating radical in polymerization of most MAMs, such as methacrylates and styrene,[47] but it is a poor initiating radical for many LAMs, such as vinyl acetate or *N*-vinylpyrrolidone.[108] Rate constants for radical addition to monomers are available.[106,109]

6.10.2   Initiator

The initiator in RAFT polymerization should be chosen such that the initiator-derived radicals are good leaving groups with reference to the propagating radical in order to avoid retardation and conversion of the RAFT agent to a relatively stable by-product.[110] Thus, lauroyl peroxide LPO (a source of undecyl radicals) is usually a poor initiator choice since the addition of an undecyl radical to a RAFT agent is irreversible. Azobis(methyl isobutyrate) (AIBMe) is a poor initiator choice in MMA polymerization for similar reasons.[53] However, it suitable for polymerizations of most monosubstituted monomers. AIBN is usually a good choice in this context since the 2-cyanoprop-2-yl radical is a very good leaving group with respect to most propagating species.

6.10.3   Impurities

Impurities in the RAFT agent have been shown to cause retardation in a number of cases. The effect of impurities in cumyl dithiobenzoate on RAFT polymerization of HEMA, styrene, and MA has been studied.[111] Samples of RAFT agent purified by column chromatography on silica gave a significant inhibition period even when used immediately following purification. Samples that had been rigorously purified by high performance liquid chromatography also gave significant retardation after storage for a short time (<3 months) under refrigeration at $-20\,^{\circ}\text{C}$. Dithiobenzoic acid was considered a likely impurity in these experiments.[111] Actual impurities in cumyl dithiobenzoate will depend on how it was synthesized and stored. We have found that cumyl dithiobenzoate purified by column chromatography on alumina with hexane as eluent and stored at $4\,^{\circ}\text{C}$ for >2 years in the absence of light remains fully effective.[19] Another potential impurity in dithiobenzoate RAFT agents is bis(thiobenzoyl)disulfide. This may form from cumyl dithiobenzoate on exposure to light and such disulfides are known to inhibit polymerization until consumed by reaction with initiator-derived radicals.[112,113]

6.10.4   Air

RAFT polymerization shows sensitivity to air and impurities in the monomer. Results for polymerizations carried out with unpurified commercial MMA at

Table 6.1 RAFT polymerization of unpurified commercial MMA with (2-cyano-4-carboxy)but-2-yl dodecyl trithiocarbonate as RAFT agent under various conditions.[a] Table reproduced from ref. 15.

Degassing	Reaction Time/h	Conversion (%)	$\bar{M}_n{}^b$	$\bar{M}_w/\bar{M}_n$
Freeze–evacuate–thaw	1.5	93	100 000	1.25
N$_2$ flush	1.5	87	90 000	1.25
In air[c]	1.5	26	38 000	1.38
In air[c]	3.0	88	112 000	1.26

[a]Polymerization of 10 g MMA with 20 mg of RAFT agent and 5 mg of AIBN at 80 °C for the times indicated.
[b]GPC molar mass in polystyrene equivalents.
[c]No degassing.

80 °C with thorough degassing by freeze-evacuate-thaw cycles, a simple nitrogen flush, or no degassing are shown in Table 6.1. All polymerizations provide poly(MMA) with relatively narrow molecular weight distributions and some degree of molecular weight control. However, polymerizations with no degassing gave substantial retardation. All polymerizations provided a lower than expected molecular weight which was attributed to the use of unpurified monomer.

6.10.5 Slow Fragmentation and/or Intermediate Radical Termination

The properties of the intermediate radicals in RAFT polymerization may lead to retardation. If fragmentation of the intermediate radical is very slow this can cause retardation directly. The estimation of RAFT equilibrium constants using *ab initio* molecular orbital calculations provides some credibility for this hypothesis. However, direct measurement of intermediate radical concentrations in RAFT polymerization as measured by EPR, indicates that these species do not have sufficient life-time for slow fragmentation, by itself, to be the cause of retardation.

If intermediate radicals are consumed in side reactions, such as combination or disproportionation with another radical species, this will also cause retardation. That so-called intermediate radical termination might complicate RAFT polymerization was first proposed by Monteiro and de Brouwer in 2001.[114] That a polystyrene intermediate radical can combine with a polystyrene propagating radical to form a stable three-armed star has been demonstrated. However, attempts to detect these species in the expected concentrations in polymerization have failed.

Monitoring the kinetics of RAFT polymerization, specifically the rate of monomer depletion, does not allow discrimination between these models or the various hybrid models that have been proposed. Klumpermann and Heuts[115] re-examined the molecular orbital calculations of Coote and coworkers and experimental determinations of rate parameters associated with RAFT polymerization using dithiobenzoate agents. They concluded that while the available data do not allow model discrimination between the slow fragmentation model and the intermediate radical termination model, the apparent incompatibility of the models was less than has been suggested in some papers.

6.11   Reaction Conditions for RAFT Polymerization

The polymerization conditions for solution or bulk RAFT polymerization closely resembles those used in conventional radical polymerization.

6.11.1   Temperature

Temperatures reported for RAFT polymerization range from ambient to 180 °C.[116] There is evidence with that retardation, when observed, is less at higher temperatures. There is also some data that show lower dispersities can be achieved at higher temperatures.[31] This is consistent with the rate coefficients for fragmentation of the RAFT intermediates and transfer constants of RAFT agents all increasing with increasing reaction temperature.

There have been several studies on the thermal stability of RAFT agents and RAFT-synthesized polymers and the influence of this on the outcome of RAFT polymerization. Cumyl dithiobenzoate appears substantially less stable than benzyl or phenylethyl dithiobenzoate and degrades rapidly at temperatures >100 °C.[117] The instability was attributed to reversible formation of α-methylstyrene (AMS) and dithiobenzoic acid. The success of high temperature polymerization (of, for example, styrene) was attributed to the fact that the RAFT agent was rapidly consumed and converted to more stable polymeric RAFT agents. It was also reported that the poor control in synthesis of poly(MMA) with dithiobenzoate RAFT agents at higher temperatures (120 °C) could be attributed to the lability of the dithiobenzoate end group.[118] More recent work,[119] while confirming that thermolysis is a suitable method for end group removal, indicates that dithiobenzoate end groups of RAFT-synthesized poly(MMA) are stable to much higher temperatures (180 °C).

6.11.2   Microwave

Two reviews on microwave-assisted polymerization have recently been published.[120,121] RAFT polymerizations of "polar" monomers (MMA,[122] MA[122] VAc[123,124] and DADMAC[125]) are reported to be substantially accelerated by microwave heating. Less, but still substantial acceleration, was observed for styrene polymerization.[123,126] It is expected that monomers with a higher dielectric constant will be more effectively heated by microwave irradiation. However, the microwave effect particularly with MMA and MA was substantially greater than expected for an effect of temperature alone.[122] An explanation for the microwave effect was not provided.

Two groups[127,128] have reported on kinetic-modeling of microwave accelerated RAFT polymerization of styrene in an attempt to gain further understanding of the process. Both came to the conclusion that there is a "microwave effect". The first study[127] attributed accelerated polymerization to an additional initiation process. The second study[128] discounted this possibility and proposed that the results were consistent with the rate constants for addition to monomer (k_p) and RAFT agent (k_{add}) both being accelerated to the same extent under microwave irradiation.

A paper by Smith *et al.*[129] examined whether a "microwave effect" on the decomposition of some radical initiators and came to the conclusion "the use of microwaves with these species cannot be expected to either increase the radical flux at a particular temperature or reduce the temperature at which these materials decompose". They suggested that a "microwave effect" on propagation in radical polymerization was more likely.

Two independent studies[130,131] compared RAFT polymerization (of MMA, BA, VAc, NIPAM or DMAM) in conventionally-heated continuous-flow tubular reactors with those in a microwave-heated batch reactor. A rapid rate of polymerization was observed for both systems. There were no significant differences in the polymerization kinetics or in the molecular weight or dispersity of the polymer produced. These results suggest that there is no "microwave effect" beyond that of rapid heating of the reaction medium.

6.11.3 Pressure

Very high pressures (> 5 kbar) have been shown to improve the effectiveness of RAFT polymerization.[116,132–134] It is known that at these very high pressures, radical–radical termination is slowed and it is thought that this allows the formation of higher molar mass polymers and higher rates of polymerization than are achievable at ambient pressure.

6.11.4 Medium

6.11.4.1 Solution Polymerization

The RAFT process is compatible with all common organic solvents, including protic solvents such as alcohols, and also with water[39,135,136] and less conventional solvents such as ionic liquids[137] and supercritical carbon dioxide.[138,139] In solution polymerization the RAFT agent should be selected for solubility in the reaction medium. In protic media some RAFT agents show hydrolytic sensitivity which increases with RAFT agent activity (dithiobenzoates > trithiocarbonates ∼ aliphatic dithioesters).[74,101,140] RAFT agents are also susceptible to nucleophilic attack by primary and secondary amines and thiols.

6.11.4.2 Heterogeneous Polymerization

Much has been written on RAFT emulsion, miniemulsion and dispersion polymerization.[141–143] The first communication on RAFT polymerization briefly mentioned the successful emulsion polymerization of butyl methacrylate with cumyl dithiobenzoate (feed emulsion polymerization at 80 °C with azo(cyanovaleric acid) initiator and sodium dodecyl sulfate surfactant to give, poly(BMA) with $M_n = 57\,700$, $Đ = 1.22$ and $> 95\%$ monomer conversion).[39] Additional examples and discussion of some of the important factors for successful use of RAFT polymerization in emulsion and miniemulsion were provided in a subsequent paper.[144] Success in RAFT emulsion polymerization depends strongly on the monomer, the choice of RAFT agent and poly- merization conditions.[144–153] The use of cumyl dithiobenzoate as a RAFT agent in *ab initio* emulsion polymerization of styrene was not recommended.[144] Many papers describe the failure of RAFT to provide control in that system.[154,155] The emulsion recipes reported in this early work[39,144] were feed processes in which the conversion of monomer to polymer was maintained at a high level (often $> 90\%$). In a first *ab initio* step a low molar mass ($Đ \sim 1.4$–2.0) macro- RAFT agent was prepared. The control obtained in this step need not substantially affect what is achieved in the later stages of polymerization. The macro-RAFT constituted a RAFT-containing seed during the subsequent feed emulsion polymerization. In general, the RAFT agents with lower transfer constants, such as xanthates, offer best control in *ab initio* batch RAFT polymerization.[156]

Many papers now describe the use of RAFT in surfactantless emulsion polymerization.[141,151,157–159] We described the formation of 'self stabilizing lattices' in macromonomer sequential feed RAFT polymerization of metha- crylic acid and non-polar methacrylates.[160] Gilbert, Hawkett and coworkers[148,151] applied a similar strategy in thiocarbonylthio RAFT poly- merization. In a first step, a water-soluble monomer, acrylic acid (AA), was polymerized to a low degree of polymerization in aqueous solution to form a poly(AA) macro RAFT agent. A hydrophobic monomer (BA) was then added under controlled feed to give oligomers which form rigid micelles. These constitute a RAFT-containing seed. Continued controlled feed of hydrophobic monomer may be used to continue the emulsion polymerization.

6.12   Polymerization of 'More-Activated' Monomers (MAMs)

Aromatic dithioesters (Z = aryl, *e.g.*, 27–29) are amongst the most active RAFT agents and have general utility in the polymerization of MAMs.[36,52] However, retardation is observed with these RAFT agents and this is particularly apparent when used in high concentrations to give low molar mass polymers. They also show greater sensitivity to hydrolysis and decomposition induced by Lewis acids than less active RAFT agents.[104,161]

The aliphatic dithioesters (Z = alkyl, *e.g.*, 33) show good control over monosubstituted MAMs (styrene, acrylates) but have lower transfer constants in polymerization of 1,1-disubstituted MAMs and provide higher dispersities. These RAFT agents are less prone to hydrolytic decomposition and give less retardation than the aromatic dithioesters.

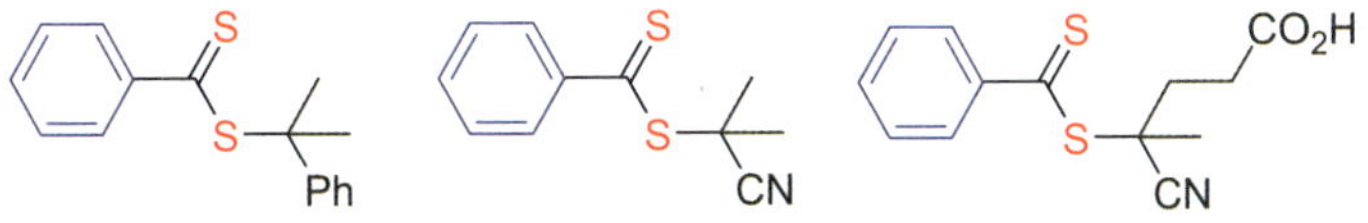

27   28   29

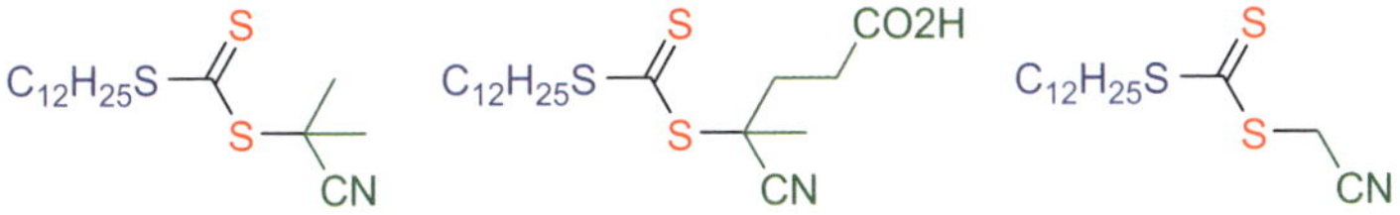

30   31   32

The trithiocarbonates ($Z = S$-alkyl *e.g.*, **30–32** and **35**, or S-aryl, *e.g.*, **34**),[36,52,162] the aromatic dithiocarbamates ($Z = N$-heteroaryl, *e.g.*, **36**)[36,52,92] and the more active form of the switchable RAFT agents (see Scheme 6.8), while less active than the aromatic dithioesters, show good control over polymerization of both mono- and 1,1-disubstituted MAMs. For the unsymmetrical trithiocarbonates, Z is preferably based on a thiol with low volatility, *e.g.*, dodecanethiol.

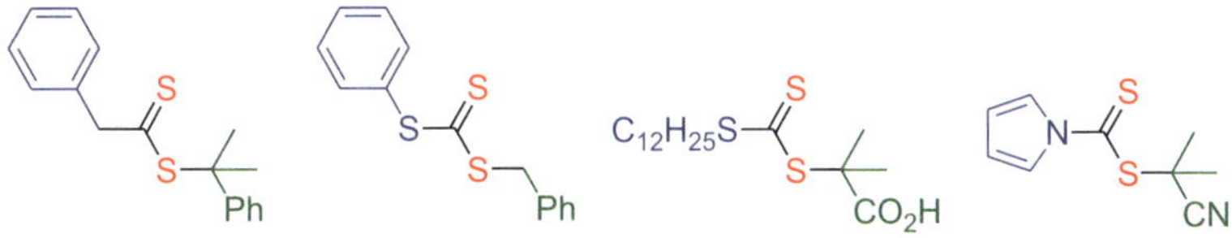

33   34   35   36

The choice of R is critical in the case of methacrylates and other 1,1-disubstituted MAMs. In some of the most effective RAFT agents R is tertiary cyanoalkyl (*e.g.*, **28**, **29**, **30**, **31**, **36**). RAFT agent **35**, where R is 2-carboxyprop-2-yl, is less suitable because the 2-carboxyprop-2-yl is a poor homolytic leaving group with reference to the methacrylate propagating radical. RAFT agents with primary or secondary R (*e.g.*, **32** where R is cyanomethyl) are ineffective with methacrylates for similar reasons.

The utility of the RAFT process in MAM polymerization is illustrated by the example shown in Figure 6.9 for RAFT polymerization of methyl methacrylate (MMA). A series of high (80–100%) conversion MMA polymerizations were carried out at 90 °C with 1,1'-azobis(1-cyclohexanenitrile) initiator, and using a ~60-fold range of initial RAFT agent concentrations.[69] The values of M_n, ranging from 2600 to 125 000, vary in accord with the concentrations of RAFT agent and initiator used.[69] All samples have narrow molecular weight distributions ($Đ < 1.2$).

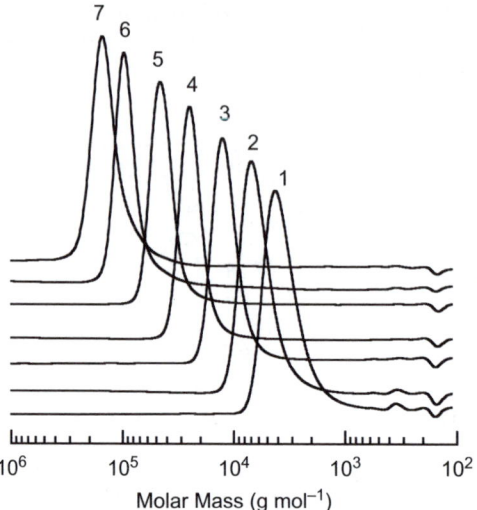

Figure 6.9 GPC traces of poly(MMA) formed by polymerization of MMA (6.55 M in benzene) with azobis(1-cyclohexanenitrile) (0.0018M) as initiator and a reaction time of 6 h at 90 °C. For trace 1, $[RAFT]_0$ 19.92×10^2 M, M_n 2870, $Đ$ 1.18, M_n(calc) 3000, 80% conversion; trace 2, $[RAFT]_0$ 9.96×10^2 M, M_n 5040, $Đ$ 1.14, M_n(calc) 5600, 80% conversion; trace 3, $[RAFT]_0$ 4.95×10^2 M, M_n 9940, $Đ$ 1.12, M_n(calc) 10 400, 79% conversion; trace 4, $[RAFT]_0$ 2.48×10^2 M, M_n 21 800, $Đ$ 1.13, M_n(calc) 22 600, 91% conversion; trace 5, $[RAFT]$ 1.24×10^2 M, M_n 41 100, $Đ$ 1.14, M_n(calc) 45 300, >99% conversion; trace 6, $[RAFT]_0$ 0.61×10^2 M, M_n 80 900, $Đ$ 1.13, M_n(calc) 80 100, >99% conversion; trace 7, $[RAFT]_0$ 0.32×10^2 M, M_n 126 000, $Đ$ 1.15, M_n(calc) 125 000, >99% conversion.
Figure reproduced from ref. 69, © Elsevier.

6.13 Polymerization of 'Less-Activated Monomers' (LAMs)

Good control over the polymerization of LAMs requires use of a less active RAFT agent such as a dithiocarbamate ($Z = NR'_2$) or a xanthates ($Z = OR'$) with $R' = $ alkyl or aryl. The more active RAFT agents $Z = R$ (dithioesters) or SR (trithiocarbonates) strongly retard or inhibit polymerization of LAMs. The choice of the R group is also critical. Inhibition periods due to slow reinitiation are expected for RAFT agents such as **37** (R = 2-cyano-2-propyl) and **38** (R = benzyl).

Some preferred RAFT agents are **39** and **40** and the less active form or the switchable RAFT agents (see Scheme 6.8).

37   **38**   **39**   **40**

6.14   Gradient Copolymers

In most copolymerization, the monomers are consumed at different rates that are dictated by the steric and electronic properties of the reactants. Consequently, both the monomer feed and copolymer composition will drift with conversion. Thus, conventional copolymers are generally not homogeneous in composition at the molecular level. In RAFT polymerization processes, where all chains grow throughout the polymerization, compositional drift is captured within the chain structure (Scheme 6.9). All chains have similar composition and the copolymers formed have a gradient or tapered structure as poly(monomer A-*grad*-monomer B).

Reactivity ratios are generally unaffected by the RAFT process. However, for very low conversions when molecular weights are low, copolymer composition may be different from that seen in conventional copolymerization because of specificity shown in the initiation step by the radical (R). The same phenomenon is observed in radical polymerization with conventional chain transfer when molecular weights are low.[56,163,164] Note that these 'low conversion' conditions correspond to those recommended for measuring reactivity ratios. It is likely that the reports of apparent dependence of reactivity ratios on the presence of RAFT agent can be attributed to this.

A wide variety of copolymers have been synthesized by RAFT polymerization and all copolymers formed by RAFT copolymerization are gradient copolymers. However, few have set out to explicitly exploit this process. RAFT copolymerization can be successful (provide molecular weight control and narrow molecular weight distributions) even when one of the monomers is not amenable to direct homopolymerization using a particular RAFT agent. For example, severe retardation is commonly observed for NVP polymerization in the presence of trithiocarbonate RAFT agents,[58] yet copolymerization of NVP with an acrylate monomer provides good control and little retardation.[165] Benzyl dithiobenzoate provides little control over MMA polymerisation, yet a low dispersity polymer can be obtained by addition of as little as 5% styrene.[15] The reactivity ratios dictate that copolymerizations of excess styrene with maleic anhydride provides poly(St-*alt*-MAH)-*block*-poly(St).[166] Many other examples can be envisaged.

6.15   Block Copolymers

RAFT polymerization is recognized as one of the most versatile methods for block copolymer synthesis and numerous examples of block synthesis have now appeared. RAFT polymerization proceeds with retention of the

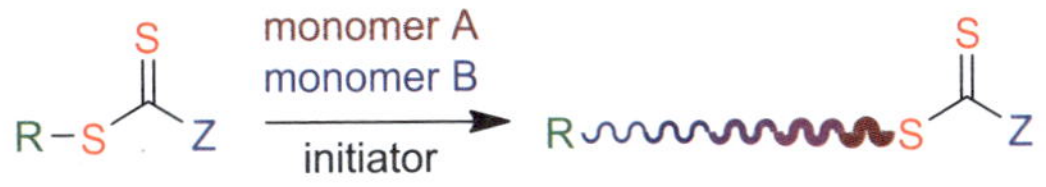

Scheme 6.9   Gradient copolymer synthesis.

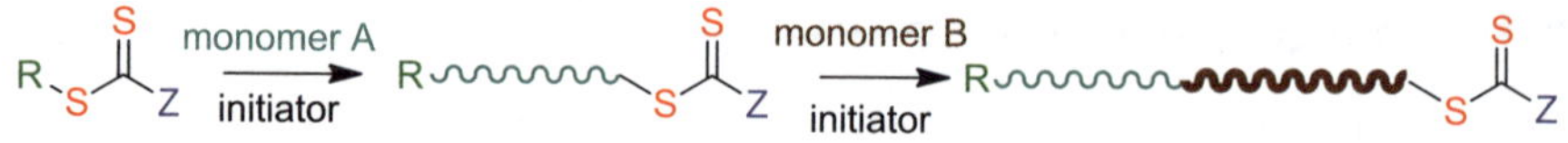

Scheme 6.10   AB diblock synthesis by sequential monomer addition.

thiocarbonylthio group. This allows an easy entry to the synthesis of **AB** diblock copolymers by the simple addition of a second monomer (Scheme 6.10).[167,168] Higher order (ABA, ABC, *etc.*) blocks are also possible by sequential addition of further monomer(s).

Of considerable interest has been the ability to use RAFT to make hydrophilic–hydrophobic or double hydrophilic block copolymers where one hydrophilic block is composed of unprotected polar monomers such as MAA or DMAEMA.

As with other RDRP, the order of constructing the blocks is very important.[47,167] In RAFT polymerization the propagating radical for the first formed block should be chosen such that it is a good homolytic leaving group with respect to that of the second block. Thus, the considerations discussed above with respect to selection of R have to be considered in designing the synthesis of block copolymers by RAFT polymerization. In the synthesis of a block copolymer comprising segments of a 1,1-disubstuted monomer and a monosubstituted monomer, the block comprising the 1,1-disubstuted monomer should be prepared first.[47] Similarly, in synthesizing a poly(MAM)-*block*-poly(LAM), using switchable RAFT,[67,68] the poly(MAM) block should be made first; both because poly(LAM) propagating radicals are relatively poor homolytic leaving groups and because poly(MAM) propagating radicals are slow to reinitiate LAM polymerization. The use of feed addition protocols, where the monomer concentration is kept low with respect to the RAFT agent concentration, can be used to circumvent this requirement.[19,169] Thus, while a polystyrene macro-RAFT agent appears essentially inert in batch solution polymerization of MMA, poly(St)S-*block*-poly(MMA) has been successfully prepared by feed emulsion polymerization starting with a poly(St) macro RAFT agent.[19] This strategy is also applied when synthesizing block copolymers from macromonomer RAFT agents. Another work-around is to maintain a small amount of an appropriate comonomer in the feed.[15] For block copolymers where the leaving group ability of the propagating species is of the same order of magnitude, the order of construction is less critical. Thus, in synthesis of block copolymers of styrene with an acrylic monomer either block can be made first.

The synthesis of block copolymers based on polymers formed by other mechanisms can be made by first preparing an end functional pre-polymer which is converted to a polymer with thiocarbonylthio groups by end group transformation. This is then used as a macro-RAFT agent in preparation of the desired block copolymer (Scheme 6.11). We first exploited this methodology to prepare poly(ethylene oxide)-*block*-poly(St) from commercially available hydroxy end-functional poly(ethylene oxide).[40,70,167]

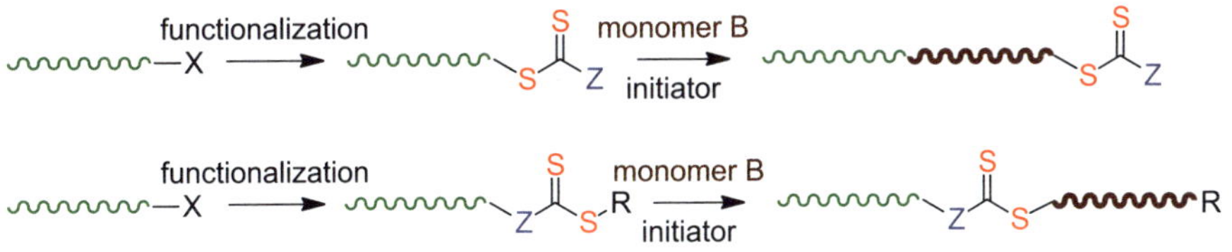

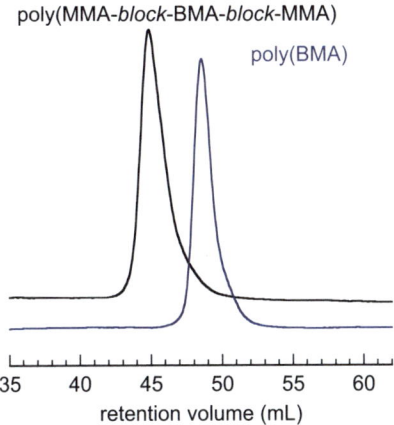

Scheme 6.11 A-B diblock synthesis from end-functional polymers *via* the RAFT process.

poly(MMA-*block*-BMA-*block*-MMA)

poly(BMA)

35   40   45   50   55   60

retention volume (mL)

Figure 6.10 GPC traces for poly(butyl methacrylate) (M_n 35 500, Đ 1.09, ——) and poly(methyl methacrylate-*block*-butyl methacrylate-*block*-methyl methacrylate) (M_n 112 200, Đ 1.14, ——) prepared by sequential monomer addition to the bis RAFT agent (**41**). For details of synthesis see ref. 167.

Use of a bis-RAFT agent (*e.g.*, **41**) allows the direct synthesis of triblock copolymers in a 'one-pot' reaction (Figure 6.10). The RAFT agent functionalities may be connected through the Z or R groups to give ABA (Scheme 6.12) or BAB blocks (Scheme 6.13), respectively. The use of multi-RAFT agents similarly allows the synthesis of star polymers. Symmetrical mono-trithiocarbonates (*e.g.*, **42**) can be considered as 'Z connected' bis-RAFT agents ($n = 0$ in Scheme 6.12) in this context.

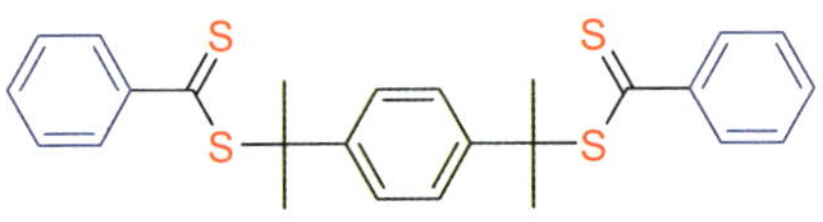

41 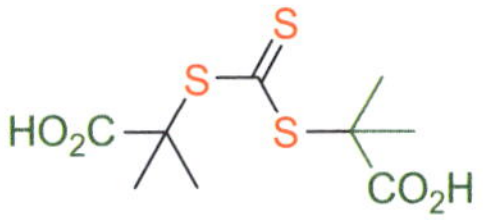   **42**

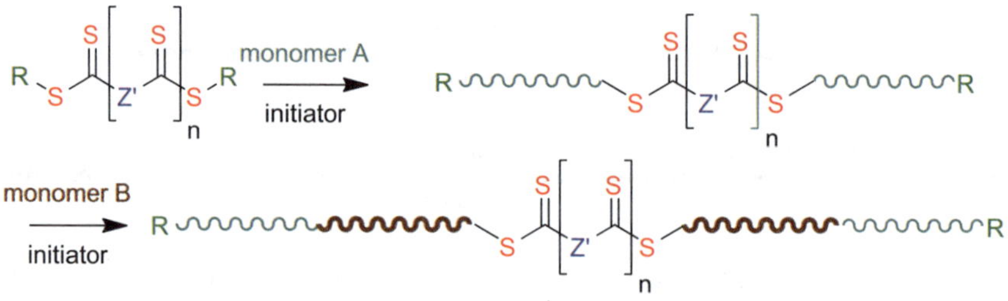

Scheme 6.12   ABA triblock synthesis from 'Z-connected' bis-RAFT agents ($n = 1$) or symmetrical trithiocarbonates ($n = 0$).

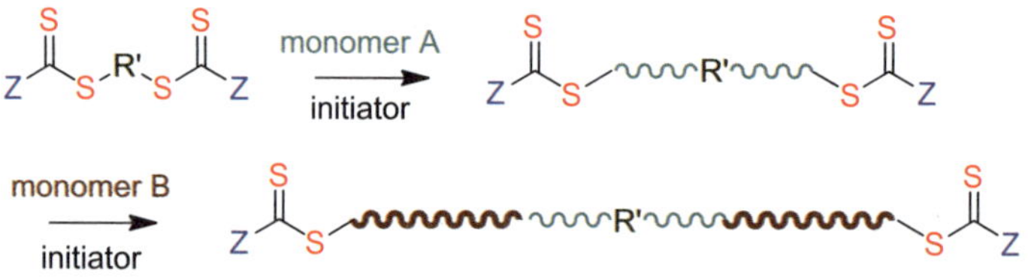

Scheme 6.13   B-A-B triblock synthesis from 'R-connected' bis-RAFT agents.

6.16   More Complex Architectures

A large number of papers address the synthesis of more complex architectures using the RAFT process. These polymers include stars, grafts, brushes, microgels, and a variety of supramolecular assemblies. A common approach to stars, grafts and brushes begins with a compound containing multiple thiocarbonylthio groups of appropriate design; a multi-RAFT agent. The synthesis of polymers from multi-RAFT agents can be seen as an extension of the triblock syntheses described above where the number of thiocarbonylthio groups exceeds two. The multi-RAFT agent may be a small organic compound,[70,144,170–174] an organometallic complex,[175] a dendrimer,[176–178] a hyperbranched species,[179] a macromolecular species,[76,180] a particle, or indeed, any moiety possessing multiple thiocarbonylthio groups. Our first RAFT patent[181] recognized two limiting forms of star growth depending on the orientation of the thiocarbonylthio group with respect to the core. We have discussed the advantages and disadvantages of the 'propagation away from core' (Scheme 6.14) and 'propagation attached to core' (Scheme 6.15) strategies in various publications.[70,144,171]

An advantage of the 'propagation away from core' strategy is that, since propagating radicals are never directly attached to the core, by-products from star–star coupling are unlikely. Radical–radical termination will involve linear propagating radicals and will produce a low molecular weight by-product that may be observed in high conversion polymerizations. All of the thiocarbonylthio functionality remains at the core of the star structure. It has been

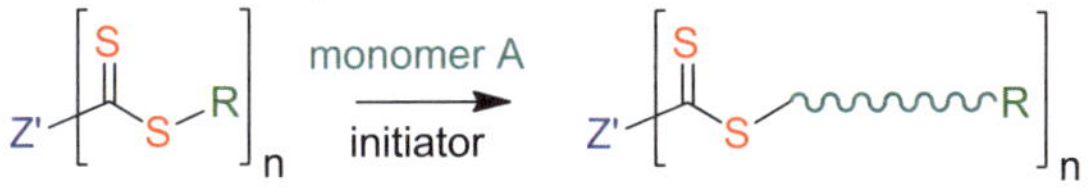

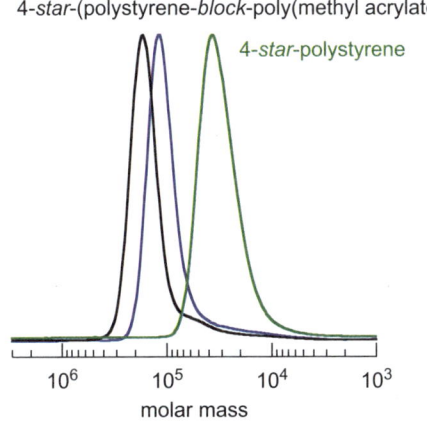

Scheme 6.14 Star polymers synthesis by the 'propagation away from core' strategy.

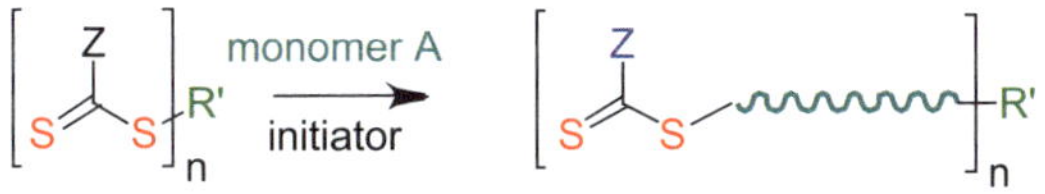

Scheme 6.15 Star polymers synthesis by the 'propagation attached to core' strategy.

4-*star*-(polystyrene-*block*-poly(methyl acrylate))

4-*star*-polystyrene

10⁶   10⁵   10⁴   10³

molar mass

Figure 6.11 Molecular weight distributions for four arm polystyrene (M_n 25 550, Đ 1.1.18, ——) and 4-*star*-(polystyrene-*block*-poly(methyl acrylate)) (M_n 87 700, Đ 1.24, —— and M_n 129 600, Đ 1.23, ——) prepared by sequential monomer addition to the tetrafunctional RAFT agent (**43**). For details of synthesis see ref. 171.

suggested that steric factors associated with attack of the propagating radical at the core of the star should be an issue particularly at high conversions. There is, however, no direct evidence of problems attributable to this cause in the examples we have reported. An example of star synthesis using this strategy and RAFT agent **43** is shown in Figure 6.11.

A potential disadvantage of the 'propagation away from core' strategy is that any reaction which cleaves the thiocarbonylthio groups (*e.g.* hydrolysis, thermolysis) results in cleavage of arms from the star structure. This property may be turned into an advantage in some circumstances; for example, in polymer supported polymer synthesis or in designing drug delivery vehicles. Arm cleavage also enables the molecular weights of the arms and the uniformity of arm growth to be readily verified.[171]

43

With the 'propagation attached to core' strategy, the thiocarbonylthio functionality remains at the periphery of the star and the majority of propagating radicals are attached to the core. Star–star coupling is therefore a potential complication. A fraction of linear dormant chain commensurate with the number of initiator-derived chains will be formed as a (low molar mass) byproduct. The likelihood of star–star coupling increases with the number of arms to the star but can be minimized by controlling the conversion and rate of initiation. Examples of star synthesis using this strategy and RAFT agents **44** is shown in Figure 6.12.

44

The so called 'arm-first' process for microgel synthesis[183,184] involves making a macro-RAFT agent then using this to initiate (co)polymerization of a di- (or higher) functional monomer (*e.g.* divinylbenzene) which results in a crosslinked

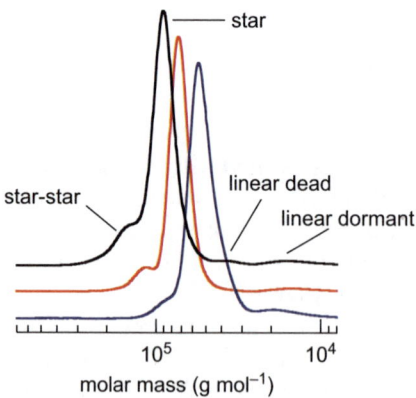

Figure 6.12 Molecular weight distributions for 8 arm (M_n 114000, Đ 1.07, 52% conversion, ——). 6 arm (M_n 92000, Đ 1.04, 50% conversion, ——) and 3 arm (M_n 55000, Đ 1.11, 59% conversion, ——) polystyrene prepared with RAFT agents (**44**).[182]

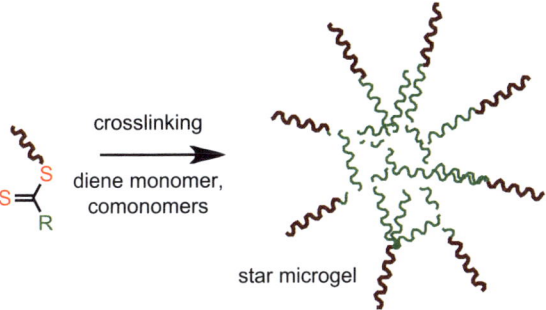

Scheme 6.16 Arm-first method for synthesis of core-crosslinked microgels.

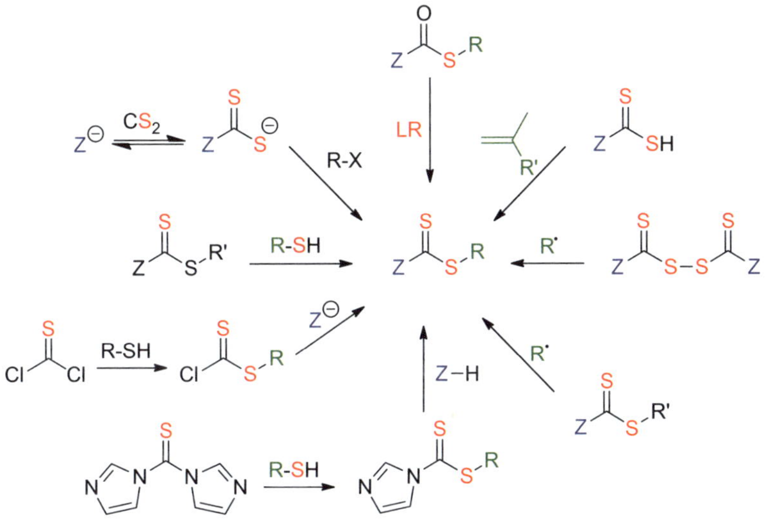

Scheme 6.17 Major methods used for RAFT agent synthesis (RX = alkylating agent, LR = Lawesson's reagent or equivalent).
Scheme reproduced from ref. 185, © American Chemical Society.

core (Scheme 6.16). The core should retain the RAFT functionality which allows for further arms to be added in a subsequent RAFT polymerization step.

6.17 Synthesis of RAFT Agents

Only a limited selection of RAFT agents is commercially available. Many procedures for the synthesis of RAFT agent have been described and the major ones are summarized in Scheme 6.17 which is taken from our recent review.[185] The most commonly used involves the reaction of a carbodithioate salt with an alkylating agent. However, other processes have important niche applications. These include various thioacylation reactions, the thiation of the corresponding carbonyl compounds, the ketoform reaction, thiol exchange, and radical substitution of bis(thioacyl) disulfides and other thiocarbonylthio compounds.

Much activity in RAFT agent synthesis involves functional RAFT agents and their use in forming macro- and surface attached-RAFT agents with particular applications in such fields as biomedicine[186] and optoelectronics.[187] Common strategies for the preparation of macro- and surface attached-RAFT agents involve esterification or amidation of a carboxy-functional RAFT agent and reactions such as azide-alkyne 1,3-dipolar cycloaddition, the active ester amine reaction or RAFT single unit monomer insertion.[185]

6.18   End Group Transformation

Removal or transformation of the thiocarbonylthio group may be required for some applications and a range of methods has been devised that can be readily incorporated into polymer syntheses.[57,58,69,119,188–190]

One major advantage of RAFT polymerization over many other RDRP techniques, such as atom transfer radical polymerization (ATRP)[8,11,191] and nitroxide mediated polymerization (NMP),[3] is its tolerance of functionality which is such that a wide range of groups can be introduced as substituents on R or Z groups. This functionality includes that for use in so-called "click" reactions. Characteristics of "click" reactions are (a) high yields with by-products (if any) that are simply removed by non-chromatographic processes, (b) high regiospecificity and stereospecificity, (c) insensitivity to oxygen and water, (d) mild, solventless reaction conditions, (e) orthogonality with other reactions, and (f) amenability to a wide variety of readily available starting materials. A number of recent reviews have focused on the combination of "click" chemistry and polymer chemistry.[192–197]

Another key feature of RAFT polymerization is that the thiocarbonylthio groups, present in the initial RAFT agent, are retained in the polymeric product. This feature is responsible for the living character of RAFT polymerization and renders the process suitable for synthesizing block copolymers and end functional polymers.

Transformation of the thiocarbonylthio group is an integral part of many polymer syntheses. The reactions of the thiocarbonylthio-group are well known from small molecule chemistry[198–201] and much of this knowledge has been shown applicable to transforming the thiocarbonylthio-groups present in RAFT-synthesized polymers.[39] Some of the methods used for thiocarbonylthio-group removal are summarized in Scheme 6.18 which is taken from our recent review.[202] Thiocarbonylthio-groups undergo reaction with nucleophiles and ionic reducing agents typically to produce a polymer with a thiol end-group. They also react with various oxidizing agents and are sensitive to UV irradiation. The thiocarbonylthio-group can be transformed directly by participation as a dieneophile in the hetero Diels Alder reaction[203,204] or as a dipolarophile in 1,3-dipolar cycloaddition.[103]

The thiocarbonylthio-group may also be transformed or, in some cases, be used directly in other forms of radical polymerization, such as atom transfer

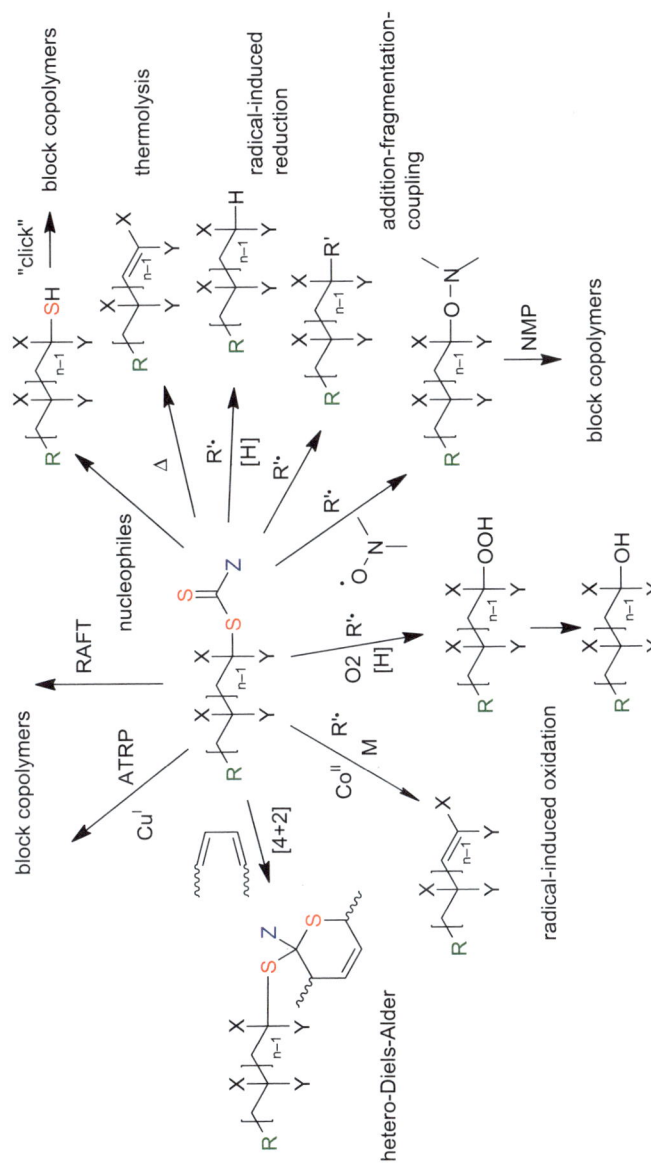

Scheme 6.18 Processes for RAFT end-group transformation (R' = radical, [H] = H atom donor, M = monomer, Co^{II} = square planar cobalt complex). Adapted from ref. 202.

radical polymerization (ATRP)[205,206] or nitroxide mediated polymerization (NMP),[207] and, as we have recently demonstrated, certain thiocarbonylthio-groups can be "switched" to enable control over polymerization of a wider range of monomers in the RAFT process.[67,68] Radical chemistry can also be used to introduce new functionality with the addition of chain transfer agents.

Reviews focussing on end-functional polymers include those by Moad *et al.*,[69,202] Willcock and O'Reilly,[208] Harvison *et al.*,[209] and Barner and Perrier.[210] Other reviews that include significant sections on end-functional polymers and end-group transformation include our previous reviews of the RAFT process,[15,16,18,36] that by Boyer *et al.*[186] on biomedical applications and that by Moad *et al.*[187] on optoelectronic applications.

6.19   Conclusions

RAFT polymerization has emerged as one of the most important methods for imparting living characteristics to radical polymerization. RAFT has been shown to be robust and has applicability to the majority of monomers that are subject to radical polymerization. However, selection of the RAFT agent for the monomers to be polymerized and the choice of reaction conditions are crucial for success. With these provisos, RAFT polymerization can be used in the synthesis of well-defined homo-, gradient, diblock and triblock polymers as well as more complex architectures which include higher order blocks, stars and polymer brushes.

References

1. A. D. Jenkins, R. I. Jones and G. Moad, *Pure Appl. Chem.*, 2010, **82**, 483–491.
2. G. Moad, E. Rizzardo and S. H. Thang, *Acc. Chem. Res.*, 2008, **41**, 1133–1142.
3. C. J. Hawker, A. W. Bosman and E. Harth, *Chem. Rev.*, 2001, **101**, 3661–3688.
4. D. Bertin, D. Gigmes, S. R. A. Marque and P. Tordo, *Chem. Soc. Rev.*, 2011, **40**, 2189–2198.
5. D. H. Solomon, E. Rizzardo and P. Cacioli *Polymerization Process and Polymers Produced Thereby*. CSIRO, US4581429, 1986.
6. M. K. Georges, R. P. N. Veregin, P. M. Kazmaier and G. K. Hamer, *Macromolecules*, 1993, **26**, 2987–2988.
7. G.Moad and D. H.Solomon, *The Chemistry of Radical Polymerization*, Elsevier, Oxford, 2006, pp. 168–176, 413–449.
8. K. Matyjaszewski and J. Xia, *Chem. Rev.*, 2001, **101**, 2921–2990.
9. K. Matyjaszewski, *Macromolecules*, 2012, **45**, 4015–4039.
10. M. Kamigaito, T. Ando and M. Sawamoto, *Chem. Rev.*, 2001, **101**, 3689–3745.
11. M. Ouchi, T. Terashima and M. Sawamoto, *Chem. Rev.*, 2009, **109**, 4963–5050.

12. J.-S. Wang and K. Matyjaszewski, *J. Am. Chem. Soc.*, 1995, **117**, 5614–5615.
13. M. Kato, M. Kamigaito, M. Sawamoto and T. Higashimura, *Macromolecules*, 1995, **28**, 1721–1723.
14. V. Percec and B. Barboiu, *Macromolecules*, 1995, **28**, 7970–7972.
15. G. Moad, E. Rizzardo and S. H. Thang, *Aust. J. Chem.*, 2005, **58**, 379–410.
16. G. Moad, E. Rizzardo and S. H. Thang, *Aust. J. Chem.*, 2006, **59**, 669–692.
17. G. Moad, E. Rizzardo and S. H. Thang, *Polymer*, 2008, **49**, 1079–1131.
18. G. Moad, E. Rizzardo and S. H. Thang, *Aust. J. Chem.*, 2009, **62**, 1402–1472.
19. G. Moad, J. Chiefari, J. Krstina, A. Postma, R. T. A. Mayadunne, E. Rizzardo and S. H. Thang, *Polym. Int.*, 2000, **49**, 993–1001.
20. E. Rizzardo, J. Chiefari, R. T. A. Mayadunne, G. Moad and S. H. Thang, *ACS Symposium Series*, 2000, **768**, 278–296.
21. T. Otsu, M. Yoshida and A. Kuriyama, *Polym. Bull.*, 1982, **7**, 45–50.
22. T. Otsu and M. Yoshida, *Makromol. Chem., Rapid Commun.*, 1982, **3**, 127–132.
23. T. Otsu, *J. Polym. Sci., Part A: Polym. Chem.*, 2000, **38**, 2121–2136.
24. B.Giese, *Radicals in Organic Synthesis: Formation of Carbon-Carbon Bonds*, Pergamon Press, Oxford, 1986.
25. W. B.Motherwell and D.Crich, *Free Radical Chain Reactions in Organic Synthesis*, Academic Press, London, 1992.
26. S. N. Lewis, J. J. Miller and S. Winstein, *J. Org. Chem.*, 1972, **37**, 1478–1485.
27. G. E. Keck, E. J. Enholm, J. B. Yates and M. R. Wiley, *Tetrahedron*, 1985, **41**, 4079–4094.
28. D. H. R. Barton and S. W. McCombie, *J. Chem. Soc., Perkin Trans. 1*, 1975, 1574–1585.
29. E. Rizzardo, Y. K. Chong, R. A. Evans, G. Moad and S. H. Thang, *Macromol. Symp.*, 1996, **111**, 1–12.
30. E. Rizzardo, G. F. Meijs and S. H. Thang, *Macromol. Symp.*, 1995, **98**, 101–123.
31. J. Chiefari and E. Rizzardo, in *Handbook of Radical Polymerization*, eds. T. P. Davis and K. Matyjaszewski, John Wiley & Sons, Hoboken, NY, 2002, pp. 263–300.
32. G. Moad and D. H. Solomon, *The Chemistry of Radical Polymerization*, Elsevier, Oxford, 2006, pp. 279–331.
33. D. Colombani and P. Chaumont, *Prog. Polym. Sci.*, 1996, **21**, 439–503.
34. D. Colombani, *Prog. Polym. Sci.*, 1999, **24**, 425–480.
35. Y. Yagci and I. Reetz, *React. Funct. Polym.*, 1999, **42**, 255–264.
36. G. Moad, E. Rizzardo and S. H. Thang, *Polymer*, 2008, **49**, 1079–1131.
37. J. Krstina, G. Moad, E. Rizzardo, C. L. Winzor, C. T. Berge and M. Fryd, *Macromolecules*, 1995, **28**, 5381–5385.
38. J. Krstina, C. L. Moad, G. Moad, E. Rizzardo, C. T. Berge and M. Fryd, *Macromol. Symp.*, 1996, **111**, 13–23.

39. J. Chiefari, Y. K. Chong, F. Ercole, J. Krstina, J. Jeffery, T. P. T. Le, R. T. A. Mayadunne, G. F. Meijs, C. L. Moad, G. Moad, E. Rizzardo and S. H. Thang, *Macromolecules*, 1998, **31**, 5559–5562.
40. T. P. Le, G. Moad, E. Rizzardo and S. H. Thang, *Polymerization with Living Characteristics* DuPont/CSIRO, WO9801478, 1998.
41. P. Corpart, D. Charmot, T. Biadatti, S. Zard and D. Michelet *Block polymer synthesis by controlled radical polymerization* (Rhodia Chimie, Fr.). WO9858974, 1998.
42. J. P. A. Heuts and N. M. B. Smeets, *Polym. Chem.*, 2011, **2**, 2407–2423.
43. A. Gridnev, *J. Polym. Sci., Part A: Polym. Chem.*, 2000, **38**, 1753–1766.
44. P. Cacioli, D. G. Hawthorne, R. L. Laslett, E. Rizzardo and D. H. Solomon, *J. Macromol. Sci., Chem.*, 1986, **A23**, 839–852.
45. C. L. Moad, G. Moad, E. Rizzardo and S. H. Thang, *Macromolecules*, 1996, **29**, 7717–7726.
46. T. Corner, *Adv. Polym. Sci.*, 1984, **62**, 95–142.
47. Y. K. Chong, J. Krstina, T. P. T. Le, G. Moad, A. Postma, E. Rizzardo and S. H. Thang, *Macromolecules*, 2003, **36**, 2256–2272.
48. J. Chiefari, J. Jeffery, R. T. A. Mayadunne, G. Moad, E. Rizzardo and S. H. Thang, *ACS Symp. Ser.*, 2000, **768**, 297–312.
49. J. Chiefari, J. Jeffery, G. Moad, R. T. A. Mayadunne, E. Rizzardo and S. H. Thang, *Macromolecules*, 1999, **32**, 7700–7702.
50. C. Barner-Kowollik and T. Junkers, *J. Polym. Sci., Part A, Polym. Chem.*, 2011, **49**, 1293–1297.
51. J. Chiefari, J. Jeffery, G. Moad, R. T. A. Mayadunne, E. Rizzardo and S. H. Thang, *Macromolecules*, 1999, **32**, 5559–5562.
52. J. Chiefari, R. T. A. Mayadunne, C. L. Moad, G. Moad, E. Rizzardo, A. Postma, M. A. Skidmore and S. H. Thang, *Macromolecules*, 2003, **36**, 2273–2283.
53. G. Moad, J. Chiefari, C. L. Moad, A. Postma, R. T. A. Mayadunne, E. Rizzardo and S. H. Thang, *Macromol. Symp.*, 2002, **182**, 65–80.
54. C. Walling, *J. Am. Chem. Soc.*, 1948, **70**, 2561–2564.
55. L. Hutson, J. Krstina, C. L. Moad, G. Moad, G. R. Morrow, A. Postma, E. Rizzardo and S. H. Thang, *Macromolecules*, 2004, **37**, 4441–4452.
56. J. Chiefari, J. Jeffery, J. Krstina, C. L. Moad, G. Moad, A. Postma, E. Rizzardo and S. H. Thang, *Macromolecules*, 2005, **38**, 9037–9054.
57. A. Postma, T. P. Davis, R. A. Evans, G. Li, G. Moad and M. O'Shea, *Macromolecules*, 2006, **39**, 5293–5306.
58. A. Postma, T. P. Davis, G. Li, G. Moad and M. O'Shea, *Macromolecules*, 2006, **39**, 5307–5318.
59. E. Bicciocchi, Y. K. Chong, L. Giorgini, G. Moad, E. Rizzardo and S. H. Thang, *Macromol. Chem. Phys.*, 2010, **211**, 529–538.
60. D. J. Keddie, C. Guerrero-Sanchez, G. Moad, R. Mulder, E. Rizzardo and S. H. Thang, *Macromolecules*, 2012, **45**, 4205–4215.
61. A. H. E. Müller and G. Litvenko, *Macromolecules*, 1997, **30**, 1253–1266.
62. A. H. E. Müller, R. Zhuang, D. Yan and G. Litvenko, *Macromolecules*, 1995, **28**, 4326–4333.

63. G.Moad and C.Barner-Kowollik, in *Handbook of RAFT Polymerization*, ed. C. Barner-Kowollik, Wiley-VCH, Weinheim, Germany, 2008, pp. 51–104.
64. M. L. Coote, E. I. Izgorodina, E. H. Krenske, M. Busch and C. Barner-Kowollik, *Macromol. Rapid. Commun.*, 2006, **27**, 1015–1022.
65. E. I. Izgorodina and M. L. Coote, *Macromol. Theory Simul.*, 2006, **15**, 394–403.
66. K. Kubo, A. Goto, K. Sato, Y. Kwak and T. Fukuda, *Polymer*, 2005, **46**, 9762–9768.
67. M. Benaglia, M. Chen, Y. K. Chong, G. Moad, E. Rizzardo and S. H. Thang, *Macromolecules*, 2009, **42**, 9384–9386.
68. M. Benaglia, J. Chiefari, Y. K. Chong, G. Moad, E. Rizzardo and S. H. Thang, *J. Am. Chem. Soc.*, 2009, **131**, 6914–6915.
69. G. Moad, Y. K. Chong, E. Rizzardo, A. Postma and S. H. Thang, *Polymer*, 2005, **46**, 8458–8468.
70. G. Moad, R. T. A. Mayadunne, E. Rizzardo, M. Skidmore and S. Thang, *Macromol. Symp.*, 2003, **192**, 1–12.
71. L. He, E. S. Read, S. P. Armes and D. J. Adams, *Macromolecules*, 2007, **40**, 4429–4438.
72. A. B. Lowe and R. Wang, *Polymer*, 2007, **48**, 2221–2230.
73. T. Y. Guo, P. Liu, J. W. Zhu, M. D. Song and B. H. Zhang, *Biomacromolecules*, 2006, **7**, 1196–1202.
74. L. Albertin, M. H. Stenzel, C. Barner-Kowollik and T. P. Davis, *Polymer*, 2006, **47**, 1011–1019.
75. L. Albertin, M. H. Stenzel, C. Barner-Kowollik, L. J. R. Foster and T. P. Davis, *Macromolecules*, 2005, **38**, 9075–9084.
76. P. Takolpuckdee, *Aust. J. Chem.*, 2005, **58**, 66–66.
77. A. Housni, H. Cai, S. Liu, S. H. Pun and R. Narain, *Langmuir*, 2007, **23**, 5056–5061.
78. L. Albertin, C. Kohlert, M. Stenzel, L. J. R. Foster and T. P. Davis, *Biomacromolecules*, 2004, **5**, 255–260.
79. P. Zhao, Q. D. Ling, W. Z. Wang, J. Ru, S. B. Li and W. Huang, *J. Polym. Sci., Part A, Polym. Chem.*, 2007, **45**, 242–252.
80. H. Mori, S. Nakano and T. Endo, *Macromolecules*, 2005, **38**, 8192–8201.
81. Y. Zhang, Z. Cheng, X. Chen, W. Zhang, J. Wu, J. Zhu and X. Zhu, *Macromolecules*, 2007, **40**, 4809–4817.
82. W. Sriprom, M. Neel, C. D. Gabbutt, B. M. Heron and S. Perrier, *J. Mater. Chem.*, 2007, **17**, 1885–1893.
83. M. Chen, K. P. Ghiggino, T. A. Smith, S. H. Thang and G. J. Wilson, *Aust. J. Chem.*, 2004, **57**, 1175–1177.
84. M. Chen, K. P. Ghiggino, S. H. Thang, J. White and G. J. Wilson, *J. Org. Chem.*, 2005, **70**, 1844–1852.
85. M. Chen, K. P. Ghiggino, A. W. H. Mau, E. Rizzardo, S. H. Thang and G. J. Wilson, *Chem. Commun.*, 2002, 2276–2277.
86. M. J. Yanjarappa, K. V. Gujraty, A. Joshi, A. Saraph and R. S. Kane, *Biomacromolecules*, 2006, **7**, 1665–1670.

87. C. M. Schilli, A. H. E. Muller, E. Rizzardo, S. H. Thang and Y. K. Chong, *ACS Symp. Ser.*, 2003, **854**, 603–618.
88. E. N. Savariar and S. Thayumanavan, *J. Polym. Sci., Part A, Polym Chem*, 2004, **42**, 6340–6345.
89. M. L. Coote and D. J. Henry, *Macromolecules*, 2005, **38**, 1415–1433.
90. M. L. Coote, E. H. Krenske and E. I. Izgorodina, in *Handbook of RAFT Polymerization*, ed. C. Barner-Kowollik, Wiley-VCH, Weinheim, Germany, 2008, pp. 5–49.
91. M. Destarac, W. Bzducha, D. Taton, I. Gauthier-Gillaizeau and S. Z. Zard, *Macromol. Rapid. Commun.*, 2002, **23**, 1049–1054.
92. R. T. A. Mayadunne, E. Rizzardo, J. Chiefari, Y. K. Chong, G. Moad and S. H. Thang, *Macromolecules*, 1999, **32**, 6977–6980.
93. M. Destarac, D. Charmot, X. Franck and S. Z. Zard, *Macromol. Rapid Commun.*, 2000, **21**, 1035–1039.
94. M. L. Coote, E. I. Izgorodina, G. E. Cavigliasso, M. Roth, M. Busch and C. Barner-Kowollik, *Macromolecules*, 2006, **39**, 4585–4591.
95. A. Theis, M. H. Stenzel, T. P. Davis, M. L. Coote and C. Barner-Kowollik, *Aust. J. Chem.*, 2005, **58**, 437–441.
96. G. Moad, M. Benaglia, M. Chen, J. Chiefari, Y. Chong, K. D. Keddie, J., E. Rizzardo and H. Thang San, *ACS Symposium Series - Non-Conventional Functional Block Copolymers*, 2011, **1066**, 81–102.
97. D. J. Keddie, C. Guerrero-Sanchez, G. Moad, E. Rizzardo and S. H. Thang, *Macromolecules*, 2011, **44**, 6738–6745.
98. M. H. Stenzel, L. Cummins, G. E. Roberts, T. R. Davis, P. Vana and C. Barner-Kowollik, *Macromol. Chem. Phys.*, 2003, **204**, 1160–1168.
99. M. L. Coote and L. Radom, *Macromolecules*, 2004, **37**, 590–596.
100. L. Nebhani, S. Sinnwell, C. Y. Lin, M. L. Coote, M. H. Stenzel and C. Barner-Kowollik, *J. Polym. Sci., Part A, Polym. Chem.*, 2009, **47**, 6053–6071.
101. D. B. Thomas, A. J. Convertine, R. D. Hester, A. B. Lowe and C. L. McCormick, *Macromolecules*, 2004, **37**, 1735–1741.
102. P. Metzner, *Top. Curr. Chem.*, 1999, **204**, 127–181.
103. M. Chen, G. Moad and E. Rizzardo, *Aust. J. Chem.*, 2011, **64**, 433–437.
104. E. Rizzardo, M. Chen, B. Chong, G. Moad, M. Skidmore and S. H. Thang, *Macromol. Symp.*, 2007, **248**, 104–116.
105. C. Z. Li and B. C. Benicewicz, *J. Polym. Sci., Part A, Polym. Chem.*, 2005, **43**, 1535–1543.
106. H. Fischer and L. Radom, *Angew. Chem., Int. Ed. Engl.*, 2001, **40**, 1340–1371.
107. C. Barner-Kowollik, M. Buback, B. Charleux, M. L. Coote, M. Drache, T. Fukuda, A. Goto, B. Klumperman, A. B. Lowe, J. B. Mcleary, G. Moad, M. J. Monteiro, R. D. Sanderson, M. P. Tonge and P. Vana, *J. Polym. Sci., Part A, Polym Chem*, 2006, **44**, 5809–5831.
108. G. Pound, J. B. McLeary, J. M. McKenzie, R. F. M. Lange and B. Klumperman, *Macromolecules*, 2006, **39**, 7796–7797.

109. G. Moad and D. H. Solomon, *The Chemistry of Radical Polymerization*, Elsevier, Oxford, 2006.
110. G. Moad, Y. K. Chong, R. Mulder, E. Rizzardo and S. H. Thang, in *Pacifichem 2010*, Honolulu, Hawaii, 2010, p. 1177.
111. R. Plummer, Y. K. Goh, A. K. Whittaker and M. J. Monteiro, *Macromolecules*, 2005, **38**, 5352–5355.
112. S. H. Thang, Y. K. Chong, R. T. A. Mayadunne, G. Moad and E. Rizzardo, *Tetrahedron Lett.*, 1999, **40**, 2435–2438.
113. E. Rizzardo, S. H. Thang and G. Moad *Synthesis of Dithioester Chain-Transfer Agents and Use of Bis(thioacyl) Disulfides or Dithioesters as Chain-Transfer Agents in Radical Polymerization* CSIRO, WO9905099, 1999.
114. M. J. Monteiro and H. de Brouwer, *Macromolecules*, 2001, **34**, 349–352.
115. B. Klumperman, E. T. A. van den Dungen, J. P. A. Heuts and M. J. Monteiro, *Macromol. Rapid. Commun.*, 2010, **31**, 1846–1862.
116. T. Arita, M. Buback and P. Vana, *Macromolecules*, 2005, **38**, 7935–7943.
117. Y. Liu, J. P. He, J. T. Xu, D. Q. Fan, W. Tang and Y. L. Yang, *Macromolecules*, 2005, **38**, 10332–10335.
118. J. Xu, J. He, D. Fan, W. Tang and Y. Yang, *Macromolecules*, 2006, **39**, 3753–3759.
119. B. Chong, G. Moad, E. Rizzardo, M. Skidmore and S. H. Thang, *Aust. J. Chem.*, 2006, **59**, 755–762.
120. K. Kempe, C. R. Becer and U. S. Schubert, *Macromolecules*, 2011, **44**, 5825–5842.
121. W. L. A. Brooks and B. S. Sumerlin, *Israel Journal of Chemistry*, 2012, **52**, 256–263.
122. S. L. Brown, C. M. Rayner and S. Perrier, *Macromol. Rapid. Commun.*, 2007, **28**, 478–483.
123. S. L. Brown, C. M. Rayner, S. Graham, A. Cooper, S. Rannard and S. Perrier, *Chem. Commun.*, 2007, 2145 –2147.
124. D. Roy and B. S. Sumerlin, *Polymer*, 2011, **52**, 3038–3045.
125. Y. Assem, A. Greiner and S. Agarwal, *Macromol. Rapid. Commun.*, 2007, **28**, 1923–1928.
126. J. Zhu, X. L. Zhu, Z. B. Zhang and Z. P. Cheng, *J. Polym. Sci., Part A, Polym. Chem.*, 2006, **44**, 6810–6816.
127. J. C. Hernández-Ortiz, G. Jaramillo-Soto, J. Palacios-Alquisira and E. Vivaldo-Lima, *Macromol. React. Eng.*, 2010, **4**, 210–221.
128. P. B. Zetterlund and S. b. Perrier, *Macromolecules*, 2011, **44**, 1340–1346.
129. A. D. Smith, E. Lester, K. J. Thurecht, J. El Harfi, G. Dimitrakis, S. W. Kingman, J. P. Robinson and D. J. Irvine, *Ind. Eng. Chem. Res.*, 2010, **49**, 1703–1710.
130. C. H. Hornung, C. Guerrero-Sanchez, M. Brasholz, S. Saubern, J. Chiefari, G. Moad, E. Rizzardo and S. H. Thang, *Organic Process Research & Development*, 2011, **15**, 593–601.
131. C. Diehl, P. Laurino, N. Azzouz and P. H. Seeberger, *Macromolecules*, 2010, **43**, 10311–10314.

132. M. J. Monteiro, R. Bussels, S. Beuermann and M. Buback, *Aust. J. Chem.*, 2002, **55**, 433–437.
133. T. Arita, M. Buback, O. Janssen and P. Vana, *Macromol. Rapid Commun.*, 2004, **25**, 1376–1381.
134. J. Rzayev and J. Penelle, *Angew. Chem. Int. Ed. Engl.*, 2004, **43**, 1691–1694.
135. C. L. McCormick and A. B. Lowe, *Acc. Chem. Res.*, 2004, **37**, 312–325.
136. A. B. Lowe, B. S. Sumerlin, M. S. Donovan and C. L. McCormick, *J. Am. Chem. Soc*, 2002, **124**, 11562–11563.
137. S. Perrier, T. P. Davis, A. J. Carmichael and D. M. Haddleton, *Chem. Commun.*, 2002, 2226–2227.
138. T. Arita, S. Beuermann, M. Buback and P. Vana, *e-Polym.*, 2004, no. 003.
139. T. Arita, S. Beuermann, M. Buback and P. Vana, *Macromol. Mater. Eng.*, 2005, **290**, 283–293.
140. M. Mertoglu, A. Laschewsky, K. Skrabania and C. Wieland, *Macromolecules*, 2005, **38**, 3601–3614.
141. M. Save, Y. Guillaneuf and R. G. Gilbert, *Aust. J. Chem.*, 2006, **59**, 693–711.
142. M. F. Cunningham, *Prog. Polym. Sci.*, 2008, **33**, 365–398.
143. P. B. Zetterlund, Y. Kagawa and M. Okubo, *Chem. Rev.*, 2008, **108**, 3747–3794.
144. E. Rizzardo, J. Chiefari, R. T. A. Mayadunne, G. Moad and S. H. Thang, *ACS Symp. Ser.*, 2000, **768**, 278–296.
145. I. Uzulina, S. Kanagasabapathy and J. Claverie, *Macromol. Symp.*, 2000, **150**, 33–38.
146. M. J. Monteiro, and J. de Barbeyrac, *Macromolecules*, 2001, **34**, 4416–4423.
147. M. J. Monteiro, M. Hodgson and H. De Brouwer, *J. Polym. Sci., Part A: Polym. Chem.*, 2000, **38**, 3864–3874.
148. M. J. Monteiro, M. Sjoberg, J. van der Vlist and C. M. Gottgens, *J. Polym. Sci., Part A, Polym Chem*, 2000, **38**, 4206–4217.
149. W. Smulders, R. G. Gilbert and M. J. Monteiro, *Macromolecules*, 2003, **36**, 4309–4318.
150. S. W. Prescott, M. J. Ballard, E. Rizzardo and R. G. Gilbert, *Aust. J. Chem.*, 2002, **55**, 415–424.
151. C. H. Such, E. Rizzardo, A. K. Serelis, B. S. Hawkett, R. G. Gilbert, C. J. Ferguson and R. J. Hughes, *Aqueous Dispersions of Polymer Particles*, University of Sydney, 03055919, 2003.
152. A. Butte, G. Storti and M. Morbidelli, *DECHEMA Monograph.*, 2001, **137**, 273–281.
153. S. Nozari and K. Tauer, *Polymer*, 2005, **46**, 1033–1043.
154. M. J. Monteiro, M. Hodgson and H. De Brouwer, *J. Polym. Sci., Part A, Polym. Chem.*, 2000, **38**, 3864–3874.
155. S. W. Prescott, M. J. Ballard, E. Rizzardo and R. G. Gilbert, *Macromolecules*, 2002, **35**, 5417–5425.

156. M. P. F. Pepels, C. I. Holdsworth, S. Pascual and M. J. Monteiro, *Macromolecules*, 2010, **43**, 7565–7576.

157. C. J. Ferguson, R. J. Hughes, B. T. T. Pham, B. S. Hawkett, R. G. Gilbert, A. K. Serelis and C. H. Such, *Macromolecules*, 2002, **35**, 9243–9245.

158. S. Freal-Saison, M. Save, C. Bui and B. Charleux, and S. Magnet, *Macromolecules*, 2006, **39**, 8632–8638.

159. J. Bernard, M. Save, B. Arathoon and B. Charleux, *J. Polym. Sci., Part A, Polym. Chem.*, 2008, **46**, 2845–2857.

160. G. Moad, C. L. Moad, J. Krstina, E. Rizzardo, C. T. Berge and T. R. Darling *Preparation of Block Copolymer with Low Dispersity by Radical Polymerization* DuPont/CSIRO, WO9615157, 1996.

161. Y. K. Chong, G. Moad, E. Rizzardo, M. A. Skidmore and S. H. Thang, *Macromolecules*, 2007, **40**, 9262–9271.

162. R. T. A. Mayadunne, E. Rizzardo, J. Chiefari, J. Krstina, G. Moad, A. Postma and S. H. Thang, *Macromolecules*, 2000, **33**, 243–245.

163. M. N. Galbraith, G. Moad, D. H. Solomon and T. H. Spurling, *Macromolecules*, 1987, **20**, 675–679.

164. T. H. Spurling, M. Deady, J. Krstina and G. Moad, *Makromol. Chem., Macromol. Symp.*, 1991, **51**, 127–146.

165. G. Moad, K. Dean, L. Edmond, N. Kukaleva, G. Li, R. T. A. Mayadunne, R. Pfaendner, A. Schneider, G. Simon and H. Wermter, *Macromol. Symp.*, 2006, **233**, 170–179.

166. D. S. Germack, S. Harrisson, G. O. Brown and K. L. Wooley, *J. Polym. Sci., Part A, Polym. Chem.*, 2006, **44**, 5218–5228.

167. Y. K. Chong, T. P. T. Le, G. Moad, E. Rizzardo and S. H. Thang, *Macromolecules*, 1999, **32**, 2071–2074.

168. E. Rizzardo, R. Mayadunne, G. Moad and S. H. Thang, *Macromol. Symp.*, 2001, **174**, 209–212.

169. M. J. Monteiro, *J. Polym. Sci., Part A, Polym. Chem.*, 2005, **43**, 5643–5651.

170. B. Y. K. Chong, T. P. T. Le, G. Moad, E. Rizzardo and S. H. Thang, *Macromolecules*, 1999, **32**, 2071–2074.

171. R. T. A. Mayadunne, J. Jeffery, G. Moad and E. Rizzardo, *Macromolecules*, 2003, **36**, 1505–1513.

172. M. H. Stenzel and T. P. Davis, *J. Polym. Sci., Part A: Polym. Chem*, 2002, **40**, 4498–4512.

173. M. Stenzel-Rosenbaum, T. P. Davis, V. Chen and A. G. Fane, *J. Polym. Sci., Part A: Polym. Chem.*, 2001, **39**, 2777–2783.

174. M. H. Stenzel, T. P. Davis and C. Barner-Kowollik, *Chem. Commun.*, 2004, 1546–1547.

175. M. Chen, K. P. Ghiggino, A. Launikonis, A. W. H. Mau, E. Rizzardo, W. H. F. Sasse, S. H. Thang and G. J. Wilson, *J Mater Chem*, 2003, **13**, 2696–2700.

176. V. Darcos, A. Dureault, D. Taton, Y. Gnanou, P. Marchand, A. M. Caminade, J. P. Majoral, M. Destarac and F. Leising, *Chem. Commun.*, 2004, 2110–2111.

177. X. J. Hao, C. Nilsson, M. Jesberger, M. H. Stenzel, E. Malmstrom and T. P. Davis, E. Ostmark and C. Barner-Kowollik, *J. Polym. Sci., Part A, Polym Chem*, 2004, **42**, 5877–5890.

178. Y. Z. You, C. Y. Hong, C. Y. Pan and P. H. Wang, *Adv. Mater.*, 2004, **16**, 1953–1957.

179. M. Jesberger, L. Barner, M. H. Stenzel, E. Malmstrom, T. P. Davis and C. Barner-Kowollik, *J. Polym. Sci., Part A, Polym Chem*, 2003, **41**, 3847–3861.

180. S. Perrier, P. Takolpuckdee, J. Westwood and D. M. Lewis, *Macromolecules*, 2004, **37**, 2709–2717.

181. T. P. Le, G. Moad, E. Rizzardo and S. H. Thang, *Polymerization with Living Characteristics*, E.I. Du Pont De Nemours and Co., USA, WO 9801478, 1998.

182. R. T. A. Mayadunne, G. Moad and E. Rizzardo, *Tetrahedron Lett.*, 2002, **43**, 6811–6814.

183. S. Abrol, P. A. Kambouris, M. G. Looney and D. H. Solomon, *Macromol. Rapid Commun.*, 1997, **18**, 755–760.

184. S. Abrol, M. J. Caulfield, G. G. Qiao and D. H. Solomon, *Polymer*, 2001, **42**, 5987–5991.

185. D. J. Keddie, G. Moad, E. Rizzardo and S. H. Thang, *Macromolecules*, 2012, doi: 10.1021/ma300410v.

186. C. Boyer, V. Bulmus, T. P. Davis, V. Ladmiral, J. Liu and S. Perrier, *Chem. Rev.*, 2009, **109**, 5402–5436.

187. G. Moad, M. Chen, M. Häussler, A. Postma, E. Rizzardo and S. H. Thang, *Polym. Chem.*, 2011, **2**, 492–519.

188. Y. K. Chong, G. Moad, E. Rizzardo and S. H. Thang, *Macromolecules*, 2007, **40**, 4446–4455.

189. A. Postma, T. P. Davis, G. Moad and M. S. O'Shea, *Macromolecules*, 2005, **38**, 5371–5374.

190. M. Chen, G. Moad and E. Rizzardo, *J. Polym. Sci., Part A, Polym. Chem.*, 2009, **47**, 6704–6714.

191. W. A. Braunecker and K. Matyjaszewski, *Prog. Polym. Sci.*, 2007, **32**, 93–146.

192. R. A. Evans, *Aust. J. Chem.*, 2007, **60**, 384–395.

193. R. K. Iha, K. L. Wooley, A. M. Nystrom, D. J. Burke and M. J. Kade, and C. J. Hawker, *Chem. Rev.*, 2009, **109**, 5620–5686.

194. W. H. Binder and R. Sachsenhofer, *Macromol. Rapid. Commun.*, 2007, **28**, 15–54.

195. W. H. Binder and R. Sachsenhofer, *Macromol. Rapid. Commun.*, 2008, **29**, 952–981.

196. B. S. Sumerlin and A. P. Vogt, *Macromolecules*, 2009, **43**, 1–13.

197. P. L. Golas and K. Matyjaszewski, *Chem. Soc. Rev.*, 2010, **39**, 1338–1354.

198. S. Kato and M. Ishida, *Sulfur Reports*, 1988, **8**, 155–323.

199. R. Mayer and S. Scheithauer, in *Methoden den Organischen Chemie*, eds. K. H. Buechel, J. Falbe, H. Hagemann and M. Hanack, Thieme, Stuttgart, 1985, pp. 891–930.

200. B. Quiclet-Sire and S. Z. Zard, *Top. Curr. Chem.*, 2006, **264**, 201–236.
201. S. Z. Zard, *Aust. J. Chem.*, 2006, **59**, 663–668.
202. G. Moad, E. Rizzardo and S. H. Thang, *Polym. Int.*, 2011, **60**, 9–25.
203. S. Sinnwell, A. J. Inglis, T. P. Davis, M. H. Stenzel and C. Barner-Kowollik, *Chem. Commun.*, 2008, 2052–2054.
204. A. J. Inglis, S. Sinnwell, M. H. Stenzel and C. Barner-Kowollik, *Angew. Chem., Int. Ed. Engl.*, 2009, **48**, 2411–2414.
205. Y. Kwak, R. Nicolay and K. Matyjaszewski, *Macromolecules*, 2008, **41**, 6602–6604.
206. Y. Kwak, R. Nicolay and K. Matyjaszewski, *Aust. J. Chem.*, 2009, **62**, 1384–1401.
207. A. Favier, B. Luneau, J. Vinas, N. Laissaoui, D. Gigmes and D. Bertin, *Macromolecules*, 2009, **42**, 5953–5964.
208. H. Willcock and R. K. O'Reilly, *Polym. Chem.*, 2010, **1**, 149–157.
209. M. A. Harvison, P. J. Roth, T. P. Davis and A. B. Lowe, *Aust. J. Chem.*, 2011, **64**, 992–1006.
210. L. Barner and S. Perrier, in *Handbook of RAFT Polymerization*, ed. C. Barner-Kowollik, Wiley-VCH, Weinheim, Germany, 2008, pp. 455–482.

CHAPTER 7

Living Radical Polymerizations with Organic Catalysts

A. GOTO,* Y. TSUJII AND H. KAJI*

Institute for Chemical Research, Kyoto University, Uji, Kyoto 611-0011, Japan
*Email: agoto@scl.kyoto-u.ac.jp; kaji@scl.kyoto-u.ac.jp

7.1 Introduction

Living radical polymerization (LRP) is a useful and powerful tool for synthesizing well-defined, low-polydispersity polymers.[1–4] The basic concept of LRP is the reversible activation of the dormant species (Polymer–X) to the propagating radical (Polymer$^{\bullet}$) (Scheme 7.1a). A sufficiently large number of activation–deactivation cycles are a requisite for good control of chain length distribution.[5–7]

Organic catalysts have extensively been studied for decades in chemistry, offering attractive approaches for organic syntheses. The recent development of organic catalysts for fine reactions such as asymmetric syntheses has particularly promoted attention to organic catalysts.[8] In the field of polymer syntheses, organic catalysts were developed for group-transfer polymerization to yield fine polymers.[9,10]

We recently developed the first LRPs using organic (non-transition-metal) catalysts.[11–16] Our LRPs use iodine as a capping agent X and a tin (Sn),[11,12] germanium (Ge),[11,12] phosphorus (P),[12] nitrogen (N),[13,16] oxygen (O),[15] or carbon (C)[15]-centered molecule (Figure 7.1) as a catalyst for reversible activation. We developed two families of LRPs with different mechanisms. We have proposed to term them reversible chain transfer catalyzed polymerization (RTCP)[11–15] and reversible complexation mediated polymerization (RCMP)[16]

RSC Polymer Chemistry Series No. 4
Fundamentals of Controlled/Living Radical Polymerization
Edited by Nicolay V Tsarevsky and Brent S Sumerlin
© The Royal Society of Chemistry 2013
Published by the Royal Society of Chemistry, www.rsc.org

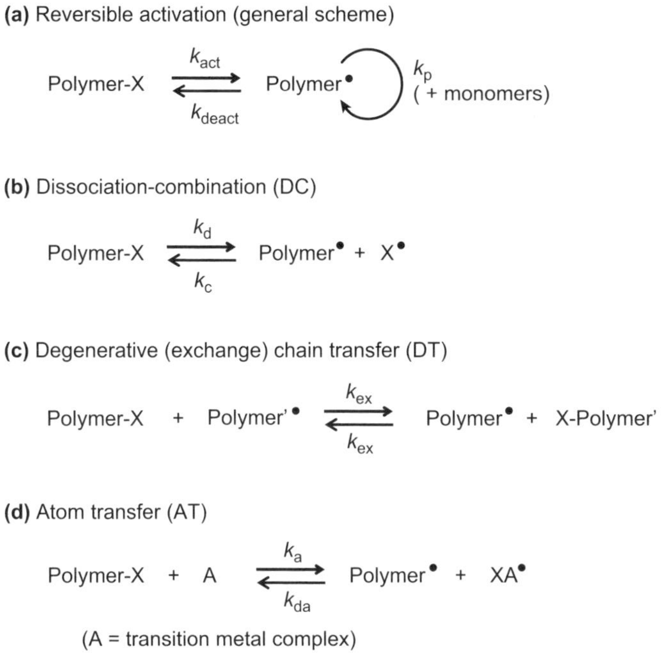

(a) Reversible activation (general scheme)

$$\text{Polymer-X} \underset{k_{deact}}{\overset{k_{act}}{\rightleftharpoons}} \text{Polymer}^{\bullet} \quad \bigcirc \quad k_p \; (+ \text{ monomers})$$

(b) Dissociation-combination (DC)

$$\text{Polymer-X} \underset{k_c}{\overset{k_d}{\rightleftharpoons}} \text{Polymer}^{\bullet} + X^{\bullet}$$

(c) Degenerative (exchange) chain transfer (DT)

$$\text{Polymer-X} + \text{Polymer'}^{\bullet} \underset{k_{ex}}{\overset{k_{ex}}{\rightleftharpoons}} \text{Polymer}^{\bullet} + \text{X-Polymer'}$$

(d) Atom transfer (AT)

$$\text{Polymer-X} + A \underset{k_{da}}{\overset{k_a}{\rightleftharpoons}} \text{Polymer}^{\bullet} + XA^{\bullet}$$

(A = transition metal complex)

Scheme 7.1 (a) General scheme of reversible activation, (b) dissociation–combination, (c) degenerative chain transfer, and (d) atom transfer.

after their new reversible activation mechanisms. In this chapter, we will describe the fundamental features of RTCP (Section 7.3) and RCMP (Section 7.4), following a brief survey of currently known LRP systems from a mechanistic point of view (Section 7.2).

7.2 Currently Known LRP Systems

7.2.1 Examples of Capping Agent X

Miscellaneous capping agents X are used for LRP.[17–53] Examples are listed in Scheme 7.2. They include sulfur compounds (Schemes 7.2a and 7.2g),[17–20] stable nitroxides (7.2b),[21–24] stable nitrogen and carbon compounds (7.2c),[25–30] transition metal complexes (7.2d),[31–34] iodine (7.2e),[35–38] halogens with transition metal catalysts (7.2f),[39–43] and tellurium, stibine, and bismuth compounds (7.2h).[44–48]

7.2.2 Mechanistic Classification of Reversible Activation Processes

The reversible activation reactions in the most successful LRPs currently known may mechanistically be classified into three types (Schemes 7.1b–7.1d), which are the dissociation–combination (DC), degenerative chain transfer (DT), and atom transfer (AT) mechanisms.

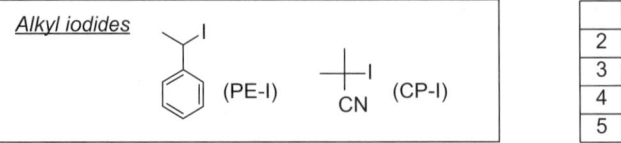

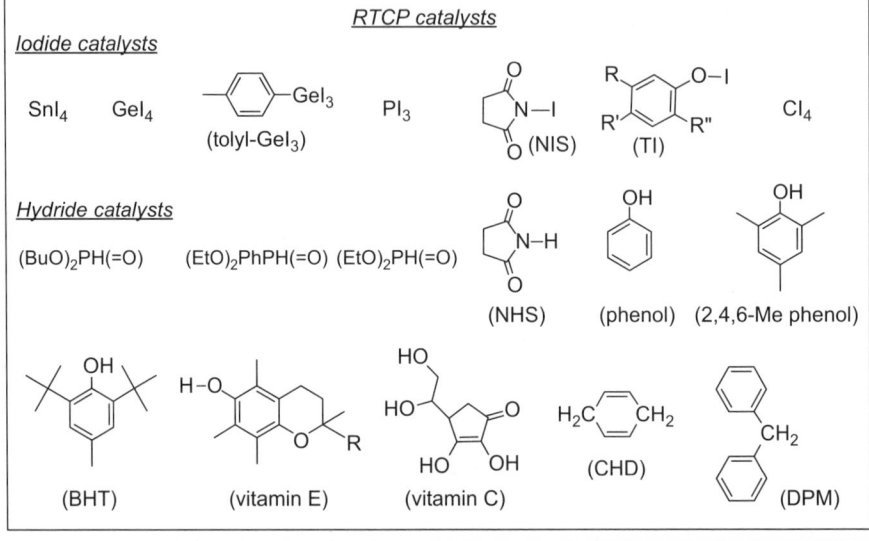

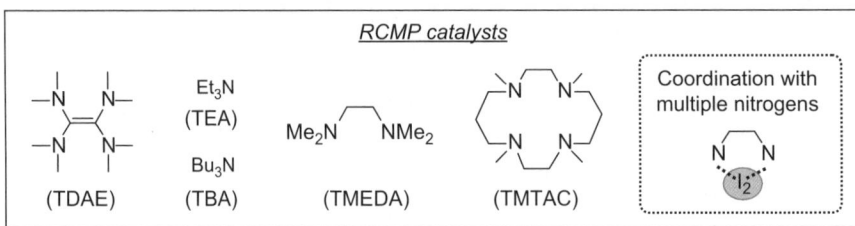

Figure 7.1 Periodic table and structures of catalysts and alkyl iodides (low-mass dormant species).

7.2.2.1 Dissociation–Combination (DC) Mechanism

In this mechanism, Polymer–X is thermally or photochemically dissociated into Polymer$^\bullet$ and a stable (persistent) radical X$^\bullet$ (Scheme 7.1b). The best-known examples of the stable radical are nitroxides such as 2,2,6,6-tetramethylpiperidinyl-1-oxy (TEMPO)[21] and N-*tert*-butyl-1-diethylphosphono-2,2-dimethylpropyl nitroxide (DEPN)[22] (Scheme 7.2b). Other examples include nitrogen compounds such as triazols[25] and verdazyls[26,27] (Scheme 7.2c), bulky carbon compounds such as triphenyl-methyls[28,29] and vinyl compounds[30] (Scheme 7.2c), and transition metal complexes such as cobalt[31–33] and titanium[34] complexes (Scheme 7.2d).

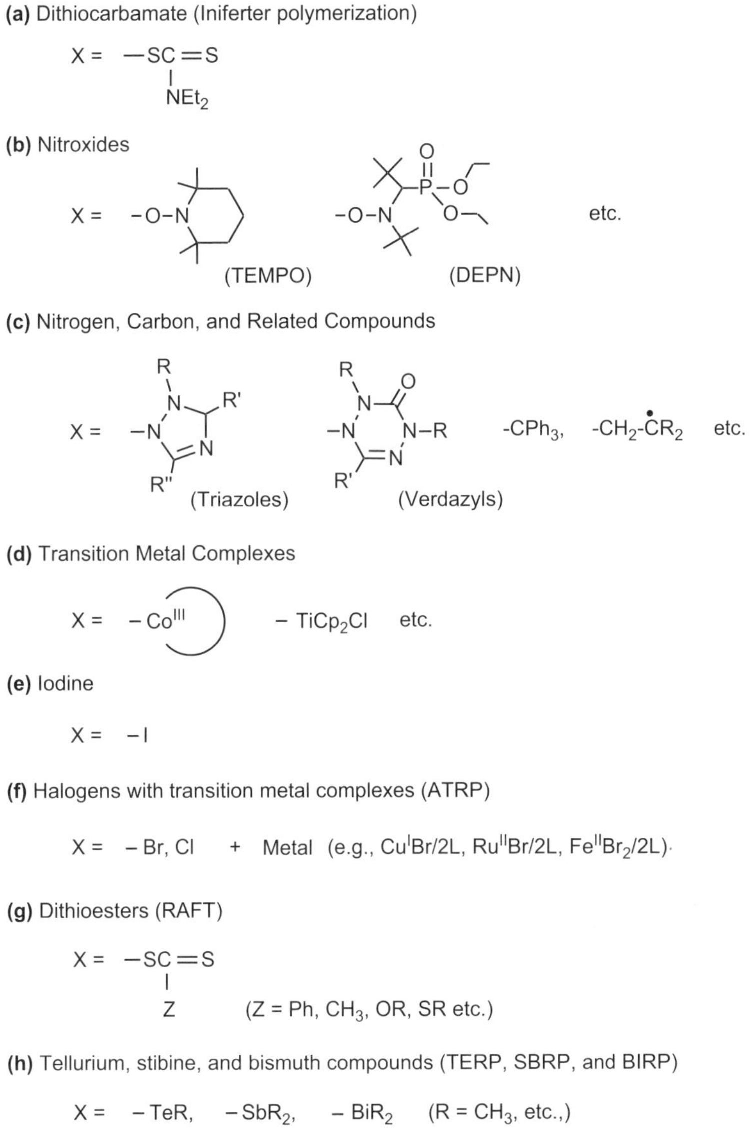

(a) Dithiocarbamate (Iniferter polymerization)

$$X = \quad -SC = S$$
$$\mid$$
$$NEt_2$$

(b) Nitroxides

X =   -O-N (TEMPO)   -O-N (DEPN)   etc.

(c) Nitrogen, Carbon, and Related Compounds

X =   -N (Triazoles)   -N (Verdazyls)   -CPh$_3$,   -CH$_2$-ĊR$_2$ etc.

(d) Transition Metal Complexes

X =   - Co^{III} - TiCp$_2$Cl   etc.

(e) Iodine

X =   -I

(f) Halogens with transition metal complexes (ATRP)

X =   - Br, Cl   +   Metal (e.g., Cu^{I}Br/2L, Ru^{II}Br/2L, Fe^{II}Br$_2$/2L).

(g) Dithioesters (RAFT)

$$X = \quad -SC = S$$
$$\mid$$
$$Z$$ (Z = Ph, CH$_3$, OR, SR etc.)

(h) Tellurium, stibine, and bismuth compounds (TERP, SBRP, and BIRP)

X =   - TeR,   - SbR$_2$,   - BiR$_2$ (R = CH$_3$, etc.,)

Scheme 7.2   Examples of X.

7.2.2.2   *Degenerative Chain Transfer (DT) Mechanism*

In this mechanism, Polymer–X is attacked by the propagating radical (Polymer′•) to form the active species (Polymer•) and the dormant species (Polymer′–X) (Scheme 7.1c). This is an exchange reaction of X. There are two types of DT. In one type, X is an atom or a simple group. X is transferred from radical to radical without forming any kinetically important intermediate.

A typical example is iodide-mediated polymerization (IMP) (X = iodine) (Scheme 7.2e). Organotellurium-mediated living radical polymerization (TERP: X = MeR)[45] (Scheme 7.2h), organostibine one[46] (SBRP: X = SbR$_2$) (7.2h), and organobismuthine one[47] (BIRP: X = BiR$_2$) (7.2h) are also based on this mechanism, while interestingly the DC mechanism coexists in TERP and BIRP with the main mechanism being DT in both cases. The cobalt and titanium systems (X = Co and Ti complexes) (Scheme 7.2d) are also assumed to include the two mechanisms. The other is the case in which X is a group with a double bond that is accessible to the addition of Polymer$^\bullet$. The exchange reaction occurs *via* the addition of Polymer$'^\bullet$ to Polymer–X to form the intermediate radical Polymer–(X$^\bullet$)–Polymer$'$ followed by its fragmentation into Polymer$^\bullet$ and Polymer$'$–X. This process was named reversible addition–fragmentation chain transfer (RAFT).[19] Dithioester compounds are representative RAFT agents (Scheme 7.2g).

7.2.2.3 Atom Transfer (AT) Mechanism

In this mechanism, Polymer–X is activated by the catalysis of activator A, and the capping agent is transferred to form a stable species XA$^\bullet$ (Scheme 7.1d). All currently known successful LRPs in this category use a halogen like Cl and Br as a capping agent X and a halide complex of transition metal like Cu[40] and Ru[39] as an activator A (Scheme 7.2f). These LRPs are often termed atom transfer radical polymerization (ATRP).[40]

7.2.2.4 Experimental Establishment of Activation Mechanism for Individual Systems

The activation mechanism can experimentally and quantitatively be established by determining the activation rate constants for the three mechanisms k_d (Scheme 7.1b), k_{ex} (Scheme 7.1c), and k_a (Scheme 7.1d) for individual systems.[6] Such an experiment was carried out for the nitroxide (Scheme 7.2b),[54] iodide (7.2e),[55] ATRP (7.2f),[56] RAFT (7.2g),[57] TERP (7.2h),[58,59] SBRP (7.2h),[46,60] and BIRP (7.2h)[47] systems. The established mechanisms for these systems were mentioned above.

7.3 RTCP

7.3.1 Concept and Mechanism

IMP (Scheme 7.2e) (X = I) is a simple LRP, as it contains only monomer, an alkyl iodide (dormant species), and a conventional radical initiator (source of Polymer$^\bullet$). In IMP, as noted above, a polymer–iodide (Polymer–I) (dormant species) is activated by Polymer$^\bullet$ (DT (Scheme 7.1c)). However, due to a low DT frequency for iodine,[16,55] the control of polydispersity is limited. To this polymerization, we added a Ge, Sn, P, N, O, or C-centered compound (Figure 7.1) such as GeI$_4$, developing a new family of LRP (Scheme 7.3a).[11–15]

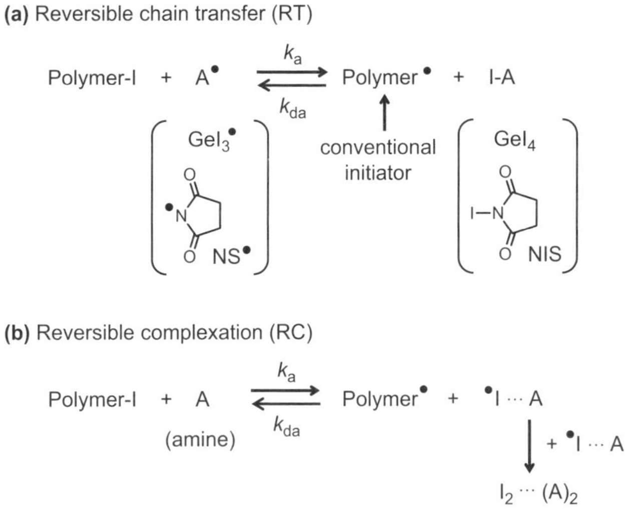

(a) Reversible chain transfer (RT)

$$\text{Polymer-I} \quad + \quad \text{A}^{\bullet} \quad \underset{k_{da}}{\overset{k_a}{\rightleftharpoons}} \quad \text{Polymer}^{\bullet} \quad + \quad \text{I-A}$$

conventional initiator

$$\begin{bmatrix} \text{GeI}_3{}^{\bullet} \\ \\ \text{NS}^{\bullet} \end{bmatrix} \qquad \begin{bmatrix} \text{GeI}_4 \\ \\ \text{NIS} \end{bmatrix}$$

(b) Reversible complexation (RC)

$$\text{Polymer-I} \quad + \quad \text{A} \quad \underset{k_{da}}{\overset{k_a}{\rightleftharpoons}} \quad \text{Polymer}^{\bullet} \quad + \quad {}^{\bullet}\text{I}\cdots\text{A}$$

(amine)

$$\downarrow + {}^{\bullet}\text{I}\cdots\text{A}$$

$$\text{I}_2\cdots(\text{A})_2$$

Scheme 7.3 Reversible activation processes.

The added GeI$_4$ works as a deactivator (I–A) of Polymer$^{\bullet}$, *in situ* producing GeI$_3{}^{\bullet}$. GeI$_3{}^{\bullet}$ works as an activator (A$^{\bullet}$) of Polymer–I, producing Polymer$^{\bullet}$ and GeI$_4$ again. This cycle allows a frequent activation of Polymer–I. Mechanistically, this process is a *reversible chain transfer* (RT) of Polymer$^{\bullet}$ with GeI$_4$ as a chain transfer agent, and Polymer–I is catalytically activated *via* the RT process. This is a new reversible activation mechanism, and we have proposed to term the related polymerization RT-catalyzed polymerization (RTCP). In this article, the reversible chain transfer agents (Ge, Sn, P, N, O, and C compounds) will be called catalysts.

7.3.2 Background – Radical Reactions of Sn, Ge, P, and N Compounds in Organic Chemistry

In organic chemistry, Sn and Ge radicals are known to abstract a halogen from an alkyl halide with a high reactivity to give the alkyl radical and their halides.[61–63] For the best example, tributyltin radical (Bu$_3$Sn$^{\bullet}$) rapidly reacts with alkyl chlorides at a rate constant 10^2–10^4 M^{-1} s^{-1},[64,65] with alkyl bromides at 10^6–10^7 M^{-1} s^{-1},[64,66] and with alkyl iodides at 10^9 M^{-1} s^{-1}.[64,66] The last value is even close to those at the diffusion controlled limit. Such high reactivity of Sn and Ge radicals urged us to use the reaction for the activation process of LRP. In organic chemistry, this reaction and the subsequent reduction of the alkyl radical with a hydride are widely utilized for the transformation of alkyl halide to alkyl hydride.[61–63] While Sn and Ge (group 14) are overwhelmingly frequently used, other group 14 elements and group 13, 15, and 16 elements (Figure 7.1) are also utilized for the transformation. This indicates that a wide variety of elements are possibly applicable to RTCP catalysts. In this article, we focus on Ge, Sn, P, N, O, and C catalysts.

As well as a high reactivity of the activator radical, a high reactivity of the deactivator halide is also essential for RTCP. However, the reactions of a carbon-centered radical to Ge, Sn, and relevant halides are generally slow.[61–63] The counterparts of the mentioned Bu$_3$Sn$^\bullet$, *i.e.*, the chloride Bu$_3$SnCl, the bromide Bu$_3$SnCl, and even the iodide Bu$_3$SnI, are also virtually unreactive (poor deactivators), and thus tributyltins are not useful for RTCP. Thus, we explored appropriate halides. We found that some Ge, Sn, P, N, O, and C iodides such as GeI$_4$ work as very effective deactivators to be used in RTCP.

7.3.3 Performance

We examined the homopolymerizations and copolymerizations of styrene (St), methyl methacrylate (MMA), acrylonitrile (AN), a functional styrene, *i.e.*, *p*-aminostyrene (ASt), and functional methacrylates, *i.e.*, 2-ethyhexyl (EHMA), benzyl (BzMA), phenyl (POMA), glycidyl (GMA), 2-hydroxyethyl (HEMA), poly(ethylene glycol) methyl ether (PEGMA), and *N*,*N*-dimethylaminoethyl (DMAEMA) methacrylates as well as methacrylic acid (MAA) (Figure 7.2), for example, by using several catalysts. The results will be summarized below.

7.3.3.1 *St with Ge, Sn, P, N, O, and C Iodides as Deactivators*

The first successful RTCP was the polymerization of St with GeI$_4$ catalyst.[11,12] We carried out the bulk polymerization of St (8 M (100 eq)) with 1-phenylethyl

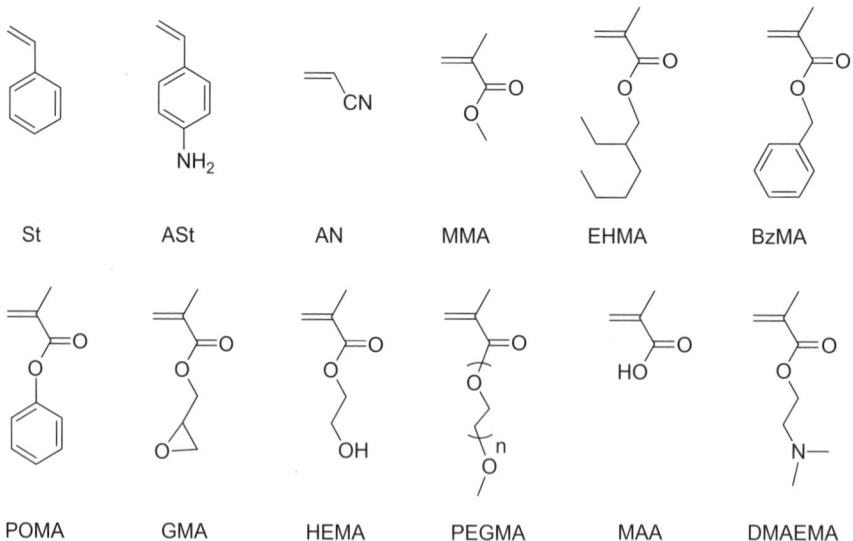

Figure 7.2 Structures of studied monomers.

iodide (PE–I (Figure 7.1)) (80 mM (1 eq.)) as a low-mass dormant species, benzoyl peroxide (BPO) (20 mM (0.25 eq.)) as a radical source, and GeI_4 (5 mM (0.0625 eq.)) as a catalyst (deactivator) at 80 °C. The theoretical degree of polymerization at a full conversion ($[St]_0/[PE–I]_0$) was 100 in this case. In this polymerization (Scheme 7.3a), Polymer•, which is originally supplied by BPO, is supposed to react with GeI_4, *in situ* producing the activator radical GeI_3• and Polymer–I. If GeI_3• effectively abstracts I from PE–I (or Polymer–I) to produce PE• (or Polymer•), cycles of activation and deactivation will be started. Table 7.1 (entry 1) and Figure 7.3 (filled circles) show the results. The first-order plot of the monomer concentration [M] (Figure 7.3a) showed a curvature in a long time range (for 24 h) due to a decrease of [BPO] with time *t*. In a relatively short time range (for ~1 h), the polymerization rate R_p was approximately constant (as will be detailed below). The M_n (Figure 7.3b) linearly increased with conversion and agreed well with the theoretical value $M_{n,theo}$. The small deviations from $M_{n,theo}$ at a later stage of polymerization are ascribed to the increase in the number of chains by the decomposition of BPO. The poly-dispersity index (PDI or M_w/M_n) (Figure 7.3b) reached a low value of about 1.2 from an early stage of polymerization, indicating a high frequency of the activation–deactivation cycle. The amount of GeI_4 catalyst was also reduced from 5 mM (above) to 2 mM (Table 7.1 (entry 2)), keeping a small PDI (1.24). The small amount of GeI_4 (2–5 mM, *i.e.*, 200–500 ppm) required to control the

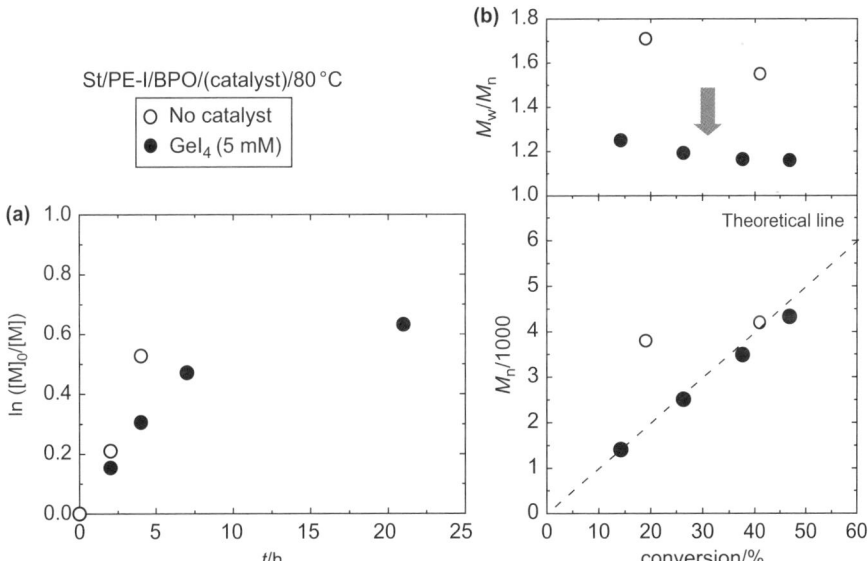

Figure 7.3 Plots of (a) $\ln([M]_0/[M])$ *vs.* *t* and (b) M_n and M_w/M_n *vs.* conversion for the St/PE–I/BPO/(GeI_4) systems in bulk (80 °C): $[St]_0 = 8$ M; $[PE–I]_0 = 80$ mM; $[BPO]_0 = 20$ mM; $[GeI_4]_0 = 0$ mM (open circles) and 5 mM (filled circles: for entry 1 in Table 7.1).
Reproduced from ref. 12 by permission of American Chemical Society.

Table 7.1 RTCPs of St and MMA with iodide (deactivator) catalysts ([Monomer]$_0$ = 8 M in bulk).

Entry	Monomer (equiv to [R-I])	Catalyst	R-I	In^a	[R-I]$_0$/[In]$_0$/[cat]$_0$ (mM)	T/°C	t/h	Conv (%)	M$_n$ (M$_{n,theo}$)	PDI
1	St (100 eq)	GeI$_4$	PE-I	BPO	80/20/5	80	21	47	4300 (4700)	1.16
2	St (100 eq)	GeI$_4$	PE-I	BPO	80/40/2	80	7	85	8200 (8500)	1.24
3	St (100 eq)	GeI$_2$	PE-I	BPO	80/40/2	80	7	85	8200 (8500)	1.24
4	St (100 eq)	SnI$_4$	PE-I	AIBN	80/40/5	60	27	72	6500 (7200)	1.21
5	St (100 eq)	SnI$_2$	PE-I	AIBN	80/20/5	60	21	50	4800 (5000)	1.23
6	St (100 eq)	PI$_3$	PE-I	DCP	80/80/0.5	100	23	52	5100 (5200)	1.25
7	St (100 eq)	NIS	PE-I	DCP	80/80/5	100	10	56	5200 (5600)	1.16
8	St (100 eq)	TI	PE-I	BPB	80/40/5	100	7	76	8500 (7600)	1.29
9	MMA (100 eq)	tolyl-GeI$_3$	CP-I	AIBN	80/20/5	70	6	84	6900 (8400)	1.19
10	MMA (100 eq)	PI$_3$	CP-I	AIBN	80/20/2	70	4	78	9600 (7800)	1.28
11	MMA (100 eq)	NIS	CP-I	AIBN	80/10/1	80	2	90	7900 (9000)	1.38
12	MMA (100 eq)	NIS	CP-I	AIBN	60/10/1^b	80	2.5	96	9400 (9600)	1.28
13	MMA (100 eq)	TI	CP-I	AIBN	80/20/15	80	2	37	3800 (3700)	1.37
14	MMA (100 eq)	CI$_4$	CP-I	AIBN	80/10/1	80	2.3	90	8000 (9000)	1.31
15	MMA (100 eq)	NIS	I$_2$	AIBN	40/80/5	80	3	90	8200 (9000)	1.28

^{a}In = conventional radical initiator, BPO = benzoyl peroxide, AIBN = azobis(isobutyronitrile), DCP = dicumyl peroxide, and BPB = t-butyl perbenzoate.
^{b}Solution polymerization with 25 vol% anisole for entry 12.

polydispersity suggests a high reactivity of GeI_4. The activation of Polymer–I occurs not only by RT (Scheme 7.3a) but also by DT (Scheme 7.1c). However, the system without GeI_4 (IMP) (Figure 7.3 (open circles)) gave much larger PDIs than those with GeI_4 (Figure 7.3 (filled circles)) (with other conditions set the same). This means that RT plays a main role in the GeI_4 system, with a small contribution of DT.

Besides GeI_4, another Ge catalyst GeI_2,[11,12] Sn catalysts SnI_4 and SnI_2,[11,12] a P catalyst PI_3,[12] an N catalyst *N*-iodosuccinimide (NIS (Figure 7.1)),[13] an O catalyst thymol iodide (TI (Figure 7.1)),[15] and a C catalyst CI_4[15] were also effective for St (Table 7.1 (entries 3–8)). Schematically (Scheme 7.3a), NIS, for example, gives the *N*-centered succinimide radical (NS•) as an activator (A•). Low polydispersity (~ 1.2) was achieved with a small amount (0.5–5 mM (50–500 ppm)) of the catalyst at 60–100 °C. NIS is a common compound and TI is a derivative of a scent component of thyme (thymol). Their commonness may be an attractive feature of RTCP.

7.3.3.2 MMA with Ge, Sn, P, N, O, and C Iodides as Deactivators

The polymerization of MMA was also successful with Ge, P, N, O, and C catalysts at 70–80 °C (Figure 7.4 and Table 7.1 (entries 9–14)).[12,13,15] To ensure a fast initiation from the alkyl iodide, we used a tertiary alkyl iodide, 2-cyanopropyl iodide (CP-I (Figure 7.1)), instead of the secondary one, PE-I. As a Ge catalyst, p-CH_3–C_6H_4–GeI_3 (tolyl-GeI_3)[12] (Table 7.1 (entry 9)) was used instead of GeI_4, which was effective for St but not active enough for MMA. With 5 mM of tolyl-GeI_3, low-polydispersity polymers were obtained. As a P catalyst, PI_3 (Table 7.1 (entry 10)) was effective for MMA as well as for St. For MMA, an N catalyst NIS exhibited particularly good performance (Figure 7.4 (filled symbols) and Table 7.1 (entries 11 and 12)).[13] Even with 1 mM (100 ppm) of NIS, which is among the lowest concentrations of the studied catalysts, PDI was small (1.15–1.4) from an early stage to a later stage of polymerization both in bulk and anisole solution. Without the catalyst (IMP), PDI was large (1.7–1.8) for MMA (Figure 7.4 (open circles)). Notably, NIS has good tolerance to functional groups, being applicable to a range of functional monomers, as shown below. The C catalyst CI_4 was also useful for MMA (Table 7.1 (entry 14)),[15] requiring only 1 mM (100 ppm) as in the case of NIS.

7.3.3.3 St and MMA with P, N, O, and C Hydrides as Precursors

Instead of adding a deactivator, a precursor of a deactivator or an activator radical (A•) may be used as a starting compound. We used compounds with hydrogen (H–A) as precursors of A•.[12,15,67,68] The precursors include such common compounds as phosphites (P catalysts),[12] imides (N),[67] phenols (O),[15,68] and dienes (C)[15] (Figure 7.1).

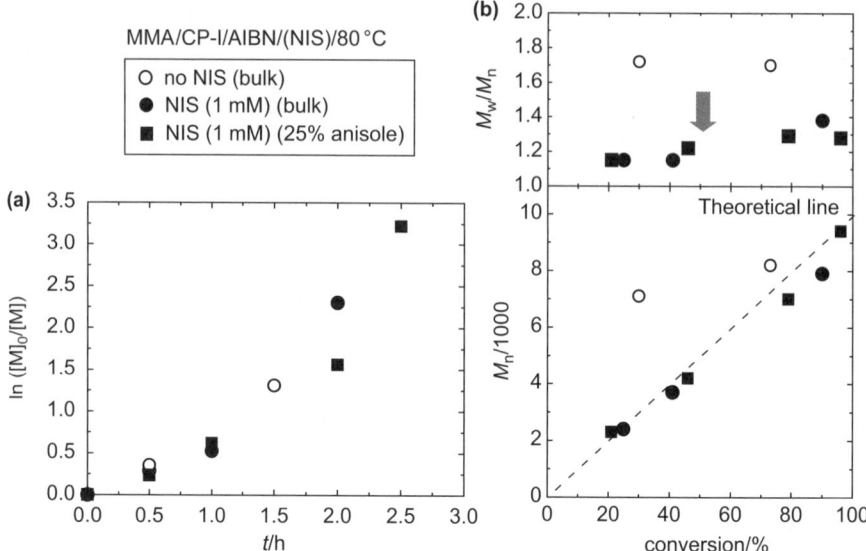

Figure 7.4 Plots of (a) $\ln([M]_0/[M])$ *vs. t* and (b) M_n and M_w/M_n *vs.* conversion for the MMA/CP–I/AIBN/ (NIS) systems in bulk and in 25 vol% anisole solution (80 °C): $[MMA]_0 = 8$ M; $[CP\text{–}I]_0 = 80$ mM; $[AIBN]_0 = 10$ mM; $[NIS]_0 = 0$ mM (open circles) and 1 mM (filled circles: for entry 11 in Table 7.1) for the bulk systems, and $[MMA]_0 = 6$ M; $[CP\text{–}I]_0 = 60$ mM; $[AIBN]_0 = 10$ mM; $[NIS]_0 = 1$ mM (filled squares: for entry 12 in Table 7.1) for the solution system.
Reproduced from ref. 13 by permission of American Chemical Society.

We first examined phosphorus hydrides $R_2PH(=O)$[12] as precursors in the presence of an alkyl iodide (dormant species) and a conventional radical initiator (radical source) as in the above deactivator systems. The conventional initiator gives a radical (Scheme 7.4a), which abstracts a hydrogen from $R_2PH(=O)$ to *in situ* produce $R_2P^\bullet(=O)$ (Scheme 7.4b). The activator radical $R_2P^\bullet(=O)$ leads to reversible activation (Scheme 7.4c). Figure 7.5 (filled symbols) and Table 7.2 (entries 1 and 2) show the polymerizations of St (8 M) with PE–I (dormant species) (80 mM), dicumyl peroxide (DCP) (radical source) (80 mM), and a precursor (10 mM) at 100 °C. With $(BuO)_2PH(=O)$ and $(EtO)PhPH(=O)$, small PDIs (≤ 1.2) were achieved from an early stage of polymerization, suggesting that the mentioned reactions (Scheme 7.4) effectively occurred. On the other hand, $Ph_2PH(=O)$ (with no alkoxy group) did not control the polydispersity, indicating the importance of the substituents. Phosphites were also effective for MMA (Figure 7.5 (open circles) and Table 7.2 (entry 9)). Cyclic phosphites with two R groups linked were also useful for St both at ambient and high (500–2000 bar) pressures.[69] The R_p became significantly higher with an increase of pressure (about three times higher at 2000 bar than at 1 bar) due to the increased propagation rate constant k_p and the decreased termination rate constant k_t.

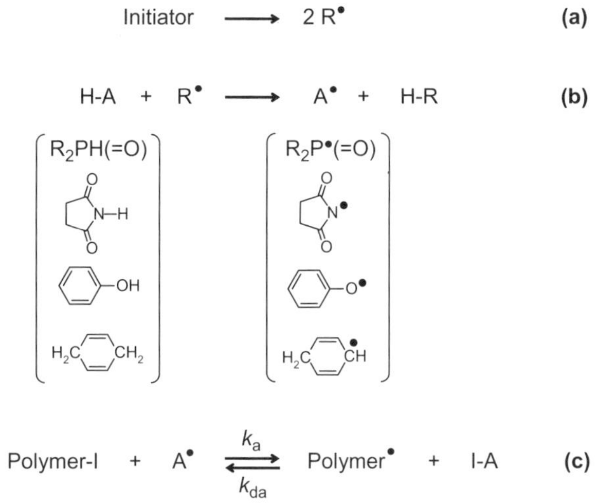

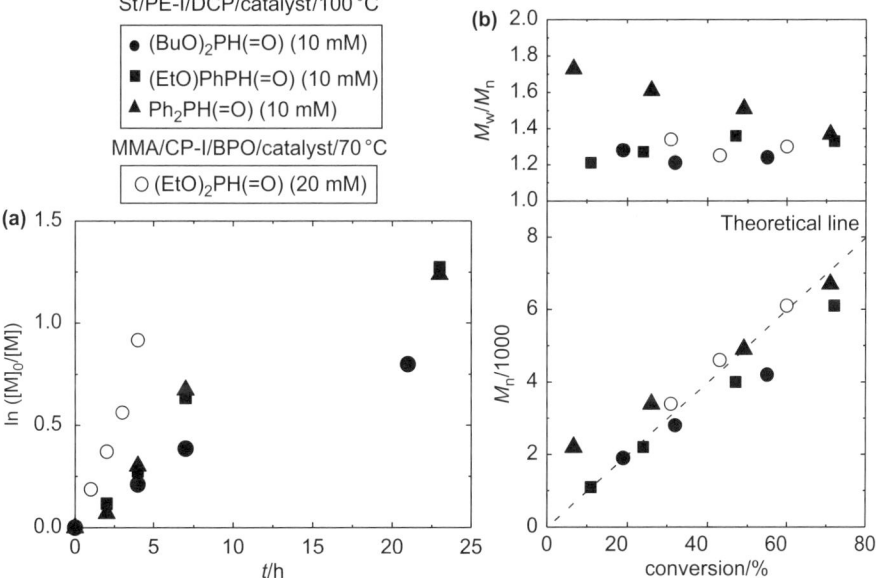

Scheme 7.4 Abstraction of hydrogen from a hydride precursor (H–A) with an initiator-derived radical (R•) to generate the activator radical (A•), which will work in the reversible activation.

Figure 7.5 Plots of (a) $\ln([M]_0/[M])$ *vs.* t and (b) M_n and M_w/M_n *vs.* conversion for the St/PE–I/DCP/catalyst systems in bulk (100 °C) and the MMA/CP-I/BPO/catalyst system in bulk (70 °C): $[St]_0 = 8$ M; $[PE–I]_0 = 80$ mM; $[DCP]_0 = 80$ mM; $[catalyst]_0 = 10$ mM for the St systems, and $[MMA]_0 = 8$ M; $[CP–I]_0 = 80$ mM; $[BPO]_0 = 40$ mM; $[catalyst]_0 = 20$ mM for the MMA systems. The catalysts and symbols are indicated in the figure. The MMA system (open circles) is for entry 1 in Table 7.2. Reproduced from ref. 12 by permission of American Chemical Society.

Table 7.2 RTCPs of St and MMA with hydride (precursor) catalysts ([Monomer]$_0$ = 8 M in bulk).

Entry	Monomer (equiv to [R-I])	Catalyst	R-I	In[a]	[R-I]$_0$/[In]$_0$/[cat]$_0$ (mM)	T/°C	t/h	Conv (%)	M$_n$ (M$_{n,theo}$)	PDI
1	St (100 eq)	(BuO)$_2$PH(=O)	PE-I	DCP	80/80/10	100	23	55	4200 (5500)	1.24
2	St (100 eq)	(EtO)PhPH(=O)	PE-I	DCP	80/80/30	100	23	69	5200 (6900)	1.19
3	St (100 eq)	2,4,6-Me phenol	CP-I	BPB	80/40/5	100	7	76	6300 (7600)	1.16
4	St (100 eq)	Phenol	CP-I	BPB	80/40/5	100	7	77	6700 (7700)	1.25
5	St (100 eq)	BHT	CP-I	BPB	80/40/5	100	7	75	6900 (7500)	1.34
6	St (100 eq)	Vitamin E	CP-I	BPB	80/40/5	100	7	76	6900 (7600)	1.18
7	St (100 eq)	Vitamin C	CP-I	BPB	80/40/5	100	7	80	8300 (8000)	1.21
8	St (100 eq)	CHD	CP-I	BPO	80/20/3	80	22	100	10 300 (10 400)	1.31
9	MMA (200 eq)	(EtO)$_2$PH(=O)	CP-I	BPO	40/20/10	70	6	74	13 000 (15 000)	1.29
10	MMA (100 eq)	BHT	CP-I	PDX	80/80/10	80	0.5	81	6300 (7900)	1.36
11	MMA (100 eq)	BHA	CP-I	PDX	80/80/10	80	0.5	52	5200 (5200)	1.21
12	MMA (100 eq)	Vitamin E	CP-I	PDX	80/80/10	80	0.5	69	6500 (6900)	1.23
13	MMA (100 eq)	Vitamin C	CP-I	PDX	80/80/10	80	0.5	87	7200 (8700)	1.33
14	MMA (100 eq)	CHD	CP-I	BPO	80/40/3	80	3	90	8800 (9000)	1.30
15	MMA (100 eq)	(EtO)$_2$PH(=O)	I$_2$	AIBN	40/80/10	80	2.2	95	10 300 (9500)	1.22
16	MMA (100 eq)	NHS	I$_2$	AIBN	40/80/30	80	3.1	90	7400 (9000)	1.38
17	MMA (100 eq)	CHD	I$_2$	AIBN	40/80/4	80	4	78	7200 (7900)	1.35
18	MMA (100 eq)	DPM	I$_2$	AIBN/BPO	40/(80/5)/30	80	4	100	8000 (10 000)	1.40

[a] In = conventional radical initiator, DCP = dicumyl peroxide, BPB = t-butyl perbenzoate, BPO = benzoyl peroxide, PDX = di(4-t-butylcyclohexyl) peroxydicarbonate, and AIBN = azobis(isobutyronitrile).

In RTCP, as requisites, (1) A$^\bullet$ should be relatively stable and easily be formed, and (2) A$^\bullet$ should undergo no or little initiation (addition to monomer) but still be active enough to abstract iodine from Polymer–I. Phenoxyl radicals (Ph-O$^\bullet$) (O-centered radical) and conjugate C-centered radicals (R$_3$C$^\bullet$) are relatively stable and do not initiate radical polymerization in many cases[2] and can be RTCP catalysts if they can activate Polymer–I. In fact, several phenols (Ph–OH)[15,68] and conjugated carbon compounds (R$_3$CH)[15] were effectively used as precursors (H–A), controlling the polymerizations of St and MMA (Figure 7.6 for early stages of polymerization and Table 7.2 (entries 3–8 and 10–14) for later stages of polymerization). Besides phenols, a conjugated alcohol, vitamin C[15,68] was also useful and one of the most effective catalysts for St (Figure 7.6 (filled circles)) in the studied condition. The phenols and alcohol include common antioxidants for foods and resins such as 2,6-di-*t*-butyl-4-hydroxytoluene (BHT) and natural compounds such as vitamins E and C. They are particularly attractive as non-toxic catalysts. The carbon compounds

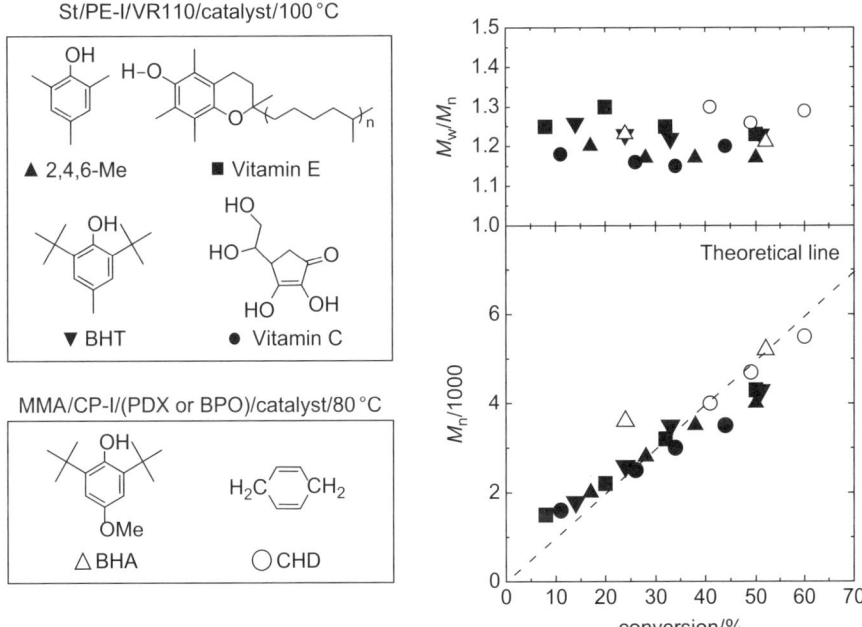

Figure 7.6 Plots of M_n and M_w/M_n vs conversion for the St/PE-I/VR110/catalyst systems in bulk (100 °C) and the MMA/CP-I/(PDX or BPO)/catalyst systems in bulk (80 °C): [St]$_0$ = 8 M; [PE-I]$_0$ = 80 mM; [VR110]$_0$ = 80 mM; [catalyst]$_0$ = 5 mM for the St systems, and [MMA]$_0$ = 8 M; [CP-I]$_0$ = 80 mM; [PDX]$_0$ = 80 mM (for BHA) or [BPO]$_0$ = 20 mM (for CHD); [BHA]$_0$ = 10 mM or [CHD]$_0$ = 3 mM for the MMA systems, where VR110 is 2,2′-azobis(2,4,4-trimethylpentane) and PDX is di(4-*t*-butylcyclohexyl) peroxydicarbonate. The catalysts and symbols are indicated in the figure. The BHA system (open triangles) is for entry 11 in Table 8.2.
Reproduced from ref. 15 by permission of American Chemical Society.

include simple hydrocarbons such as 1,4-cyclohexadiene (CHD). CHD has been used in organic syntheses taking advantage of the easy generation of the CHD radical,[70,71] while it was not used as a polymerization catalyst.

The amount of the precursor (*e.g.*, Table 7.2 and Figures 7.5 and 7.6) was about 3–30 mM in the studied cases, *i.e.*, about 10 mM for phosphites (St and MMA),[12] 30 mM for imides (MMA),[67] and 5–10 mM for phenols (St and MMA).[15,68] Among the precursors, CHD is more active over others, requiring 3 mM, for example.[15] CHD also has good tolerance to functional groups (like NIS as a deactivator catalyst), being useful for widening monomer versatility, as shown below.

7.3.3.4   Various Monomers

RTCP was applicable to acrylonitrile (AN) and various functional monomers,[13,15] for which NIS[13] (deactivator catalyst) and CHD[15] (precursor catalyst) were particularly effective, as Table 7.3 shows the examples. The compatible functional groups included hydrophobic and hydrophilic groups such as 2-ethylhexyl (EHMA), benzyl (BzMA), phenyl (POMA), epoxy (GMA), hydroxyl (HEMA), poly(ethylene glycol) (PEGMA), dimethylamino (DMAEMA), non-protected amino (ASt), and non-protected carboxylic acid (MAA) groups. In the homopolymerizations and random copolymerizations, M_n well agreed with $M_{n,theo}$, and PDI was small (1.1–1.4) from low to high conversions, with a small amount (1–10 mM) of the catalyst (as in the case of St and MMA) in many cases.

BzMA and GMA are widely applied for photoresists (electronic materials) and HEMA and PEGMA are biomaterials. No conductivity and low toxicity of NIS and CHD may be attractive for those applications. Also notably, the catalysts are applicable to both amino and acid functionalities (DMAEMA, ASt, and MAA) and even to unprotected amino and acid groups (ASt and MAA). Such dual compatibility to amino and acid functionalities is intriguing. The acid and amino functional polymers listed in Table 7.3 are water soluble and may find a range of applications. (The random copolymer of MAA (entry 18) with a relatively large MAA composition (40 mol%) is water soluble.)

The polymer produced in RTCP possesses iodine at the chain end. In the case of methacrylate polymers (tertiary alkyl chain-end polymers), upon heating, the decomposition (elimination of HI) (Scheme 7.5) occurs to produce a dead chain, and this side reaction is accelerated in the presence of acid. A mild temperature like 40–60 °C can suppress this side reaction. Thus, the polymerizations of acid monomers were carried out at a mild temperature 40 °C, leading to satisfactory control.

7.3.3.5   Block Copolymerizations

Exploiting the living character, block copolymerizations were conducted with a PMMA-I macroinitiator as the first block (Table 7.3 (entries 19–21)),[13,15] where PMMA is poly(methyl methacrylate), for example. Using BzMA (entry 19),

Table 7.3 RTCPs of AN and functional monomers ([Monomer]$_0$ = 8 M in bulk).

Entry	Monomer (equiv to [R-I])	Catalyst	R-I	In[a]	[R-I]$_0$/[In]$_0$/[cat]$_0$ (mM)	T/°C	t/h	Conv (%)	M$_n$[b] (M$_{n,theo}$)	PDI[b]
1	AN (100 eq)	CHD	CP-I	BPO	40/20/2.5[c]	80	2	73	11 800 (3900)	1.29
2	EHMA (100 eq)	NIS	CP-I	AIBN	80/10/1	80	2	87	13 200 (17 000)	1.26
3	EHMA (100 eq)	CHD	CP-I	BPO	80/20/3	80	2	87	13 200 (17 000)	1.20
4	BzMA (100 eq)	NIS	CP-I	AIBN	80/10/1	80	0.8	70	13 300 (13 000)	1.26
5	BzMA (100 eq)	CHD	CP-I	AIBN/BPO	80/(40/3)/1	80	3	94	13 300 (16 500)	1.27
6	POMA (100 eq)	CHD	CP-I	V70	40/40/2.5[c]	40	2	100	10 000 (16 200)	1.24
7	GMA (100 eq)	NIS	CP-I	AIBN	60/7.5/0.75[c]	80	1	71	9700 (9100)	1.29
8	GMA (100 eq)	CHD	CP-I	V70	60/15/2.25[c,d]	50	0.67	90	7300 (12 800)	1.36
9	HEMA (100 eq)	NIS	CP-I	AIBN	40/15/10[c]	80	1.5	57	7900 (7400)	1.36
10	PEGMA (100 eq)[e]	NIS	CP-I	AIBN	80/10/2	80	1	39	8600 (9600)	1.33
11	PEGMA (100 eq)[e]	CHD	CP-I	V70	80/80/3[d]	50	2	49	11 500 (22 500)	1.29
12	DMAEMA (100 eq)	CHD	CP-I	AIBN	80/40/10	80	1	90	12 300 (14 000)	1.40
13	ASt (100 eq)	CHD	CP-I	none	80/0/30	50	1	78	8000 (9300)	1.32
14	HEMA/BzMA (22/78 eq)	NIS	CP-I	AIBN	80/20/2	80	1.25	90	18 500 (15 200)	1.37
15	HEMA/MMA (50/50 eq)	CHD	CP-I	V70	40/10/10[c,d]	50	1	100	10 300 (11 900)	1.48
16	DMAEMA/BzMA (15/85 eq)	NIS	CP-I	AIBN	80/10/1	80	1.25	90	19 000 (15 600)	1.34
17	MAA/BzMA (15/85 eq)	NIS	CP-I	AIBN	80/40/1	80	1	60	6400 (9700)	1.21
18	MAA/MMA (40/60 eq)	CHD	CP-I	V70	40/40/5[c]	40	1.67	100	8300 (9400)	1.17
19	BzMA (100 eq)	NIS	PMMA-I[f]	AIBN	80/10/1	80	1	63	11 000 (15 000)	1.20
20	DMAEMA (100 eq)	CHD	PMMA-I[f]	AIBN	80/40/10	80	0.5	95	18 700 (15 000)	1.32
21	MAA/MMA (16/24 eq)	CHD	PMMA-I[f]	V70	80/80/5	40	2	80	6600 (5700)	1.31
22	MMA (100 eq) + BzMA (100 eq)	NIS	CP-I	AIBN	40/10/1[c]	80	1.5 / +0.5 / +1.0	54 / 108 / 145	5400 (5400) / 9700 (13 600) / 13 100 (19 300)	1.31 / 1.37 / 1.42
23	MMA (100 eq) + BzMA (100 eq)	CHD	CP-I	AIBN	60/15/3.75[c]	80	1.5 / +0.5 / +1.0	60 / 102 / 154	6200 (6000) / 10 100 (12 500) / 16 300 (20 500)	1.29 / 1.35 / 1.43

[a] In = conventional radical initiator, BPO = benzoyl peroxide, AIBN = azobis(isobutyronitrile), V70 = 2,2'-azobis(4-methoxy-2,4-dimethyl valeronitrile).

[b] Determined by GPC with a multiangle laser ?light-??scattering? (MALLS) detector for entry 1, and poly(methyl methacrylate)-calibration for entries 1–13, and poly(methyl methacrylate)-calibration for entries 14–23.

[c] Solution polymerization with 50 vol% ethylene carbonate for entry 1, 50 vol% dipropylene glycol monomethyl ether for entries 6 and 18, 25 vol% anisole for entries 7 and 23, 25 vol% toluene for entry 8, 15 vol% 1-propanol and 35 vol% methyl ethyl ketone for entries 9 and 15, and 50 vol% anisole for entry 22 ([monomer]$_0$ = 4 M and 6 M for 50 and 25 vol% solvent, respectively).

[d] Addition of I$_2$ (1 mM for entries 8 and 11, and 2 mM for entry 15).

[e] Molecular weight = 246.

[f] M$_n$ = 3900 and PDI = 1.14 for entry 19, and M$_n$ = 2700 and PDI = 1.15 for entries 20 and 21.

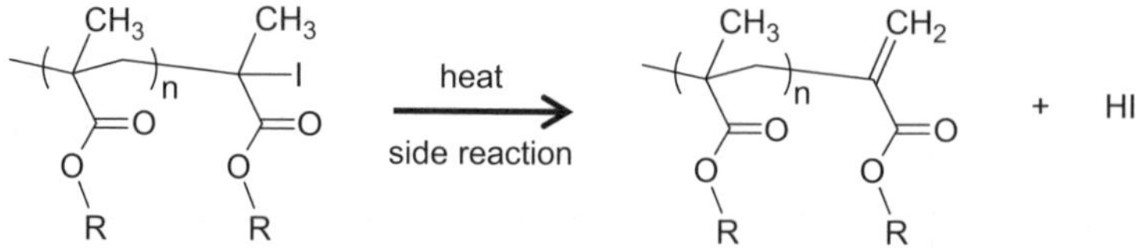

Scheme 7.5 Elimination of HI from Polymer–I for methacrylates.

DMAEMA (entry 20), and the mixture of MMA and MAA (entry 21) as the second block monomers, the PMMA chain successfully extended, yielding low-polydispersity block copolymers (PDI = 1.2–1.3). Instead of using an isolated macroinitiator, successive addition of two monomers was also successful.[13,15] To a polymerization of MMA (first monomer) with CP-I (low-mass alkyl iodide) at an about 60% conversion, the addition of BzMA (second monomer) yielded PMMA-*block*-(PMMA-*random*-PBzMA)s with relatively small PDIs (~1.4) (entries 22 and 23), where PBzMA is poly(benzyl methacrylate).

For the studied homopolymerizations and copolymerizations of St, MMA, AN, and functional monomers (Tables 7.1, 7.2, and 7.3), the polymerization was fairly fast in many cases. The conversion reached 70–90% in 1–7 h in the studied conditions.

7.3.3.6 *Aqueous Microsuspension Polymerization*

Okubo *et al.*[72] successfully achieved an aqueous microsuspension RTCP of MMA with NIS catalyst. The particle phase included MMA (8 M), CP-I (80 mM), AIBN (10 mM), and NIS (1 mM), as in the mentioned bulk polymerization (Table 7.1 (entry 11)). The water content was 90% with the use of a cationic surfactant, dodecyltrimethyl ammonium bromide. The polymerization smoothly proceeded and the conversion reached 93% for 2 h at 80 °C. The PDI was small (1.3–1.5) throughout the polymerization, demonstrating the successful polymerization in water continuous phase.

7.3.4 **Alkyl Iodide Dormant Species**

This section focuses on alkyl iodide dormant species (not catalysts).

7.3.4.1 *Use of Alkyl Iodide Formed* In Situ

Instead of a preformed alkyl iodide (used in the above mentioned systems), we may add molecular iodine I_2 and an azo compound (R–N = N–R) to the system as starting compounds and use the alkyl iodide (R–I) *in situ* formed in the polymerization. This (I_2/azo) method was originally invented by Lacroix-Desmazes *et al.* for IMP[73–75] and may be practically useful due to the general lack of the long-term stability of alkyl iodides upon storage. The

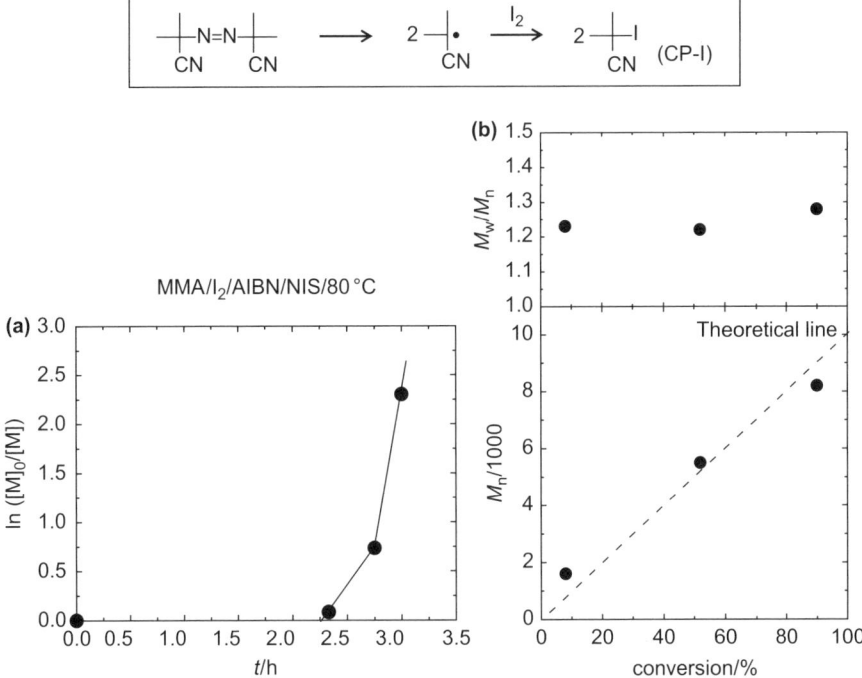

Figure 7.7 Plots of (a) $\ln([M]_0/[M])$ *vs. t* and (b) M_n and M_w/M_n *vs.* conversion for the MMA/I_2/AIBN/NIS system in bulk (80 °C): $[MMA]_0 = 8$ M; $[I_2]_0 = 40$ mM; $[AIBN]_0 = 80$ mM; $[NIS]_0 = 5$ mM (for entry 15 in Table 7.1). Reproduced from ref. 13 by permission of American Chemical Society.

polymerization of MMA (8 M) was examined with I_2 (40 mM), AIBN (80 mM), and NIS (5 mM) at 80 °C (Figure 7.7 and Table 7.1 (entry 15)).[13] AIBN gives 2-cyanopropyl radical CP•, and CP• reacts with I_2 to form CP–I (Figure 7.7).[76] AIBN also works as a radical source to run the polymerization after the completion of CP–I formation. In view of the rather low efficiency ($f \sim 0.6$) of AIBN to produce free CP•, we used an excess of it (2 equivalents to I_2). Virtually no polymerization occurred for 2.25 h, during which time CP• predominantly reacted with I_2 (rather than monomer) and CP–I (theoretically 80 mM from 40 mM of I_2) accumulated. After this period, the polymerization smoothly proceeded. The M_n well agreed with $M_{n,theo}$, and PDI was 1.2-1.3 from a low conversion to a high conversion.

This method was also applicable to precursor catalysts (Table 7.2 (entries 15–18)) such as $(EtO)_2PH(=O)$ (P precursor),[67] succinimide (NHS) (N precursor),[67] CHD (C precursor),[15] and diphenylmethane (DPM) (C precursor).[15]

7.3.4.2 Direct Initiation from Alcohol

A hydroxyl (OH) group, one of the most common functional groups, was used as an initiating moiety.[77] The system consists of an alcohol (RO–H), an

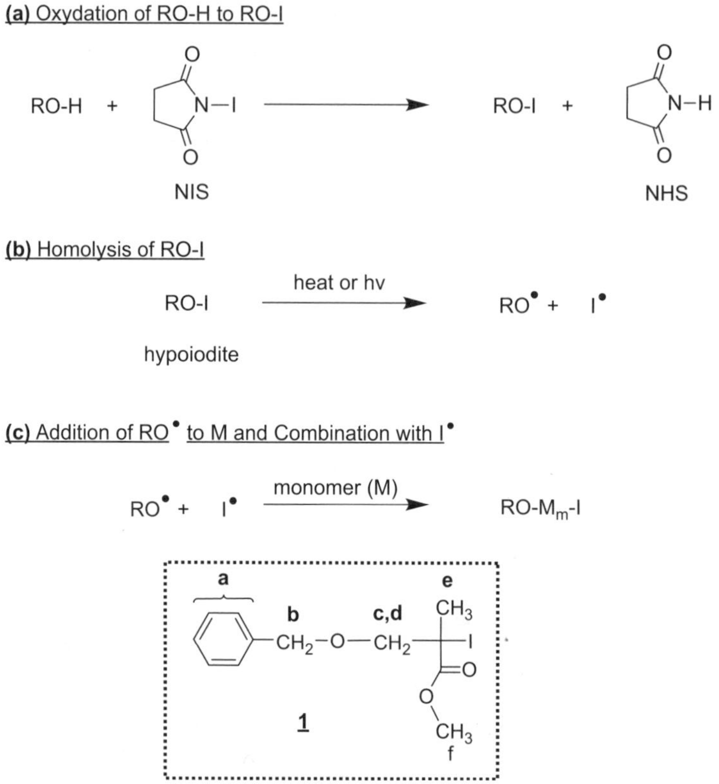

(a) Oxydation of RO-H to RO-I

NIS   NHS

(b) Homolysis of RO-I

$$RO\text{-}I \xrightarrow{\text{heat or hv}} RO^{\bullet} + I^{\bullet}$$

hypoiodite

(c) Addition of RO$^{\bullet}$ to M and Combination with I$^{\bullet}$

$$RO^{\bullet} + I^{\bullet} \xrightarrow{\text{monomer (M)}} RO\text{-}M_m\text{-}I$$

Scheme 7.6   Generation of alkyl iodide RO–M_m–I from alcohol (RO–H).

iodinating agent (NIS, for example), and a monomer (M), giving an alkyl iodide dormant species RO–M–I (I = iodine) (Scheme 7.6). The dormant species may be used to the subsequent polymerization with or without purification. The latter case can lead to a one-pot RTCP from an alcohol.

Mechanistically, RO–H is oxidized with NIS to form the hypoiodite (RO–I) (Scheme 7.6a).[78] RO–I thermally or photochemically dissociates to generate the alkoxy radical (RO$^{\bullet}$) and the iodine radical (I$^{\bullet}$) (Scheme 7.6b).[78] This reaction sequence (the radical formation from alcohol *via* hypoiodite) is called the Barton reaction.[78–80] When RO$^{\bullet}$ is non-conjugated (R = alkyl *etc.*), it can add to M and then is capped by I$^{\bullet}$ to form RO–M–I (Scheme 7.6c). RO–I can also directly react with M (*via* an insertion mechanism) to form RO–M–I.[81]

Benzyl alcohol was used as RO–H, for example.[77] The mixture of benzyl alcohol (240 mM), NIS (240 mM), and MMA (8 M) (in bulk) heated at 80 °C for 3 h gave RO–M–I (compound **1** in Scheme 7.6) in a moderate 40% yield. The yield was improved to 55% with a doubled amount of NIS (480 mM). Despite the non-quantitative generation of RO-M-I for benzyl alcohol, one-pot RTCP from the alcohol would be interesting and was attempted in two manners. In the first manner, to the mentioned mixture (after generation of **1** (yield = 40%)), AIBN (120 mM) was added as a radical source to run the polymerization at 80 °C (in the same pot and without purification of **1**). In this system, the remaining NIS after the generation of **1** worked as a catalyst for the polymerization. NIS thus plays two roles on the initiation (generation of **1**) and polymerization stages. After the polymerization for 3 h, a low-polydispersity polymer with $M_n = 7400$ and PDI = 1.27 was obtained (including alcohol-derived and AIBN-derived polymers). In the second manner, we mixed benzyl alcohol, NIS, MMA, and AIBN altogether at the onset of reaction and heated it at 80 °C. The all-mixed system also gave a relatively low-polydispersity polymer with $M_n = 17\,000$ and PDI = 1.41, although the polydispersity control was less satisfactory than that in the first manner.

This method would be applicable to various alcohols, while the yield of the dormant species would depend on alcohols. This method would be useful for graft polymerizations from materials with many OH groups, even if not all of the OH groups can initiate. Star and comb polymers and polymer brushes on solid surfaces may easily be obtained without prior attachment of preformed LRP dormant species.

7.3.4.3 Surface Initiated Polymerization

Surfaces play crucial roles in many important properties, including optical, mechanical, thermodynamic, and chemical properties, and thus surface modification is an important issue.[82–84] One of the effective surface modification methods is surface-initiated graft polymerization. Fukuda *et al.*[83,85] used LRP for graft polymerization for the first time to obtain a polymer brush with a high surface density, which was one order of magnitude higher than those of the conventional brushes. The surface occupancy was as large as 40%. Such a so-called concentrated polymer brush has an extremely extended conformation, as extended as 80% of the length of the all-*trans* conformation in good solvents.[83] Such a conformation affords many new properties, such as high elasticity, ultra-low friction, and excellent repellency of proteins and cells, which are not attainable by conventional semi-dilute and dilute brushes, and thus can find useful applications.[82–84]

In this regard, surface-initiated RTCP was examined, the details of which will be reported in a future publication. In brief, an immobilizing dormant species IHE[86] (Figure 7.8) consisting of an alkyl iodide moiety and a triethoxysilyl (TEOS) group was synthesized. IHE was fixed on a silicon wafer *via* the TEOS group, and the polymerizations of some monomers yielded well-defined (PDI ∼ 1.3) concentrated polymer brushes with about 40% surface occupancy (*e.g.*, about 0.5 chains/nm² for PMMA brushes). An example is shown in Figure 7.8, in which IHE (small circles) was fixed on the surface in a patterned manner. In this example, for a demonstration purpose, the polymerization of BzMA was carried out, using vitamin E as a non-toxic catalyst, without prior bubbling with inert gas (argon), and at 85 °C for a very short time 5 min. After the polymerization, the polymer brush is observed as black square spots

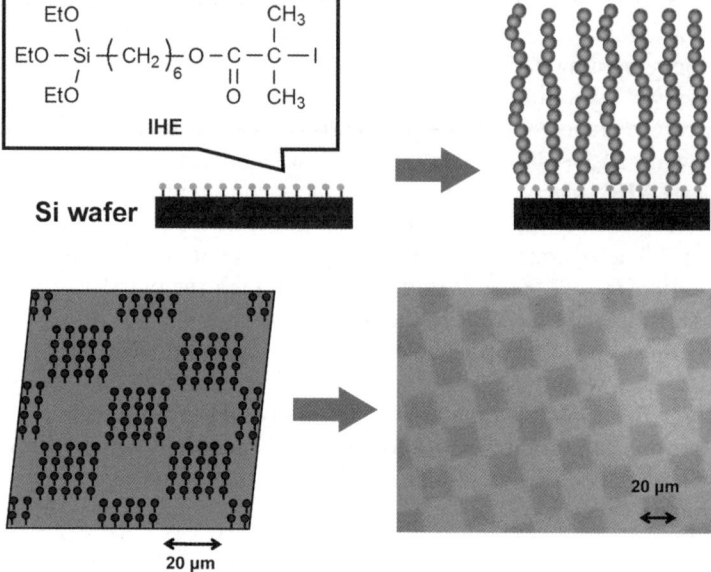

Figure 7.8 Surface-initiated RTCP: Graft polymerization from IHE on surface (top) and microscope image of polymer brush obtained by the graft polymerization from IHE (small circles) patterned on surface (bottom).

(Figure 7.8), demonstrating the patterning and the robust and quick polymerization.

7.3.5 Kinetic Features

Kinetic studies were carried out on the activation process (Section 7.3.5.1) and R_p (Section 7.3.5.2) in the polymerizations of St with a polystyrene iodide (PSt-I) ($M_n = 2000$; PDI $= 1.20$) (dormant species), BPO (radical source), and a catalyst (GeI$_4$, tolyl-GeI$_3$, SnI$_4$, PI$_3$ or NIS).[12,87] The catalysts are the deactivators shown in section 7.3.3.1. The polymeric adduct PSt-I was used as a starting alkyl iodide to focus on the kinetics of polymer region. The mechanistic proof of the RT process (Scheme 7.3a) was also obtained in the actual polymerizations and model systems (Section 7.3.5.3).[12,15,87]

7.3.5.1 Reversible Activation

In the presence of a deactivator I–A, Polymer–I can be activated *via* the RT process (Scheme 7.3a) with an activator radical A$^\bullet$ (rate constant k_a) and the DT process (Scheme 7.1c) (rate constant k_{ex}). Thus, k_{act} (Scheme 7.1a) is generally given by

$$k_{act} = k_{ex}[\text{Polymer}^\bullet] + k_a[\text{A}^\bullet] \qquad (7.1)$$

In the quasi-equilibrium of the RT process (Scheme 7.3a), eqn (7.1) takes the form

$$k_{act} = k_{ex}[Polymer^\bullet] + k_{da}[Polymer^\bullet] \left(\frac{[I-A]}{[Polymer-I]} \right) \qquad (7.2)$$

where k_{da} is the deactivation rate constant with I–A (Scheme 7.3a). Thus, k_{act} increases with the ratio [I–A]/[Polymer–I].

Figure 7.9 shows the plots of k_{act} *vs.* the ratio for the St/PSt–I/BPO/catalyst systems at 80 °C, where k_{act} was experimentally determined at a fixed value of [Polymer$^\bullet$].[12,87] In all cases, k_{act} linearly increased with the ratio, as expected from eqn (7.2). The system at the ratio = 0 (without catalyst) is IMP, where k_{act} is too small to achieve low polydispersity because of the small k_{ex}.[55] The plot for GeI$_4$ suggests that in typical RTCP conditions with [GeI$_4$]/[PSt–I] = 5 mM/ 80 mM = 0.0625 (Table 7.1 (entry 1)), k_{act} is about 20 times larger than in the absence of GeI$_4$ (IMP). Namely, PSt–I is about 20 times more frequently activated due to the contribution of RT with GeI$_4$, along with the small contribution of DT. This explains why low-polydispersity polymers were obtained from an early stage of polymerization for the GeI$_4$ system. SnI$_4$, NIS, tolyl–GeI$_3$, and PI$_3$ (Figure 7.9) also had large k_{act} values at the ratio [catalyst]/[PSt–I] = 0.0625, which were about 13, 10, 5, and 3 times larger than in their absence, respectively. These k_{act} values are also so large as to achieve low polydispersity from an early stage of polymerization. Of the five studied catalysts, GeI$_4$ was the most active for St at 80 °C.

It should be noted that the activity depends on temperature and monomer. For St at a lower temperature 60 °C, SnI$_4$ was the most active, for example.[87]

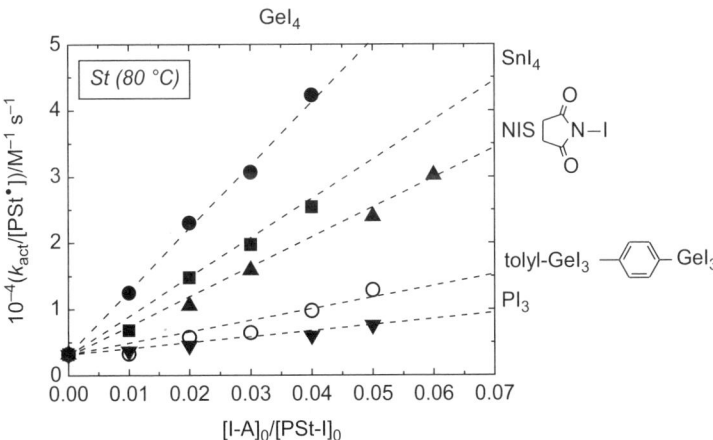

Figure 7.9 Plot of k_{act} *vs.* [I–A]$_0$/[PSt–I]$_0$ for the St/PSt–I/BPO/I–A systems in bulk (80 °C): [PSt-I]$_0$ = 10 mM; [BPO]$_0$ = 1 mM; [I–A]$_0$ = 0–0.6 mM. The I–A and symbols are indicated in the figure.
Reproduced from ref. 87 by permission of Wiley-VCH Verlag GmbH & Co. KGaA, Weinheim.

For the monomer, k_{act} was about 120 times larger for MMA (PMMA–I) than St (PSt–I) with the same catalyst tolyl–GeI$_3$,[87] indicating that the catalytic activity strongly depends on monomer. The order of the catalytic activity among the catalysts may also be different between St and MMA.

For GeI$_4$, the slope of the line (Figure 7.9) gave k_{da} of 6.0×10^5 M^{-1} s^{-1}. This is large, nearly as large as the deactivation rate constant (5.7×10^5 M^{-1} s^{-1} (120 °C))[22] for a PSt radical with a representative nitroxide DEPN, a good deactivator for LRP. As shown below, the equilibrium constant K ($= k_a/k_{da}$) for the GeI$_4$ system was estimated to be about 10^0. It follows that, for typical conditions, [PSt–I] $= 10^{-2}$ M, [PSt$^\bullet$] $= 10^{-8}$ M, and [GeI$_4$] $= 10^{-3}$ M, [GeI$_3$$^\bullet$] is 10^{-9} M (nanomolar). Namely, the observed large k_{act} is achieved by a nanomolar level concentration of the activator, meaning a remarkably large k_a. The k_a was calculated to be about 10^6 M^{-1} s^{-1} with the mentioned K and k_{da}. Such a large k_a is a unique kinetic feature of RTCP. The k_a and k_{da} were also similarly large for other examined catalysts (Table 7.4).

7.3.5.2 Polymerization Rate

In the presence of the RTCP catalysts, R_p is somewhat smaller than in their absence (IMP), as shown in Figure 7.10a (and Figure 7.3a) for the St/PSt–I/BPO with and without GeI$_4$ at 80 °C.[12] This is because A$^\bullet$ undergoes irreversible cross-termination with Polymer$^\bullet$ (rate constant k_t').[12] This mechanism is analogous to the one for the rate retardation in the RAFT polymerization, where the intermediate radical (Polymer–(X$^\bullet$)–Polymer) undergoes the relevant cross-termination.[88–93] In theory, when the quasi-equilibrium of RT holds and the radical concentrations [Polymer$^\bullet$] and [A$^\bullet$] are stationary, R_p is given by eqn (7.3).[12]

$$R_p = R_{p,0}\left(1 + 2\left(\frac{k_t'}{k_t K}\right)\frac{[I - A]}{[\text{Polymer} - I]}\right)^{-1/2} \qquad (7.3)$$

where $R_{p,0}$ is the R_p without I–A and k_t is the self-termination rate constant of Polymer$^\bullet$. Eqn (7.3) means that R_p decreases with increasing the ratio [I–A]/[Polymer–I].

In the mentioned system (Figure 7.10a), R_p (hence [Polymer$^\bullet$]) was stationary in the studied range of time in all cases and decreased with increasing ratio of [GeI$_4$]$_0$/[PSt–I]$_0$ as the theory demands. Figure 7.10b shows the plot of R_p vs. [GeI$_4$]$_0$/[PSt-I]$_0$. The plot was linear, confirming the validity of eqn (7.3) for [GeI$_4$]$_0$/[PSt–I]$_0$ = 0–0.25. Thus, when this ratio is relatively small ($\sim$0.0625), as in entry 1 in Table 7.1, the cross-termination is the main cause for the retardation. The existence of the cross-termination brings about an increase of the termination rate and hence the number of dead chains by, e.g., about 20% in a typical condition (Table 7.1 (entry 1)). However, the effect of this much of dead chains on polydispersity is rather minor in this as well as other LRP systems. The cross-termination also results in a loss of GeI$_4$, but it is also minor, at least

Table 7.4 Summary of k_a, k_{da}, k_{ex}, and k_{act} values.

Entry	System	Dormant species	Catalyst	T/°C	$10^3 \, k_{act}/s^{-1}$ [a]	$k_a/M^{-1} \, s^{-1}$	$k_{da}/M^{-1} \, s^{-1}$	$k_{ex}/M^{-1} \, s^{-1}$	Reference
1	RTCP	PSt-I	GeI$_4$	80	5.3	$\sim 10^6$	9.0×10^5	—	12,87
2		PSt-I	SnI$_4$	80	3.5	$\sim 10^6$	5.7×10^5	—	87
3		PSt-I	NIS	80	2.7	$\sim 10^5$	4.3×10^5	—	87
4		PSt-I	Tolyl-GeI$_3$	80	1.6	$\sim 10^5$	2.5×10^5	—	87
5		PSt-I	PI$_3$	80	0.9	$\sim 10^5$	1.2×10^5	—	87
6		PMMA-I	Tolyl-GeI$_3$	70	110	$\sim 10^7$	310×10^5	—	87
7	RCMP	PMMA-I	TMEDA	90	6.6	0.13	7.2×10^5	—	16
8	IMP	PSt-I	—	80	0.22	—	—	0.024×10^5	55
9	IMP	PMMA-I	—	90	0.096	—	—	0.042×10^5	16
10	ATRP	PSt-Br	CuBr/dHbiby	110	23	0.45	110×10^5	—	56

[a]Value approximately estimated for $R_p = 4.8 \times 10^{-4} \, M \, s^{-1}$ (50% conversion for 3 h).

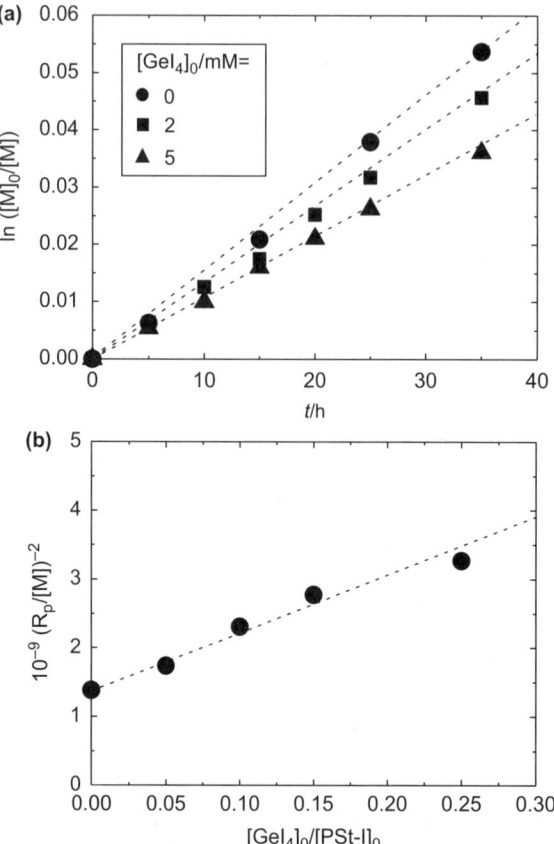

Figure 7.10 Plots of (a) ln([M]$_0$/[M]) *vs.* t and (b) (R_p/[M])$^{-2}$ *vs.* [GeI$_4$]$_0$/[PSt-I]$_0$ for the St/PSt–I/BPO/(GeI$_4$) systems in bulk (80 °C): [PSt–I]$_0$ = 20 mM; [BPO]$_0$ = 10 mM; [GeI$_4$]$_0$ = 0–5 mM. The [GeI$_4$]$_0$ and symbols are indicated in the figure.
Reproduced from ref. 12 by permission of American Chemical Society.

at an early stage of polymerization. In any case, the cross-termination products, such as PSt–GeI$_3$, may still work as I–A, controlling the polydispersity. The loss of catalyst can be important at a later stage of polymerization when the cross-termination product Polymer–A is inactive (such as with NIS to generate Polymer–NS with no iodine). The effects of the catalyst loss on molecular weight distribution and R_p were predicted by computer simulation.[94]

The slope of the line (Figure 7.10b) gave $k_t'/(k_tK) = 3$. The k_t' between a polymer radical and a low-mass radical (A$^\bullet$) would be about 10^9 M^{-1} s^{-1}, and the k_t between polymer radicals would be about 10^8 M^{-1} s^{-1},[95,96] at the studied low conversions. Thus, K is estimated to be on the order of 10^0. The rate retardation was also observed for other catalysts because of the cross-termination.[87] The same analysis was also carried out for other catalysts (Table 7.4).[87]

7.3.5.3 Experimental Proof of RT Process

The existence of RT (Scheme 7.3a) was examined in three ways. (a) If RT exists, A• is mediated. Thus, we attempted to detect A• by electron spin resonance (ESR) in the St polymerizations. The Sn, Ge, P, and N-centered A• radicals could not be detected due to their low equilibrium concentrations, while TI (O deactivator), BHT (O precursor), and CHD (C precursor) derived A• radicals were in fact observed (Figure 7.11).[15] The spectrum for TI was broad, probably due to the aggregation of TI in the St medium, and that for BHT clearly split into four peaks (1:3:3:1 ratio) by the methyl group at the para position and further split into three peaks (1:2:1 ratio) by the protons at the meta positions. The spectrum for CHD was somewhat noisy due to the lower concentration of A• than those for TI and BHT. The spectrum split into three peaks (1:2:1 ratio) by the protons at the meta positions and further split into two peaks (1:1 ratio) by the proton at the central position. In these experiments, a large amount (100–400 mM) of the catalyst was used to facilitate the detection of A•. Nevertheless, the concentration of A• was low, in the order of 10^{-6} M, in the three cases, indicating that in typical polymerization conditions (with 3–10 mM of the catalyst), it is very low, on the order of 10^{-8}–10^{-7} M (nanomolar), and the activation process effectively occurs at such a low A• concentration (as indicated above). Polymer• was not apparently detected (Figure 7.11) due to its much lower concentration (10^{-8}–10^{-7} M estimated from the polymerization rate) than that of A• (10^{-6} M). (b) If RT exists, the propagating species is a free radical (Polymer•). This was confirmed from the tacticity of the product polymer, namely, the tacticity was virtually the same as that without the catalyst (IMP) in all studied cases (St and MMA with various

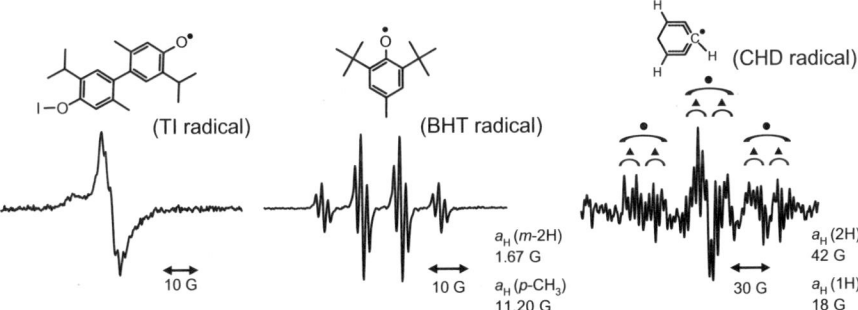

Figure 7.11   ESR spectra for the St/PE–I/AIBN/TI system (85 °C), the St/PE–I/ BPB/BHT system (100 °C), and the St/PE-I/PDX/CHD system (60 °C), in bulk in all cases: $[St]_0 = 8$ M; $[PE–I]_0 = 400$ mM; $[AIBN]_0 = 200$ mM; $[TI]_0 = 150$ mM for the TI system, $[St]_0 = 8$ M; $[PE–I]_0 = 400$ mM; $[BPB]_0 = 200$ mM; $[BHT]_0 = 100$ mM for the BHT system, and $[St]_0 = 8$ M; $[PE–I]_0 = 200$ mM; $[PDX]_0 = 200$ mM; $[CHD]_0 = 400$ mM for the CHD system, where BPB is *t*-butyl perbenzoate and PDX is di(4-*t*-butylcyclohexyl)peroxydicarbonate.
Reproduced from ref. 15 by permission of American Chemical Society.

catalysts).[12,15] (c) While the activation process (reaction of alkyl iodide and A•) is known to be fast for Sn, Ge, and P-centered A• in organic chemistry,[62,63] the deactivation process (reaction of alkyl radical and I-A) had not been well studied. The existence of the deactivation process was confirmed for SnI_4, GeI_4, PI_3, and NIS in a model experiment using a unimer methacrylate radical and a deactivator.[87] The three experiments (a, b, and c) strongly support the existence of RT. The results would commonly hold for St and MMA with various catalysts.

7.4 RCMP

7.4.1 Background, Concept, and Mechanism

The background of RCMP is ATRP[39–43] (Scheme 7.1d) that uses a halogen as X and a transition metal compound such as $Cu(I)X$[40,41,43] as an activator catalyst. $Cu(I)X$ reversibly activates Polymer–X to generate Polymer• and $Cu(II)X_2$. This is a redox reaction of copper. As an initial motivation, we got an idea that even organic compounds may work as catalysts, if they have redox ability. We tested this concept, using iodine as X (an alkyl iodide dormant species) and a well-known organic reducing agent tetra(dimethylamino)ethylene (TDAE) (Figure 7.1) as an activator catalyst instead of $Cu(I)X$. This in fact worked well. After this initial study, we found that even a simple amine, triethylamine (TEA), which has little or a much weaker redox ability, also well worked as a catalyst.

As a possible mechanism (Scheme 7.3b), the amine abstracts iodine from Polymer–I to generate Polymer• and a complex of the iodine radical and amine (I•/amine complex). Since the iodine radical is not a stable radical, it recombines with another iodine radical to form a complex of the iodine molecule and amine (I_2/amine complex).[97–99] Polymer• reacts with these complexes (deactivators) to form Polymer–I and the amine. In this process, electron transfer from the amine to iodine would occur to a range of different degrees including full (redox), partial (coordination) and nearly no transfer, depending on the kind of amines. The process is *reversible complexation* (RC) of iodine and catalyst, and the polymerization is termed RC mediated polymerization (RCMP).[16,100] Clearly, it is mechanistically distinguished from both ATRP and RTCP.

Technically, RCMP is like ATRP in that they include only a dormant species and an activator catalyst. Unlike RTCP, it requires no conventional radical initiator, even though a conventional radical initiator would, as in other LRPs, work to decrease the concentration of deactivator and thus increase R_p.[5–7] A small amount of the deactivator $Cu(II)X_2$ is sometimes added in ATRP. In the same way, the deactivator I_2/amine complex is sometimes added in RCMP.

In the literature, the formation of an alkyl radical from an alkyl halide with an amine was studied in organic chemistry[101,102] and employed as a conventional radical initiation in polymer chemistry.[103,104] However, in all cases, the reaction was irreversible. We found the reversible reaction, which is new in chemistry.

7.4.2 Performance

7.4.2.1 MMA with TEA Catalyst

We studied the polymerizations of MMA with amine catalysts at 60–90 °C.[16,100] The polydispersity control was essentially the same at these temperatures. Figure 7.12 (open circles) and Table 7.5 (entry 1) show the bulk polymerization of MMA (8 M) with CP–I (80 mM) as a dormant species and TEA (40 mM) as an activator catalyst at 90 °C. The polymerization proceeded up to, *e.g.*, a 66% monomer conversion in 2.3 h, confirming the generation of the propagating radical from CP–I and TEA. Without CP–I or TEA, no polymerization proceeded, again confirming that the radical was generated by the combination of CP–I and TEA. The first order plot of [M] was linear in the studied range of time (Figure 7.12a (open circle)). However, M_n deviated from $M_{n,theo}$ and PDI was larger than 1.5 at an early stage of polymerization (Figure 7.12b (open circles)). This is because a sufficient amount of deactivator (I_2/amine complex) was not accumulated at an early stage of polymerization, at which many monomers added to Polymer•. Thus, we added a small amount of iodine molecule (I_2) as a starting compound, which will form the I_2/amine

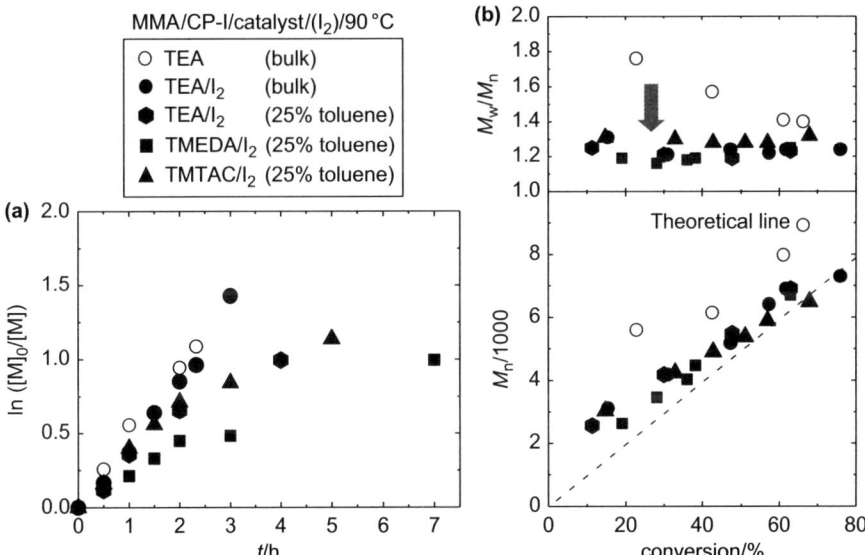

Figure 7.12 Plots of (a) $\ln([M]_0/[M])$ *vs.* t and (b) M_n and M_w/M_n *vs.* conversion for the MMA/CP–I/TEA/(I_2) systems in bulk (90 °C) and the MMA/CP–I/catalyst/I_2 systems in 25 vol% toluene (90 °C): $[MMA]_0 = 8$ M; $[CP–I]_0 = 80$ mM; $[TEA]_0 = 40$ mM; $[I_2]_0 = 0$ mM (open circles: for entry 1 in Table 7.5) and 1 mM (filled circles: for entry 2 in Table 7.5) for the bulk systems, and $[MMA]_0 = 6$ M; $[CP–I]_0 = 60$ mM; [N atom of catalyst]$_0 = 30$ mM; $[I_2]_0 = 0.75$ mM for the solution systems. The catalysts and symbols for the solution systems are indicated in the figure.
Reproduced from ref. 16 by permission of American Chemical Society.

Table 7.5 RCMPs of some monomers ($[\text{Monomer}]_0 = 8$ M in bulk).

Entry	Monomer (equiv to [R-I])	Catalyst	R-I	$[R\text{-}I]_0/[cat]_0/[I_2]_0$ (mM)	T/°C	t/h	Conv (%)	$M_n{}^a$ ($M_{n,\text{theo}}$)	PDI^a
1	MMA (100 eq)	TEA	CP-I	80/40/0	90	2.3	66	8900 (6600)	1.40
2	MMA (100 eq)	TEA	CP-I	80/40/1	90	3	76	7300 (7600)	1.24
3	MMA (100 eq)	TBA	CP-I	80/20/5	90	2.3	72	6400 (7200)	1.35
4	MMA (100 eq)	TMTAC	CP-I	80/40/2	90	0.75	67	7800 (6700)	1.30
5	MMA (100 eq)	TEA	CP-I	80/40/2	80	5	80	8500 (8000)	1.25
6	MMA (100 eq)	TMEDA	CP-I	80/40/2	80	6	68	7900 (6800)	1.32
7	MMA (100 eq)	TMTAC	CP-I	80/40/2	80	2	80	9200 (8000)	1.39
8	MMA (100 eq)	TEA	CP-I	80/80/20	60	24	81	10000 (8100)	1.26
9	MMA (200 eq)	TEA	CP-I	40/80/20	60	24	67	17000 (14000)	1.29
10	MMA (400 eq)	TEA	CP-I	20/80/20	60	24	62	30000 (25000)	1.42
11	St (100 eq)	TMTAC	CP-I	80/40/20	120	24	74	10000 (7700)	1.45
12	AN (100 eq)	TMTAC	CP-I	40/10/5^b	75	4	100	14900 (5700)	1.49
13	BzMA (100 eq)	TMTAC	CP-I	80/405	90	1	84	13400 (14800)	1.35
14	HEMA (100 eq)	TBA	CP-I	80/80/20	90	1	70	6700 (9100)	1.28
15	HEMA/MMA (13/87 eq)	TDME	CP-I	80/40/5	80	2	53	9700 (5600)	1.28
16	DMAEMA/MMA (5/95 eq)	none	CP-I	80/0/20	90	24	37	5300 (5800)	1.39
17	MMA (110 eq)	TEA	AIBN/I$_2$	60/10/40^c	80	5.8	90	11000 (9900)^d	1.42
18	MMA (110 eq)	TEA	AIBN/I$_2$	60/40/40^c	80	2.1	87	15000 (9600)^d	1.45

^{a}Determined by GPC with poly(methyl methacrylate)-calibration for entries 1–10 and 15–18, polystyrene-calibration for entry 11, and a multiangle laser light-scattering (MALLS) detector for entries 12–14.

^{b}Solution polymerization with 50 vol% ethylene carbonate (hence $[AN]_0 = 4$ M) for entry 12.

^{c}$[AIBN]_0/[TEA]_0/[I_2]_0 = 60/10/40$ for entry 17 and 60/40/40 for entry 18.

^{d}Theoretical M_n calculated with $[MMA]_0$, $[I_2]_0$, and conversion, in which 72 mM of CP-I is assumed to be formed (as noted in the text).

complex with TEA (Scheme 7.3b).[73,105,106] In fact (Figure 7.12 (filled circles) and Table 7.5 (entry 2)), with the addition of 1 mM of I_2 (as small as 1/40 equivalent to TEA), M_n agreed with $M_{n,theo}$ and PDI was about 1.3 from an early stage of polymerization. Also importantly, PDI remained small (about 1.2) up to high conversions, *e.g.*, 80%, demonstrating the success of RCMP. The R_p with I_2 was slightly lower than that without it (Figure 7.12a), as expected from the equilibrium (Scheme 7.3b).

Mechanistically, RCMP includes DT as well as RC. However, as mentioned above, the rate constant k_{ex} of DT is too small,[16] and the good polydispersity control observed in RCMP is mainly due to the work of the catalyst (RC) with a small contribution of DT.

7.4.2.2 Amines with Multiple Nitrogens as Catalysts

Iodine can be stabilized by using amines with multiple nitrogens (Figure 7.1), which may work as good activators. The amines thus examined and their abbreviated names are shown in Figure 7.1. At the same concentration of the nitrogen atom (30 mM), compared with TEA with one nitrogen (Figure 7.12b (pentagons)), TMEDA with two nitrogens (Figure 7.12b (squares)) afforded lower polydispersity at an early stage of polymerization.[16,100] The PDI for TMEDA also kept small (about 1.2) up to a high conversion (about 60%). TMTAC with four nitrogens (Figure 7.12b (triangles)) was also a good catalyst, although it did not further lower polydispersity, possibly because the coordination is divalent and thus the remaining two nitrogens did not effectively work (other factors may also be involved).[16,100]

The number of alkyl groups attached to the nitrogen atom was also important.[16] PDI became larger in the order of tertiary amines < secondary ones < primary ones, meaning that tertiary amines are the most effective catalysts. This could be due to a higher electron density on the nitrogen atom with a more number of alkyl groups and hence a higher coordination ability. For primary amines, another cause for the larger PDI is a side reaction, *i.e.*, chain-end transformation of Polymer–I (dormant species) to Polymer–NH–R (dead chain) by amine R–NH$_2$, which is particularly significant for primary amines.

The studied amines such as TEA and TMEDA are among the simplest and cheapest amines, which is an attractive feature. The small PDI achievable even at high conversions would also be useful. Other examples of the MMA polymerizations are also listed in Table 7.5 (entries 3–7).

7.4.2.3 Higher Molecular Weights

As mentioned above, in the case of methacrylate polymers, upon heating, elimination of HI (Scheme 7.5) from Polymer–I occurs to produce a dead chain. A mild temperature like 60 °C can suppress this side reaction, and we may have relatively high molecular weight polymers. At 60 °C with TEA as a catalyst (Table 7.5 (entries 8–10)), when the target degree of polymerization at a full conversion was set to 100, 200, and 400, we obtained low polydispersity polymers (PDI = 1.26–1.42) up to high conversions (up to M_n = 30 000 in the studied case),[100] which was difficult at higher temperatures. With TEA, the polymerization was rather slow and took 24 h for high conversions (*e.g.*, 60%) in the studied conditions. We are currently exploring more active catalysts which can lead to faster polymerization.

7.4.2.4 Some Other Monomers

Besides MMA, RCMP was applicable to some other monomers[16] including St, AN, and functional methacrylates with a benzyl group (BzMA) and a hydroxyl group (HEMA). The PDI was 1.28–1.49 even at high conversions (>70%) (Table 7.5 (entries 11–15)). Interestingly, DMAEMA monomer possesses a tertiary amine on the side chain and hence may work as a catalyst at the same time. In fact, the random copolymerization of DMAEMA (5 mol%) and MMA (95 mol%) proceeded in a controlled manner without the addition of catalyst (Table 7.5 (entry 16)).

7.4.2.5 Use of In Situ Formed Alkyl Iodide

The I_2/azo method was adopted to RCMP.[100] Figure 7.13 (filled circles) and Table 7.5 (entry 17) show the polymerization of MMA (8 M) with I_2 (40 mM),

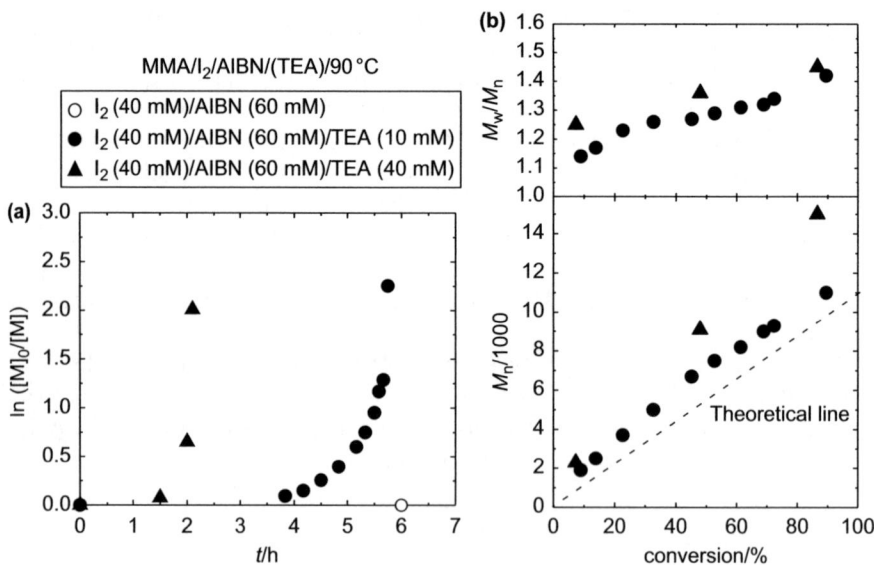

Figure 7.13   Plots of (a) $\ln([M]_0/[M])$ *vs. t* and (b) M_n and M_w/M_n *vs.* conversion for the MMA/I_2/AIBN/(TEA) systems in bulk (80 °C): $[MMA]_0 = 8$ M; $[I_2]_0 = 40$ mM; $[AIBN]_0 = 60$ mM; $[TEA]_0 = 0$ (open circle), 10 mM (filled circles, for entry 17 in Table 7.5), and 40 mM (filled triangles, for entry 18 in Table 7.5). Reproduced from ref. 100 by permission of American Chemical Society.

AIBN (60 mM), and TEA (10 mM) at 80 °C. In view of the efficiency (about 0.6)[2] of AIBN to produce free CP•, 60 mM of AIBN can give about 72 mM of free CP• and hence about 72 mM (theoretical amount) of CP–I. Virtually no polymerization occurred for 3.5 h, during which time CP–I was formed, and after this period, the polymerization proceeded. The M_n almost agreed with $M_{n,theo}$, and PDI was 1.2–1.4 throughout the polymerization. Without TEA (catalyst) (Figure 7.13a (open circle)), no polymerization occurred. With a much larger amount (40 mM) of TEA (Figure 7.13 (triangles) and Table 7.5 (entry 18)), the polymerization began before all of the AIBN was consumed to yield CP–I, indicating the importance of the amounts of AIBN and TEA for RCMP.

7.4.3 Kinetic Studies

Kinetic studies were carried out on the reversible activation and R_p for the MMA polymerizations with TMEDA (with two nitrogens), a representative good catalyst, at 90 °C.[16] (The existence of the RC process (Scheme 7.3b) was also supported by some experiments.[16])

7.4.3.1 Activation and Deactivation Rate Constants

We determined k_{act} for the activation of PMMA–I with various concentrations of TMEDA at 90 °C. According to Schemes 7.1a, 7.1c, 7.3b, k_{act} takes the form:

$$k_{act} = k_{ex}[\text{Polymer}^\bullet] + k_a]\text{TMEDA}] \qquad (7.4)$$

where k_a is the activation rate constant in Scheme 7.3b. Figure 7.14 shows the plot of k_{act} vs. [TMEDA]$_0$. The plot was linear, and the slope of the line gave k_a

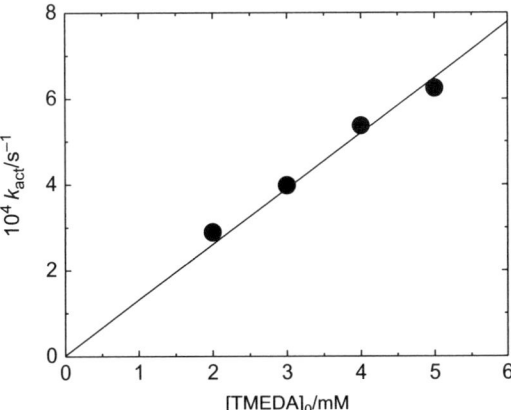

Figure 7.14 Plot of k_{act} vs. [TMEDA]$_0$ for the MMA/PMMA–I/TMEDA systems in bulk (90 °C): [MMA]$_0$ = 8 M; [PMMA–I]$_0$ = 0.5 mM; [TMEDA]$_0$ = 2–5 mM. Reproduced from ref. 16 by permission of American Chemical Society.

to be $0.13\,\mathrm{M}^{-1}\,\mathrm{s}^{-1}$. (The plot virtually passed through the origin due to a small contribution of DT in the studied condition.) This k_a ($0.13\,\mathrm{M}^{-1}\,\mathrm{s}^{-1}$) is similar in magnitude to that ($0.45\,\mathrm{M}^{-1}\,\mathrm{s}^{-1}$)[56] in a representative ATRP system of styrene with PSt–Br and CuBr/dHbipy catalyst (110 °C) (Table 7.4). Of course, since the monomer, halogen, and temperature are different for the two systems, this does not directly compare the catalytic activities of TMEDA and CuBr/dHbipy, but this suggests that the MMA/TMEDA system is virtually as effective as this good ATRP system as a system.

The equilibrium constant K defined as $K=[\mathrm{PMMA}^{\bullet}][\mathrm{I}_2/\mathrm{TMEDA}]/([\mathrm{PMMA{-}I}][\mathrm{TMEDA}])$ was determined to be 1.8×10^{-7}. The obtained K and k_a give the deactivation rate constant $k_{da}=7.2\times10^5\,\mathrm{M}^{-1}\mathrm{s}^{-1}$ according to $K=k_a/k_{da}$. This k_{da} value is about 20 times smaller than that ($1.1\times10^7\,\mathrm{M}^{-1}\mathrm{s}^{-1}$)[56] in the mentioned ATRP system, explaining why the RCMP system requires a small amount of deactivator added at the beginning of polymerization while the ATRP system does not.

7.4.3.2 Polymerization Rate

In the quasi-equilibrium of RC (Scheme 7.3b), R_p should follow the relation

$$R_p / [\mathrm{M}] = k_p[\mathrm{polymer}^{\bullet}] = k_p K([\mathrm{polymer-I}][\mathrm{amine}]) / [\mathrm{I}_2/\mathrm{amine}] \qquad (7.5)$$

This was confirmed to hold in the polymerizations of MMA with CP–I, TMEDA, and I_2/TMEDA at 90 °C. The R_p was proportional to $[\mathrm{CP{-}I}]_0$ and $[\mathrm{TMEDA}]_0$ and reciprocal to $[\mathrm{I}_2/\mathrm{TMEDA}]_0$.

7.5   Conclusions

This chapter summarized the fundamental features of RTCP and RCMP that use Ge, P, N, O, and C-centered molecules as catalysts. They are based on new reversible activation mechanisms (RT and RC). Low-polydispersity polymers were obtained in the homopolymerizations and random and block copolymerizations of St, MMA, AN, and several functional monomers. Attractive features of the catalysts include their good polydispersity controllability, good tolerance to functional groups, inexpensiveness, low toxicity, and ease of handling (robustness). With these advantages, RTCP and RCMP may serve as useful tools for synthesizing well-defined polymers.

Acknowledgments

The authors acknowledge the financial support from Grants-in-Aid for Scientific Research, the Ministry of Education, Culture, Sports, Science and Technology, Japan, New Energy and Industrial Technology Development Organization (NEDO) of Japan, and Japan Science and Technology Agency (JST).

References

1. For reviews on LRP, see refs 1–4: K. Matyjaszewski and T. P. Davis (ed.), *Handbook of Radical Polymerization*, Wiley & Sons, New York, 2002.
2. G. Moad and D. H. Solomon (ed.), *The Chemistry of Radical Polymerization*, Elsevier, Amsterdam, 2006.
3. K. Matyjaszewski (ed.), Controlled/Living Radical Polymerization, *ACS Symp. Ser.*, 2009, **1023** & **1024**.
4. W. A. Braunecker and K. Matyjaszewski, *Prog. Polym. Sci.*, 2007, **32**, 93.
5. For reviews on kinetics, see refs 5–7: T. Fukuda, *J. Polym. Sci.: Part A: Polym. Chem.*, 2004, **42**, 4743.
6. A. Goto and T. Fukuda, *Prog. Polym. Sci.*, 2004, **29**, 329.
7. H. Fischer, *Chem. Rev.*, 2001, **101**, 3581.
8. L. S. Hegedus, *J. Am. Chem. Soc.*, 2009, **131**, 17997.
9. J. Raynaud, Y. Gnanou and D. Taton, *Macromolecules*, 2009, **42**, 5996.
10. R. Kakuchi, K. Chiba, K. Fuchise, R Sakai, T. Satoh and T. Kakuchi, *Macromolecules*, 2009, **42**, 8747.
11. A. Goto, H. Zushi, Y. Kwak and T. Fukuda, *ACS Symp. Ser.*, 2006, **944**, 595.
12. A. Goto, H. Zushi, N. Hirai, T. Wakada, Y. Tsujii and T. Fukuda, *J. Am. Chem. Soc.*, 2007, **129**, 13347.
13. A. Goto, N. Hirai, T. Wakada, K. Nagasawa, Y. Tsujii and T. Fukuda, *Macromolecules*, 2008, **41**, 6261.
14. For a review on RTCP: A. Goto, Y. Tsujii and T. Fukuda, *Polymer*, 2008, **49**, 5177 (Feature Article).
15. A. Goto, N. Hirai, K. Nagasawa, Y. Tsujii, T. Fukuda and H. Kaji, *Macromolecules*, 2010, **43**, 7971.
16. A. Goto, T. Suzuki, H. Ohfuji, M. Tanishima, T. Fukuda, Y. Tsujii and H. Kaji, *Macromolecules*, 2011, **44**, 8709.
17. T. Otsu and M. Yoshida, *Makromol. Chem. Rapid Commun.*, 1982, **3**, 127.
18. T. Otsu, *J. Polym. Sci., Polym. Chem.*, 2000, **38**, 2121.
19. J. Chiefari, Y. K. Chong, F. Ercole, J. Krstina, J. Jeffery, T. P. T. Le, R. T. A. Mayadunne, G. F. Meijs, C. L. Moad, G. Moad, E. Rizzardo and S. H. Thang, *Macromolecules*, 1998, **31**, 5559.
20. G. Moad, E. Rizzardo and S. H. Thang, *Polymer*, 2008, **49**, 1079.
21. M. K. Georges, R. P. N. Veregin, P. M. Kazmaier and G. K. Hamer, *Macromolecules*, 1993, **26**, 2987.
22. D. Benoit, S. Grimaldi, S. Robin, J. P. Finet, P. Tordo and Y. Gnanou, *J. Am. Chem. Soc.*, 2000, **122**, 5929.
23. D. Benoit, V. Chaplinski, R. Braslau and C. J. Hawker, *J. Am. Chem. Soc.*, 1999, **121**, 3904.
24. V. Sciannamea, R. Jérôme and C. Detrembleur, *Chem. Rev.*, 2008, **108**, 1104.
25. D. Colombani, M. Steenbock, M. Klapper and K. Müllen, *Macromol. Rapid Commun.*, 1997, **18**, 243.
26. B. Yamada, Y. Nobukane and Y. Miura, *Polym. Bull.*, 1998, **41**, 539.

27. E. K. Y. Chen, S. J. Teertstra, D. Chan-Seng, P. O. Otieno, R. G. Hicks and M. K. Georges, *Macromolecules*, 2007, **40**, 8609.
28. A. Bledzki and D. Braun, *Polym. Bull.*, 1986, **16**, 19.
29. E. V. Chernikova, Z. A. Pokataeva, E. S. Garina, M. B. Lachinov and V. B. Golubev, *Macromol. Chem. Phys.*, 2001, **202**, 188.
30. P. C. Wieland, B. Raether and O. Nuyken, *Macromol. Rapid Commun.*, 2001, **22**, 700.
31. B. B. Wayland, G. Poszmik, S. L. Mukerjee and M. Fryd, *J. Am. Chem. Soc.*, 1994, **116**, 7943.
32. A. Debuigne, J. R. Caille and R. Jérôme, *Angew. Chem. Int. Ed.*, 2005, **44**, 1101.
33. A. Debuigne, R. Poli, C. Jérôme, R. Jérôme and C. Detrembleur, *Prog. Polym. Sci.*, 2009, **34**, 211.
34. A. D. Asandei and I. W. Moran, *J. Am. Chem. Soc.*, 2004, **126**, 15932.
35. Y. Yutani and M. Tatemoto, *Eur. Pat. Appl.*, 1992, EP489370 A1.
36. M. Kato, M. Kamigaito, M. Sawamoto and T. Higashimura, *Polym. Prepr., Jpn.*, 1994, **43**, 225.
37. K. Matyjaszewski, S. Gaynor and J.S. Wang, *Macromolecules*, 1995, **28**, 2093.
38. G. David, C. Boyer, J. Tonnar, B. Ameduri, P. Lacroix-Desmazes and B. Boutevin, *Chem. Rev.*, 2006, **106**, 3936.
39. M. Kato, M. Kamigaito, M. Sawamoto and T. Higashimura, *Macromolecules*, 1995, **28**, 1721.
40. J. S. Wang and K. Matyjaszewski, *J. Am. Chem. Soc.*, 1995, **117**, 5614.
41. N. V. Tsarevsky and K. Matyjaszewski, *Chem. Rev.*, 2007, **107**, 2270.
42. M. Ouchi, T. Terashima and M. Sawamoto, *Chem. Rev.*, 2009, **109**, 4963.
43. F. Lena and K. Matyjaszewski, *Prog. Polym. Sci.*, 2010, **35**, 959.
44. K. Takagi, A. Soyano, T. S. Kwon, H. Kunisada and Y. Yuki, *Polym. Bull.*, 1999, **43**, 143.
45. S. Yamago, K. Iida and J. Yoshida, *J. Am. Chem. Soc.*, 2002, **124**, 2874.
46. S. Yamago, B. Ray, J. Yoshida, T. Tada, K. Yoshizawa, Y. Kwak, A. Goto and T. Fukuda, *J. Am. Chem. Soc.*, 2004, **126**, 13908.
47. S. Yamago, E. Kayahara, M. Kotani, B. Ray, Y. Kwak, A. Goto and T. Fukuda, *Angew. Chem. Int. Ed.*, 2007, **46**, 1304.
48. S. Yamago, *Chem. Rev.*, 2009, **109**, 5051.
49. P. B. Zetterlund, Y. Kagawa and M. Okubo, *Chem. Rev.*, 2008, **108**, 3747.
50. K. Satoh and M. Kamigaito, *Chem. Rev.*, 2009, **109**, 5120.
51. B. M. Rosen and V. Percec, *Chem. Rev.*, 2009, **109**, 5069.
52. C. Boyer, V. Bulmus, T. P. Davis, V. Ladmiral, J. Liu and S. Perrier, *Chem. Rev.*, 2009, **109**, 5402.
53. M. A. Tasdelen, M. U. Kahveci and Y. Yagci, *Prog. Polym. Sci.*, 2011, **36**, 455.
54. A. Goto and T. Fukuda, *Macromolecules*, 1997, **30**, 5183.
55. A. Goto, K. Ohno and T. Fukuda, *Macromolecules*, 1998, **31**, 2809.
56. K. Ohno, A. Goto, T. Fukuda, J. Xia and K. Matyjaszewski, *Macromolecules*, 1998, **31**, 2699.

57. A. Goto, K. Sato, Y. Tsujii, T. Fukuda, G. Moad, E. Rizzardo and S. H. Thang, *Macromolecules*, 2001, **34**, 402.
58. A. Goto, Y. Kwak, T. Fukuda, S. Yamago, K. Iida, M. Nakajima and J. Yoshida, *J. Am. Chem. Soc.*, 2003, **125**, 8720.
59. Y. Kwak, A. Goto, T. Fukuda, Y. Kobayashi and S. Yamago, *Macromolecules*, 2006, **39**, 4671.
60. Y. Kwak, A. Goto, T. Fukuda and S. Yamago, *Z. Chem. Phys.*, 2005, **219**, 283.
61. W. P. Neumann, *Synthesis*, 1987, 665.
62. C. Chatgilialoglu and M. Newcomb, *Adv. Organomet. Chem.*, 1999, **44**, 67.
63. A. Studer and S. Amrein, *Synthesis*, 2002, 835.
64. D. P. Curran, C. P. Jasperse and M. J. Totleben, *J. Org. Chem.*, 1991, **56**, 7169.
65. A. L. J. Beckwith and P. E. Pigou, *Aust. J. Chem.*, 1986, **39**, 77.
66. K. U. Ingold, J. Lusztyk and J. C. Scaiano, *J. Am. Chem. Soc.*, 1984, **106**, 343.
67. A. Goto, K. Nagasawa, A. Shinjo, Y. Tsujii and T. Fukuda, *Aust. J. Chem.*, 2009, **62**, 1492.
68. A. Goto, N. Hirai, T. Wakada, K. Nagasawa, Y. Tsujii and T. Fukuda, *ACS Symp. Ser.*, 2009, **1023**, 159.
69. A. Wolpers, L. Ackermann and P. Vana, *Macromol. Chem. Phys.*, 2011, **212**, 259.
70. G. Binmore, J. C. Walton and L. Cardellin, *J. Chem. Soc., Chem. Commun.*, 1995, 27.
71. G. Binmore, L. Cardellin and J. C. Walton, *J. Chem. Soc., Perkin Trans.*, 1997, **2**, 757.
72. M. Yorizane, T. Nagasuga, Y. Kitayama, A. Tanaka, H. Minami, A. Goto, T. Fukuda and M. Okubo, *Macromolecules*, 2010, **43**, 8703.
73. P. Lacroix-Desmazes, R. Severac and B. Boutevin, *Macromolecules*, 2005, **38**, 6299.
74. C. Boyer, P. Lacroix-Desmazes, J. J. Robin and B. Boutevin, *Macromolecules*, 2006, **39**, 4044.
75. For another method for in situ producing an alkyl iodide: J. Tonnar and P. Lacroix-Desmazes, *Angew. Chem., Int. Ed.*, 2008, **47**, 1294
76. P. Balczewski and M. Mikolajczyk, *New J. Chem.*, 2001, **25**, 659.
77. J. Kim, A. Nomura, T. Fukuda, A. Goto and Y. Tsujii, *Macromol. React. Eng.*, 2010, **4**, 272.
78. C. E. McDonald, T. R. Beebe, M. Beard, D. McMillen and D. Selski, *Tetrahedron Lett.*, 1989, **30**, 4791.
79. D. H. R. Barton, J. M. Beaton, L. E. Geller and M. M. Pechet, *J. Am. Chem. Soc.*, 1960, **82**, 2640.
80. J. Allen, R. B. Boar, J. F. McGhie and D. H. R. Barton, *J. Chem. Soc., Perkin Trans.*, 1973, **1**, 2402.
81. S. A. Glover and A. Goosen, *Tetrahedron Lett.*, 1980, **21**, 2005.
82. S. Edmondson, V. L. Osborne and W. T. S. Huck, *Chem. Soc. Rev.*, 2004, **33**, 14.

83. Y. Tsujii, K. Ohno, S. Yamamoto, A. Goto and T. Fukuda, *Adv. Polym. Sci.*, 2006, **197**, 1.

84. R. Barbey, L. Lavanant, D. Paripovic, N. Schuwer, C. Sugnaux, S. Tugulu and H. Klok, *Chem. Rev.*, 2009, **109**, 5437.

85. M. Ejaz, S. Yamamoto, K. Ohno, Y. Tsujii and T. Fukuda, *Macromolecules*, 1998, **31**, 5934.

86. A. Nomura, A. Goto, K. Ohno, E. Kayahara, S. Yamago and Y. Tsujii, *J. Polym. Sci. Part A: Polym. Chem.*, 2011, **49**, 5284.

87. A. Goto, T. Wakada, Y. Tsujii and T. Fukuda, *Macromol. Chem. Phys.*, 2010, **211**, 594.

88. H. de Brouwer, M. A. J. Schellekens, B. Klumperman, M. J. Monteiro and A. L. German, *J. Polym. Sci. Part A: Polym. Chem.*, 2000, **38**, 3596.

89. Y. Kwak, A. Goto, Y. Tsujii, Y. Murata, K. Komatsu and T. Fukuda, *Macromolecules*, 2002, **35**, 3026.

90. F. M. Calitz, M. P. Tonge and R. D. Sanderson, *Macromolecules*, 2003, **36**, 5.

91. Y. Kwak, A. Goto and T. Fukuda, *Macromolecules*, 2004, **37**, 1219.

92. W. Meiser, J. Barth, M. Buback, H. Kattner and P. Vana, *Macromolecules*, 2011, **44**, 2474.

93. C. Barner-Kowollik, M. Buback, B. Charleux, M. L. Coote, M. Drache, T. Fukuda, A. Goto, B. Klumperman, A. B. Lowe, J. B. Mcleary, G. Moad, M. J. Monteiro, R. D. Sanderson, M. P. Tonge and P. Vana, *J. Polym. Sci. Part A: Polym. Chem.*, 2006, **44**, 5809.

94. P. Vana and A. Goto, *Macromol. Theory Simul.*, 2010, **19**, 24.

95. M. Buback, M. Egorov, R. G. Gilbert, V. Kaminsky, O. F. Olaj, G. T. Russell, P. Vana and G. Zifferer, *Macromol. Chem. Phys.*, 2002, **203**, 2570.

96. C. Barner-Kowollik, M. Buback, M. Egorov, T. Fukuda, A. Goto, O. F. Olaj, G. T. Russell, P. Vana, B. Yamada and P. B. Zetterlund, *Prog. Polym. Sci.*, 2005, **30**, 605.

97. For the combination of two free I•s (not I•/amine complexes) to generate an iodine molecule I$_2$, see ref. 97–99H. Rosman and R. M. Noyes, *J. Am. Chem. Soc*, 1958, **80**, 2410.

98. J. A. LaVerne and L. Wojnarovits, *J. Phys. Chem.*, 1994, **98**, 12635.

99. P. Lacroix-Desmazes, J. Tonnar and B. Boutevin, *Macromol. Symp.*, 2007, **248**, 150.

100. A. Goto, Y. Tsujii and H. Kaji, *ACS Symp. Ser.*, 2012, **1100**, 305.

101. D. P. Stevenson and G. M. Coppinger, *J. Am. Chem. Soc.*, 1962, **84**, 149.

102. H. Ishibashi, S. Haruki, M. Uchiyama, O. Tamura and J. Matsuo, *Tetrahedron Lett.*, 2006, **47**, 6263.

103. S. R. Sen and N. N. Dass, *Eur. Polym. J.*, 1982, **18**, 477.

104. P. C. Dwivedi, S. D. Pandey, A. Tandon and S. Singh, *Colloid. Polym. Sci.*, 1995, **273**, 822.

105. H. Yada, J. Tanaka and S. Nagakura, *Bull. Chem. Soc. Jpn.*, 1960, **33**, 1660.

106. J. K. S. Wan and J. N. Jitts Jr., *Tetrahedron Lett.*, 1964, **44**, 3245.

CHAPTER 8

Atom Transfer Radical Polymerization (ATRP)

NICOLAY V. TSAREVSKY*[a] AND KRZYSZTOF MATYJASZEWSKI[b]

[a] Department of Chemistry and Center for Drug Design, Discovery and Delivery in Dedman College, Southern Methodist University, Dallas, TX 75275, USA; [b] Department of Chemistry, Carnegie Mellon University, Pittsburgh, PA 15213, USA
*Email: nvt@smu.edu

8.1 Introduction

Radical polymerization is one of the most useful methods for synthesis of homo- and copolymers derived from a large number of monomers, due to its tolerance toward many solvents, functional groups, and additives or impurities. Although a true "living" radical polymerization cannot be achieved (owing to the ever-present rapid bimolecular radical termination), it is possible to design controlled/"living" radical polymerization (CRP) if a dynamic equilibrium can be established between the propagating radicals and a larger amount of dormant species, which can be reactivated.[1-3] In the first living radical poly-merizations, trityl (triphenylmethyl),[4,5] nitroxide, or various S-centered radicals were employed, which could not efficiently initiate polymerization but were able to form relatively labile bonds with the propagating radicals to afford the dormant species. The latter could be further reactivated, either photochemically or thermally, and the generated polymeric radicals could continue to grow in the presence of the monomer. Additionally, the product consisting of dormant polymer molecules could be isolated from the reaction mixture, purified, and

RSC Polymer Chemistry Series No. 4
Fundamentals of Controlled/Living Radical Polymerization
Edited by Nicolay V Tsarevsky and Brent S Sumerlin
© The Royal Society of Chemistry 2013
Published by the Royal Society of Chemistry, www.rsc.org

used as macroinitiator (after suitable reactivation) in the presence of another monomer or mixture of monomers, yielding segmented copolymers.

A number of CRP methods[6–10] have been developed that allow for the preparation of a multitude of well-defined polymers, *i.e.*, polymers with predetermined molecular weight, narrow (or in some cases even controlled) molecular weight distribution, and high degree of chain-end functionalization. The use of multifunctional initiators, including initiators containing also a polymerizable moiety, or monomers with more than one polymerizable group under CRP conditions has enormously expanded the number of accessible molecular architectures. The most widely used CRP methods are (i) stable free radical polymerization (which includes iniferter- and nitroxide-mediated polymerization (NMP),[11–15] organometallic radical polymerization (OMRP),[16–19] and several others), (ii) atom transfer radical polymerization (ATRP),[20–23] and reversible chain transfer catalyzed polymerizations;[24] and (iii) degenerative transfer-based polymerization[25] (with reversible addition–fragmentation chain transfer (RAFT) polymerization [26–30] being the most successful technique but also including iodine-mediated polymerizations and polymerizations in the presence of tellurium or antimony compounds[31]). Most of these techniques are covered in detail in other chapters of this book.

8.2   Main Mechanistic Aspects of ATRP and Origin of Polymerization Control

ATRP is based on the reversible homolytic cleavage of the C–X bond (activation) of an alkyl halide or pseudohalide,[32] RX, by a redox-active low-oxidation state metal complex Mt^zL_m (activator; Mt^z represents the metal ion in oxidation state z, and L is a ligand; throughout this text, the charges of ionic species are omitted for simplicity). The reaction affords a radical and the corresponding high-oxidation state metal complex with a coordinated halide ligand $XMt^{z+1}L_m$ (termed deactivator owing to its ability to transfer the atom or group X back to the radicals transforming them into the corresponding (pseudo)halides). ATRP mechanistically resembles the synthetically useful radical addition of alkyl halides across an unsaturated carbon–carbon bond, termed *atom transfer radical addition*[33–35] (ATRA) or the closely related *atom transfer radical cyclization;*[35,36] the latter is observed in molecules containing both C–X and carbon–carbon double bonds. ATRP can be viewed as a special case of ATRA, which involves the reactivation of the alkyl halide adduct of the unsaturated compound (monomer) and the further reaction of the formed radical with the monomer (propagation).[37] Activation and deactivation occur throughout the polymerization and ideally the deactivation should be efficient and fast, so that the propagating radicals are converted to the dormant species after just a few monomer additions. The ATRA and ATRP processes are presented in Figure 8.1 along with all relevant rate constants. In this text, the symbol $R_n^{\bullet}$ is used to designate growing radicals with degree of polymerization n, and P_n-X denotes the corresponding dormant polymeric alkyl halide or pseudohalide.

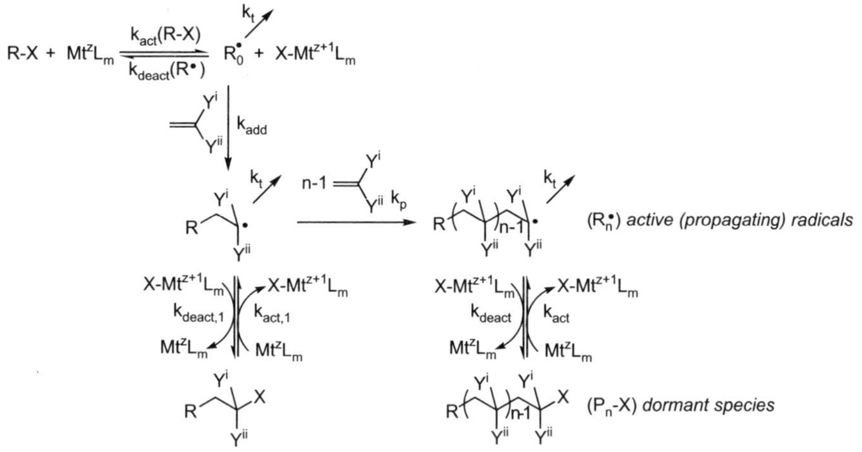

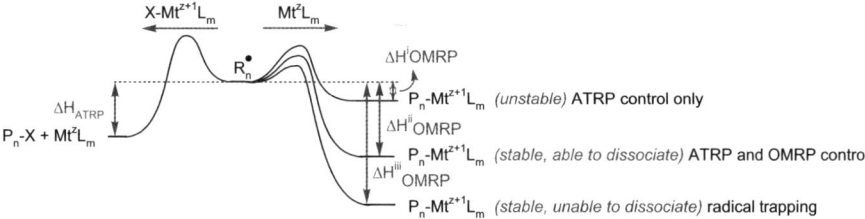

Figure 8.1 Mechanism of metal complex-mediated ATRA and ATRP.

Figure 8.2 Polymerization control in systems where ATRP mechanism is operational but also an organometallic compound can be formed.
Adapted from ref. 39.

Metal complexes can affect all the elementary steps of radical polymerizations *via* interaction with the initiator, monomer, or polymer,[38] and can therefore affect the reaction kinetics or the structure of the final product. This chapter only deals with redox-active metal complexes that can reversibly activate organic (pseudo)halides to form radicals able to initiate polymerization.

It should be mentioned that metal complexes can mediate controlled polymerization *via* two mechanisms (Figure 8.2):[39] (i) reversible transfer of an atom, typically halogen, or a group from the polymer chain end to the metal center (ATRP), which is the subject of this chapter, or (ii) reversible formation of a metal–carbon bond upon reaction of the complex Mt^zL_m with the propagating radical (a process termed organometallic radical polymerization (OMRP)). In both cases, the oxidation state of the central metal atom or ion changes. In principle, depending on the stability of the formed dormant species, both mechanisms can contribute to the polymerization control simultaneously. For example, in a situation where a redox active metal complex Mt^zL_m is able to reversibly abstract the atom (group) X from an initiating molecule RX or an X-terminated polymer chain P_n-X with the formation of the higher oxidation state complex $XMt^{z+1}L_m$ and the corresponding radical ($R_0^\bullet$ or $R_n^\bullet$), ATRP will operate. However, if the lower oxidation state metal complex can react with radicals with the formation of the corresponding organometallic species $R\text{-}Mt^{z+1}L_m$ or $P_n\text{-}Mt^{z+1}L_m$, OMRP equilibrium can also be established. If the organometallic species is unstable and readily dissociates, the OMRP mechanism will not have a significant contribution to the overall polymerization control. In the other extreme case, where the organometallic compound is very stable (and is formed rapidly), the activation of RX or $P_n\text{-}X$ will be followed by formation of the organometallic species and, due to the inability of the latter to dissociate, the reaction may stop before the formation of polymer; in this case, radical trapping occurs. However, in the intermediate situation, where the formation of a relatively stable organometallic species is reversible, OMRP mechanism will operate along with ATRP.[39] For example, in polymerizations mediated by Mo^{40} and $Os^{41,42}$ compounds, both ATRP and OMRP mechanisms contribute to the overall polymerization control.

The ATRP process is characterized by the ATRP equilibrium constant, which can be defined as either the ratio of the rate constants of activation and deactivation or as the ratio of concentrations of all involved species.

$$K_{ATRP} = \frac{k_{act}}{k_{deact}} = \frac{[R^{\bullet}][X - Mt^{z+1}L_m]}{[PX][Mt^z L_m]}$$

$$(R^{\bullet} = R_0^{\bullet} + R_1^{\bullet} + R_2^{\bullet} + \ldots + R_n^{\bullet}; \ PX = RX + P_1X + P_2X + \ldots + P_nX)$$

$$(8.1)$$

The rate of the process R_p, like that of any other radical polymerization, is determined by the product of the (total) radical and monomer concentrations, and the former can be expressed using eqn (8.1).

$$R_p = k_p[M][R^{\bullet}] = k_p K_{ATRP} \frac{[M][PX][Mt^z L_m]}{[X - Mt^{z+1}L_m]} \quad (8.2)$$

The "livingness" of ATRP can be ascertained from a linear first-order kinetic plot of monomer consumption (which by itself only indicates a constant number of propagating chains and is not a sufficient criterion), accompanied by a linear increase of polymer molecular weights with conversion, with a value of the number-average degree of polymerization (DP_n) determined by the ratio of reacted monomer to initially introduced initiator (eqn (8.3)).

$$DP_n = \text{conv.} \times \frac{[M]_0}{[RX]_0} \quad (8.3)$$

$$M_n = MW(RX) + MW(M) \times DP_n$$
$$= MW(RX) + MW(M) \times \text{conv.} \times \frac{[M]_0}{[RX]_0} \quad (8.4)$$

In equations (8.3) and (8.4), conv. is the monomer conversion, and MW(RX) and MW(M) are the molecular weights of the initiator (*i.e.*, the total molecular weight of the two end groups in the polymer) and the monomer, respectively. The ratio $[M]_0/[RX]_0$ is often referred to as targeted degree of polymerization at complete conversion, $DP_{n,targ}$.

Provided that all polymer chains start growing within a narrow time period (fast initiation) and that the growth is relatively slow prior to each deactivation, the formed polymers are characterized by a narrow molecular weight distribution (MWD) or low, close to unity, MWD dispersity ($Đ = M_w/M_n$, where M_w and M_n are the polymer weight- and number-average molecular weights, respectively).

It has been shown for classical "living" (ionic) polymerizations that in order to obtain polymers with narrow MWDs, the initiation (k_i) should be fast with respect to polymerization (k_p), particularly when low molecular weight polymers are targeted. The important parameter that determines the width of the MWD for living ionic polymerizations is the initiation efficiency, f, which is defined as the fraction of reacted initiator (designated as in) at a very long reaction time, *i.e.*, at the end of the polymerization.[43]

$$f \equiv \frac{[in]_0 - [in]_\infty}{[in]_0} \tag{8.5}$$

$$Đ = \frac{M_w}{M_n} = \frac{\frac{(\ln(1-f))^2}{f + \ln(1-f)} + 2}{\frac{\ln(1-f)}{f} + 1} \tag{8.6}$$

The value of $Đ = M_w/M_n$ reaches unity as the initiation efficiency approaches unity. When the polymerization is slow with respect to the initiation, high initiation efficiencies (and therefore narrow MWDs) can be achieved even when relatively low molecular weight polymers are targeted, *i.e.*, when the values of $DP_{n,targ} = [M]_0/[in]_0$ are low. As the propagation rate constant increases with respect to the initiation rate constant, high initiation efficiencies can be reached only when higher molecular weight polymers are being synthesized. In other words, the ratio k_p/k_i determines the size of the macromolecules that can be targeted to achieve values of the initiation efficiency as close as possible to unity (which in turn determines values of $Đ$ close to unity), as shown by eqn (8.7) and Figure 8.3.[43,44]

$$DP_{n,targ} = \frac{[M]_0}{[in]_0} = f - \frac{k_p}{k_i}(f + \ln(1-f)) \tag{8.7}$$

Just as in anionic polymerization, efficient initiation is of an utmost importance in ATRP for the formation of polymers with narrow MWDs. According to a relation, firstly derived for the case of living ionic polymerization,[45,46] later[47] modified to describe the somewhat simpler case of living radical polymerization, and finally generalized for all polymerizations involving exchange reactions,[48] the MWD dispersity of polymers made by ATRP is directly related to the rate of deactivation, the targeted degree of polymerization (*i.e.*, $[RX]_0$, which is large for low values of $DP_{n,targ}$), and the monomer conversion.

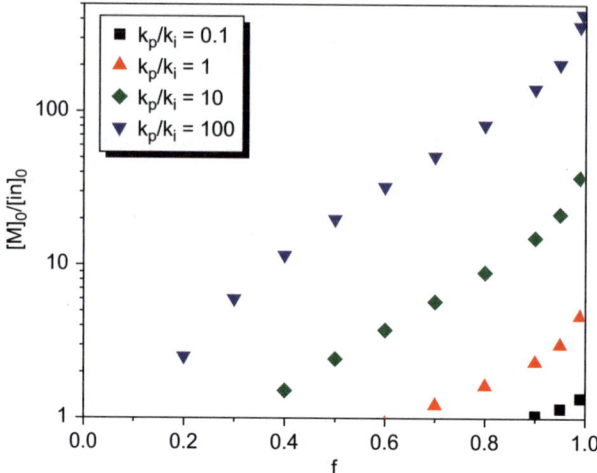

Figure 8.3   Relation between targeted degrees of polymerization ($DP_{n,targ} = [M]_0/[in]_0$) and efficiency of initiation at various ratios of the propagation and initiation rate constants (k_p/k_i).[43]

$$\DJ = \frac{M_w}{M_n} \approx 1 + \left(\frac{k_p[RX]_0}{k_{deact}[XMt^{z+1}L_m]} \right) \left(\frac{2}{conv.} - 1 \right) \qquad (8.8)$$

In ATRP and other CRP reactions, as in any other radical process, continuous termination takes place. If, for simplicity, chain length independent termination and propagation coefficients are considered, and if a CRP reaction is characterized with the same rate as a conventional radical polymerization, the concentration of propagating radicals is identical, as are the rate of termination and the concentration of terminated chains. However, in conventional radical polymerization, virtually all chains in the reaction mixture are "dead", with the exception of very small amounts of growing chains that are continuously generated and that terminate after a very fast growth (usually in less than 1 s). In contrast, in CRP, the large majority (typically, > 90–95%) of chains in the mixture are "living", mostly in the dormant state. However, in CRP the ever-present termination yields "dead" chains that cannot be reactivated and grow any further, due to a loss of their ω-chain end functionality, and it is important to quantify the amount or fraction of dead chains. The fractions of ω-chain end functionalized (P_n-X) and terminated (T) chains depend upon the reaction conditions, *e.g.*, the temperature, the nature of the monomer (through the reaction coefficients k_p and k_t), the targeted degree of polymerization ($DP_{n,targ}$, *i.e.*, the ratio $[M]_0/[RX]_0$), and the monomer conversion. In ATRP, the fraction of T (or the dead chain fraction, DCF, defined as the ratio $[T]/[RX]_0$) is given by eqn (8.9), in which it is assumed that the termination *via* disproportionation occurs exclusively (in order to keep the number of chains constant).

Table 8.1 Minimal time required for polymerization of MA, MMA and St with preserved 90% functionality (DCF = 0.1).[a]

Monomer	$k_p/M^{-1}s^{-1}$ [b]	$k_t/M^{-1}s^{-1}$ [b]	$DP_{n,targ} = 500$ conv. = 60%	$DP_{n,targ} = 500$ conv. = 90%	$DP_{n,targ} = 100$ conv. = 60%	$DP_{n,targ} = 100$ conv. = 90%
Methyl acrylate	47 400	1.10×10^8	37 s	234 s	7 s	47 s
Methyl methacrylate	1300	9.00×10^7	13.3 h	83.9 h	2.7 h	16.8 h
Styrene	665	1.10×10^8	2.8 d	17.5 d	0.6 d	3.5 d

[a]Conditions: 80 °C, bulk polymerization.
[b]k_p and k_t values from literature data.[49,50] The value for k_t is the sum of $k_{t,c}$ (rate constant of radical termination by combination) and $k_{t,d}$ (rate constant of termination by disproportionation).

$$DCF \equiv \frac{[T]}{[RX]_0} = \frac{2DP_{n,targ}k_t[\ln(1 - \text{conv.})]^2}{[M]_0 k_p^2 t} = \frac{2k_t[\ln(1 - \text{conv.})]^2}{[RX]_0 k_p^2 t} \quad (8.9)$$

According to eqn (8.9), slower polymerizations (longer t), particularly at comparatively low monomer conversion, afford a higher fraction of ω-end functionalized chains, *i.e.*, lower fraction of dead chains. Table 8.1 shows the minimal reaction times required to polymerize methyl acrylate, methyl methacrylate, and styrene in bulk, for two different targeted degrees of polymerization, in order to obtain polymers with 90% of preserved ω-chain end functionality (*i.e.*, DCF = 0.1). It is possible to prepare poly(methyl acrylate) with only 10% DCF for $DP_{n,targ} = 500$ at 60% monomer conversion in only 37 s. The same level of control requires 13 h for poly(methyl methacrylate) and 2.8 days for polystyrene. The effect of the targeted chain length is also important. For instance, although 13.3 h are needed to obtain poly(methyl methacrylate) with only 10% of dead chains at 60% conversion for $DP_{n,targ} = 500$, only 2.7 h will suffice if the targeted degree of polymerization is lower, $DP_{n,targ} = 100$. Finally, the monomer conversion also affects the degree of chain end functionality. The same DCF of 0.1 (90% end-functionalized chains) for polystyrene at $DP_{n,targ} = 100$ requires 3.5 days at 90% monomer conversion but only 0.6 days at 60% monomer conversion. This indicates that to preserve the same functionality in synthesis of polymers with specific final DP, it is best to target a higher $DP_{n,targ}$ and stop the reaction at lower conversion. For example, to prepare poly(methyl methacrylate) with $DP_n = 60$ within 5 h, it is better to target a $DP_{n,targ} = 150$ and stop the reaction at 40% monomer conversion (the DCF in this case will be as low as 0.025) than to target a $DP_{n,targ} = 100$ and stop the reaction at 60% monomer conversion (DCF = 0.053), which is in turn better than targeting a $DP_{n,targ}$ of 75 and stopping the reaction at 80% conversion (DCF = 0.123). Reducing the ratio $k_t/(k_p)^2$, which can be accomplished by increasing temperature and pressure, should lead to better ω-chain end functionality, and therefore to higher molecular weights that can be reached. However, it should be remembered that increasing the temperature also causes increased chain transfer (particularly important in the polymerization of acrylates), self-initiation (styrene), or depropagation (methacrylates) and other side reactions involving mediating agents in CRP, such as catalyst dissociation.

8.3 Kinetics and Thermodynamics of ATRP Structure–Reactivity Correlations

8.3.1 The Activation Process

Reactions of organic halogenated compounds RX with redox-active transition metals, metal salts, or complexes that generate radicals *via* the homolytic cleavage of a C–X bond (X = Cl, Br, I) have been known for a long time. The formed radicals have often been used to add across the C=C double bond of an unsaturated compound, *i.e.*, in ATRA reactions. For instance, in 1962, Asscher and Vofsi[51] reported that both chloroform and carbon tetrachloride can transfer a chlorine atom to either Cu^I or Fe^{II} compounds with the generation of the respective dichloromethyl or trichloromethyl radicals and the higher oxidation state metal compound with an associated chloride ligand. Radical addition reactions were reported with various olefins. It was also shown[52] that carbon tetrachloride could be activated by Cu^{II} in the presence of amines (which reduced Cu^{II} to Cu^I), and that, in the presence of ethylene, oligomers were formed. A number of polymerization reactions were reported in the 1960s by Bamford and coworkers, in which the initiating radicals were generated *via* the activation of halogenated compounds (CCl_4,[53–60] CCl_3COONH_4,[61] or trichloroacetate esters, *e.g.*, poly(vinyl trichloroacetate)[54]) by various redox active complexes of Cu,[61] Cr,[54] Mo,[54,56,57,59] W,[54] Mn,[53,58] Re,[55] Ni,[60] and Pt.[60] Free metals such as Cu, Fe, Co, and Ni were employed by Otsu and Yamaguchi in 1968[62] to activate a variety of alkyl halides in the polymerization of styrene. The yield of polymer reached after a fixed time was used to estimate the activity of the alkyl halide initiator and it was shown that the activity increased in the order $CH_2Cl_2 < CHCl_3 < CCl_4 < CHBr_3$. Further, it was shown that organic halides with one halogen atom that would not yield stabilized radicals upon cleavage of the C–X bond, for instance, n-butyl chloride and chlorobenzene, showed very low activity. In contrast, benzyl chloride, allyl chloride, and even *t*-butyl chloride were active.[62] Some of the early work on the activation of organic halides by transition metal compounds has been summarized in a review by Bamford.[63] The ability of the formed higher oxidation state metal compounds with a coordinated halide ligand to deactivate radicals, or the evolution of the polymer molecular weights with monomer conversion were not studied. Therefore, it is not clear if some of the early processes had some of the characteristics of CRP.

With the development of ATRP in the mid 1990s,[20,21,64] many metal complexes, including those of Ti,[65] Mo,[40,66–68] Re,[69–71] Ru,[21,72–81] Fe,[82–90] Rh[72,91–93] (for a review on the use of Rh-containing catalysts see ref. 94), Ni,[92,95–102] Pd,[103] Co,[104] Os,[41] and Cu,[20,64,105–113] have been employed as polymerization mediators leading to excellent control with a range of monomers. Some other examples of useful metal catalysts are summarized in recent reviews.[114–116] Additionally, dual metallic ATRP catalysts, in which, the complexes of two metals are simultaneously employed have also been reported, including the combination of two Cu complexes with different ligands,[117] or $Sn^{II}Cl_2$ with the complex of $Fe^{III}Cl_3$ and N-substituted diethylenetriamine (DETA),[118] or $Sn^{II}Cl_2$, $Mn^{II}Cl_2$, $Ni^{II}Cl_2$, and $Co^{II}Cl_2$ with the complex of $Fe^{III}Cl_3$ and Ph_3P.[119]

Thorough understanding was needed of the structural characteristics of both the alkyl halide initiators and metal complexes that determine both the activity and the degree of polymerization control. The effects of the alkyl group structure and the nature of the halogen atom on the activity of alkyl halides in activation by redox-active metal complexes began to be studied in a systematic fashion. To quantify the activity of alkyl halide initiators, the development of reliable methods for the measurement of the rate constant of activation k_{act} was crucial. One approach is to react an alkyl halide RX with an excess of the lower oxidation state metal complex, *i.e.*, the activator Mt^zL_m, followed by fast and irreversible trapping of the formed radicals (Figure 8.4). The consumption of alkyl halide can be monitored by spectroscopic or chromatographic techniques, and, under the mentioned conditions, at which the concentration of the activating complex can be considered constant, is given by eqn (8.10).[120–128] Stable radicals, *e.g.*, TEMPO (shown in Figure 8.4) are used to trap the radicals. Obviously, it is important to select a radical trap that does not participate in redox reactions with the metal complexes present in the system. Eqn (8.10) can be directly used to determine k_{act} as the slope of the first order kinetic plot

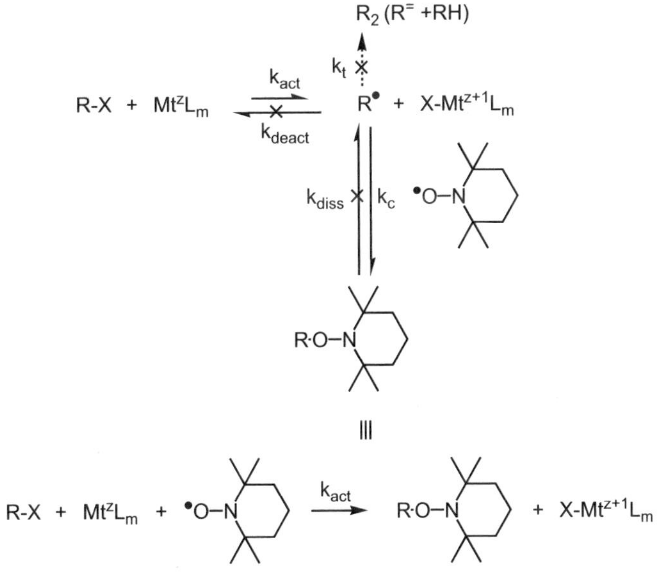

Figure 8.4 Model reactions for determination of k_{act} values.

describing the disappearance of the initiator, divided by the initial concentration of activator.

$$\ln\frac{[RX]_0}{[RX]} = -\ln(1 - \text{conv.}(RX)) = k_{act}[Mt^zL_m]_0 t \qquad (8.10)$$

Of particular relevance to ATRP are alkyl halides that structurally resemble the halogen-terminated polymer chains, *i.e.*, the dormant species. Some of these alkyl halides (see Figure 8.5) include 1-phenylethyl halides (1-PhEtX), 2-halopropionate esters (R-XP, where R = alkyl or aryl), 2-halopropionitriles (XPN), and 2-haloisobutyrate esters (R-XiB, R = alkyl or aryl) that mimic the chain ends of halogen-capped polystyrenes, polyacrylates, polyacrylonitrile, and polymethacrylates, respectively (in all cases, X = Cl, Br). Some other important initiators include methacrylate-type dimers ((R-MA)$_2$X, which are better models of X-capped polymethacrylates than R-XiB), arenesulfonyl halides (ArSO$_2$X, *e.g.*, 4-methylbenzenesulfonyl (tosyl) chloride, which is often used in the polymerization of methacrylates[129] and acrylonitrile[130]), 2-halophenylacetate esters (R-XPA), and 2-halo-2-methylmalonate esters (RR′XMM, suitable in the polymerization of methacrylates under low-catalyst-concentration conditions[131,132]).

The reactivities of alkyl halides and pseudohalides in the activation process depend upon the structure of both the alkyl group and the nature of the transferable (pseudo)halogen. An early study[123] demonstrated that the activity of alkyl halides that mimic halogen-capped dormant polymer chains in reactions of activation mediated by the Cu^I complex of di(5-nonyl)-2,2′-bipyridne (dNbpy) in acetonitrile decreased in the order Et-BriB ($k_{act} = 0.60 \, M^{-1}s^{-1}$) > 1-PhEtBr ($k_{act} = 8.5 \times 10^{-2} \, M^{-1}s^{-1}$) > Me-BrP ($k_{act} = 5.2 \times 10^{-2} \, M^{-1}s^{-1}$). The methacrylate dimer-type alkyl bromides, (RMA)$_2$Br, are characterized by values of k_{act} that are 7–8 times higher than simple 2-bromoisobutyrates R-BriB.[133,134] This is the reason why the latter are not the best, *i.e.*, the most efficient, initiators for the ATRP of methacrylates (after the first addition of the isobutyryl radical followed by deactivation, the formed alkyl bromides are more likely to be reactivated than the

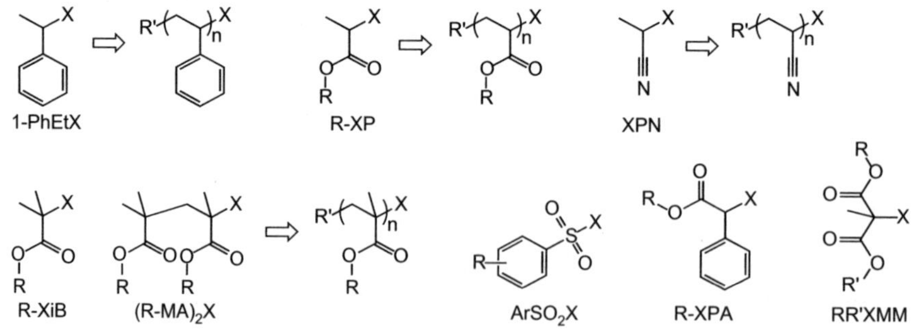

Figure 8.5 Structures of some alkyl halides that structurally resemble halogen-terminated polymers and some other alkyl halides used as ATRP initiators.

2-bromoisobutyrate initiator, which, as described above (Figure 8.3), leads to inefficient initiation and formation of polymers of broad molecular weight distributions, particularly at relatively low $DP_{n,targ}$). The activation rate constant of the alkyl halide resembling the chlorine-terminated polystyrene chain, 1-PhEtCl, was markedly lower than for the corresponding bromo-derivative, 1-PhEtBr. Alkyl (pseudo)halides, in which the (pseudo)halogen is attached to a tertiary carbon atom, are more reactive than secondary, which are in turn more reactive than primary alkyl (pseudo)halides, in agreement with the bond dissociation energies of the respective C–X bonds. If the cleavage of the C–X bond yields a stabilized radical, the value of k_{act} increases. For instance, alkyl (pseudohalides) bearing an α-cyano group are more easily activated than those with α-phenyl or α-carbonyl (ester) groups, which is expected due to the highest radical stabilization ability of the α-cyano moiety. When more than one stabilizing group is present at the α-carbon, very high activity is observed. For instance, in Et-BrPA, the radical stabilization effects of an α-phenyl and an α-carbonyl (ester) groups are combined, which results in activity that is >10 000-fold higher than that of 1-PhEtBr (with only α-phenyl group), and >100 000-fold higher than Me-BrP (containing only an α-carbonyl (ester) group). The activities of 2-bromo-2-methylmalonate initiators (RR'XMM) have not been determined but are also expected to be high due to the presence of two radical-stabilizing α-carbonyl (ester) substituents in addition to the fact that these halides are tertiary. As far as the nature of the X group is concerned, the reactivity decreases in the order I > Br > Cl > pseudohalogen (with the same alkyl group), which, once again, is related to the bond dissociation energies. Figure 8.6 summarizes the studies on the influence of both the R and the X groups on the activity of RX initiators in the activation reaction mediated by the N,N,N',N''N''-pentamethyldiethylenetriamine (PMDETA) complex of Cu^I at 35 °C in acetonitrile.[135]

Most systematic studies on the influence of the catalyst on the value of k_{act} have been determined in reactions mediated by Cu complexes. Thus, the role of the ligand in alkyl (pseudo)halide activation reactions will be discussed here, and not the role of the central atom. The nature of the ligand with N-donor atoms in Cu^I complex-based activators affects markedly the value of k_{act}. As shown in Figure 8.7, the activation rate constant of Et-BriB varies by more than 6 orders of magnitude as the ligand of the activating complex is altered.[136] Currently, it is difficult to predict the activity of a complex in the homolytic cleavage of a C–X bond or to correlate this activity to specific characteristics (*e.g.*, reducing power, degree of "unsaturation" of the coordination sphere). Such predictions and correlations are only possible when the values of the ATRP equilibrium constant are concerned, not the values of k_{act}. However, some general remarks can be made. Typically, bidentate ligands form Cu complexes of relatively low activity, although with many of them (derivatives of 2,2'-bipyridyl (bpy), for instance) the polymerization control is excellent, indicating relatively high values of k_{deact} (*vide infra*). Tridentate ligands with both aliphatic and pyridine-type N donor atoms have been studied as well. In aliphatic amine-type ligands, the distance between the donor atoms influences

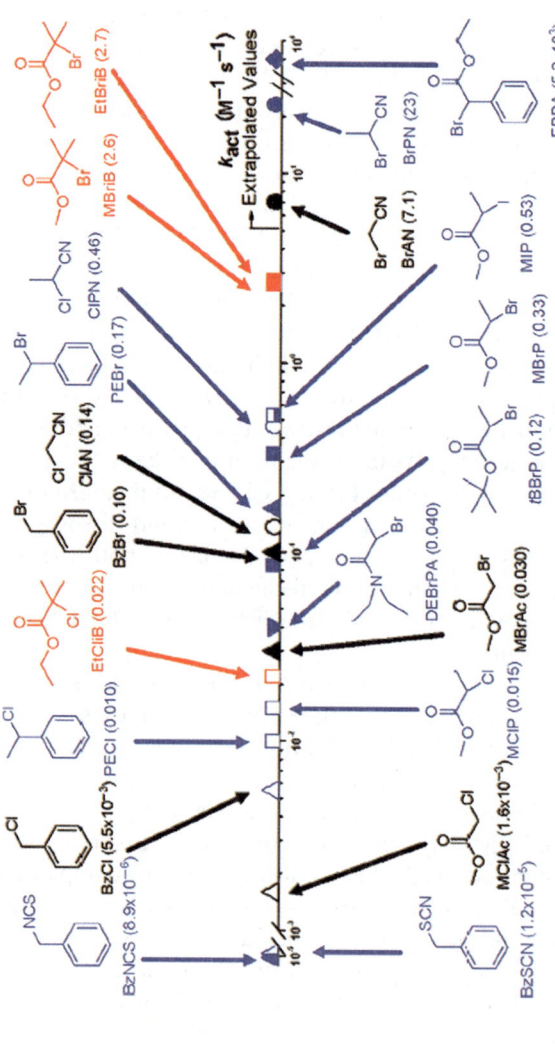

Figure 8.6 Activation rate constants for various alkyl (pseudo)halide initiators with $Cu^IX/PMDETA$ in MeCN at 35 °C. Reprinted with permission from ref. 135.

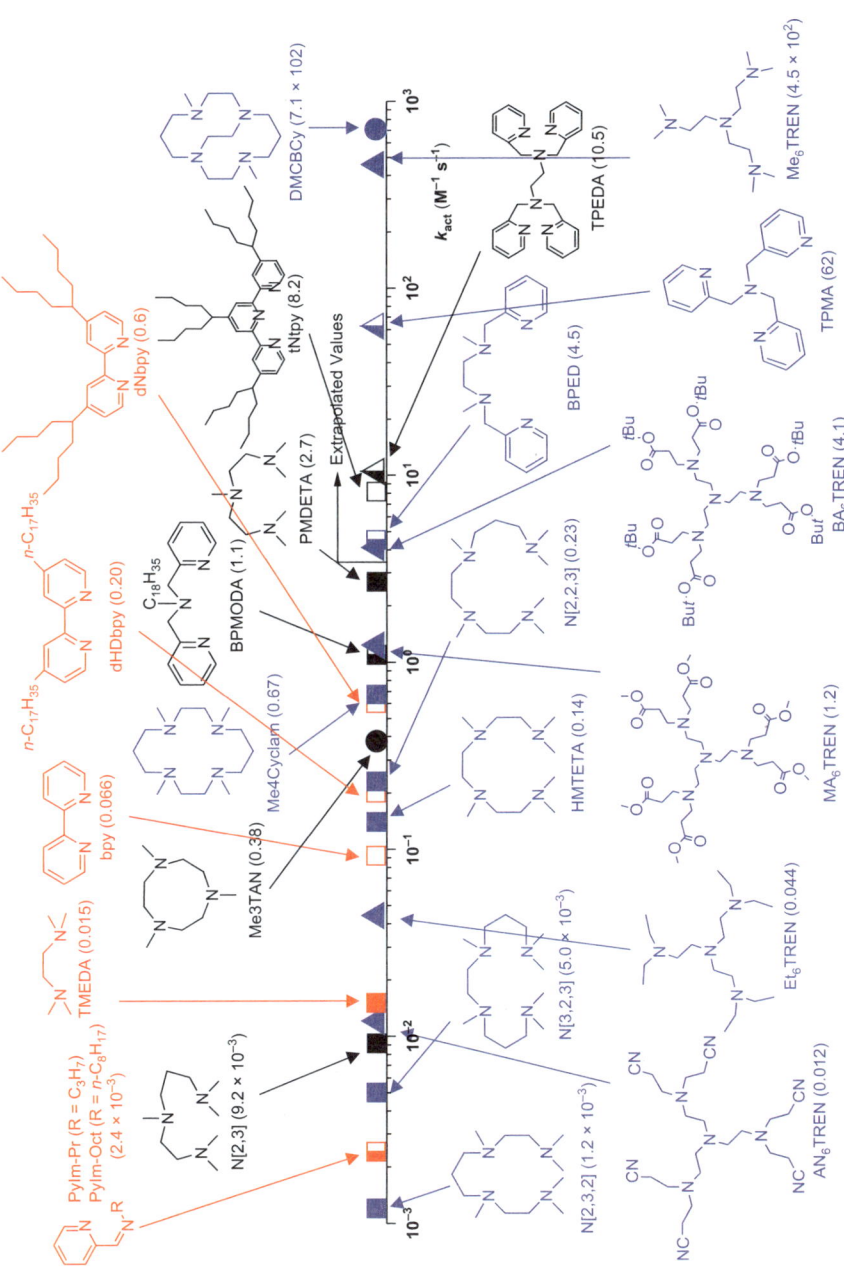

Figure 8.7 Activation rate constants of Cu^{I} complexes derived from various N-donor ligands in the reaction with Et-2BriB in MeCN at 35 °C. Reprinted from ref. 136 with permission.

the activity of the corresponding Cu^I complex. Aliphatic amines with a C_3 bridge between the N atoms form less active Cu^I-based catalysts compared to ligands with only C_2 bridges. A similar trend is observed with tetradentate aliphatic amine ligands. The number of C atoms between any two adjacent N donor atoms determines the coordination angle and strain in each chelate ring, and also the mutual arrangement of contiguous chelate rings; it has been known that these parameters are related to the stability of the complexes.[137] Branched ligands form very active catalysts. In some instances, seemingly minor alterations of substituents at the N atoms in certain amine-type ligands (for instance, replacement of a methyl with an ethyl group) can dramatically influence the catalytic activity, as shown in Figure 8.7 for the derivatives of tris(2-aminoethyl)amine (TREN). As seen, the hexaethyl derivative of TREN, Et_6TREN, forms a Cu^I complex, which activates Et-BriB about 10 000 times slower than the hexamethyl derivative, Me_6TREN, plausibly partially due to steric effects. This observation may also be related to the reported effect of hydrophobic groups attached to N atoms in aliphatic amine ligands, which lead to increased redox potentials of the Cu^{II}/Cu^I couple compared to complexes of ligands with a similar "frame", but with more hydrophilic groups attached to the N donor atoms.[138] (The reducing power of a complex is related to its activity, or more precisely – to the value of K_{ATRP} as will be shown below.) The most active Cu-based ATRP activators are derived from the tetradentate cyclic ligand cyclam and particularly from the dimethyl-cross-bridged cyclam DMCBCy (Figure 8.7).[136,139]

Linear free energy relationships have been established between the values of k_{act} and parameters characterizing the activity of the catalyst (C) and of the alkyl halide initiator (I) (eqn (8.11), where s_C is the slope, which is specific for each catalyst).[140]

$$\log k_{act} = C + s_C \times I \tag{8.11}$$

8.3.2 The ATRP Equilibrium

Logically, this section should discuss the other rate constant characterizing the ATRP process, namely the deactivation rate constant, k_{deact}. However, the values of k_{deact} are typically rather large and difficult to determine experimentally. They are often calculated as the ratio $k_{deact} = k_{act}/K_{ATRP}$ of the much easier to determine rate constant of activation and equilibrium constant of ATRP. This is why this section is dedicated to the experimental determination of K_{ATRP} as well as to the factors (initiator and catalyst structure, solvent, *etc.*) that influence its values. As seen from eqn (2), the rate of polymerization under "classical" ATRP conditions depends on the value of the equilibrium constant.

8.3.2.1 Experimental Determination of K_{ATRP}

K_{ATRP} can be determined experimentally using several different approaches. Polymerization kinetics yields an apparent value, $K_{ATRP}/[XMt^{z+1}L_m]$ as the slope of the time dependence of $\ln([M]_0/[M])/(k_p[Mt^zL_m]_0[RX]_0.^{141}$ In ATRP, the persistent radical effect (PRE) is operational, *i.e.*, the deactivator ($XMt^{z+1}L_m$, which is the persistent radical) accumulates with time as a result of radical termination.[142–144] Assuming that the concentrations of both the activator and alkyl halide initiator do not change significantly (in other words, that $k_{act}[Mt^zL_m]_0[RX]_0 = k_{deact}[R^\bullet][XMt^{z+1}L_m]$, *i.e.*, the product $[R^\bullet][XMt^{z+1}L_m]$ does not change), the accumulation of deactivator due to the PRE is a function of the cube root of time as shown by eqn (8.12).[142–144]

$$[XMt^{z+1}L_m] = (3K_{ATRP}^2 k_t[RX]_0^2[Mt^zL_m]_0^2)^{1/3}t^{1/3} \qquad (8.12)$$

To determine K_{ATRP}, the activator, Mt^zL_m, is reacted with an alkyl halide, and the deactivator concentration (experimentally accessible through ESR or electronic spectroscopy) is monitored as a function of time. Then, a plot of $[XMt^{z+1}L_m]$ *vs.* $t^{1/3}$ is constructed, and K_{ATRP} is determined from the slope provided that the termination rate constant k_t is known.[145] The equilibrium must be established rapidly and, due to the assumptions mentioned above, this strategy is only applicable for comparatively low conversions of activator and initiator. If these conditions are not met, the dependence (8.12) is not observed. This is a severe limitation when the values of K_{ATRP} for active catalysts should be determined, where both the concentrations of activator and initiator decrease rapidly.

More accurate equations describing the PRE were derived that take into account that the concentrations of both the activator and initiator change during the experiment.[146] If the activator and initiator are mixed in a 1:1 molar ratio, the reaction stoichiometry requires that $[RX]_{reacted} = [RX]_0 - [RX] = [Mt^zL_m]_{reacted} = [Mt^zL_m]_0 - [Mt^zL_m] = [XMt^{z+1}L_m]$. Using the assumption (justified by simulations) that the rate of generation of deactivator significantly exceeds the rate of consumption of radicals, the time-dependence of accumulation of $[XMt^{z+1}L_m]$ can be calculated. For the simple 1:1 stoichiometry ($[Mt^zL_m]_0 = [RX]_0$), a function $F([XMt^{z+1}L_m])$ is defined (eqn (8.13)), whose values can be experimentally calculated if the concentration of deactivator can be monitored. K_{ATRP} is obtained as the square root of the slope of the linear dependence of the function $F([XMt^{z+1}L_m])$ on time, divided by $2k_t$ (the value of the termination rate constant should be known). The accuracy of the analysis can be checked from the intercept, which should have a value equal to $1/3[Mt^zL_m]_0$, which is known for each experiment.

$$F([XMt^{z+1}L_m]) \equiv \frac{[Mt^zL_m]_0^2}{3([Mt^zL_m]_0 - [XMt^{z+1}L_m])^3} - \frac{[Mt^zL_m]_0}{([Mt^zL_m]_0 - [XMt^{z+1}L_m])^2}$$

$$+ \frac{1}{[Mt^zL_m]_0 - [XMt^{z+1}L_m]}$$

$$= 2k_t K_{ATRP}^2 t + \frac{1}{3[Mt^zL_m]_0}$$

$$(8.13)$$

In the case when $[Mt^zL_m]_0 \neq [RX]_0$, the time dependence of deactivator accumulation is more complex (eqn (8.14)).[146]

$$
\begin{aligned}
F([XMt^{z+1}L_m]) &\equiv \left(\frac{[RX]_0[Mt^zL_m]_0}{[Mt^zL_m]_0-[RX]_0}\right)^2 \left(\frac{1}{[Mt^zL_m]_0^2([RX]_0-[XMt^{z+1}L_m])}\right. \\
&\quad + \frac{2}{[RX]_0[Mt^zL_m]_0([Mt^zL_m]_0-[RX]_0)} \ln\left(\frac{[RX]_0-[XMt^{z+1}L_m]}{[Mt^zL_m]_0-[XMt^{z+1}L_m]}\right) \\
&\quad \left.+ \frac{1}{[RX]_0^2([Mt^zL_m]_0-[XMt^{z+1}L_m])}\right) \\
&= 2k_t K_{ATRP}^2 t + \left(\frac{[RX]_0[Mt^zL_m]_0}{[Mt^zL_m]_0-[RX]_0}\right)^2 \left(\frac{1}{[Mt^zL_m]_0^2[RX]_0}\right. \\
&\quad \left.+ \frac{2}{[RX]_0[Mt^zL_m]_0([Mt^zL_m]_0-[RX]_0)} \ln\frac{[RX]_0}{[Mt^zL_m]_0} + \frac{1}{[RX]_0^2[Mt^zL_m]_0}\right)
\end{aligned}
$$

(8.14)

The concentration of deactivator accumulated in the system is often easy to monitor using spectroscopic techniques but in some cases it may be more convenient to determine the concentration of the alkyl halide initiator at any given time (for example, by GC or NMR). An equation describing the time dependence of the initiator consumption (for the case when $[Mt^zL_m]_0 \neq [RX]_0$) has also been derived (eqn (8.15))[146] and has been used to experimentally access K_{ATRP}.[146,147] In these cases, the values of a function F(RX) are calculated from known concentrations of reagents and are plotted against time.

$$
\begin{aligned}
F(RX) &\equiv \left(\frac{[RX]_0[Mt^zL_m]_0}{[Mt^zL_m]_0 - [RX]_0}\right)^2 \times \left(\frac{1}{[Mt^zL_m]_0^2[RX]_0}\right. \\
&\quad + \frac{2}{[Mt^zL_m]_0[RX]_0([Mt^zL_m]_0 - [RX]_0)} \ln\left(\frac{[RX]}{[Mt^zL_m]_0 - [RX]_0 + [RX]}\right) \\
&\quad \left.+ \frac{1}{[RX]_0^2([Mt^zL_m]_0 - [RX]_0 + [RX])}\right) \\
&= 2k_t K_{ATRP}^2 t + \left(\frac{[Mt^zL_m]_0[RX]_0}{[Mt^zL_m]_0 - [RX]_0}\right)^2 \left(\frac{1}{[Mt^zL_m]_0^2[RX]_0}\right. \\
&\quad \left.+ \frac{2}{[Mt^zL_m]_0[RX]_0([Mt^zL_m]_0 - [RX]_0)} \ln\frac{[RX]_0}{[Mt^zL_m]_0} + \frac{1}{[RX]_0^2[Mt^zL_m]_0}\right)
\end{aligned}
$$

(8.15)

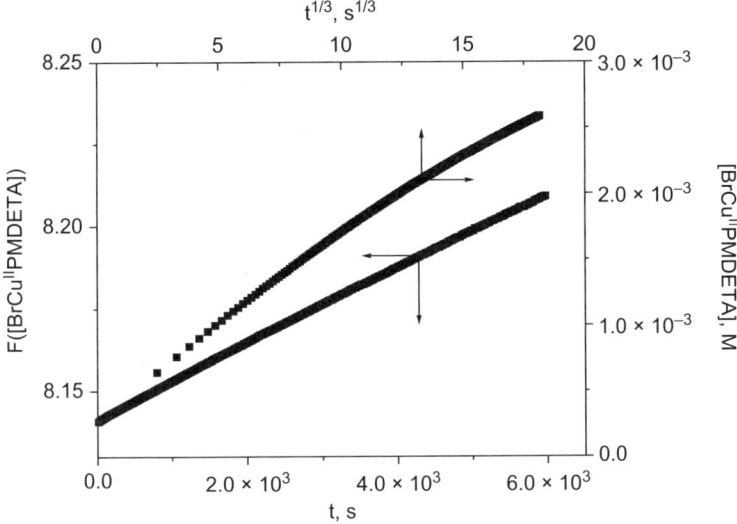

Figure 8.8 Experiment for the determination of K_{ATRP} *via* plotting F([XCu^{II}L$_m$]) against time. Reaction conditions: [Cu^{I}Br/PMDETA]$_0$ = 5 mM, [1-PhEtBr]$_0$ = 100 mM, in MeCN at 22 ± 2 °C.
Reprinted from ref. 146 with permission.

Figure 8.8 shows both the dependence of deactivator concentration on $t^{1/3}$ (approximated equation describing the PRE) and also the time dependence of the function F([XCu^{II}L$_m$]) from eqn (8.14) for the reaction of Cu^{I}Br/PMDETA complex with 1-PhEtBr (non-stoichiometric conditions) in MeCN.[146] As seen, the plot of deactivator concentration *vs.* $t^{1/3}$ is curved due to the fast consumption of activator and alkyl halide, but the plot of F([XCu^{II}L$_m$]) *vs.* t is linear and can be used to determine the value of K_{ATRP} for this system.

Table 8.2 lists selected experimentally determined values of K_{ATRP} for various Cu complexes with various alkyl halide initiators in acetonitrile using the approaches outlined above (eqns (8.13)–(8.15)).

As seen, as the catalyst and/or initiator are changed, the values of K_{ATRP} change by many orders of magnitude. To compare the catalytic activity of two complexes, the values of K_{ATRP} in a reaction with the same alkyl halide and in the same solvent should be compared. The ATRP catalytic activity of Cu^I complexes increases in the order bpy < HMTETA < PMDETA < TPMA < Me$_6$TREN < DMCBCy. The most active complex known to date is derived from the cross-bridged cyclam ligand DMCBCy.[139] The catalytic activity (K_{ATRP}) of a large number of Cu complexes in a reaction with the same alkyl halide, Et-BriB is compared in Figure 8.9.[151] Even for ligands that are not extremely diverse structurally (all shown ligands have N donor atoms and are aliphatic amines, pyridine derivatives, or imines), the values of K_{ATRP} vary substantially (by over 8 orders of magnitude).

When the same catalyst is used (see the entries in Table 8.1), the activity of various alkyl halide initiators can be compared. The values of K_{ATRP} for alkyl

Table 8.2 Experimental values of Cu-mediated K_{ATRP}.[a]

No.	Ligand	Initiator	K_{ATRP}	Ref.
1	**bpy**	Et-BriB	3.93×10^{-9}	146
2	,,	1-PhEtBr	8.5×10^{-10}	148
3	,,	Allyl-Br	3.0×10^{-9}	149
4	**HMTETA**	Et-BriB	8.38×10^{-9}	150
5	,,	1-PhEtBr	2.9×10^{-9}	148
6	,,	1-PhEtCl	7.9×10^{-10}	148
7	**PMDETA**	BrPN	5.89×10^{-7}	146
8	,,	Et-BriB	$(6.06{-}7.46) \times 10^{-8}$	146
9	,,	1-PhEtBr	$(3.27{-}3.68) \times 10^{-8}$	146
10	,,	Me-BrP	3.95×10^{-9}	146
11	**TPMA**	Et-BriB	9.65×10^{-6}	146
12	,,	1-PhEtBr	4.58×10^{-6}	146
13	,,	1-PhEtCl	8.60×10^{-7}	146
14	,,	BnBr	6.78×10^{-7}	146
15	,,	Me-BrP	3.25×10^{-7}	146
16	,,	Me-ClP	$(4.07{-}4.28) \times 10^{-8}$	146
17	,,	Allyl-Br	1.72×10^{-5}	149
18	,,	Allyl-Cl	2.34×10^{-6}	149
19	**Me$_6$TREN**	Et-BriB	1.54×10^{-4}	146
20	,,	Methyl chloroacetate	3.3×10^{-6}	139
21	**DMCBCy**	Methyl chloroacetate	9.9×10^{-5}	139

Table 8.2 (*Continued*)

No.	Ligand	Initiator	K_{ATRP}	Ref.
22	BPMPrA	Et-BriB	6.2×10^{-8}	147
23	TPEDA	Et-BriB	2.0×10^{-6}	147

[a]Cu[I]Br was used for alkyl bromide and Cu[I]Cl was used for alkyl chloride initiators, respectively. All values were measured in MeCN at $22 \pm 2\,^{\circ}C$.

halides increase in the order primary < secondary < tertiary, similarly to the values of the rate constant k_{act}. Alkyl halides with α-functional groups that stabilize radicals (such as cyano or carbonyl (ester)) are more active (larger K_{ATRP}) compared to those with no such substituents. Also, alkyl bromides are more active than the corresponding alkyl chlorides by a factor of 2–5 (generally, less than an order of magnitude).

Not only the nature of the ligand and the alkyl halide but also the solvent influences the values of K_{ATRP}.[152] Table 8.3 lists the values of K_{ATRP} for the reaction between Cu[I]Br/HMTETA and Me-BriB (a small molecule analogue of Br-capped polymethacrylate chain). The values of K_{ATRP} were determined by estimating the values of k_t of methyl 2-isobutyryl radicals in each individual solvent. For this purpose, the diffusion coefficients D of methyl 2-isobutyrate (a compound with almost the same size and mass as the methyl 2-isobutyryl radical) were measured in all studied solvents. The values of D were a function of the solvent viscosity. For a diffusion controlled termination, the termination rate constant was estimated[153] from eqn (8.16), where N_A is Avogadro's number and ρ is the reaction distance, which, according to the model proposed by Gorrell and Dubois,[154] is related to the molar volume V_m (*i.e.*, to the molecular weight MW and the bulk density d) of the studied compound (methyl 2-isobutyryl radical or, in this case, the model compound, methyl 2-isobutyrate) as shown by eqn (8.17).

$$2k_t^{Diff} = 2 \times 10^{-3} \times \pi \times N_A \times D \times \rho = 3.78 \times 10^{21} \times D \times \rho \qquad (8.16)$$

$$\rho^{GD} = \left(\frac{V_m}{N_A}\right)^{1/3} = \left(\frac{MW}{d \times N_A}\right)^{1/3} \qquad (8.17)$$

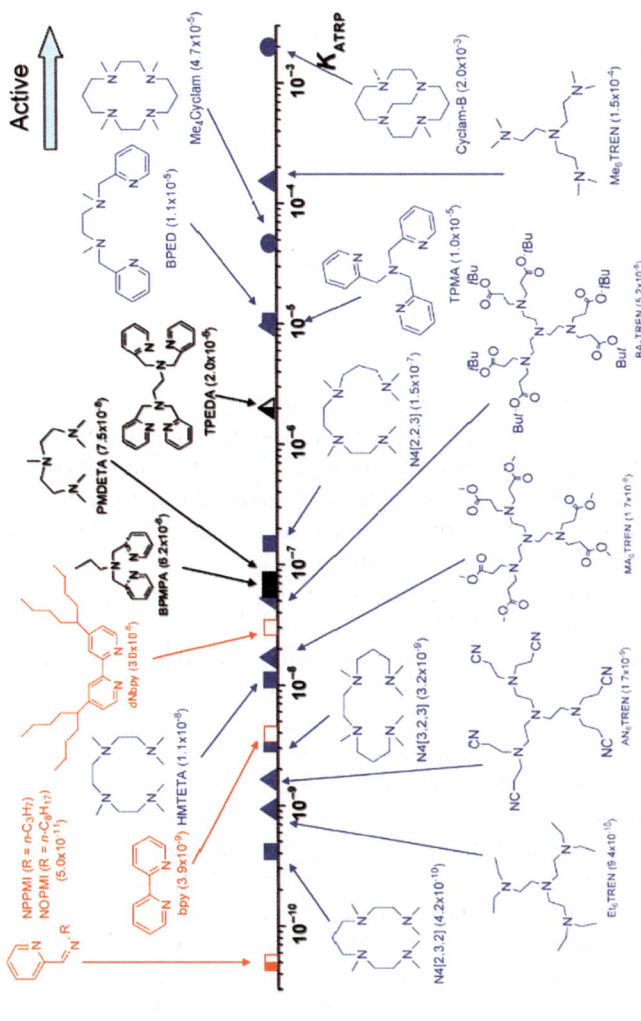

Figure 8.9 ATRP equilibrium constants for ligands with N donor atoms in a reaction of the corresponding Cu^{I}Br complexes with Et-BriB in MeCN at 22 ± 2 °C. Reprinted from ref. 151 with permission.

Table 8.3 Values of K_{ATRP} for the reaction of Me-BriB with $Cu^I Br/HMTETA$ in various solvents at 25 °C.[152]

Solvent	K_{ATRP}	$K_{ATRP,rel}$
Acetone	3.14×10^{-9}	1.0
2-Propanol	6.89×10^{-9}	2.2
Ethanol	7.46×10^{-9}	2.4
Acetonitrile	7.49×10^{-9}	2.4
1-Propanol	8.52×10^{-9}	2.7
Methanol	9.27×10^{-9}	3.0
N,N-Dimethylacetamide	1.91×10^{-8}	6.1
Propylene carbonate	3.45×10^{-8}	11
N,N-Dimethylformamide	4.60×10^{-8}	15
N-Methylpyrrolidone	1.39×10^{-7}	44
Dimethylsulfoxide	2.61×10^{-7}	83

As the solvent changes from acetone to dimethylsulfoxide, the values of K_{ATRP} for the mentioned catalyst–initiator system increase by nearly two orders of magnitude. For a more extended selection of solvents, the changes in K_{ATRP} are likely to be much more pronounced.

8.3.2.2 Theoretical Treatment

It is essential to determine the important ligand characteristics that are responsible for the activity of the corresponding Cu^I (or other metal) complexes. Likewise, a theoretical treatment that explains the effect of the transferable atom on the value of K_{ATRP} is very desirable. This theoretical treatment should also be able to explain any observed solvent effects. It is convenient to formally represent the ATRP equilibrium constant (Figure 8.10) as a product of two equilibrium constants – that of C–X bond homolysis (K_{BH}) and that of the activator halogenophilicity (K_{Halo}, which is the inverse of the equilibrium constant of homolytic dissociation of the $X\text{-}Mt^{z+1}L_m$ bond). The latter equilibrium constant is a measure of the catalyst activity and can in turn be represented as the product of three additional equilibrium constants – electron transfer (K_{ET}) to the higher oxidation metal complex $Mt^{z+1}L_m$, *i.e.*, the ease of its reduction ($1/K_{ET}$ is a measure of the reducing power of the activator $Mt^z L_m$), electron affinity of the halogen atom X (K_{EA}), and halido-philicity of the higher oxidation state complex $Mt^{z+1}L_m$ (K_{Halido}). It is important to note that the "contributing" reversible reactions do not necessarily take place in the reaction system and the representation is purely formal. The "splitting" of the overall equilibrium into "contributing" equilibria is useful because it is relatively easy to understand, or even quantify, the effect of the nature of the atom X and the ligand L, and of the reaction medium on each one of them.[155]

$$K_{ATRP} = K_{BH} K_{Halo} = \frac{K_{BH} K_{EA} K_{Halido}}{K_{ET}} \quad (8.18)$$

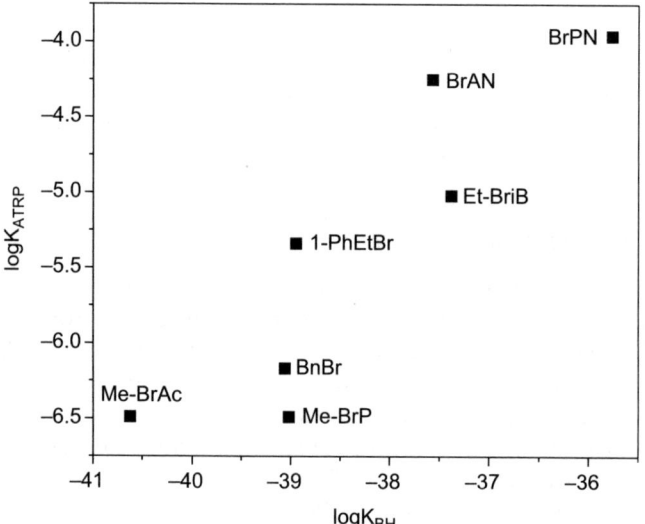

Figure 8.10 Atom transfer as a formal combination of a C–X bond homolysis of alkyl halide (RX) and halogenophilicity of the deactivator, which, in turn can be represented as a product of two redox processes, and association (coordination) of halide anion to a higher oxidation state metal complex $Mt^{z+1}L_m$.

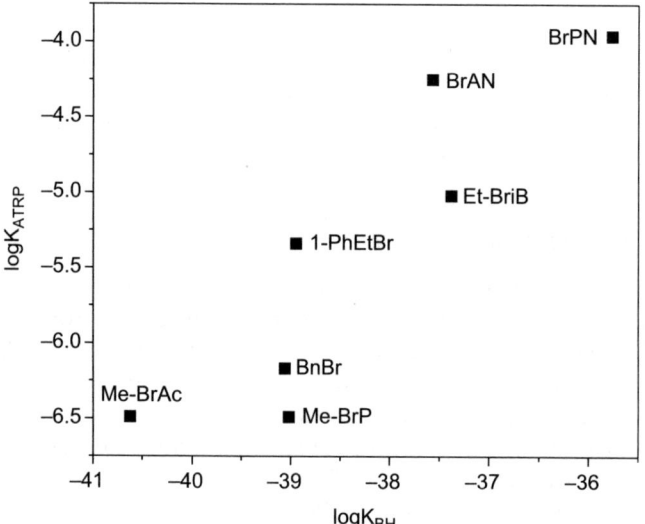

Figure 8.11 K_{ATRP} of various alkyl halide initiators measured at $22 \pm 2\,°C$ with $Cu^IBr/$ $TPMA^{146,157}$ against the calculated values of K_{BH} of RBr initiators.[151]

   The constant K_{BH} is related to the nature of the alkyl halide initiator and the structure of the dormant species; its values are accessible both experimentally and through DFT calculations.[156] Figure 8.11 shows the correlation between the experimentally determined values of K_{ATRP} for the reaction of the same catalyst (to ensure the value of K_{Halo} on the left hand side of eqn (8.18) remains constant), $Cu^IBr/TPMA$, with several alkyl bromide ATRP initiators *vs.* the calculated values of K_{BH} for the C–Br bonds of these initiators.[151]

8.3.2.2.1 The Importance of the Transferable Atom. As seen in Table 8.2, the values of K_{ATRP} for alkyl bromides are larger than those for the alkyl chlorides with the same alkyl substituent (for reactions mediated by the same catalyst under identical conditions). This is partially related to the higher bond dissociation energy or lower K_{BH} of the C–Cl bond. However, C–Cl bonds are stronger than C–Br bonds by *ca.* 8–9 kcal mol^{-1} and if the bond dissociation energy was the sole factor determining the values of K_{ATRP}, those for alkyl bromides should be about 6 orders of magnitude larger than for the chlorides. The experimentally determined differences are less than an order of magnitude, which can be attributed to the higher electron affinity of chlorine compared to bromine (see Figure 8.10 and the right hand side of eqn (8.18)). Indeed, the electron affinity of the atoms Cl, Br, and I in the absence of a solvent have been reported as 3.61, 3.36, and 3.06 eV.[158] On the other hand, the values of the redox potentials of the couples $X_2/2X^-$, which are also related to the electron affinity of the corresponding halogen atoms for X = Cl, Br, and I are 1.12, 0.83, and 0.30 V (*vs.* standard calomel electrode (SCE)), respectively.[159] A difference of 0.29 V between the redox potentials of the couples $Cl_2/2Cl^-$ and $Br_2/2Br^-$ corresponds to a difference between the corresponding equilibrium constants of reduction of the two halogens to halides of about 5 orders of magnitude, *i.e.*, one can estimate the ratio $K_{EA}(Cl)/K_{EA}(Br) \approx 10^5$. Therefore, the ratio of the ATRP equilibrium constants, which, if the catalyst is not altered (*i.e.*, both K_{ET} and K_{Halido} in Figure 8.6 and eqn (8.18) are constant) for alkyl bromides and alkyl chlorides with identical alkyl groups should be given by $K_{ATRP}(RBr)/K_{ATRP}(RCl) = K_{BH}(RBr)K_{EA}(Br)/K_{BH}(RCl)K_{EA}(Cl) \approx 10^6 \times 10^{-5} = 10$, which is close to the observed difference.

On the other hand, as shown by the left-hand side of eqn (8.18), K_{ATRP} depends upon the relative values of the C–X and Cu^{II}–X dissociation energies, in other words to K_{BH} and K_{Halo} (K_{Halo} is, as seen from its definition in Figure 8.6, the inverse of the Cu^{II}–X bond dissociation energy). The lower than expected difference between $K_{ATRP}(RCl)$ and $K_{ATRP}(RBr)$, based solely on the ease of homolytic cleavage (K_{BH}) of C–X bonds, can be explained also by the greater stability of the Cu^{II}–Cl compared to Cu^{II}–Br bond, *i.e.*, higher chloro- than bromophilicity of Cu^I complexes.[151,156]

8.3.2.2.2 Catalytic Activity. The value of K_{Halo} is a measure of the catalyst activity and depends upon the structure of the ATRP catalyst. Therefore, the catalyst structure is related to the equilibrium constants that "contribute" to the value of K_{Halo}, namely, K_{ET} and K_{Halido}. The constant K_{ET} is directly related to the easy to measure redox potential E of the couple $Mt^{z+1}L_m/Mt^zL_m$ according to eqn (8.19).

$$E = -\frac{RT}{F}\ln K_{ET} \qquad (8.19)$$

How the ligand determines the values of these equilibrium constants and therefore the value of K_{ATRP} is discussed next.

The ATRP catalysts, like all coordination compounds, are characterized by *stability constants*, also often named *formation* or *association constants*. Since many of the following mathematical expressions will contain formation constants, their definition is given here. If the metal or metal ion Mt (charges are not shown in the expressions) can react with the ligand Y yielding a series of complexes with different stoichiometries (with ratios of Mt to $Y = 1 : 1$, $1 : 2$, ..., $1 : m$), a mixture of complexes is usually formed in solution containing both Mt and Y and their respective concentrations are governed by the *absolute* (*i.e.*, *initial* or *total*) concentrations of the reagents as well as by the values of the formation constants. All complex-formation equilibria can be characterized by two sets of stability constants: *stepwise*, denoted with K_j^L ($j = 1, 2, ..., m$), and *overall* (also termed *cumulative* or *gross*) stability constants, β_j^L, as shown by the set of eqn (8.20). The overall constants β_j^L in fact define formal equilibria, which do not necessarily take place in real systems (where step processes occur, characterized by the constants K_j^L), but can rather be considered a useful mathematical model.

$$Mt + Y \Leftrightarrow MtY \quad K_1 = \frac{[MtY]}{[Mt][Y]} = \beta_1$$

$$MtY + Y \Leftrightarrow MtY_2 \quad K_2 = \frac{[MtY_2]}{[MtY][Y]} \; ; \quad \beta_2 = \frac{[MtY_2]}{[Mt][Y]^2} = K_1 K_2$$

$$...$$

$$MtY_{m-1} + Y \Leftrightarrow MtY_m \quad K_m = \frac{[MtY_m]}{[MtY_{m-1}][Y]} \quad \beta_m = \frac{[MtY_m]}{[Mt][Y]^m} = K_1 K_2...K_m = \prod_{j=1}^{m} K_j$$

$$(8.20)$$

In addition, the equilibria on the left hand side of eqn (8.20) can sometimes be defined by the dissociation constants $K_{diss,j}$ ($j = 1, 2, ..., m$), that are very familiar from acid–base reactions (the acidity constants $K_{a,j}$ for polyprotic acids are in fact dissociation equilibrium constants). The relationships between all these constants are given by the sets of eqn (8.21). In certain cases, it is more convenient to use one type of equilibrium constants or another, and the choice is typically made based on the simplicity of the mathematical equations describing a system of interest. Definitions and uses of all equilibrium constants in determination of concentrations of all species present in a complex system are given in many monographs on coordination chemistry.[160–166]

$$\beta_r = \prod_{j=1}^{r} K_j = \frac{1}{\prod_{j=1}^{r} K_{diss,n-j+1}}$$

$$K_r = \frac{\beta_r}{\beta_{r-1}} = \frac{1}{K_{diss,n-r+1}} \qquad (8.21)$$

$$K_{diss,r} = \frac{1}{K_{n-r+1}} = \frac{\beta_{n-r}}{\beta_{n-r+1}}$$

The activity of the ATRP catalyst is related to its reducing power, *i.e.* to the value of K_{ET} (eqn (8.19)) or the redox potential of the couple $Mt^{z+1}L_m/Mt^zL_m$, which in turn is related to the relative stabilization of the Mt^{z+1} *vs.* the Mt^z state upon complexation with the ligand L, according to eqn (8.22), valid for one-electron transfer processes between two complexes with the same coordination number of the central metal or metal ion (E^0 is the standard potential of the Mt^{z+1}/Mt^z couple).[167–173]

$$E = E^0 + \frac{RT}{F}\ln\frac{[Mt^{z+1}]}{[Mt^z]} = E^0 + \frac{RT}{F}\ln\frac{[Mt^{z+1}]_{tot}}{[Mt^z]_{tot}} - \frac{RT}{F}\ln\frac{1 + \sum_{j=1}^{m}\beta_j^{z+1,L}[L]^j}{1 + \sum_{j=1}^{m}\beta_j^{z,L}[L]^j}$$

(8.22)

For ligands forming complexes with 1 : 1 stoichiometry, the redox potential depends upon the ratio of the stability constants of the $Mt^{z+1}L$ and Mt^zL complexes ($\beta^{z+1,L}$ and $\beta^{z,L}$, respectively; the lower case index "1", indicating the stoichiometry of the complexes is temporarily omitted for simplicity), according to eqn (8.23).

$$E = E^0 + \frac{RT}{F}\left(\ln\frac{[Mt^{z+1}]_{tot}}{[Mt^z]_{tot}} - \ln\frac{1 + \beta^{z+1,L}[L]}{1 + \beta^{z,L}[L]}\right)$$

(8.23)

$$\approx E^0 + \frac{RT}{F}\left(\ln\frac{[Mt^{z+1}]_{tot}}{[Mt^z]_{tot}} - \ln\frac{\beta^{z+1,L}}{\beta^{z,L}}\right)$$

Knowledge of the readily measurable stability constants of the higher and the lower oxidation state metal complexes can be used to predict the ATRP catalytic activity of a given complex. Ideally, the value of K_{Halido} should be known as well, for it also contributes to the value of K_{ATRP} (*vide infra*). Table 8.4 shows experimental values of $\beta^{II,L}/\beta^{I,L}$ (measured in aqueous media) for Cu complexes used as ATRP catalysts along with the measured values of K_{ATRP} in the reaction of those complexes with Et-BriB in acetonitrile. Although the two sets of numbers were determined in two different solvents, the trend that higher values

Table 8.4 Correlation between the ratio $\beta^{II,L}/\beta^{I,L}$ and K_{ATRP} for various Cu^I complexes used as ATRP catalysts.

Catalyst	β^I [a]	β^{II} [a]	β^{II}/β^I [a]	Ref.	K_{ATRP} [b]
Cu^IBr/bpy	8.9×10^{12} [c]	4.5×10^{13} [c]	5.0	175	3.93×10^{-9} [146]
$Cu^IBr/HMTETA$	1×10^{11}	3.98×10^{12}	39.8	176	8.38×10^{-9} [150]
$Cu^IBr/PMDETA$	$<1 \times 10^8$	1.45×10^{12}	$>1.45 \times 10^4$	176	7.46×10^{-8} [146]
$Cu^IBr/TPMA$	7.94×10^{12}	3.89×10^{17}	4.90×10^4	177	9.65×10^{-6} [146]
Cu^IBr/Me_6TREN	6.3×10^8	2.69×10^{15}	4.3×10^6	178,179	1.54×10^{-4} [146]

[a] Measured in aqueous solution.
[b] Reaction with Et-BriB in MeCN at $22 \pm 2\,°C$.
[c] The values of β_2^{II} and β_2^{II} and their ratio are reported.

of β^{II}/β^{I} correspond to higher values of K_{ATRP} is clearly seen. The stability and the speciation in systems used in ATRP and consisting of Cu^{II} or Cu^{I}, ligand, and halide ions in acetonitrile have also been reported.[174]

In practice, redox potentials are easily accessible *via* techniques such as cyclic voltammetry. Also, a variety of methods has been developed for the determination of stability constants (including electrochemical methods,[162,165,180–184] potentiometric (pH) titration,[185,186] spectroscopy,[187–189] and titration calorimetry[190–192]). These constants determine the redox potential as shown by eqn (8.22) and (8.23) or their more complex forms useful when more than one ligand are present that can form mixed complexes. A typical ATRP deactivator contains both a halide and an organic ligand, and for cases with 1 : 1 stoichiometry, the redox potential of the couple $XMt^{z+1}L/(Mt^{z}L + X^{-})$ should be considered as determining the catalytic activity in ATRP, not of the couple $Mt^{z+1}L/Mt^{z}L$. The expression for the redox potential of the former couple will be similar to eqn (8.23), but will contain an extra term $-\ln((1 + K_{Halido}[X^{-}])/(1 + \beta_{1}^{I,X}[X^{-}]))$, where $\beta_{1}^{I,X}$ is the stability constant of the $Cu^{I}L$ halide complex (which is often negligible). (Note that the halidophilicity of the Cu^{II} complex, K_{Halido}, can also be expressed, using the definitions in eqn (8.20), as $\beta_{1}^{II,X}$; in this text, the simpler K_{Halido} is used.)

In a detailed study, $Cu^{I}Cl$ and $Cu^{I}Br$ complexes of several N-based ligands were characterized by cyclic voltammetry and the measured redox potentials were correlated with the activity of the complexes in the ATRP of methyl acrylate initiated by Et-BrP.[141] The values of either k_{act} or K_{ATRP} and the redox potentials of a series of Cu complexes with tridentate N-based ligands (where the nitrogen atom was an amine-, imine-, or pyridine-type) were well-correlated.[124] It has been shown that although the measured redox potentials of various Cu complexes with non-coordinating counterions such as trifluoromethanesulfonate are correlated to the values of the equilibrium constants of ATRP that those complexes mediate, the correlation becomes much better if bromide ions are added to the system or, simply, if the redox properties of complexes with halide ligands/counterions are determined (Figure 8.12).[152] The reason is that the Cu^{II} complexes with different ligands have different halidophilicities (and some of the Cu^{I} complexes may be, to some extent, halidophilic as well). For complexes with ligands that are able to saturate the Cu^{I} coordination sphere (tetradentate ligands), where the halidophilicity of $Cu^{I}L$ is very low, the halidophilicity of the Cu^{II} complex can be estimated from the difference between the redox potentials of complexes with trifluoromethanesulfonate (or other non-coordinating counterions) and halide counterions. For $Me_{6}TREN$, bpy, and TPEDA the differences in the $E_{1/2}$ values of the bromide and trifluoromethanesulfonate Cu complexes are 230, 160, and 10 mV, respectively. Thus, the relative bromidophilicities of the three corresponding Cu^{II} complexes differ by approximately $10^{3.9}$, $10^{2.7}$, and $10^{0.17}$. The value for bpy is consistent with the bromidophilicity of $[Cu^{II}(bpy)_{2}]^{2+}$ measured in another aprotic solvent, DMF, for which ($K_{Bromido} \approx 10^{5}$ M^{-1}).[193]

The redox potential of complexes with common electrochemistry and spin state can be predicted based on electrochemical parametrization methods.[194,195]

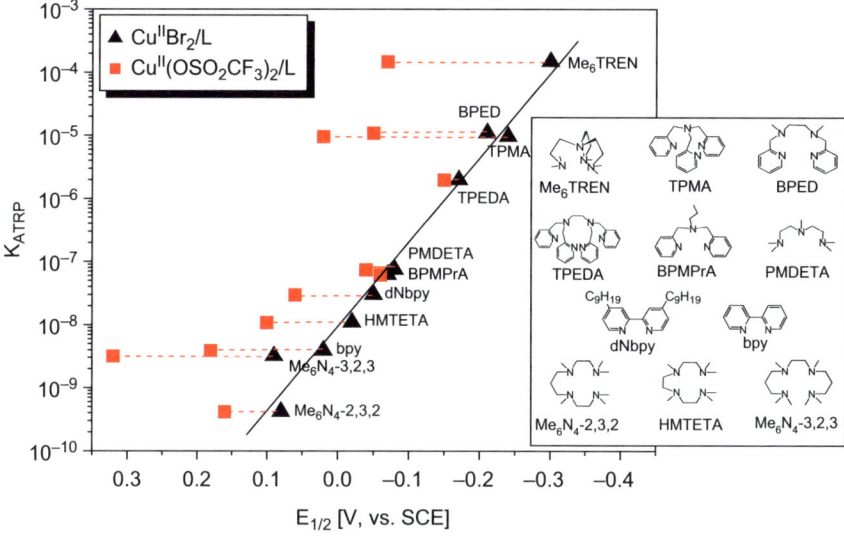

Figure 8.12 Comparison of redox potentials of Cu complexes with various N-based ligands with two different counterions – trifluoromethanesulfonate and bromide. The values of K_{ATRP} are for the reaction of the Cu^I complexes with Et-BriB in MeCN at r.t.
Reprinted from ref. 152 with permission.

Unfortunately, the application of such methodologies is rather limited when multi-dentate ligands are employed that introduce varying degrees of distortion from ideal geometries. In general, complexes of Cu^I are typically tetrahedral while Cu^{II} species are more stable in tetragonal or square pyramidal geometries,[196] and ligand morphologies that force a more planar geometry tend to stabilize Cu^{II} relative to the Cu^I state.[197] The result is a higher ratio of β^{II}/β^I and hence a more reducing Cu^I-based activator, which is a more active ATRP catalyst. This is especially true of constrained macrocyclic ligands and it was recently demonstrated with a derivative of cross-bridged cyclam, DMCBCy, which, as shown in Table 8.2 and Figure 8.9, forms an exceptionally active ATRP catalyst.[139] The attachment of electron donating substituents on the ligand will also result in a more reducing complex,[198–200] with the possible exception that added steric constraints may distort the geometry and destabilize a particular oxidation state of the complex.[197] Subtle changes in the ligand substituents can have a dramatic effect on the redox properties of the complexes, *i.e.*, changes in the $E_{1/2}$ values by more than 60 mV, which corresponds to changes in the values of K_{ET} exceeding 1 order of magnitude. For example, ligands with the generic structures shown in Figure 8.13 form Cu complexes with very different redox properties as the substituents in the ligands change. Some measured redox potentials of Cu complexes are presented in Table 8.5.[198–200] Similarly, as the substituents in the 4- and 4′-positions of 2,2′-bipyridyls changed in the order Cl, H, Me, 5-nonyl, MeO, and Me$_2$N, the

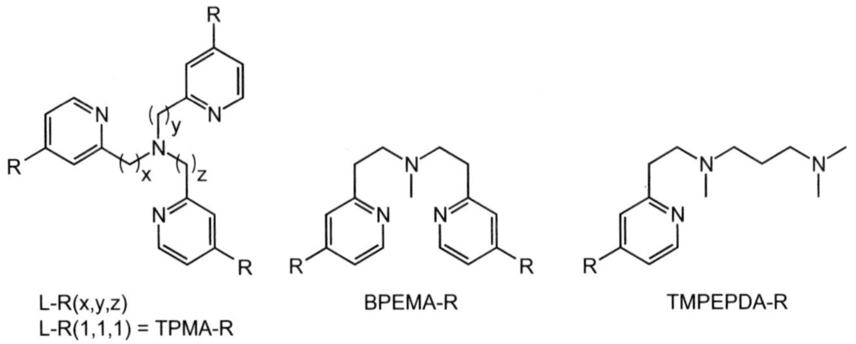

L-R(x,y,z)   BPEMA-R   TMPEPDA-R
L-R(1,1,1) = TPMA-R

Figure 8.13 Substituted multidentate N-containing ligands forming O_2-binding copper complexes.

Table 8.5 Redox potentials (*vs.* the ferricinium/ferrocene redox couple) of the Cu complexes with the ligands shown in Figure 8.9.

Ligand	$E_{1/2}$ (*vs.* Fc/Fc$^+$, in V) of $Cu^IL/Cu^{II}L$ for R =				
	Cl	*H*	*tBu*	*MeO*	*Me$_2$N*
TPMA-R[198]		−0.40^a	−0.46^a	−0.49^a	−0.70^a
BPEMA-R[199]	−0.27^b	−0.31^b		−0.36^b	−0.44^b
TMPEPDA-R[200]	−0.32^b	−0.33^b		−0.38^b	−0.40^b

^{a}In MeCN.
^{b}In DMF.

redox potentials of the corresponding Cu^{II}/Cu^I complex redox couples decreased in the order 270, 55, −48, −55, −88, and −313 mV (*vs.* SCE), respectively, which was accompanied by increased ATRP activity of the complexes.[201] The electrochemical properties of various Cu complexes of N-based macrocyclic ligands have also been shown to change as the electronic effects of the substituents is altered.[202,203]

As mentioned, the size of the chelate ring (determined by the number of C atoms bridging two neighboring N-donor atoms) also affects the redox properties of the Cu complexes. For example, when a 5-membered chelate ring is replaced by a 6-membered one in ligands L-R(x,y,z) (Figure 8.13) and as the number of 6-membered chelate rings increases at the expense of 5-membered ones, the Cu^I complexes become less reducing. The $E_{1/2}$ values (*vs.* NHE in DMF) of the Cu^I/L-H(x,y,z) complexes are −0.386, −0.300, −0.200, and +0.115 V as the values of x, y, and z change in the order 1, 1, 1 (only 5-membered chelate rings); 1, 1, 2 (two 5- and one 6-membered rings); 1, 2, 2 (one 5-membered and two 6-membered rings); and 2, 2, 2 (only 6-membered chelate rings).[204] In other words, the difference in the K_{ET} values for the complexes containing only 5-membered and only 6-membered rings is about 8.5 orders of magnitude. Similar trends have been observed for Cu complexes with macrocyclic ligands.[203]

As far as the donor atoms of the ligands are concerned, the stability of Cu^{II} complexes typically increases according to $S < O < N$.[205] S- and also P-based ligands form rather stable Cu^{I} complexes, but relatively weak Cu^{II} complexes[206] and such ligands are unlikely to form active ATRP catalysts, due to the low β^{II}/β^{I} ratio and the weak reducing power of the corresponding Cu^{I} complexes. For example, it was shown that in ligands analogous to TPMA with mixed donor atoms (N and S; Figure 8.14), the increase of the number of sulfur atoms led to formation of less reducing Cu^{I} complexes.[177] Consequently, S-only-containing ligands are not appropriate as components of active ATRP catalysts, but "heterodonor" (with N in addition to S donor atoms) ligands are expected to form more catalytically active complexes. Such complexes are of interest because they may be sufficiently stable in the presence of acids (due to the low basicity of the S atoms in thioethers) and therefore may be used to mediate the ATRP of acidic monomers. As in the case of ligands with only N donor atoms, the redox potential increases, *i.e.*, the reducing power decreases as an additional methylene group is added between the donor atoms, causing the formation of 6- instead of 5-membered chelate rings.[177] A detailed review on Cu complexes with ligands containing the N_2S_2 donor set[207] demonstrates that the electrochemical properties can serve as a "probe" for the structural characteristics of the complexes. The electrochemistry of Cu^{II} complexes with N_2O_2-type heterodonor ligands has also been shown to depend upon the molecular geometry[208] and the electronic effects of the substituents.[209] Redox potentials of numerous Cu complexes with ligands with N, O, and S donor atoms have been reviewed.[197] A thorough monograph by Zanello[210] summarizes all the important concepts related to the redox properties of coordination compounds.

The redox potentials and the ATRA catalytic activity of Ru complexes with *p*-substituted triphenylphosphine ligands were correlated.[211] Also, the ATRP catalytic activity of a series of pentacoordinated $Fe^{II}Cl_2$ complexes with tridentate N-based ligands was shown to correlate to their reducing strength.[212] However, in general, it is difficult to predict the catalytic activity of Fe complexes based only on redox potentials because, depending on the structure of the ligand, either low- or high-spin complexes can be formed, and the spin state can influence not only the ATRA or ATRP catalytic activity, but even the

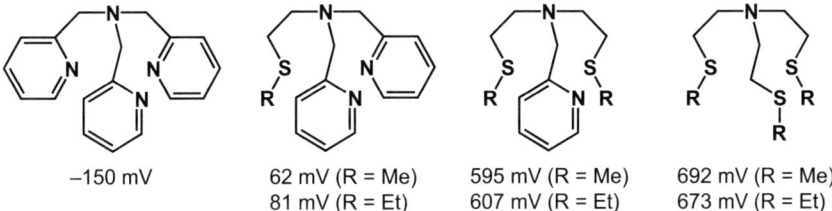

−150 mV   62 mV (R = Me)   595 mV (R = Me)   692 mV (R = Me)
          81 mV (R = Et)   607 mV (R = Et)   673 mV (R = Et)

Figure 8.14 Homo- and heterodonor ligands with similar structural features forming only 5-membered chelate rings and redox potentials of their Cu complexes *vs.* NHE.[177] As the number of S donor groups increases, the reducing power of the Cu^{I} complexes decreases gradually.

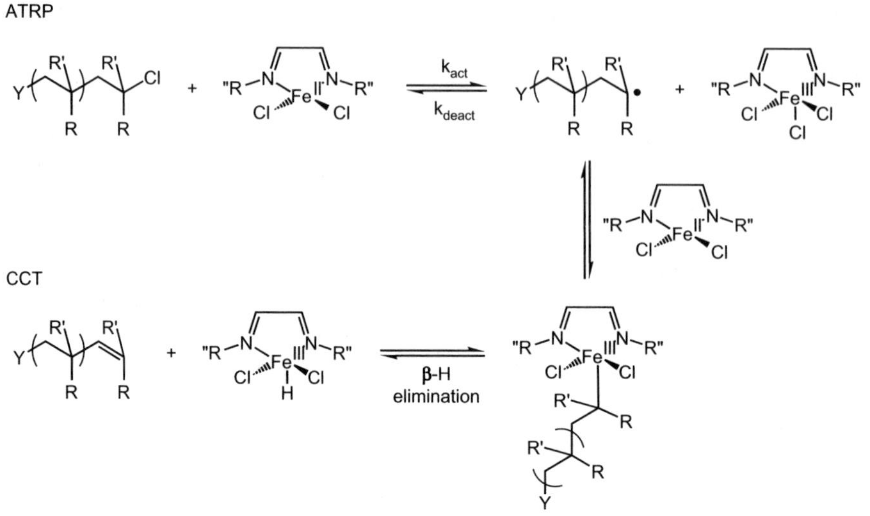

ATRP

CCT

Figure 8.15 Competing ATRP and CCT mediated by Fe complexes.

mechanism of the reactions these complexes mediate. It was shown, for instance, that depending on the substituents at the N atoms of bisimine-type ligands, either ATRP or catalytic chain transfer (CCT, Figure 8.15) could be mediated by the Fe complexes. Both the reversibility of Fe^{II}-to-Fe^{III} transition[89,212,213] and the spin state of the Fe^{III} center[214] play a role in determining whether the ATRP or CCT mechanism will dominate, the former being more pronounced for high-spin complexes with a reversible redox process.

It should be remembered that the halogen atom transferred from the alkyl halide RX to the activator Mt^zL_m becomes a ligand (halide) in the formed deactivating complex $XMt^{z+1}L_m$. For metals different from Cu (*e.g.*, Ru, Os, *etc.*), the activity in ATRP reactions sometimes cannot be explained based solely on the reducing power of the activator, which is often low. However, for those complexes that mediate controlled ATRP reactions, the halidophilicity of the higher oxidation state metal complex is very large (often orders of magnitude larger than for Cu^{II} complexes), which compensates for the low reducing activity (*i.e.*, large value of K_{ET}), and leads to high ATRP catalytic activity (*i.e.*, high K_{Halido} determines high K_{ATRP}).[42] It has been shown that K_{ET} contributes largely (>85%) to the overall value of K_{ATRP} for Cu^I-based activators such as $Cu^IBr/TPMA$ or Cu^IBr/Me_6TREN, whereas it contributes less than 50% to the overall K_{ATRP} for $Os^{II}(PPh_3)_3Br_2$ or $Os^{II}(P(i-Pr)_3)(Cp^*)Br$ (Cp^* = pentamethylcyclopentadienyl anion); the rest of the activity in these cases originates from the halidophilicity.[42]

Halidophilicity is related not only to the catalyst activity, as discussed above, but also to the fraction of homolytically dissociated $Mt^{z+1}L_m$-X bonds (with formation of the complex $Mt^{z+1}L_m$ and halide anions) in a given solvent at a specific total catalyst concentration. In solvents where the halidophilicity is low (*e.g.*, protic solvents such as water or alcohols, or in their mixtures with other solvents), a fraction, sometimes significant, of the deactivating $XMt^{z+1}L_m$ complex can effectively "lose" its halide ligand, and, since the produced $Mt^{z+1}L_m$ complex is unable to deactivate radicals, the polymerization control can be largely lost.[215] This is discussed below, after the discussion dedicated to the process of transfer of the X atom (group) from the complex $XMt^{z+1}L_m$ to the propagating radicals.

The theoretical treatment shown above can also be used to understand and explain the observed pronounced solvent effects on the values of K_{ATRP} (Table 8.3). It is not always immediately obvious how K_{ATRP} will change from one solvent to another, but, in general, it should be understood how each of the "contributing" equilibria in Figure 8.10 is affected by the solvent. For example, solvents that solvate halide anions well would increase K_{EA} but at the same time will contribute to a decrease in K_{Halido} (*vide infra*). Solvents that tend to stabilize the Mt^zL_m complex relative to $Mt^{z+1}L_m$ *via* preferential solvation or coordination, will lower the redox potential of the $Mt^{z+1}L_m/Mt^zL_m$ couple, leading to a decreased ATRP catalytic activity (high K_{ET}). The effect of several solvents on K_{ET}, K_{EA}, and K_{Halido} (K_{BH} is thought not to be affected significantly by the solvent) have been quantitatively explained.[152] Importantly, the solvent effects were also quantitatively analyzed in terms of Kamlet–Taft parameters,[216,217] and linear solvation energy relationships were employed to extrapolate catalyst activity over seven orders of magnitude in a large selection of organic solvents and water (Figure 8.16).[152] Prior to that work, redox properties of Cu^I complexes in various solvents were analyzed using the Kamlet–Taft parameters approach.[218] The very high predicted value of K_{ATRP} in water has also been experimentally estimated for the $Cu^IBr/$ TPMA–bromoisobutyrate system.[219]

8.3.3 The Deactivation Process

The ability of high oxidation state metal halides to transfer a halogen atom to polymeric radicals, which was accompanied with reduction to the corresponding lower oxidation state compounds has been known for a long time. For instance, Bamford[220,221] reported on the transfer of a chlorine atom from $Fe^{III}Cl_3$ to radicals derived from azobisisobutyronitrile and the propagating radicals derived from several monomers – acrylonitrile, methacrylonitrile, methyl acrylate, and methyl methacrylate. The transfer of a halogen atom from $Cu^{II}Cl_2$[222–225] and $Cu^{II}Br_2$[226] to propagating radicals was extensively investigated in the 1960s and 1970s by Bengough. However, these halogen transfer reactions were not fully utilized in the synthesis of polymers until the development of ATRP.

The values of k_{act} and K_{ATRP} can be determined independently, as described above, and the values of k_{deact} can be calculated from the ratio k_{act}/K_{ATRP}, which is one of the commonly used approaches. Alternative experimental methods for determination of k_{deact} include the clock reaction wherein radicals are simultaneously trapped by TEMPO and the deactivator $XMt^{z+1}L_m$,[124] or analysis of the initial degrees of polymerization with no reactivation, end groups, and molecular weight distributions.[227–229] Additionally, k_{deact} can be

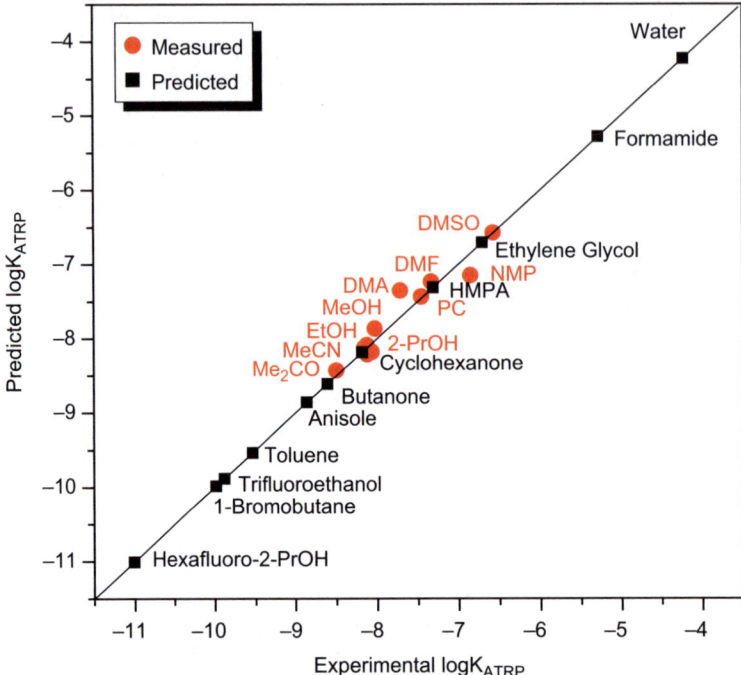

Figure 8.16   Plot of log K_{ATRP} values for $Cu^I Br/HMTETA + Me\text{-}BriB$ against values predicted by the Kamlet–Taft relationship where $XYZ = -11.54 - 0.80\alpha + 1.83\beta + 1.72\pi^* + 0.011(\delta_H)^2$. The line represents values predicted by the Kamlet–Taft relationship. Predicted values of K_{ATRP} for fifteen organic solvents and water are also provided based on solvent-independent coefficients and the appropriate solvatochromic parameters. Reprinted from ref. 152 with permission.

measured by applying laser single pulses for radical production in conjunction with subsequent time-resolved detection of the decay of radical concentration in the absence and then in the presence of deactivator.[230]

The efficiency of deactivation, as shown by eqn (8.8), is responsible for the degree of polymerization control and the width of the attained MWDs. The rate of deactivation depends on the value of k_{deact} and also on the concentration of available deactivator. As the latter normally increases due to the persistent radical effect, ATRP can be viewed as a "self-regulating" system. The importance of available deactivator concentration, related to the halido-philicity of the higher oxidation state complex $Mt^{z+1}L_m$, which is very solvent-dependent, is discussed in the next sections dedicated to competitive equilibria in ATRP. In this section, only the values of k_{deact} are discussed. The values of k_{deact} depend strongly on the ligand. The appropriate ligands for ATRP are those that form active complexes but also yield higher oxidation state halide complexes able to rapidly deactivate the propagating radicals. Figure 8.17 shows a "map" for rational selection of Cu-based catalysts, in

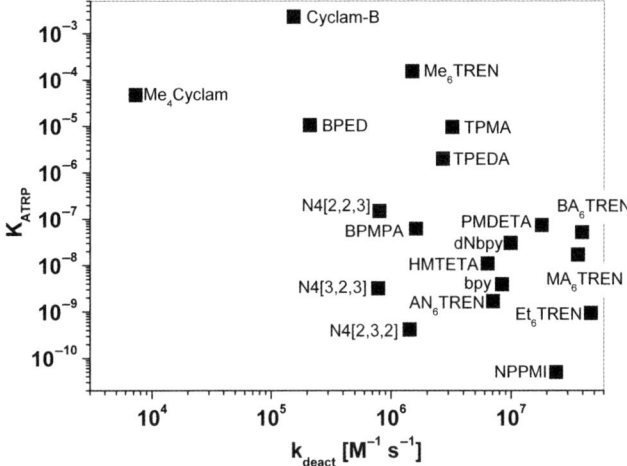

Figure 8.17 "Map" for selection of catalysts with best performance in ATRP. The values are for the reaction of $Cu^I Br$ complexes with Et-BriB in MeCN at $22 \pm 2 \, ^\circ C$.
Reprinted from ref. 151 with permission.

which K_{ATRP} in the reaction of various Cu complexes with Et-BriB in aceto-nitrile is plotted against k_{deact}. The ligands can be divided into four groups. Ligands in the lower right part are typical ligands used in ATRP, which yield moderately active catalysts (*i.e.*, mediate moderately fast polymerizations) and also provide good control due to sufficiently high values of k_{deact}. Most "traditional" ATRP ligands belong to this group, such as bpy, dNbpy, PMDETA and HMTETA. The ligands in the upper right section form complexes mediating fast (large K_{ATRP}) and well-controlled (high k_{deact}) polymerizations. Ligands from the lower left area of the map can form catalysts that mediate relatively slow and not well-controlled polymerizations. These are inefficient ligands for ATRP, and therefore only a few of them are included in Figure 8.17, such as N4[3,2,3] and N4[2,3,2]. Finally, the ligands in the top left part of the map can form catalysts that afford very fast but uncontrolled (owing to low deactivation rate constant) polymerizations.

Like k_{act}, k_{deact} depends upon the nature of the transferable atom. For instance, the deactivation of 1-phenylethyl radicals (structurally similar to the propagating polystyrene radicals) with the bromide complex $Cu^{II}(dNbpy)_2 Br$ ($k_{deact} = 2.5 \times 10^7 \, M^{-1} \, s^{-1}$) is almost 6 times faster than for the analogous chloride complex $Cu^{II}(dNbpy)_2 Cl$ ($k_{deact} = 4.3 \times 10^6 \, M^{-1} \, s^{-1}$).[123] It is consequently expected and indeed observed that polymerization control is better if alkyl bromides and copper bromide-containing catalysts are used in ATRP provided that no "side" reactions take place in the system.[215] However, if the alkyl halide initiator or polymeric dormant state can easily participate in nucleophilic substitution reactions, the use of chloride-based initiator and/or catalyst is advantageous due to lower reactivity of alkyl chlorides in nucleophilic substitution. For instance, the ATRP of 4-vinylpyridine using

2-bromoisobutyrate ester as the initiator and Cu^IBr/HMTETA or Cu^IBr/ TPMA as the catalyst yielded polymers with polymodal MWDs, most likely due to reaction of the bromine-terminated poly(4-vinylpyridine) with either the monomer or polymer yielding pyridinium salts, therefore leading to branching. This reaction was suppressed when the corresponding Cu^ICl-based catalysts were used and polymers of narrow and symmetrical MWDs were produced.[148] Cu^ICl/TPMA was also used as the ATRP catalyst in the synthesis of polymeric brushes with 4-vinylpyridine-derived arms.[231]

8.4   Halogen Exchange

The importance of knowing how the nature of the halogen atom and the alkyl group of the ATRP initiator affects the values of k_{act} and K_{ATRP} is nicely illustrated by the synthesis of block copolymers. In order to synthesize a polyA-*block*-polyB-type segmented copolymer by ATRP (A and B are the corresponding monomers), a bromine-terminated macroinitiator (polyA-Br) is usually dissolved in a monomer B, and the appropriate catalyst is added. However, to obtain polymers, in which the second, B-derived, segment has a narrow MWD, fast initiation from the polyA-Br macroinitiator is needed, particularly for the cases when a low molecular weight second (polyB) segment is targeted. In cases when the polymeric alkyl bromide derived from the monomer B (polyB-Br) is more active than polyA-Br ($K_{ATRP}^{polyB-Br} \gg K_{ATRP}^{polyA-Br}$), poorly defined block copolymers are obtained. This is the case for the chain extension of Br-capped polystyrene or polyacrylate macro-initiators with methacrylates or acrylonitrile. The values of k_p and k_i are also important, and in systems, for which the addition of the polyA$^\bullet$ radical to the monomer B is slow, a poorly defined mixture of polyA-*block*-polyB copolymers with varying lengths of the polyB block and possibly even unreacted polyA-Br macroinitiator are obtained. To achieve complete consumption of the macro-initiator polyA-Br before the chains that already contain one or more units of the monomer B had propagated too much (due to the higher propensity of the polyB-Br-type chains to get activated compared to polyA-Br), it is important to slow down the growth of the B-containing segment. One very useful approach is to employ halogen exchange,[232,233] shown in Figure 8.18. The ligands in the complexes are not shown for the sake of simplicity, but it should be remembered that they are of utmost importance for the catalyst performance.

In a typical halogen exchange experiment, a polyA-Br macroinitiator is used but the catalyst is formed from ligated Mt^zCl (Cl can be a counterion and not necessarily a ligand in the lower oxidation state complex) rather than Mt^zBr. The activation yields the mixed halide higher oxidation state complex $BrMt^{z+1}Cl$ and the macroradical polyA$^\bullet$, which can add to the double bond of the monomer B yielding the radical polyA-*block*-polyB$^\bullet$. The deactivator $BrMt^{z+1}Cl$ can transfer either a bromine or chlorine atom to the propagating radical (some rearrangement may be necessary), yielding respectively polyA-*block*-polyB-Br- or polyA-*block*-polyB-Cl-type dormant species. If the halogen

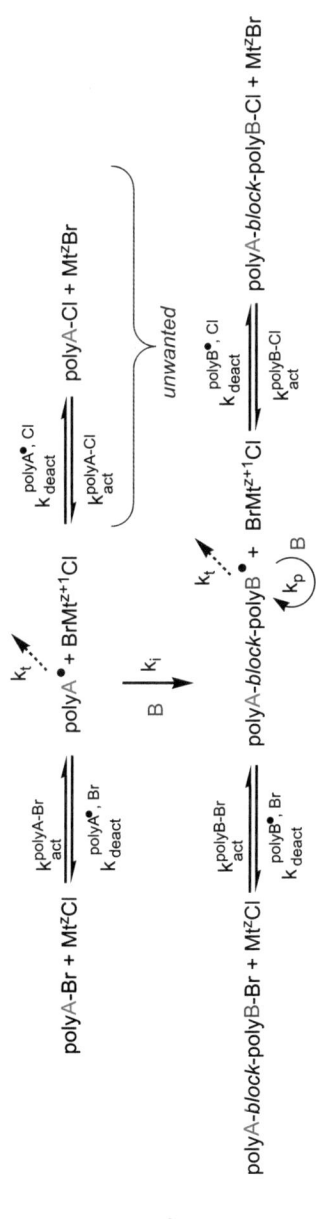

Figure 8.18 Halogen exchange.

exchange is efficient and the majority of the polyA-*block*-polyB• radicals are converted to polyA-*block*-polyB-Cl, while most of the polyA-type dormant species are in the form polyA-Br, well-defined polyA-*block*-polyB copolymer (*i.e.*, with narrow MWD of the polyB-type segment) will be afforded, even in the case when $K_{ATRP}^{polyB-Br} \times k_p(B) > K_{ATRP}^{polyA-Br} \times k_i$. The reason is that the value of the ATRP equilibrium constant for alkyl chlorides is several times (up to 1 order of magnitude) lower than for alkyl bromides with the same alkyl group. As a result, it often happens that $K_{ATRP}^{polyB-Cl} \times k_p(B) \leq K_{ATRP}^{polyA-Br} \times k_i$, and efficient initiation (cross-propagation) takes place. It is important that the halogen exchange occurs at chains containing at least one terminal B-unit rather than at the macroinitiator polyA-Br. If the Br-capped macroinitiator is rapidly converted to the corresponding chloride, polyA-Cl, before addition of the polyA• radicals to the monomer B has taken place, again, just as in the absence of halogen exchange, the crosspropagation (*i.e.*, the initiation efficiency) will be poor and block copolymers with a polyB block of wide MWD will be produced.

It has been shown[234] that the copper-mediated halogen exchange is more efficient for alkyl bromides with higher values of K_{ATRP}, for instance, Et-BriB is converted faster to the corresponding chloride, Et-CliB, than 1-PhEtBr is converted to 1-PhEtCl. This is the reason why chain extension of Br-capped polystyrene macroinitiator with methyl methacrylate works efficiently under halogen exchange conditions. Likewise, as k_{act} (or K_{ATRP}) increases for a given alkyl bromide by selecting more active catalysts, the rate of halogen exchange also increases, *e.g.*, in the order $Cu^I Cl/bpy < Cu^I Cl/HMTETA < Cu^I Cl/PMDETA$. The addition of deactivator ($Cu^{II}Cl_2$ complex with the same ligand) does not affect the rate of halogen exchange, although it may prove beneficial in polymer synthesis, since it increases the rate of deactivation – a prerequisite for good polymerization control. In addition to the radical pathway shown in Figure 8.18, nucleophilic substitution by chloride ions weakly associated to the Cu^I and/or Cu^{II} complexes also contributes to the conversion of polymeric alkyl bromides to chlorides, *i.e.*, halide exchange takes place in addition to halogen exchange.[235]

8.5   Initiation Techniques in ATRP

8.5.1   Normal, Reverse, Simultaneous reverse and Normally Initiated ATRP, and ATRP with Activators Generated by Electron Transfer

In the original reports on ATRP, a combination of alkyl halide and a lower oxidation state transition metal complex or, more often, a mixture of a metal halide or other salt and a ligand (activator) were used to reversibly generate radicals that initiated the polymerization of a monomer, which was also present in the reaction mixture. This experimental setup is suitable for small-scale reactions, but it suffers from an important drawback, namely the sensitivity of the lower oxidation state metal complex to air. The oxidation of the catalyst became an even larger concern as more reducing (and more active) catalysts were developed. Polymerization systems in large vessels can be difficult to deoxygenate, which can lead to irreversible oxidation of part of the ATRP activator, and therefore to slow and difficult to reproduce (due to the unknown extent of oxidation in each batch) reactions. Considering the handling of the oxidatively unstable catalysts, an alternative initiation technique was developed, termed *reverse ATRP*, in which instead of the lower oxidation state metal complex, the deactivator was added to the reaction mixture in conjunction with a radical source, typically a conventional radical initiator. Since all components are stable toward air, the systems of this type were easier to handle and the reaction mixture can be prepared and stored until needed, provided that care is taken not to decompose the radical initiator. Then, the mixture is deoxygenated and the reaction is started with the decomposition of the radical initiator, usually upon heating. The produced radicals reduce the higher oxidation state deactivator and the ATRP activator is generated *in situ*.[84,236,237] The method has been successfully employed in aqueous heterogeneous (miniemulsion)[238,239] systems as well as in a variety of solvents, including ionic liquids.[240] The fact that reverse ATRP operates, demonstrates that the equilibrium shown in Figure 8.1 can be reached from both directions (starting either with an activator and an alkyl halide or with a deactivator and a radical source), and can be viewed as one of several ways to prove the radical nature of this polymerization method.[108] As a drawback of the reverse ATRP technique, it should be pointed out that the transferable atom or group, which eventually becomes the end group of each polymer chain, originates from the deactivator, and therefore the concentration of catalyst needed should be equal to or higher than the concentration of generated polymer chains, *i.e.*, the catalyst concentration cannot be lowered regardless of the catalyst activity. Further, block copolymers cannot be formed with this technique, unless a polymeric precursor of radicals (*e.g.*, polymeric azo compound) can be prepared. Similarly, the introduction of a large selection of α-end functional groups is not possible if a special radical source containing the functionality of interest (or its precursor) is not available.

An important improvement was the development of another initiation strategy, dubbed *simultaneous reverse and normally initiated (SR&NI) ATRP*.[241] In this technique, a low concentration of active catalyst can be used and block copolymers, although not absolutely free of homopolymer impurities, can be synthesized. The reaction mixture contains an alkyl halide (which may be polymeric or functionalized), a smaller amount of radical source and the oxidatively stable higher oxidation state metal halide complex (deactivator). The reaction starts with the decomposition of the conventional radical source. The formed radicals initiate some chains and reduce the higher oxidation state metal halide complex to afford an ATRP activator, which can then activate the alkyl halide and concurrently mediate normal ATRP. The amount of catalyst used in this technique is determined not by the number of chains but only by its activity. SR&NI ATRP has been successfully conducted in aqueous heterogeneous media such as mini-emulsion,[242,243] where addition of the catalyst in its oxidatively stable deactivator form simplifies large-scale and commercial procedures. A drawback of this reaction protocol is the use of radical initiators, which leads to the formation of a fraction of chains that are not initiated by the alkyl halide. In other words, not all chains have the same α-functionality, which is a problem when the preparation of highly functionalized or pure segmented copolymers is desired.

The next major step toward reducing the amount of catalyst needed to mediate polymerizations, which yield highly functionalized polymers, was the development another technique named *activators generated by electron transfer* (*AGET*) ATRP.[244,245] The method is in some sense a logical continuation of work showing that zero-valent metals could reduce the deactivator accumulated in the system,[246] and that other reducing agents such as mono-saccharides[247] or phenols[248] could be employed for the same purpose. AGET ATRP uses a combination of an alkyl halide functional initiator or macro-initiator with an active ATRP catalyst in its higher oxidation state (deactivator) in conjunction with a reducing agent. The role of the reducing agent is to generate the activator *in situ*, but unlike SR&NI ATRP, where radicals play the same role, the reducing agent (or the product of its oxidation by the ATRP deactivating complex) is not able to initiate new chains. A variety of reducing compounds have been employed in AGET ATRP including Sn^{II} compounds,[244,249] ascorbic acid,[245,250] phenols,[251] or thiophenols.[252] AGET ATRP can also be successfully carried out in the presence of limited amounts of air, both in bulk and in miniemulsion, provided that a larger amount of reducing agent is added to the system.[253]

All described initiation techniques are schematically represented in Figure 8.19.

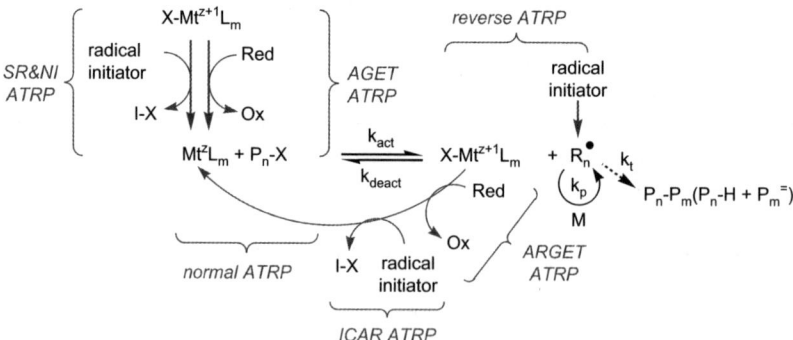

Figure 8.19   ATRP initiation techniques. I-X is a halide produced by the deactivation of radicals originating from the conventional radical initiator, Red and Ox respectively designate a non-radical reducing agent and the product of its oxidation, and M is a monomer.

8.5.2 ATRP in the Presence of Excess of Reducing Agents: ARGET and ICAR ATRP

In "classical" ATRP, relatively large amounts of catalyst were used, often comparable or even equal to the amount of the initiator. Although very active ATRP catalysts were developed, they could not be used at very low concentrations because in ATRP, as in any other radical polymerization, radical termination occurs leading to the irreversible accumulation of the deactivator, $XMt^{z+1}L_m$, at the expense of the activating complex Mt^zL_m, *i.e.*, due to the persistent radical effect. The amount of activator, which is "lost" owing to termination, equals the amount of terminated chains, as shown by eqn (8.24).

$$-\Delta[Mt^zL_m] = \Delta[XMt^{z+1}L_m] = \Delta[P_{dead}] = k_t[R^\bullet]^2t \qquad (8.24)$$

As a consequence, when the concentration of terminated chains starts to approach the concentration of activator initially present in the system, the polymerization slows down and, when all the activator is consumed, the reaction stops. This can happen at a relatively early stage of the polymerization, *i.e.*, at low monomer conversion. As indicated by eqn (8.2), the rate of ATRP does not depend upon the absolute concentration of catalyst but only upon the concentration ratio of activator and deactivator. In other words, if there was a way to keep that ratio sufficiently high (or constant) throughout the polymerization, the reaction rate should remain relatively high and will decrease with time only due to the consumption of monomer. With that in mind, it was realized that if an additional redox cycle could operate in the system that would continuously reduce the higher oxidation state metal complex $XMt^{z+1}L_m$ to the lower oxidation state activator Mt^zL_m, the persistent radical effect would not lead to slowing down the polymerization. It was demonstrated that indeed, in the presence of reducing agents, ATRP could be successfully conducted to high monomer conversion at very low (often single-digit ppm) amounts of catalyst.[254] It is important to use a reducing agent that does not reduce all the deactivator, because, if that were to happen, polymerization control would be lost. As eqn (8.8) shows, a sufficient amount of deactivator should be present in the system to ensure that during the active state of each chain, only a few monomer units can be added prior to conversion of the growing macroradical to the dormant species. Otherwise, polymers with broad MWDs will be formed. The average number of added monomer units per activation–deactivation cycle is given by the ratio of the rates of propagation and deactivation, *i.e.*, by $k_p[M]/k_{deact}[XMt^{z+1}L_m]$.

Both non-radical (Sn^{II} compounds, amines, hydrazines, *etc.*) and radical source-based (*e.g.*, radicals generated by the decomposition of conventional radical initiators) reducing agents have been successfully used and the corresponding processes have been named *activators regenerated by electron transfer* (ARGET)[255,256] and *initiators for continuous activator regeneration* (ICAR) ATRP[257] (Figure 8.19). As in SR&NI ATRP, in ICAR ATRP, some polymer chains are generated that do not originate from the alkyl halide initiator (which can contain a specific functional group or be polymeric), but rather from the radical source used as a reducing agent. These additional chains may amount to 5–15% of all the chains, depending on the amount of used radical-based reducing agent. In cases where pure α-end functionalized or block copolymers are desired, ICAR ATRP is not the most appropriate synthetic technique. Scaling-up ICAR ATRP may be challenging due to the large amounts of radical initiator that need to be used, which, if the temperature is not controlled precisely, may decompose rather quickly and lead to fast and exothermic polymerization. To solve this last problem, slow dosing of radical initiator throughout the polymerization has been proposed and utilized on relatively large scales.[258] In ARGET ATRP, non-initiating reducing agents are used, and the number of chains is determined solely by the amount of alkyl halide initiator, and therefore the method is suitable for the synthesis of α-end functionalized or block copolymers that are not contaminated by non-functionalized or homopolymers.[259] However, the reducing agent and the products of its oxidation have to be removed at the end of the reaction, and this purification step increases the cost of the final product. Additionally, some reducing agents are reactive and can participate in side reactions with certain solvents or functional monomers. For example, the ARGET ATRP of glycidyl methacrylate, a functional monomer that is used to prepare reactive epoxide-containing polymers (almost universal precursors of a plethora of functional macromolecules[260–264]) afforded poorly defined polymers with polymodal MWDs. This was attributed to side reactions between the epoxide groups in the monomer and polymer (ring opening) and the reducing agent (Sn^{II} octanoate).[131] When ICAR ATRP was employed, well-defined poly(glycidyl methacrylate) with narrow MWD and high molecular weight was obtained.[131] Some reducing agents are rather basic or nucleophilic (examples include amines and hydrazines) and are able to participate in side reactions with the alkyl halide chain ends. For instance, the nucleophilic reaction between 1-PhEtBr, a low molecular weight analogue of Br-capped polystyrene, with both hydrazine and phenylhydrazine in DMSO was rather fast even at ambient temperature, and significant part of the alkyl bromide functional groups were "lost" within time periods shorter than those required for a typical polymerization.[132] Finally, some reducing agents, or the products of their oxidation, may not be tolerant to a wide range of solvents, for instance, protic or coordinating ones. As an alternative to both ICAR and ARGET ATRP, where the drawbacks associated with the use of additional reagents (reducing agents) are virtually eliminated, electrochemical reduction of the deactivator has been successfully achieved.[265]

Synthetically useful organic reactions similar to ATRP, which are mediated by redox-active transition metal complexes, *e.g.*, atom transfer radical addition or cyclization, can also be carried out successfully at low catalyst concentrations in the presence of both radical-based[266,267] and non-radical (ascorbic acid)[268] reducing agents. The continuous activator regeneration throughout the process *via* reduction has made these reactions more environmentally friendly than the traditionally used protocols.[269–271]

Since in both ICAR and ARGET ATRP the amount of catalyst needed to mediate the polymerization is substantially reduced compared to the more traditional initiation techniques, the methods have contributed to making ATRP a "green", environmentally friendly, process. ARGET and ICAR ATRP can be performed in the presence of limited amounts of air, provided that a sufficient amount of reducing agent is added to the reaction mixtures.[272] Moreover, higher degrees of polymer ω-chain-end functionalization[273] and higher molecular weights[273–275] than in the traditional, high catalyst concentration, ATRP can be attained. An additional benefit of these techniques is that the width of the polymer molecular weight distribution can be controlled by adjusting the amount of catalyst.[276,277] On the other hand, because ICAR and ARGET ATRP use a lower amount of catalyst compared to the amount of polymer chains, halogen exchange is not possible. However, chain extensions of macroinitiators of relatively low activity (low K_{ATRP}) with monomers yielding more active polymeric alkyl halides is still possible by adding a comonomer forming dormant chains of low activity (*e.g.*, styrene) to the main monomer of the second segment.[259]

The rate of ICAR ATRP is catalyst concentration- and structure-independent. It is determined by the concentration of radicals, which in turn depends on the concentration of radical source (initiator, in) and the rate constant of its decomposition (which is temperature-dependent). The steady-state radical concentration, assuming slow radical generation, is given by eqn (8.25), where k_{diss} is the rate constant of dissociation (decomposition) of the radical source. Values of k_{diss} and their temperature dependence for various radical initiators are collected in several sources.[278–280]

$$[R^{\bullet}]_{st} = \sqrt{\frac{k_{diss}[in]}{k_t}} \approx \sqrt{\frac{k_{diss}[in]_0}{k_t}} \qquad (8.25)$$

Simulations and experiments indicate that the polymerization rate can be tuned by adjusting the rate of radical initiator decomposition rate, *i.e.*, by adjusting k_{diss} (which depends upon the temperature and the initiator structure) and concentration. As mentioned, efficient deactivation is needed to ensure that the average number of monomer units added to a propagating radical before deactivation is small. The rate of deactivation depends on k_{deact} (*i.e.*, on the nature of the catalyst) and the concentration of deactivator present in the system, which is given by eqn (8.26).[257]

$$[XMt^{z+1}L_m] = [XMt^{z+1}L_m]_0 \left(1 - \frac{1}{K_{ATRP}\frac{[RX]_0}{[R^{\bullet}]_{st}} + 1} \right) \qquad (8.26)$$

It has been shown that only ligands forming catalysts with a high value of K_{ATRP} were successful in mediating a well-controlled ICAR ATRP. The stability of the higher oxidation state metal halide complex towards heterolytic dissociation of the Mt^{z+1}–X bond (halidophilicity) determines to a large degree the amount of deactivator present in the system at high dilution. The importance of complex stability and competitive complex formation reactions are discussed in the next section.

8.6   Competitive Complex Formation Reactions in ATRP

ATRP is a metal complex-mediated process and dissociation or any ligand-exchange reaction affecting the activator or the deactivator may influence the outcome (rate or degree of control) of the polymerization. This section discusses the importance of the mentioned reactions.

8.6.1   Dissociation of the ATRP Catalyst

The degree of dissociation of a complex with the simplest, 1:1, stoichiometry, MtY (Mt represents either the lower or the higher oxidation state metal ion and Y is the ligand or the halide ion, which is part of the higher oxidation state deactivator), depends upon the analytical (initial) concentration of the complex and its stability constant $\beta_1^{Mt,Y}$. When the complex MtY is dissolved in the reaction medium at concentration $[MtY]_{tot}$, part of it will dissociate with the formation of Mt and Y. The products of dissociation are generally not active catalytically, and the dissociation is detrimental in ATRP. From the mass balance, $[MtY]_{tot} = [Mt] + [MtY]$ (and, if no free Y had been added to the system, $[MtY]_{tot} = [Y] + [MtY]$), the fraction of remaining, non-dissociated complex is given by eqn (8.27).

$$\frac{[MtY]}{[MtY]_{tot}} = \frac{[Mt]}{[Mt] + [MtY]} = \frac{-1 + \sqrt{1 + 4\beta_1^{Mt,Y}[MtY]_{tot}}}{4\beta_1^{Mt,Y}[MtY]_{tot}} \qquad (8.27)$$

The above dependence is presented in Figure 8.20. At low complex concentrations, only complexes with large stability constants will not dissociate appreciably. For instance, if the analytical concentration of the ATRP catalytically active complex is $[MtY]_{tot} = 1 \times 10^{-5}$ M, and if the stability constant is $\beta_1^{Mt,Y} = 10^2$, 10^4, 10^6, or 10^8 M^{-1}, *ca.* 1, 8.4, 73, or 97% of the complex will "survive" (will not dissociate), respectively. Clearly at this concentration, complexes with stability constants larger than 10^6 M^{-1} will be suitable to mediate ATRP, if it is desired that at least 70% of the catalyst does not dissociate. In order for about 70% of an unstable complex with stability constant of only 10^2 M^{-1} to remain in the system after the dissociation–association equilibrium has been established, its initial concentration should be of the order of 0.1 M, which is impractical. On the other hand, a very stable complex with stability constant of 10^{12} M^{-1} will not dissociate appreciably even at very high dilutions: over 90% of the complex will remain intact at total concentration as low as 10^{-10} M. In other words, at high dilutions, the stability of the complexes should be high to ensure insignificant dissociation.

It should be remembered that polymerization reactions are typically carried out at elevated temperatures but the majority of the stability constants reported

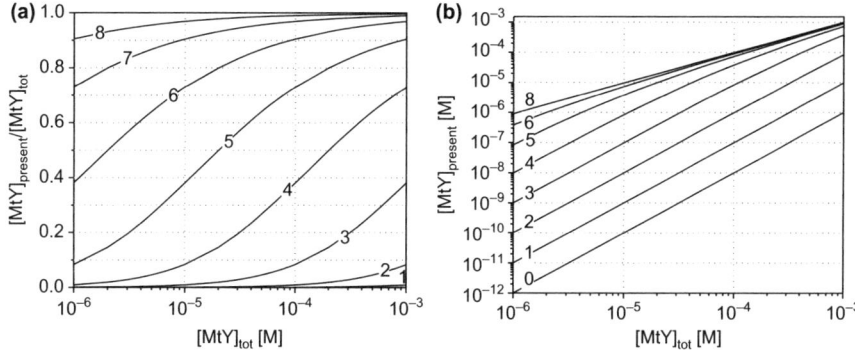

Figure 8.20 Fraction (a) and concentration (b) of remaining, non-dissociated, complex MtY as a function of the analytical concentration of initially present complex at various values of the stability constant $\beta_1^{Mt,Y}$ (the logarithm of which is shown at each curve).

in the literature have been determined at room temperature. The stability of complexes generally decreases as the temperature is raised, which means that dissociation becomes more prominent, according to eqn (8.27). Therefore, thermochemical paramters (*e.g.*, ΔH or ΔG) of complex formation should be known in order to evaluate the stability constant of the ATRP catalyst at high temperatures. The thermochemistry of polyamine complexes of metal ions, including those of Cu^{II}, which are employed in ATRP, has been extensively studied.[281] The enthalpies of formation of Cu^{II} complexes of polyamines in aqueous solution are in the range of -40 to -80 kJ mol^{-1}, and consequently a temperature increase from 25 to 110 °C leads to a decrease in the stability constant by 2–3 orders of magnitude. This is indeed the case[282] for the Cu^{II} complexes of PMDETA,[283] TREN,[284] Me$_6$TREN,[283] TPMA,[285,286] and TPEN.[285,286]

8.6.1.1 Dissociation of the Higher Oxidation State Metal Halide Complex (Deactivator)

ATRP reactions in aqueous solvents are usually fast, even at ambient temperature, and the polymerizations are accelerated as the amount of water in the solvent is increased. This could be caused by the effect of solvent on k_p, K_{ATRP}, and/or the halidophilicity of the $Mt^{z+1}L_m$ complex, which is related to the degree of heterolytic dissociation and therefore to the concentration of the deactivator present in the system (eqn (8.27)). It is known that the specific solvation of some polar monomers able to form hydrogen bonds with protic solvents, particularly with water, does indeed lead to a small increase in k_p.[49,287,288] As discussed above, the value of K_{ATRP} can also change profoundly as the solvent is changed, and in water, very high values of K_{ATRP} have been predicted.[152] In addition to the high polymerization rate, it has been been observed that ATRP reactions in water are generally characterized by relatively poor control, which is most likely due to a low rate of deactivation.[215] The coordination of halide ions to Cu^{II} complexes is comparatively weak in protic (particularly in aqueous) media and, as a result, in some cases a significant part of the deactivator $XCu^{II}L_m$ can dissociate to $Cu^{II}L_m$ and halide anions. The concentration of deactivator actually present in the system depends upon the value of the $Cu^{II}L_m$ halidophilicity complex, K_{Halido}, and upon the total concentrations of Cu^{II} complexes ($[Cu^{II}]_{tot} = [Cu^{II}L_m] + [XCu^{II}L_m]$) and halide ions ($[X]_{tot} = [X] + [XCu^{II}L_m]$), according to eqn (8.28).

$$[XCu^{II}L_m] = \frac{F - \sqrt{F^2 - 4K_{Halido}^2[Cu^{II}]_{tot}[X]_{tot}}}{2K_{Halido}} \qquad (8.28)$$

$$\times \left(F \equiv 1 + K_{Halido}[Cu^{II}]_{tot} + K_{Halido}[X]_{tot}\right)$$

The equilibrium constant of halide anion coordination can be measured by spectroscopic means,[187] as described in the literature for bpy-based ATRP deactivators in several protic solvents.[215] Many values of bromido- and chloridophilicity of the Cu^{II} bpy complex in various water-organic solvent mixtures have been reported[132,215,282] and it has been shown that as the amount of water in the system increases, the halidophilicity decreases markedly. The curves of K_{Halido} *vs.* $[H_2O]$ can, in fact, be used to estimate the halidophilicity in the pure (water-free) solvent by extrapolation. Dissociation of the $XCu^{II}L_m$ complex is pronounced in protic media, particularly in water-rich solvents, and the lower deactivator concentration causes the formation of polymers of relatively wide molecular weight distributions.[46] There are three ways to improve the polymerization control in protic or aqueous media: (i) select ATRP catalysts, for which the higher oxidations state complex is very halidophilic (K_{Halido} depends upon the nature of the ligand L and the central metal); (ii) in classical (high catalyst concentration) ATRP systems, employ catalyst containing large initial amounts of deactivator (in some cases, up to 80 mol% of the total catalyst); or (iii) add extra halide salts to the system. The use of complexes with highly halidophilic metals (*e.g.*, Ru and Os) may be very beneficial in aqueous solvents, but such systems have yet to be studied. The utility of the last two approaches in "classical" ATRP (at high catalyst concentration) has been demonstrated.[215,289,290] In organic solvents, however, halide ions coordinate relatively strongly to both Cu^I and Cu^{II} and, when used in a large excess relative to the catalyst (*e.g.*, in ICAR or ARGET ATRP), may displace the ligand L and form catalytically inactive halide complexes. This has been observed when ICAR ATRP of a halide-containing monomer, 4-vinylbenzyltriphenylphosphonium chloride was attempted. To achieve good polymerization control, the counterion of the monomer had to be exchanged with tetrafluoroborate.[291] The inability of $[Cu^IX_2]^-$ complexes, which are present in the system Cu^IX/Me_6TREN in acetonitrile, particularly at high halide concentrations, to activate alkyl halides has been demonstrated.[292]

To determine if ICAR ATRP can be carried out in protic media, one has to know the value of the halidophilicity for the Mt^{z+1} complex in this media. Low halidophilicity, particularly at high dilutions, would lead to significant dissociation of the halide complex $XMt^{z+1}L_m$ and therefore to poor deactivation efficiency and poor polymerization control. The halidophilicities measured in ethanol–water mixtures are higher than for methanol–water mixtures[132] and it can therefore be expected that ethanol is a more appropriate reaction medium for ICAR or ARGET ATRP than methanol.

8.6.2 Competitive Equilibria and Conditional Stability Constants

In ATRP systems, competitive complex equilibria may take place, which destabilize the catalytically active complex MtL. Side reactions may include (i) competitive complexation of the Mt center with a coordinating group-bearing reaction component, *e.g.*, monomer (M), polymer (P), solvent (S), reducing agent (RA) or its oxidation product (Ox); (ii) protonation of the ligand L, which often happens in ARGET ATRP in the presence of either acidic reducing agents (*e.g.*, ascorbic acid) or acidic oxidation products of the reducing agents (*e.g.*, gluconic acid formed in the oxidation of glucose); and (iii) competitive complexation of L with other metal ions present in the system, *e.g.*, Sn^{II} or the product of its oxidation Sn^{IV}. Some of the side reactions are presented in Figure 8.21.

In such systems, the degree of surviving catalyst MtL is governed by the analytical concentration of the complex of interest, the concentrations of all reagents participating in the side reactions, and the stability constants of all products of the side reactions. It is convenient to introduce a conditional stability constant of the complex MtL, designated as $\beta^{*Mt,L}_1$, which depends upon the extent to which each side reaction occurs. In the most general form, the conditional stability constant can be expressed by eqn (8.29), where the alpha coefficients take into account all possible competitive equilibria leading to destabilization of the complex of interest.[163,293]

$$\beta^{*Mt,L}_1 = \frac{\beta^{Mt,L}_1}{\alpha^{Mt,M}\alpha^{Mt,P}\alpha^{Mt,S}\alpha^{Mt,RA}\alpha^{Mt,Ox}\alpha^{L,H}\alpha^{L,Mt'}\alpha^{L,Mt''}}$$

$$\alpha^{Mt,Z} = 1 + \beta^{Mt,Z}_1[Z] \quad (Z = M, P, S, RA, Ox, \ldots)$$

$$\alpha^{L,H} = 1 + \beta^{L,H}_1[H^+] + \beta^{L,H}_2[H^+]^2 + \cdots \tag{8.29}$$

$$\alpha^{L,Mt^j} = 1 + \beta^{Mt^j,L}_1[Mt^j] \quad (Mt^j = Mt', Mt'')$$

If a reagent participating in a side reaction is present at a very low concentration and/or the stability of the formed product is low, the alpha coefficient will be close to unity, and the corresponding side reaction will not contribute significantly to the catalyst destabilization.

The stability constants $\beta^{Mt,M}_1$, $\beta^{Mt,P}_1$, $\beta^{Mt,S}_1$, $\beta^{Mt,RA}_1$, and $\beta^{Mt,Ox}_1$ characterize the formation of complexes between Mt and the monomer, polymer, solvent, reducing agent, and its oxidized form, respectively, $\beta^{H,L}_1$, $\beta^{H,L}_2$, ..., are the protonation constants of the ligand L (inverse of the acid dissociation

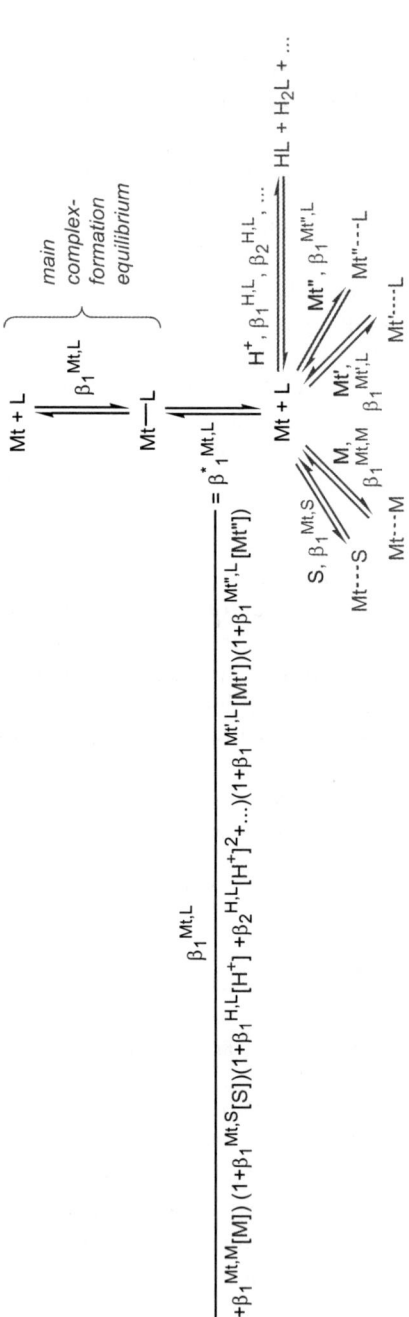

Figure 8.21   Formation of a catalytically active metal complex MtL (characterized with the stability constant $\beta_1^{Mt,L}$) and side reactions of the central metal ion (Mt) and the ligand (L) leading to destabilization of MtL. The conditional stability constant of MtL ($\beta_1^{*\,Mt,L}$), which takes into account side reactions is lower than $\beta_1^{Mt,L}$.

constants of its protonated forms), and $\beta_1^{Mt',L}$, and $\beta_1^{Mt'',L}$ are the formation constants of the complex between the ligand L and present metal ions Mt′ and Mt″, which can be a metal-based reducing agent and its oxidized form (*e.g.*, Sn^{II} and Sn^{IV}). Organic compounds with double or triple carbon–carbon bonds are known to form complexes with various transition metals, including copper.[294–297] The coordination of several monomers such as styrene, 1-octene, methyl acrylate, and methyl methacrylate to the Cu^I/PMDETA complex with non-coordinating anions was reported.[298,299] Although some of the Cu^I – olefin complexes, which have been studied and characterized, are rather stable,[300] the coordination of most monomers that do not contain special coordinating (donor) groups is relatively weak. However, when a catalytically active complex is used at low concentration, the excess of monomer relative to the ligand L (which is part of the catalyst) can become significant, and ligand displacement may take place, affecting the catalyst performance. Some monomers coordinate strongly to metal ions, and even at high catalyst concentrations, they can displace ligands from the ATRP catalyst, often forming a catalytically inactive complex. This has been observed in the Cu/bpy-mediated ATRP of 4-vinylpyridine,[148] in which good polymerization control was achieved only with strongly binding ligands.

When ATRP catalysts derived from relatively basic ligands (*e.g.*, aliphatic amines) are used, particularly in the presence of acidic compounds, *e.g.*, under ARGET ATRP conditions or when ATRP of acidic monomers is attempted, significant protonation may take place. For example, the stability of the Cu^{II} complexes of Me$_4$cyclam,[178] HMTETA,[178] Me$_6$TREN,[178] PMDETA,[176] and TPMA[177] can decrease by many (>6) orders of magnitude as the pH of the medium decreases from 6 to 1.[282] However, it is noteworthy that the Cu^{II} complex of TPMA still possesses considerable stability ($log\beta^{II,TPMA} > 5$) even at pH 1.[282] To avoid protonation of the ligand by an acidic reducing agent (*e.g.*, ascorbic acid) or its oxidation product (*e.g.*, dehydroascorbic or gluconic acid), the ligand should be used in excess in order to "trap" the acidic reaction component. Similarly, if the reducing agent employed in an ARGET ATRP process is a metal compound able to coordinate to the ligand of the catalyst (*e.g.*, Sn^{II} compound), an excess of the ligand with respect to Cu^I and Cu^{II} ought to be added to prevent decomplexation of the ligand from the Cu-based catalyst. Knowing the stability constant of a catalytically active complex, one can predict whether or not the complex is suitable for ATRP under dilute conditions, *i.e.*, for ARGET and ICAR ATRP.

8.7 Concurrent Electron Transfer Reactions in ATRP

Three important concurrent reactions involving electron transfer may occur in ATRP (Figure 8.22; E is the redox potential of the couple shown in parentheses): (i) disproportionation of the ATRP (usually Cu^I-based) activator, (ii) oxidation or reduction of organic radicals to carbocations or carbanions, respectively, and (iii) formation of organometallic species *via* a reaction between radicals and the (usually lower oxidation state) complexes catalyzing the ATRP process.[179]

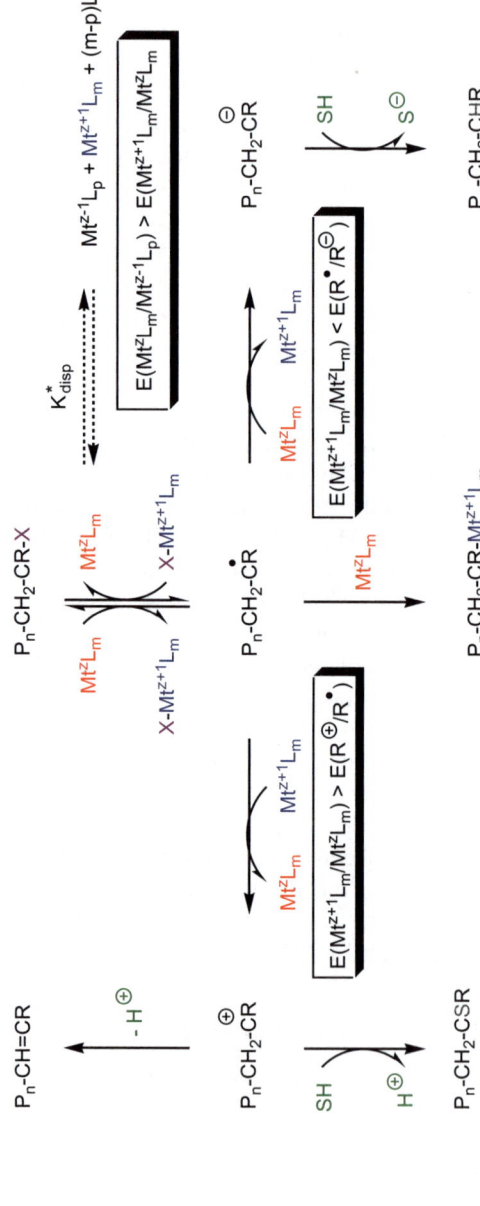

Figure 8.22 Possible redox reactions in ATRA and ATRP.

8.7.1 Disproportionation of Cu-based Activating Complexes in Aqueous Media

The compounds of Cu^I can participate in a bimolecular redox process termed *disproportionation*, which yields a Cu^{II} compound and elemental copper (eqn (8.30)). The process is reversible and is characterized by the equilibrium constant K_{disp}. The process of reduction of Cu^{II} by Cu^0 (the reverse of Cu^I disproportionation) is named *comproportionation*.

$$Cu^I + Cu^I \overset{K_{disp}}{\rightleftharpoons} Cu^{II} + Cu^0 \tag{8.30}$$

The extent to which a Cu^I compound disproportionates in a certain medium is related to both the value of K_{disp} and the concentration.[179] The equilibrium constant K_{disp} is related to the difference of the redox potentials of the couples Cu^I/Cu^0 and Cu^{II}/Cu^I, according to eqn (8.31), where R is the universal gas constant ($8.314\,J\,mol^{-1}\,K^{-1}$), T is the absolute temperature, and F is the Faraday constant ($96\,500\,C\,mol^{-1}$).

$$\log K_{disp} = \frac{E^0(Cu^I\,/\,Cu^0) - E^0(Cu^{II}\,/\,Cu^I)}{2.303\,RTF^{-1}} \tag{8.31}$$

Since the redox potentials of the two couples depend very strongly upon the nature of the medium (*i.e.*, its polarity and ability to solvate or coordinate to all involved species), the values of K_{disp} can change dramatically as the solvent is altered. This is illustrated by the data in Table 8.6. For a narrow range of solvents that do not coordinate strongly to Cu^I or Cu^{II}, the value of the disproportionation constant depends upon the dielectric constant of the medium ε (*i.e.*, $\log K_{disp}$ decreases linearly as $1/\varepsilon$ increases).[301] For solvents that coordinate strongly to copper ions such as pyridine, sulfoxides,[302] and nitriles,[303,304] the value of K_{disp} depends on their coordinating ability. In the presence of solvents that preferentially solvate and stabilize Cu^I ions compared

Table 8.6 Disproportionation equilibrium constant of Cu^I ions in various solvents.

Solvent	log $K_{disp}{}^a$	Ref.
H_2O	5.9–6.2 (0.05 – 3)	309–311
DMF	4.26	306
MeOH	3.6 (0.08)–3.8 (0.1)	312,313
AcOH	2.7 (0.11–0.26 (0.01–0.16 M $HClO_4$))	314
EtOH	0.56 (0.01–0.08)	312
Me_2SO	0.2–0.3 (0.1–1)	305,315,316
Me_2CO	−1.50	301
Pyridine	−13.97 (0.1)	317
EtCN	−19.59 (0)	318
PhCN	−19.97 (0)	318
MeCN	~ −21 (0–0.1)	318,319

^{a}The number in parentheses is the ionic strength of the medium (in mol L^{-1}).

to Cu^{II} (*e.g.*, nitriles), the disproportionation reaction is suppressed. In solvents that solvate or coordinate to Cu^{II} ions stronger than to Cu^{I} ions (*e.g.*, DMF and particularly water), on the other hand, the disproportionation becomes more prominent (*i.e.*, $K_{disp} > 1$). In mixed solvents containing water and DMSO,[305] DMF,[306] MeOH,[307] or MeCN,[308] the extent of disproportionation increases with the concentration of water.

Upon addition of a ligand L that forms complexes with Cu^{I} and Cu^{II} with different stabilities, the disproportionation process is affected and is characterized by a new equilibrium constant, K_{disp}^{*}, which can be named *conditional disproportionation constant*, similarly to the conditional stability constants discussed above and other conditional equilibrium constants. If the ligand L forms $1:1$ complexes with both Cu^{I} and Cu^{II}, the disproportionation equilibrium is expressed by eqn (8.32).

$$Cu^{I}L + Cu^{I}L \;\overset{K_{disp}^{*}}{\rightleftharpoons}\; Cu^{II}L + Cu^{0} + L \tag{8.32}$$

There are two ways to express the conditional disproportionation constant K_{disp}^{*} as a function of $K_{disp} = [Cu^{II}]/[Cu^{I}]^2$ (the constant for "bare" Cu^{I} ions in the same solvent) and the stability constants of the Cu^{I} and Cu^{II} complexes ($\beta_1^{I,L}$ and $\beta_1^{II,L}$, respectively), each of which may be more or less convenient to use in specific situations.

$$K_{disp,1}^{*} = \frac{[Cu^{II}L][L]}{[Cu^{I}L]^2} = \frac{\beta_1^{II,L}[Cu^{II}][L][L]}{(\beta_1^{I,L})^2[Cu^{I}]^2[L]^2} = \frac{\beta_1^{II,L}}{(\beta_1^{I,L})^2} K_{disp} \tag{8.33}$$

or

$$K_{disp,2}^{*} = \frac{[Cu^{II}]_{tot}}{[Cu^{I}]_{tot}^2} = \frac{[Cu^{II}] + [Cu^{II}L]}{([Cu^{I}] + [Cu^{I}L])^2} = \frac{(1 + \beta_1^{II,L}[L])[Cu^{II}]}{(1 + \beta_1^{I,L}[L])^2[Cu^{I}]^2} \approx \frac{\beta_1^{II,L}}{(\beta_1^{I,L})^2[L]} \cdot K_{disp} \tag{8.34}$$

In cases of complex stoichiometries different from 1:1, the disproportionation equilibrium is expressed by eqn (8.35) and the general expression for $K_{disp,2}^{*}$ is given by eqn (8.36).

$$Cu^{I}L_m + Cu^{I}L_m \;\overset{K_{disp}^{*}}{\rightleftharpoons}\; Cu^{II}L_m + Cu^{0} + m\,L \tag{8.35}$$

$$K_{disp,2}^{*} = \frac{[Cu^{II}]_{tot}}{[Cu^{I}]_{tot}^2} = \frac{1 + \sum_{j=1}^{m} \beta_j^{II,L}[L]^j}{\left(1 + \sum_{j=1}^{m} \beta_j^{I,L}[L]^j\right)^2} K_{disp} \tag{8.36}$$

The disproportionation constant $K_{disp,2}^{*}$, which is defined by the total concentrations of Cu^{I}- and Cu^{II}-containing species, in analogy with other conditional equilibrium constants, is ligand concentration-dependent. The addition of free ligand to a solution containing the corresponding Cu^I and Cu^{II} complexes leads to a decrease in the conditional disproportionation constant $K^*_{disp,2}$ and also to a decrease of the ratio $[Cu^{II}]_{tot}/[Cu^I]_{tot}^2$. In systems that have already reached the disproportionation–comproportionation equilibrium, the addition of ligand L will have the effect of "dissolving" part of the metal Cu^0 present and reduction of Cu^{II} with the formation of 2 equivalents of Cu^I for each equivalent of reacted Cu^{II} and Cu^0. Eqn (8.33), (8.34), and the more general eqn (8.36) can be used to predict whether a ligand is suitable for the formation of a Cu^I complex, which is stable with respect to disproportionation in a given reaction medium, provided that the stability constants of Cu^I and Cu^{II} complexes with L are known in the same medium. It should be borne in mind that the values of both β^I and β^{II} change with temperature,[281] likely to a different degree due to the different enthalpies of formation of the Cu^I and Cu^{II} complexes with the ligand L. As a result, even if the disproportionation of a Cu^I complex is negligible at ambient temperature in a given solvent, it may become appreciable upon altering the temperature.

Both the Cu-based catalyst activity, *i.e.*, K_{ATRP} and the tendency of the Cu^I complex to disproportionate (K^*_{disp}) depend upon the stability constants of the Cu^I and Cu^{II} complexes. For ligands forming 1 : 1 complexes with copper ions, K_{ATRP} is proportional to the ratio $\beta^{II,L}/\beta^{I,L}$ whereas the propensity of the Cu^I complex to disproportionate depends on the ratio $\beta^{II,L}/(\beta^{I,L})^2[L]$. A "map" of Cu-based ATRP catalysts that are both active and stable towards disproportionation can be constructed (Figure 8.23),[150,320–322] which is particularly useful for the rational selection of catalysts for aqueous ATRP reactions. The

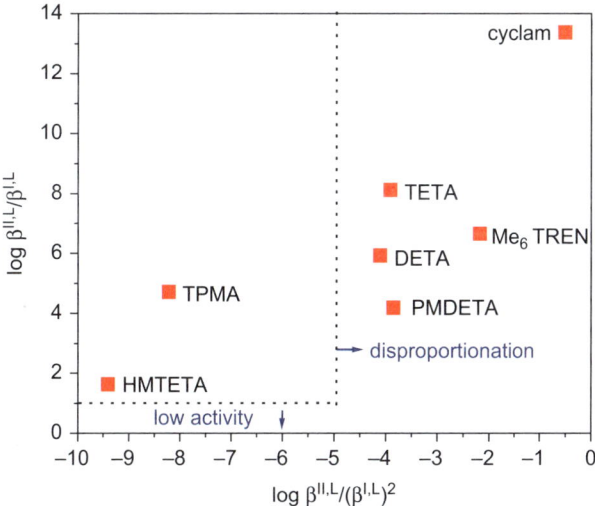

Figure 8.23 Correlation between ATRP catalytic activity and disproportionation ability for several Cu^I complexes.
Reprinted from ref. 322 with permission.

performance in aqueous solution of catalysts derived from various tri- and tetradentate N-based ligands, all forming 1:1 complexes with copper ions, is presented (TETA[176] = triethylenetetramine, DETA[176] = diethylenetriamine, cyclam[323] = 1,4,8,11-tetraazacyclotetradecane or [14]aneN$_4$). The cyclam-derived Cu^I catalyst is rather active but disproportionates easily. Both PMDETA and TPMA form catalysts that are active, but only the latter forms a Cu^I complex that does not disproportionate significantly in aqueous media. It should be remembered that K_{Halido} is also responsible for the observed polymerization rate and control, and its values may also be plotted on a separate axis to yield a three-dimensional catalyst selection map. Real polymerization systems contain monomer (with a concentration which changes throughout the polymerization), the presence of which may significantly affect the stability of the lower and the higher oxidation state metal complexes. In other words, the behavior of the catalyst expected in a pure solvent may be, and often is, very different from that observed in an ATRP reaction system. If the system contains two complex-forming compounds, say the ligand L that is part of the ATRP catalyst, and an additional coordinating compound S (monomer, solvent, additive, *etc.*), the conditional disproportionation constant is given by eqn (8.37). In fact, the catalysts used in ATRP contain coordinating halide anions (originating either from the original catalyst or the alkyl halide initiator), and eqn (8.37) should be employed to evaluate the extent of disproportionation, where S represents halide ions, and $\beta_1^{II,S}$ is actually $\beta_1^{II,X}$, *i.e.*, the halidophilicity K_{Halido}. If more than two types of coordinating compounds are present in the system, more terms should be added to eqn (8.37), which have the same form as the shown sums, but in which the concentrations of the additional compounds are used as well as the corresponding stability constants for the complexes in both oxidation states.

$$K_{disp,2}^{*} = \frac{1 + \sum_{j=1}^{m} \beta_j^{II,L}[L]^j + \sum_{k=1}^{p} \beta_k^{II,S}[S]^k}{\left(1 + \sum_{j=1}^{m} \beta_j^{I,L}[L]^j + \sum_{k=1}^{q} \beta_k^{I,S}[S]^k\right)^2} K_{disp,2} \qquad (8.37)$$

The catalyst disproportionation in water can be suppressed by using an appropriate cosolvent that stabilizes Cu^I *vs.* Cu^{II} (such as pyridine). It has been demonstrated that the use of pyridine as a cosolvent for aqueous ATRP of ionic monomers such as sodium 4-styrenesulfonate and 2-(N,N,N-trialkyl-ammonio)ethyl methacrylate salts completely suppressed the catalyst disproportionation and well-defined polyelectrolytes were obtained.[324]

In systems where Cu^0 and Cu^I can coexist due to disproportionation, activation of alkyl halides by copper in both oxidation states can take place. From a thermodynamic point of view, the relative activity of Cu^0 and Cu^I as ATRP activators can be correlated to the ability of the Cu^I complex to disproportionate.[325] The disproportionation of a Cu^I complex yields the corresponding Cu^{II} complex and Cu^0. Even if one assumes that the formed Cu^0 is so finely dispersed that it can be considered as "soluble," and therefore participating in the expressions of all equilibrium constants, the relative values of K_{ATRP} with activation from Cu^I (designated as K_{ATRP}^I) and from Cu^0 (K_{ATRP}^0), which are presented in Figure 8.24, will be given by eqn (8.38), where K_{Halido}^{II} and K_{Halido}^I are the respective halidophilicities of the Cu^{II} and Cu^I complexes.

$$\frac{K_{ATRP}^I}{K_{ATRP}^0} = \frac{K_{Halido}^{II}}{K_{Halido}^I} \frac{K_{ET}(Cu^0/Cu^IL \ldots X)}{K_{ET}(Cu^IL \ldots X/XCu^{II}L)} = \frac{K_{Halido}^{II}}{K_{Halido}^I} K_{disp}^* \qquad (8.38)$$

Cu^{II} complexes with N-based ligands, particularly in cases where four N donor atoms are attached to the metal center, are significantly (by several orders of magnitude) more halidophilic than Cu^I complexes, *i.e.*, the ratio $K_{Halido}^{II}/K_{Halido}^I$ exceeds unity. Therefore, in systems with significant dispro-portionation ($K_{disp}^* \gg 1$) where significant amounts of Cu^0 are present, acti-vation by the small amounts of present Cu^I, and not by Cu^0, will be the dominating process, even if it is assumed that Cu^0 is so finely dispersed that it is "soluble." In other words, K_{ATRP}^I will significantly exceed K_{ATRP}^0.[325]

The complexes of other transition metals used to mediate ATRP can also disproportionate. For example, Ru^{III} complexes with N-based ligands dispro-portionate to a mixture of Ru^{II} and Ru^{IV} complexes.[326–328] Disproportionation of carbonyl compounds of monovalent metals such as Fe^I,[329] Ru^I,[330] and Os^I,[330] with phosphine ligands, has also been reported.

8.7.2 Electron Transfer Between Organic Radicals and ATRP Catalysts

Radicals are redox-active species and can either be oxidized to cations or reduced to anions (Figure 8.22). Although radicals are very reactive and their electrochemical properties are not easy to study, redox potentials for both the reduction and oxidation of C-centered radicals have been reported.[331–333] The redox potentials of several small radicals that structurally resemble polymeric radicals are collected in Table 8.7. Computed (*ab-initio*) values of redox potentials of alkyl radicals and halogen atoms as well as alkyl halides used as ATRP initiators have been reported.[334] Radicals with strong electron with-drawing α-substituents such as cyano or carboxylate are oxidizing and are therefore prone to accept electrons from strong reducing agents, *e.g.*, from active ATRP activating complexes. On the other hand, radicals with electron donating α-substituents (such as those derived from vinyl ethers or vinyl-amines) can be easily oxidized to carbocations by the higher oxidation state metal complex that serves as the ATRP deactivator. Both mentioned electron transfer reactions effectively "kill" the growing radicals.

The carbanions formed by reduction of radicals $R^\bullet$, *e.g.*, propagating polymeric radicals $R_n^\bullet$, can quickly react with any protic compound (solvent or moisture) yielding RH, *e.g.*, "dead" (unable to be reactivated) polymer chains P_nH. The ratio of the rate of propagation to the rate of reduction (eqn (8.39))

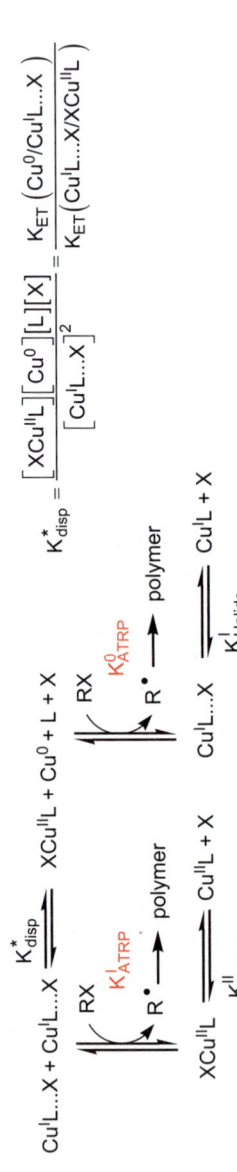

Figure 8.24 ATRP in systems where Cu^0 and Cu^I can coexist (due to disproportionation), and where both oxidation states can serve as alkyl halide activators. It is assumed that the formed higher oxidation state complexes (Cu^I or Cu^{II}, respectively) can, at least to some degree, deactivate radicals.

Table 8.7 Redox potentials of selected radicals in MeCN.

Radical	$E^{ox}_{1/2}$ (vs. SCE)/V	$E^{red}_{1/2}$ (vs. SCE)/V	Ref.
MeC$^{\bullet}$(CO$_2$Et)$_2$		-0.03^{a}–$0.45^{a,b}$	335,336
MeC$^{\bullet}$(CN)$_2$		$0.16^{a,b}$	335
NCCH$_2$$^{\bullet}$		-0.69^{c}	337
MeCH$_2$$^{\bullet}$	<0.99	-1.36–-1.10	332
PhCH$_2$$^{\bullet}$	0.73	-1.45	338
Me$_2$CH$^{\bullet}$	0.47	-1.44	332
PhC$^{\bullet}$HMe	0.37	-1.60	338
EtOC$^{\bullet}$HMe	-0.45	(-1.2)	338
Et$_2$NC$^{\bullet}$HMe	-1.12	(-2.0)	338

^{a}E$^{red}_{irr}$.
^{b}In DMSO.
^{c}In DMF.

determines the molecular weights that can be attained. Note that this ratio is inversely proportional to the concentration of the lower oxidation state activating complex, which serves as the reducing agent in the mentioned side reaction.

$$\frac{R_p}{R_{red}} = \frac{k_p[M]}{k_{red}[Mt^zL_m]} \qquad (8.39)$$

The reaction of various alkyl halides RX with the Cu^{I} complex of TPMA in acetonitrile yielded, in addition to the expected radical coupling product R$_2$, the corresponding product of substitution of the halogen atom with hydrogen, RH. The latter was formed by the reduction of the radicals R$^{\bullet}$, produced after activation of RX, to the corresponding anions R^{-}, followed by protonation by any protic impurity, *e.g.*, moisture (Figure 8.22). For instance, one of the products of the reaction of benzyl chloride with Cu^{I} complexes derived from aliphatic amine- and pyridine-based ligands was toluene,[339] which could be formed *via* the reduction of the initially formed benzyl radicals to carbanions followed by protonation. The reaction of 9-bromofluorene with the Cu^{I} complex of TPMA afforded 3% fluorene, and that of phenacyl chloride afforded 9% acetophenone, the yield of which increased to 36% when a mixture of acetonitrile and water (9:1) was used as the reaction medium instead of pure acetonitrile.[340] Further, the products of the reaction of benzyl bromide with acrylonitrile included, along with the products of addition (ATRA or oligomerization) and radical coupling, compounds that could be formed by the reduction of the electrophilic and relatively oxidizing α-cyano-radicals to anions followed by protonation, as shown in Figure 8.25.[340]

The low conversion and limited molecular weights attainable in the ATRP of acrylonitrile[341] might be attributed to reduction of the electrophilic propagating radicals with α-cyano groups. Limited conversions and molecular weights were also observed when very reducing Cu-based catalysts derived from dimethyl cross-bridged cyclam were employed in the ATRP of *n*-butyl acrylate (a monomer forming electrophilic radicals due to the presence of an α-carboxylate group).[139]

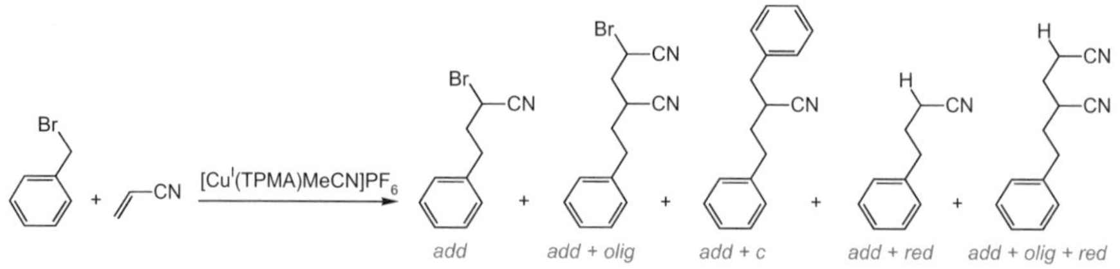

Figure 8.25 Products of the reaction between benzyl bromide and acrylonitrile (add = addition; olig = oligomerization; c = radical coupling; red = reduction).

If the propagating radicals are relatively reducing and the higher oxidation state halide complex $XMt^{z+1}L_m$ (the ATRP deactivator) or the potentially present product of its heterolytic dissociation $Mt^{z+1}L_m$ are oxidizing, oxidation of the radicals to cations may occur. This is likely to occur when weakly reducing ATRP catalysts are used to mediate the polymerization. Each cation, as shown on the left-hand side of Figure 8.22, either quickly reacts with a nucleophile (*e.g.*, halide ion or solvent, monomer, *etc.*) or eliminates a proton, yielding unsaturated compounds. In fact, the oxidation of radicals by transition metal complexes can also proceed by ligand-transfer mechanism.[342,343] Alkyl radicals have been oxidized by Cu^{II} compounds with both coordinating (carboxylate, halide, or pseudohalide) and non-coordinating or weakly coordinating (tetrafluoroborate, trifluoromethanesulfonate, or pechlorate) counterions in various solvents.[344,345] The oxidation of phenyl radicals formed in the thermal decomposition of benzoyl peroxide by various Cu^{II} compounds in acetic acid and its mixtures with acetonitrile was also studied.[346] The exact mechanism of oxidation depends upon many factors including the nature of the ligand, counterion, and the solvent.

The radical oxidation and radical reduction are chain "killing" reactions, which limits the molecular weights of polymers prepared by ATRP. As with reduction, the molecular weights that can be reached in systems where radical oxidation is possible are determined by the ratio of the rates of propagation and oxidation, which is inversely proportional to the amount of oxidant (eqn (8.40)).

$$\frac{R_p}{R_{ox}} = \frac{k_p[M]}{k_{ox}[XMt^{z+1}L_m]} \qquad (8.40)$$

It has been shown[347] that the reaction of 1-PhEtBr with the Cu^I complex of dNbpy yields styrene, which could plausibly be formed by oxidation of the radical PhC•HMe by the deactivator (Cu^{II} halide complex of dNbpy) to the cation PhC⁺HMe, followed by elimination of a proton. In the polymerization of styrene, mediated by the Cu complexes of dNbpy, the chain end functionality decreases at monomer conversions exceeding 60–70%, to a large degree due to β-elimination with the formation of polymers with terminal double bonds.[348,349] In the ATRP of styrene mediated by relatively high concentrations of Cu complexes of dNbpy, the molecular weights were limited to $<50\,000\,\text{g}\,\text{mol}^{-1}$, to a large degree due to oxidation of the nucleophilic propagating radicals by the relatively oxidizing Cu^{II} complex of dNbpy.

In order to reach high molecular weights, it is necessary to minimize the relative rates of the chain "killing" redox process relative to propagation. As shown by eqn (8.39) and (8.40), this can be accomplished by employing low-catalyst concentration techniques such as ICAR and ARGET ATRP. Indeed, higher molecular weight styrene–acrylonitrile copolymers can be prepared by ARGET ATRP compared to classical ATRP with high catalyst concentration ($\sim 200\,000$ and $\sim 50\,000$ g mol^{-1}, respectively).[274]

8.7.3 Radical Coordination to the Metal Catalyst and the Interplay of Controlled Radical Polymerization Mechanisms

One-electron oxidative addition between alkyl halides RX and reducing transition metal compounds Mt^zL_m with the formation of a higher oxidation state metal halide complex $XMt^{z+1}L_m$ and an organometallic species $RMt^{z+1}L_m$ have found many applications in organic synthesis. Examples include the Cr-catalyzed Nozaki–Hiyama–Kishi additions of organic halides to aldehydes that proceed *via* organometallic intermediates.[350] Organocopper species may be formed in the reactions of organic radicals with Cu^I and Cu^{II}.[351,352] Currently, there is no experimental evidence that suggests that organocopper(II) species $RCu^{II}L_m$ are formed in Cu-mediated ATRP.[353] However, organomolybdenum species are likely involved in polymerizations initiated by alkyl halides under ATRP conditions.[40,354] In polymerizations in the presence of transition metal compounds, the polymerization control can be either due to the reversible formation of organometallic compounds or due to the transfer of an atom (group) from a higher oxidation state metal compound to the propagating radical. As mentioned above (see Figure 8.2), both organometallic polymerization and ATRP can operate simultaneously and which process dominates depends upon the relative bond dissociation energies of the the the R–X, Mt–X, and Mt–R bonds. The identification of any $RMt^{z+1}L_m$ species generated during ATRP would have implications on optimization of the polymerization conditions, and as more and more non-Cu based ATRP catalysts are developed, such studies will become increasingly important.

References

1. D. Greszta, D. Mardare and K. Matyjaszewski, *Macromolecules*, 1994, **27**, 638.
2. K. Matyjaszewski, *J. Phys. Org. Chem.*, 1995, **8**, 197.
3. K. Matyjaszewski, S. Gaynor, D. Greszta, D. Mardare and T. Shigemoto, *J. Phys. Org. Chem.*, 1995, **8**, 306.
4. T. Otsu, M. Yoshida and T. Tazaki, *Makromol. Chem., Rapid Commun.*, 1982, **3**, 133.

5. T. Otsu and T. Tazaki, *Polym. Bull.*, 1986, **16**, 277.
6. K. Matyjaszewski (ed.), *Controlled Radical Polymerization (ACS Symp. Ser. 685)*, ACS, Washington, DC, 1998.
7. K. Matyjaszewski (ed.), *Controlled/Living Radical Polymerization. Progress in ATRP, NMP, and RAFT (ACS Symp. Ser. 768)*, ACS, Washington, DC, 2000.
8. K. Matyjaszewski and T. P. Davis, (ed.), *Handbook of Radical Polymerization*, Wiley, Hoboken, 2002.
9. K. Matyjaszewski (ed.), *Advances in Controlled/Living Radical Polymerization (ACS Symp. Ser. 854)*, ACS, Washington, DC, 2003.
10. K. Matyjaszewski (ed.), *Controlled/Living Radical Polymerization. From Synthesis to Materials (ACS Symp. Ser. 944)*, ACS, Washington, DC, 2006.
11. M. K. Georges, R. P. N. Veregin, P. M. Kazmaier and G. K. Hamer, *Macromolecules*, 1993, **26**, 2987–2988.
12. D. Benoit, V. Chaplinski, R. Braslau and C. J. Hawker, *J. Am. Chem. Soc.*, 1999, **121**, 3904.
13. D. Benoit, S. Grimaldi, J. P. Finet, P. Tordo, M. Fontanille and Y. Gnanou, *ACS Symp. Ser.*, 1998, **685**, 225.
14. C. J. Hawker, A. W. Bosman and E. Harth, *Chem. Rev.*, 2001, **101**, 3661–3688.
15. C. Le Mercier, S. Acerbis, D. Bertin, F. Chauvin, D. Gigmes, O. Guerret, M. Lansalot, S. Marque, F. Le Moigne, H. Fischer and P. Tordo, *Macromol. Symp.*, 2002, **182**, 225–247.
16. B. B. Wayland, G. Poszmik, S. L. Mukerjee and M. Fryd, *J. Am. Chem. Soc.*, 1994, **116**, 7943.
17. B. B. Wayland, L. Basickes, S. Mukerjee, M. Wei and M. Fryd, *Macromolecules*, 1997, **30**, 8109.
18. B. B. Wayland, S. Mukerjee, G. Poszmik, D. C. Woska, L. Basickes, A. A. Gridnev, M. Fryd and S. D. Ittel, *ACS Symp. Ser.*, 1998, **685**, 305–315.
19. L. E. N. Allan, M. R. Perry and M. P. Shaver, *Prog. Polym. Sci.*, 2012, **37**, 127–156.
20. J.-S. Wang and K. Matyjaszewski, *J. Am. Chem. Soc.*, 1995, **117**, 5614–5615.
21. M. Kato, M. Kamigaito, M. Sawamoto and T. Higashimura, *Macromolecules*, 1995, **28**, 1721–1723.
22. K. Matyjaszewski and J. Xia, *Chem. Rev.*, 2001, **101**, 2921–2990.
23. M. Kamigaito, T. Ando and M. Sawamoto, *Chem. Rev.*, 2001, **101**, 3689–3745.
24. A. Goto, Y. Tsujii and T. Fukuda, *Polymer*, 2008, **49**, 5177–5185.
25. K. Matyjaszewski, S. Gaynor and J.-S. Wang, *Macromolecules*, 1995, **28**, 2093.
26. G. Moad, J. Chiefari, Y. K. Chong, J. Krstina, R. T. A. Mayadunne, A. Postma, E. Rizzardo and S. H. Thang, *Polym. Int.*, 2000, **49**, 993–1001.
27. E. Rizzardo, J. Chiefari, R. Mayadunne, G. Moad and S. Thang, *Macromol. Symp.*, 2001, **174**, 209–212.

28. C. Barner-Kowollik, T. P. Davis, J. P. A. Heuts, M. H. Stenzel, P. Vana and M. Whittaker, *J. Polym. Sci.: Part A: Polym. Chem.*, 2003, **41**, 365–375.

29. S. Perrier and P. Takolpuckdee, *J. Polym. Sci.: Part A: Polym. Chem.*, 2005, **43**, 5347–5393.

30. G. Moad, Y. K. Chong, A. Postma, E. Rizzardo and S. H. Thang, *Polymer*, 2005, **46**, 8458–8468.

31. S. Yamago, *J. Polym. Sci.: Part A: Polym. Chem.*, 2005, **44**, 1.

32. R. Nicolay and Y. Kwak, *Isr. J. Chem.*, 2012, **52**, 288–305.

33. F. Minisci, *Acc. Chem. Res.*, 1975, **8**, 165.

34. D. Bellus, *Pure Appl. Chem.*, 1985, **57**, 1827–1838.

35. J. M. Munoz-Molina, T. R. Belderrain and P. J. Perez, *Eur. J. Inorg. Chem.*, 3155–3164.

36. A. J. Clark, *Chem. Soc. Rev.*, 2002, **31**, 1–11.

37. K. Matyjaszewski, *Curr. Org. Chem.*, 2002, **6**, 67.

38. J. Barton and E. Borsig, *Complexes in Free-Radical Polymerization*, Elsevier, Amsterdam, 1988.

39. R. Poli, *Angew. Chem. Int. Ed.*, 2006, **45**, 5058–5070.

40. E. Le Grognec, J. Claverie and R. Poli, *J. Am. Chem. Soc.*, 2001, **123**, 9513–9524.

41. W. A. Braunecker, Y. Itami and K. Matyjaszewski, *Macromolecules*, 2005, **38**, 9402–9404.

42. W. A. Braunecker, W. C. Brown, B. C. Morelli, W. Tang, R. Poli and K. Matyjaszewski, *Macromolecules*, 2007, **40**, 8576–8585.

43. M. Litt, *J. Polym. Sci.*, 1962, **58**, 429.

44. M. Szwarc, *Carbanions, Living Polymers, and Electron Transfer Processes*, Wiley, New York, 1968.

45. R. V. Figini, *Makromol. Chem.*, 1964, **71**, 193.

46. K. Matyjaszewski and C.-H. Lin, *Makromol. Chem., Macromol. Symp.*, 1991, **47**, 221.

47. K. Matyjaszewski, *Macromol. Symp.*, 1996, **111**, 47.

48. G. Litvinenko and A. H. E. Mueller, *Macromolecules*, 1997, **30**, 1253.

49. S. Beuermann and M. Buback, *Prog. Polym. Sci.*, 2002, **27**, 191–254.

50. C. Barner-Kowollik, M. Buback, M. Egorov, T. Fukuda, A. Goto, O. F. Olaj, G. T. Russell, P. Vanna, B. Yamada and P. B. Zetterlund, *Prog. Polym. Sci.*, 2005, **30**, 605–643.

51. M. Asscher and D. Vofsi, *Chem. Ind.*, 1962, 209–210.

52. M. Asscher, E. Levy, H. Rosin and D. Vofsi, *Ind. Eng. Chem. Prod. Res. Devel.*, 1963, **2**, 121–126.

53. C. H. Bamford and C. A. Finch, *Trans. Faraday Soc.*, 1963, **59**, 540–547.

54. C. H. Bamford, G. C. Eastmond and V. J. Robinson, *Trans. Faraday Soc.*, 1964, **60**, 751–758.

55. C. H. Bamford, G. C. Eastmond and W. R. Maltman, *Trans. Faraday Soc.*, 1965, **61**, 267–273.

56. C. H. Bamford, R. Denyer and G. C. Eastmond, *Trans. Faraday Soc.*, 1965, **61**, 1459–1469.

57. C. H. Bamford, R. Denyer and G. C. Eastmond, *Trans. Faraday Soc.*, 1966, **62**, 688–700.
58. C. H. Bamford and R. Denyer, *Trans. Faraday Soc.*, 1966, **62**, 1567–1574.
59. C. H. Bamford, G. C. Eastmond and W. R. Maltman, *Trans. Faraday Soc.*, 1966, **62**, 2531–2543.
60. C. H. Bamford, G. C. Eastmond and K. Hargreaves, *Trans. Faraday Soc.*, 1968, **64**, 175–184.
61. C. H. Bamford, G. C. Eastmond and J. A. Rippon, *Trans. Faraday Soc.*, 1963, **59**, 2548–2559.
62. T. Otsu and M. Yamaguchi, *Bull. Chem. Soc. Jpn.*, 1968, **41**, 2931–2935.
63. C. H. Bamford, in Reactivity, *Mechanism and Structure in Polymer Chemistry*, ed. A. D. Jenkins and A. Ledwith, Wiley, London, 1974, pp. 52–116.
64. J. S. Wang and K. Matyjaszewski, *Macromolecules*, 1995, **28**, 7901–7910.
65. Y. A. Kabachii, S. Y. Kochev, L. M. Bronstein, I. B. Blagodatskikh and P. M. Valetsky, *Polym. Bull.*, 2003, **50**, 271.
66. J. A. M. Brandts, P. van de Geijn, E. E. van Faassen, J. Boersma and G. Van Koten, *J. Organomet. Chem.*, 1999, **584**, 246.
67. F. Stoffelbach, J. Claverie and R. Poli, *Compt. Rend. Chim.*, 2002, **5**, 37.
68. F. Stoffelbach, D. M. Haddleton and R. Poli, *Eur. Polym. J.*, 2003, **39**, 2099.
69. Y. Kotani, M. Kamigaito and M. Sawamoto, *Macromolecules*, 1999, **32**, 2420.
70. Y. Kotani, M. Kamigaito and M. Sawamoto, *Macromolecules*, 2000, **33**, 6746.
71. H. Uegaki, Y. Kotani, M. Kamigaito and M. Sawamoto, *ACS Symp. Ser.*, 2000, **760**, 196.
72. V. Percec, B. Barboiu, A. Neumann, J. C. Ronda and M. Zhao, *Macromolecules*, 1996, **29**, 3665.
73. T. Ando, M. Kamigaito and M. Sawamoto, *Tetrahedron*, 1997, **53**, 15445.
74. F. Simal, A. Demonceau and A. F. Noels, *Angew. Chem. Int. Ed.*, 1999, **38**, 538.
75. F. Simal, A. Demonceau and A. F. Noels, *Tetrahedron Lett.*, 1999, **40**, 5689–5693.
76. F. Simal, S. Sebille, L. Hallet, A. Demonceau and A. F. Noels, *Macromol. Symp.*, 2000, **161**, 73.
77. B. De Clercq and F. Verpoort, *Tetrahedron Lett.*, 2002, **43**, 4687.
78. T. Opstal and F. Verpoort, *Angew. Chem. Int. Ed.*, 2003, **42**, 2876.
79. M. Haas, E. Solari, Q. T. Nguyen, S. Gautier, R. Scopelliti and K. Severin, *Adv. Synth. Catal.*, 2006, **348**, 439.
80. B. De Clercq and F. Verpoort, *Macromolecules*, 2002, **35**, 8943.
81. B. De Clercq and F. Verpoort, *J. Mol. Catal. A: Chem.*, 2002, **180**, 67.
82. T. Ando, M. Kamigaito and M. Sawamoto, *Macromolecules*, 1997, **30**, 4507.
83. K. Matyjaszewski, M. Wei, J. Xia and N. E. McDermott, *Macromolecules*, 1997, **30**, 8161.

84. G. Moineau, P. Dubois, R. Jerome, T. Senninger and P. Teyssie, *Macromolecules*, 1998, **31**, 545–547.

85. J. Xia, H.-j. Paik and K. Matyjaszewski, *Macromolecules*, 1999, **32**, 8310.

86. M. Teodorescu, S. G. Gaynor and K. Matyjaszewski, *Macromolecules*, 2000, **33**, 2335.

87. S. Zhu and D. Yan, *Macromolecules*, 2000, **33**, 8233.

88. S. Zhu, D. Yan, G. Zhang and M. Li, *Macromol. Chem. Phys.*, 2000, **201**, 2666.

89. V. C. Gibson, R. K. O'Reilly, W. Reed, D. F. Wass, A. J. P. White and D. J. Williams, *Chem. Commun.*, 2002, 1850.

90. V. C. Gibson, R. K. O'Reilly, D. F. Wass, A. J. P. White and D. J. Williams, *Macromolecules*, 2003, **36**, 2591.

91. G. Moineau, C. Granel, P. Dubois, R. Jerome and P. Teyssie, *Macromolecules*, 1998, **31**, 542.

92. D. Mecerreyes, G. Moineau, P. Dubois, R. Jerome, J. L. Hedrick, C. J. Hawker, E. E. Malmstrom and M. Trollsas, *Angew. Chem., Int. Ed.*, 1998, **37**, 1274–1276.

93. T. Opstal, J. Zednik, J. Sedlacek, J. Svoboda, J. Vohlidal and F. Verpoort, *Coll. Czech. Chem. Commun.*, 2002, **67**, 1858.

94. J. Vohlidal, M. Pacovska, J. Sedlacek, J. Svoboda, J. Zednik and H. Balcar, *NATO Sci. Ser., II: Math., Phys. and Chem.*, 2003, **122** , 131.

95. C. Granel, P. Dubois, R. Jerome and P. Teyssie, *Macromolecules*, 1996, **29**, 8576.

96. H. Uegaki, Y. Kotani, M. Kamigaito and M. Sawamoto, *Macromolecules*, 1997, **30**, 2249.

97. H. Uegaki, Y. Kotani, M. Kamigaito and M. Sawamoto, *Macromolecules*, 1998, **31**, 6756.

98. G. Carrot, J. Hilborn, J. L. Hedrick and M. Trollsas, *Macromolecules*, 1999, **32**, 5171–5173.

99. M. Husseman, E. E. Malmstrom, M. McNamara, M. Mate, D. Mecerreyes, D. G. Benoit, J. L. Hedrick, P. Mansky, E. Huang, T. P. Russell and C. J. Hawker, *Macromolecules*, 1999, **32**, 1424.

100. G. Moineau, M. Minet, P. Dubois, P. Teyssie, T. Senninger and R. Jerome, *Macromolecules*, 1999, **32**, 27.

101. C. Moineau, M. Minet, P. Teyssie and R. Jerome, *Macromolecules*, 1999, **32**, 8277.

102. H. Uegaki, M. Kamigaito and M. Sawamoto, *J. Polym. Sci.: Part A: Polym. Chem.*, 1999, **37**, 3003.

103. P. Lecomte, I. Drapier, P. Dubois, P. Teyssie and R. Jerome, *Macromolecules*, 1997, **30**, 7631.

104. B. Wang, Y. Zhuang, X. Luo, S. Xu and X. Zhou, *Macromolecules*, 2003, **36**, 9684.

105. V. Percec and B. Barboiu, *Macromolecules*, 1995, **28**, 7970.

106. J. Xia and K. Matyjaszewski, *Macromolecules*, 1997, **30**, 7697.

107. D. M. Haddleton, C. B. Jasieczek, M. J. Hannon and A. J. Shooter, *Macromolecules*, 1997, **30**, 2190.

108. K. Matyjaszewski, *Macromolecules*, 1998, **31**, 4710.

109. V. Percec, B. Barboiu and M. van der Sluis, *Macromolecules*, 1998, **31**, 4053.

110. D. M. Haddleton, M. C. Crossman, B. H. Dana, D. J. Duncalf, A. M. Heming, D. Kukulj and A. J. Shooter, *Macromolecules*, 1999, **32**, 2110.

111. J. Xia and K. Matyjaszewski, *Macromolecules*, 1999, **32**, 2434.

112. M. Teodorescu and K. Matyjaszewski, *Macromolecules*, 1999, **32**, 4826.

113. T. E. Patten and K. Matyjaszewski, *Acc. Chem. Res.*, 1999, **32**, 895.

114. M. Ouchi, T. Terashima and M. Sawamoto, *Chem. Rev.*, 2009, **109**, 4963–5050.

115. L. Fetzer, V. Toniazzo, D. Ruch and F. di Lena, *Isr. J. Chem.*, 2012, **52**, 221–229.

116. F. di Lena and K. Matyjaszewski, *Prog. Polym. Sci.*, 2010, **35**, 959–1021.

117. Y. Inoue and K. Matyjaszewski, *Macromolecules*, 2003, **36**, 7432.

118. J. Chen, J. Chu and K. Zhang, *Polymer*, 2004, **45**, 151.

119. C. Jian, J. Chen and K. Zhang, *J. Polym. Sci.: Part A: Polym. Chem.*, 2005, **43**, 2625.

120. K. Ohno, A. Goto, T. Fukuda, J. Xia and K. Matyjaszewski, *Macromolecules*, 1998, **31**, 2699.

121. A. Goto and T. Fukuda, *Macromol. Rapid Commun.*, 1999, **20**, 633.

122. G. Chambard, B. Klumperman and A. L. German, *Macromolecules*, 2000, **33**, 4417.

123. K. Matyjaszewski, H.-j. Paik, P. Zhou and S. J. Diamanti, *Macromolecules*, 2001, **34**, 5125–5131.

124. K. Matyjaszewski, B. Goebelt, H.-j. Paik and C. P. Horwitz, *Macromolecules*, 2001, **34**, 430.

125. M. A. J. Schellekens, F. de Wit and B. Klumperman, *Macromolecules*, 2001, **34**, 7961.

126. A. K. Nanda and K. Matyjaszewski, *Macromolecules*, 2003, **36**, 599.

127. A. K. Nanda and K. Matyjaszewski, *Macromolecules*, 2003, **36**, 1487–1493.

128. W. Tang, A. K. Nanda and K. Matyjaszewski, *Macromol. Chem. Phys.*, 2005, **206**, 1171.

129. V. Percec, H.-J. Kim and B. Barboiu, *Macromolecules*, 1997, **30**, 6702–6705.

130. B. Barboiu and V. Percec, *Macromolecules*, 2001, **34**, 8626.

131. N. V. Tsarevsky and W. Jakubowski, *J. Polym. Sci.: Part A: Polym. Chem.*, 2011, **49**, 918–925.

132. S. R. Woodruff, B. J. Davis and N. V. Tsarevsky, *ACS Symp. Ser.*, 2012, **1100**, 99–113.

133. A. K. Nanda and K. Matyjaszewski, *Macromolecules*, 2003, **36**, 8222–8224.

134. C. Y. Lin, M. L. Coote, A. Petit, P. Richard, R. Poli and K. Matyjaszewski, *Macromolecules*, 2007, **40**, 5985–5994.

135. W. Tang and K. Matyjaszewski, *Macromolecules*, 2007, **40**, 1858–1863.
136. W. Tang and K. Matyjaszewski, *Macromolecules*, 2006, **39**, 4953.
137. A. E. Martell and R. D. Hancock, *Metal Complexes in Aqueous Solutions*, Plenum Press, New York, 1995.
138. G. Golub, H. Cohen, P. Paoletti, A. Bencini, L. Messori, I. Bertini and D. Meyerstein, *J. Am. Chem. Soc.*, 1995, **117**, 8353.
139. N. V. Tsarevsky, W. A. Braunecker, W. Tang, S. J. Brooks, K. Matyjaszewski, G. R. Weisman and E. H. Wong, *J. Mol. Catal. A: Chem.*, 2006, **257**, 132–140.
140. K. Matyjaszewski, *Macromolecules*, 2012, **45**, 4015–4039.
141. J. Qiu, K. Matyjaszewski, L. Thouin and C. Amatore, *Macromol. Chem. Phys.*, 2000, **201**, 1625–1631.
142. H. Fischer, *J. Polym. Sci.: Part A: Polym. Chem.*, 1999, **37**, 1885–1901.
143. H. Fischer, *Chem. Rev.*, 2001, **101**, 3581–3610.
144. A. Goto and T. Fukuda, *Prog. Polym. Sci.*, 2004, **29**, 329–385.
145. T. Pintauer, B. McKenzie and K. Matyjaszewski, *ACS Symp. Ser.*, 2003, **854**, 130.
146. W. Tang, N. V. Tsarevsky and K. Matyjaszewski, *J. Am. Chem. Soc.*, 2006, **128**, 1598–1604.
147. H. Tang, N. Arulsamy, J. Sun, M. Radosz, Y. Shen, N. V. Tsarevsky, W. A. Braunecker, W. Tang and K. Matyjaszewski, *J. Am. Chem. Soc.*, 2006, **128**, 16277–16285.
148. N. V. Tsarevsky, W. A. Braunecker, S. J. Brooks and K. Matyjaszewski, *Macromolecules*, 2006, **39**, 6817–6824.
149. W. Jakubowski, N. V. Tsarevsky, T. Higashihara, R. Faust and K. Matyjaszewski, *Macromolecules*, 2008, **41**, 2318–2323.
150. N. V. Tsarevsky, W. Tang, S. J. Brooks and K. Matyjaszewski, *ACS Symp. Ser.*, 2006, **944**, 56–70.
151. W. Tang, Y. Kwak, W. Braunecker, N. V. Tsarevsky, M. L. Coote and K. Matyjaszewski, *J. Am. Chem. Soc.*, 2008, **130**, 10702–10713.
152. W. A. Braunecker, N. V. Tsarevsky, A. Gennaro and K. Matyjaszewski, *Macromolecules*, 2009, **42**, 6348–6360.
153. H.-H. Schuh and H. Fischer, *Helv. Chim. Acta*, 1978, **61**, 2130–2164.
154. J. H. Gorrell and J. T. Dubois, *Trans. Faraday Soc.*, 1967, **63**, 347.
155. N. V. Tsarevsky, W. A. Braunecker, W. Tang and K. Matyjaszewski, *ACS Symp. Ser.*, 2009, **1023**, 85–96.
156. M. B. Gillies, K. Matyjaszewski, P.-O. Norrby, T. Pintauer, R. Poli and P. Richard, *Macromolecules*, 2003, **36**, 8551–8559.
157. W. Tang, Y. Kwak, N. V. Tsarevsky and K. Matyjaszewski, *Polym. Prep. (Am. Chem. Soc., Div. Polym.Chem.)*, 2007, **48**, 392–393.
158. R. S. Berry, C. W. Reimann and G. N. Spokes, *J. Chem. Phys.*, 1962, **37**, 2278.
159. P. Vanysek, in *Handbook of Chemistry and Physics*, CRC Press, Boca Raton, 1994.
160. G. Schwarzenbach, *Die Komplexometrische Titration*, 2nd edn, Enke, Stuttgart, 1956.

161. L. G. Sillen, in *Treatise on Analytical Chemistry: Part I: Theory and Practice*, ed. I. M. Kolthoff and P. J. Elving, Interscience Encyclopedia, New York, 1959, vol. 1, pp. 277–317.

162. F. J. C. Rossotti and H. Rossotti, *The Determination of Stability Constants*, McGraw Hill, New York, 1961.

163. A. Ringbom, *Complexation in Analytical Chemistry*, Interscience, New York, 1963.

164. J. N. Butler, *Ionic Equilibrium: A Mathematical Approach*, Addison–Wesley, Reading, 1964.

165. M. T. Beck, *Chemistry of Complex Equilibria*, Van Nostrand Reinhold, London, 1970.

166. E. Hoegfeldt, in *Treatise on Analytical Chemistry: Part I: Theory and Practice*, eds. I. M. Kolthoff and P. J. Elving, Wiley, New York, 1979, vol. 2, pp. 1–61.

167. J. J. Lingane, *Chem. Rev.*, 1941, **29**, 1.

168. F. J. C. Rossotti and H. Rossotti, in *The Determination of Stability Constants*, McGraw–Hill, New York, Editon edn., 1961, pp. 127–170.

169. F. J. C. Rossotti and H. Rossotti, in *The Determination of Stabiltiy Constants*, McGraw–Hill, New York, Editon edn., 1961, pp. 171–202.

170. A. A. Vlcek, *Prog. Inorg. Chem.*, 1963, **5**, 211.

171. D. A. Buckingham and A. M. Sargeson, in *Chelating Agents and Metal Chelates,* eds. F. P. Dwyer and D. P. Mellor, Academic press, New York, Editon edn., 1964, pp. 237–282.

172. D. R. Crow and J. V. Westwood, *Quart. Rev. (London)*, 1965, **19**, 57.

173. R. Tamamushi and G. P. Sato, *Prog. Polarography*, 1972, **3**, 1.

174. N. Bortolamei, A. A. Isse, V. B. Di Marco, A. Gennaro and K. Matyjaszewski, *Macromolecules*, 2010, **43**, 9257–9267.

175. *Stability Constants of Metal-Ion Complexes. Supplement No 1. Special Publication No 25*, The Chemical Society, London, 1971.

176. N. Navon, G. Golub, H. Cohen, P. Paoletti, B. Valtancoli, A. Bencini and D. Meyerstein, *Inorg. Chem.*, 1999, **38**, 3484.

177. E. A. Ambundo, M.-V. Deydier, A. J. Grall, N. Aguera-Vega, L. T. Dressel, T. H. Cooper, M. J. Heeg, L. A. Ochrymowycz and D. B. Rorabacher, *Inorg. Chem.*, 1999, **38**, 4233–4242.

178. G. Golub, A. Lashaz, H. Cohen, P. Paoletti, A. Bencini, B. Valtancoli and D. Meyerstein, *Inorg. Chim. Acta*, 1997, **255**, 111–115.

179. N. V. Tsarevsky, W. A. Braunecker and K. Matyaszewski, *J. Organomet. Chem.*, 2007, **692**, 3212–3222.

180. H. Irving, in *Advances in Polarography*, ed. I. S. Longmuir, Pergamon Press, New York, Editon edn., 1960, vol. 1, pp. 42–67.

181. H. L. Schlaefer, *Komplexbildung in Losung*, Springer, Berlin, 1961.

182. S. Fronaeus, *Techn. Inorg. Chem.*, 1963, **1**, 1.

183. J. Biernat, in *Theory and Structure of Complex Compounds*, ed. B. Jezowska-Trzebiatowska, Pergamon, Oxford, Editon edn., 1964, pp. 627–636.

184. F. R. Hartley, C. Burgess and R. M. Alcock, *Solution Equilibria*, Ellis Horwood, Chichester, 1980.

185. J. Bjerrum, *Metal Ammine Formation in Aqueous Solution. Theory of the Reversible Step Reactions*, P. Haase and Son, Copenhagen, 1957.

186. A. E. Martell and R. J. Motekaitis, *The Determination and Use of Stability Constants*, VCH, New York, 1988.

187. W. A. E. McBryde, *Talanta*, 1974, **21**, 979.

188. L. Fielding, *Tetrahedron*, 2000, **56**, 6151–6170.

189. P. Thordarson, *Chem. Soc. Rev.*, 2011, **40**, 1305–1323.

190. J. J. Christensen, J. Ruckman, D. J. Eatough and R. M. Izatt, *Thermochim. Acta*, 1972, **3**, 203.

191. D. J. Eatough, J. J. Christensen and R. M. Izatt, *Thermochim. Acta*, 1972, **3**, 219.

192. D. J. Eatough, R. M. Izatt and J. J. Christensen, *Thermochim. Acta*, 1972, **3**, 233.

193. S. Ishiguro, L. Nagy and H. Ohtaki, *Bull. Chem. Soc. Jpn.*, 1987, **60**, 2053.

194. A. B. P. Lever, *Inorg. Chem.*, 1990, **29**, 1271.

195. A. B. P. Lever, in *Comprehensive Coordination Chemistry II*, eds. J. A. McCleverty and T. J. Meyer, Elsevier, Amsterdam, Editon edn., 2004, vol. 2, pp. 251–268.

196. T. Pintauer and K. Matyjaszewski, *Coord. Chem. Rev.*, 2005, **249**, 1155.

197. P. Zanello, in *Stereochemical Control, Bonding and Steric Rearrangements*, ed. I. Bernal, Elsevier, Amsterdam, Editon edn., 1990, vol. 4, pp. 181–366.

198. C. X. Zhang, S. Kaderli, M. Costas, E.-i. Kim, Y.-M. Neuhold, K. D. Karlin and A. D. Zuberbuehler, *Inorg. Chem.*, 2003, **42**, 1807.

199. C. X. Zhang, H.-C. Liang, E.-i. Kim, J. Shearer, M. E. Helton, E. Kim, S. Kaderli, C. D. Incarvito, A. D. Zuberbuehler, A. L. Rheingold and K. D. Karlin, *J. Am. Chem. Soc.*, 2003, **125**, 634.

200. L. Q. Hatcher, M. A. Vance, A. A. N. Sarjeant, E. I. Solomon and K. D. Karlin, *Inorg. Chem.*, 2006, **45**, 3004.

201. A. J. D. Magenau, Y. Kwak, K. Schroeder and K. Matyjaszewski, *Macro Lett.*, 2012, **1**, 508–512.

202. S. Chandra, S. Thakur and S. Thakur, *Trans. Metal Chem.*, 2004, **29**, 925.

203. R. Xifra, X. Ribas, A. Llobet, A. Poater, M. Duran, M. Sola, T. D. P. Stack, J. Benet-Buchholz, B. Donnadieu, J. Mahia and T. Parella, *Chem. Eur. J.*, 2005, **11**, 5146.

204. M. Schatz, M. Becker, F. Thaler, F. Hampel, S. Schindler, R. R. Jacobson, Z. Tyeklar, N. N. Murthy, P. Ghosh, Q. Chen, J. Zubieta and K. D. Karlin, *Inorg. Chem.*, 2001, **40**, 2312.

205. A. W. Addison, *Inorg. Chim. Acta*, 1989, **162**, 217.

206. A. E. Martell and R. M. Smith, *Critical Stability Constants, Vol. 3: Other Organic Ligands*, Plenum, New York, 1977.

207. P. Zanello, *Comments Mod. Chem., Part A, Inorg. Chem.*, 1988, **8**, 45.

208. M. K. Taylor, J. Reglinski, L. E. A. Berlouis and A. R. Kennedy, *Inorg. Chim. Acta*, 2006, **359**, 2455.

209. S. Zolezzi, E. Spodine and A. Decinti, *Polyhedron*, 2002, **21**, 55.
210. P. Zanello, *Inorganic Electrochemistry: Theory, Practice and Application*, RSC, Cambridge, 2003.
211. A. Richel, A. Demonceau and A. F. Noels, *Tetrahedron Lett.*, 2006, **47**, 2077.
212. R. K. O'Reilly, V. C. Gibson, A. J. P. White and D. J. Williams, *Polyhedron*, 2004, **23**, 2921.
213. V. C. Gibson, R. K. O'Reilly, D. F. Wass, A. J. P. White and D. J. Williams, *Dalton Trans.*, 2003, 2824.
214. M. P. Shaver, L. E. N. Allan, H. S. Rzepa and V. C. Gibson, *Angew. Chem., Int. Ed.*, 2006, **45**, 1241.
215. N. V. Tsarevsky, T. Pintauer and K. Matyjaszewski, *Macromolecules*, 2004, **37**, 9768–9778.
216. M. J. Kamlet, J. L. M. Abboud, M. H. Abraham and R. W. Taft, *J. Org. Chem.*, 1983, **48**, 2877–2887.
217. Y. Marcus, M. J. Kamlet and R. W. Taft, *J. Phys. Chem.*, 1988, **92**, 3613–3622.
218. G. Coullerez, E. Malmstrom and M. Jonsson, *J. Phys. Chem. A*, 2006, **110**, 10355–10360.
219. N. Bortolamei, A. A. Isse, A. J. D. Magenau, A. Gennaro and K. Matyjaszewski, *Angew. Chem., Int. Ed.*, 2011, **50**, 11391–11394.
220. C. H. Bamford, A. D. Jenkins and R. Johnston, *Proc. Roy. Soc. London, Ser. A*, 1957, **239**, 214–229.
221. C. H. Bamford, A. D. Jenkins and R. Johnston, *Trans. Faraday Soc*, 1962, **58**, 1212–1225.
222. W. I. Bengough and W. H. Fairservice, *Trans. Faraday Soc.*, 1965, **61**, 1206–1215.
223. W. I. Bengough and W. H. Fairservice, *Trans. Faraday Soc.*, 1967, **63**, 382–391.
224. W. I. Bengough and T. O'Neill, *Trans. Faraday Soc.*, 1968, **64**, 1014–1021.
225. W. I. Bengough and T. O'Neill, *Trans. Faraday Soc.*, 1968, **64**, 2415–2422.
226. W. I. Bengough and W. H. Fairservice, *Trans. Faraday Soc.*, 1971, **67**, 414–419.
227. D. Greszta and K. Matyjaszewski, *Macromolecules*, 1996, **29**, 7661.
228. J. Gromada and K. Matyjaszewski, *Macromolecules*, 2002, **35**, 6167.
229. G. Chambard, B. Klumperman and A. L. German, *Macromolecules*, 2002, **35**, 3420.
230. N. Soerensen, J. Barth, M. Buback, J. Morick, H. Schroeder and K. Matyjaszewski, *Macromolecules*, 2012, **45**, 3797–2801.
231. J. Pietrasik and N. V. Tsarevsky, *Eur. Polym. J.*, 2010, **46**, 2333–2340.
232. K. Matyjaszewski, D. A. Shipp, J.-L. Wang, T. Grimaud and T. E. Patten, *Macromolecules*, 1998, **31**, 6836.
233. D. A. Shipp, J.-L. Wang and K. Matyjaszewski, *Macromolecules*, 1998, **31**, 8005.

234. N. V. Tsarevsky, B. M. Cooper, O. J. Wojtyna, N. M. Jahed, H. Gao and K. Matyjaszewski, *Polym. Prepr.*, 2005, **46**(2), 249.

235. C.-H. Peng, J. Kong, F. Seeliger and K. Matyjaszewski, *Macromolecules*, 2011, **44**, 7546–7557.

236. J. Xia and K. Matyjaszewski, *Macromolecules*, 1997, **30**, 7692–7696.

237. J. Xia and K. Matyjaszewski, *Macromolecules*, 1999, **32**, 5199.

238. K. Matyjaszewski, J. Qiu, N. V. Tsarevsky and B. Charleux, *J. Polym. Sci.: Part A: Polym. Chem.*, 2000, **38**, 4724–4734.

239. M. Li and K. Matyjaszewski, *Macromolecules*, 2003, **36**, 6028.

240. H. Ma, X. Wan, X. Chen and Q.-F. Zhou, *J. Polym. Sci.: Part A: Polym. Chem.*, 2002, **41**, 143.

241. J. Gromada and K. Matyjaszewski, *Macromolecules*, 2001, **34**, 7664.

242. M. Li, K. Min and K. Matyjaszewski, *Macromolecules*, 2004, **37**, 2106.

243. M. Li, N. M. Jahed, K. Min and K. Matyjaszewski, *Macromolecules*, 2004, **37**, 2434.

244. W. Jakubowski and K. Matyjaszewski, *Macromolecules*, 2005, **38**, 4139.

245. K. Min, H. Gao and K. Matyjaszewski, *J. Am. Chem. Soc.*, 2005, **127**, 3825.

246. K. Matyjaszewski, S. Coca, S. G. Gaynor, M. Wei and B. E. Woodworth, *Macromolecules*, 1997, **30**, 7348.

247. A. de Vries, B. Klumperman, D. de Wet-Roos and R. D. Sanderson, *Macromol. Chem. Phys.*, 2001, **202**, 1645.

248. Y. Gnanou and G. Hizal, *J. Polym. Sci.: Part A: Polym. Chem.*, 2004, **42**, 351.

249. J. K. Oh and K. Matyjaszewski, *J. Polym. Sci.: Part A: Polym. Chem.*, 2006, **44**, 3787.

250. K. Min, H. Gao and K. Matyjaszewski, *J. Am. Chem. Soc.*, 2006, **128**, 10521.

251. G. Hizal, U. Tunca, S. Aras and H. Mert, *J. Polym. Sci.: Part A: Polym. Chem*, 2005, **44**, 77.

252. H. Mert, U. Tunca and G. Hizal, *J. Polym. Sci., Part A: Polym. Chem.*, 2006, **44**, 5923–5932.

253. K. Min, W. Jakubowski and K. Matyjaszewski, *Macromol. Rapid Commun.*, 2006, **27**, 594.

254. T. Pintauer and K. Matyjaszewski, *Chem. Soc. Rev.*, 2008, **37**, 1087–1097.

255. W. Jakubowski and K. Matyjaszewski, *Angew. Chem., Int. Ed.*, 2006, **45**, 4482–4486.

256. W. Jakubowski, K. Min and K. Matyjaszewski, *Macromolecules*, 2006, **39**, 39–45.

257. K. Matyjaszewski, W. Jakubowski, K. Min, W. Tang, J. Huang, W. A. Braunecker and N. V. Tsarevsky, *Proc. Natl. Acad. Sci. USA*, 2006, **103**, 15309–15314.

258. W. Jakubowski, *ACS Symp. Ser.*, 2012, **1100**, 203–216.

259. L. Mueller, W. Jakubowski, W. Tang and K. Matyjaszewski, *Macromolecules*, 2007, **40**, 6464–6472.
260. J. Kalal, F. Svec and V. Marousek, *J. Polym. Sci., Polym. Symp.*, 1974, **47**, 155–166.
261. G. Li, X. Zhu, J. Zhu, Z. Cheng and W. Zhang, *Polymer*, 2005, **46**, 12716–12721.
262. P. McCarthy, N. V. Tsarevsky, L. Bombalski, K. Matyjaszewski and C. Pohl, *ACS Symp. Ser.*, 2006, **944**, 252–268.
263. N. V. Tsarevsky, S. A. Bencherif and K. Matyjaszewski, *Macromolecules*, 2007, **40**, 4439–4445.
264. P. McCarthy, M. Chattopadhyay, G. L. Millhauser, N. V. Tsarevsky, L. Bombalski, K. Matyjaszewski, D. Shimmin, N. Avdalovic and C. Pohl, *Anal. Biochem.*, 2007, **366**, 1–8.
265. A. J. D. Magenau, N. C. Strandwitz, A. Gennaro and K. Matyjaszewski, *Science*, 2011, **332**, 81–84.
266. M. N. C. Balili and T. Pintauer, *Dalton Trans.*, 2011, **40**, 3060–3066.
267. W. T. Eckenhoff and T. Pintauer, *Dalton Trans.*, 2011, **40**, 4909–4917.
268. M. J. W. Taylor, W. T. Eckenhoff and T. Pintauer, *Dalton Trans.*, 2010, **39**, 11475–11482.
269. T. Pintauer, *ACS Symp. Ser.*, 2009, **1023**, 63–84.
270. W. T. Eckenhoff and T. Pintauer, *Catal. Rev. - Sci. Eng.*, 2010, **52**, 1–59.
271. T. Pintauer, *Eur. J. Inorg. Chem.*, 2010, 2449–2460.
272. K. Matyjaszewski, H. Dong, W. Jakubowski, J. Pietrasik and A. Kusumo, *Langmuir*, 2007, **23**, 4528–4531.
273. W. Jakubowski, B. Kirci-Denizli, R. R. Gil and K. Matyjaszewski, *Macromol. Chem. Phys.*, 2008, **209**, 32–39.
274. J. Pietrasik, H. Dong and K. Matyjaszewski, *Macromolecules*, 2006, **39**, 6384–6390.
275. H. Dong, W. Tang and K. Matyaszewski, *Macromolecules*, 2007, **40**, 2974–2977.
276. J. Listak, W. Jakubowski, L. Mueller, A. Plichta, K. Matyjaszewski and M. R. Bockstaller, *Macromolecules*, 2008, **41**, 2974–2977.
277. W. Jakubowski, N. V. Tsarevsky and P. McCarthy, *ACS Symp. Ser.*, 2009, **1023**, 343–355.
278. J. Brandrup, E. H. Immergut and E. A. Grulke, eds., *Polymer Handbook*, 4th edn., Wiley, New York, 1999.
279. G. Moad and D. H. Solomon, *The Chemistry of Radical Polymerization*, 2nd edn., Elsevier, Amsterdam, 2006.
280. E. T. Denisov, T. G. Denisova and T. S. Pokidova, *Handbook of Free Radical Initiators*, Wiley, Hoboken, 2003.
281. P. Paoletti, L. Fabbrizzi and R. Barbucci, *Inorg. Chim. Acta Rev.*, 1973, **7**, 43.
282. N. V. Tsarevsky, W. A. Braunecker, A. Vacca, P. Gans and K. Matyaszewski, *Macromol. Symp.*, 2007, **248**, 60–70.
283. P. Paoletti and M. Ciampolini, *Inorg. Chem.*, 1967, **6**, 64.
284. P. Paoletti, M. Ciampolini and L. Sacconi, *J. Chem. Soc.*, 1963, 3589.

285. F. Wenk and G. Anderegg, *Chimia*, 1970, **24**, 427.

286. G. Anderegg, E. Hubmann, N. G. Podder and F. Wenk, *Helv. Chim. Acta*, 1977, **60**, 123.

287. V. F. Gromov and P. M. Khomikovskii, *Russ. Chem. Rev.*, 1979, **48**, 1040.

288. V. F. Gromov, E. V. Bune and E. N. Teleshov, *Russ. Chem. Rev.*, 1994, **63**, 507.

289. S. Perrier, S. P. Armes, X. S. Wang, F. Malet and D. M. Haddleton, *J. Polym. Sci.: Part A: Polym. Chem.*, 2001, **39**, 1696–1707.

290. D. Paneva, L. Mespouille, N. Manolova, P. Degee, I. Rashkov and P. Dubois, *Macromol. Rapid Commun.*, 2006, **27**, 1489–1494.

291. Y. Borguet and N. V. Tsarevsky, *Polym. Chem.*, 2012, **3**, 2487.

292. P. De Paoli, A. A. Isse, N. Bortolamei and A. Gennaro, *Chem. Commun.*, 2011, **47**, 3580–3582.

293. A. Ringbom and E. Still, *Anal. Chim. Acta*, 1972, **59**, 143.

294. M. A. Bennett, *Chem. Rev.*, 1962, **62**, 611.

295. R. Jones, *Chem. Rev.*, 1968, **68**, 785.

296. F. R. Hartley, *Chem. Rev.*, 1973, **73**, 163.

297. S. D. Ittel and J. A. Ibers, *Adv. Organomet. Chem.*, 1976, **14**, 33.

298. W. A. Braunecker, T. Pintauer, N. V. Tsarevsky, G. Kickelbick and K. Matyjaszewski, *J. Organometal. Chem.*, 2005, **690**, 916–924.

299. W. A. Braunecker, N. V. Tsarevsky, T. Pintauer, R. R. Gil and K. Matyjaszewski, *Macromolecules*, 2005, **38**, 4081–4088.

300. Q. Ye, X.-S. Wang, H. Zhao and R.-G. Xiong, *Chem. Soc. Rev.*, 2005, **34**, 208.

301. D. Datta, *Ind. J. Chem.*, 1987, **26A**, 605.

302. M. Calligaris and O. Carugo, *Coord. Chem. Rev.*, 1996, **153**, 83.

303. R. A. Walton, *Quart. Rev. Chem. Soc.*, 1965, **19**, 126.

304. B. N. Storhoff and H. C. Lewis, jr., *Coord. Chem. Rev.*, 1977, **23**, 1.

305. J. Malyszko and M. Scendo, *Monatsh. Chem.*, 1987, **118**, 435.

306. J. Malyszko and M. Scendo, *J. Electroanal. Chem.*, 1989, **269**, 113.

307. L. A. Yanov and A. I. Molodov, *Elektrokhimiia*, 1975, **11**, 1112.

308. P. Singh, I. D. MacLeod and A. J. Parker, *J. Solution Chem.*, 1982, **11**, 495.

309. F. Fenwick, *J. Am. Chem. Soc.*, 1926, **48**, 860.

310. G. Anderegg, *Helv. Chim. Acta*, 1963, **46**, 2397.

311. L. Ciavatta, D. Ferri and R. Palombari, *J. Inorg. Nucl. Chem.*, 1980, **42**, 593.

312. J. E. B. Randles, *J. Chem. Soc.*, 1941, 802.

313. J. P. Desmarquest, C. Trinh-Dinh and O. Bloch, *J. Electroanal. Chem.*, 1970, **27**, 101.

314. A. I. Molodov, L. A. Yanov and D. V. Golodnitskaya, *Elektrokhimiia*, 1977, **13**, 300.

315. A. Foll, M. Le Demezet and J. Courtot-Coupez, *J. Electroanal. Chem.*, 1972, **35**, 41.

316. S. Ahrland, P. Blauenstein, B. Tagesson and D. Tuhtar, *Acta Chem. Scand. A*, 1980, **34**, 265.

317. S. Ahrland, S. Ishiguro and I. Persson, *Acta Chem. Scand. A*, 1986, **40**, 418.
318. A. Lewandowski and J. Malinska, *Electrochim. Acta*, 1989, **34**, 333.
319. S. Ahrland, K. Nilsson and B. Tagesson, *Acta Chem. Scand. A*, 1983, **37**, 193.
320. N. V. Tsarevsky and K. Matyjaszewski, *J. Polym. Sci.: Part A: Polym. Chem.*, 2006, **44**, 5098–5112.
321. N. V. Tsarevsky and K. Matyjaszewski, *ACS Symp. Ser.*, 2006, **937**, 79–94.
322. N. V. Tsarevsky and K. Matyjaszewski, *Chem. Rev.*, 2007, **107**, 2270–2299.
323. M. M. Bernardo, M. J. Heeg, R. R. Schroeder, L. A. Ochrymowycz and D. B. Rorabacher, *Inorg. Chem.*, 1992, **31**, 191.
324. N. V. Tsarevsky, T. Pintauer and K. Matyjaszewski, *Polym. Prepr.*, 2002, **43**(2), 203.
325. N. V. Tsarevsky, *Isr. J. Chem.*, 2012, **52**, 276–287.
326. D. P. Rudd and H. Taube, *Inorg. Chem.*, 1971, **10**, 1543.
327. R. A. Binstead, L. K. Stultz and T. J. Meyer, *Inorg. Chem.*, 1995, **34**, 546.
328. E. Masllorens, M. Rodriguez, I. Romero, A. Roglans, T. Parella, J. Benet-Buchholz, M. Poyatos and A. Llobet, *J. Am. Chem. Soc.*, 2006, **128**, 5306.
329. M. J. Therien, C. L. Ni, F. C. Anson, J. G. Osteryoung and W. C. Trogler, *J. Am. Chem. Soc.*, 1986, **108**, 4037.
330. L. Song and W. C. Trogler, *J. Am. Chem. Soc.*, 1992, **114**, 3355.
331. D. Griller and D. D. M. Wayner, *Pure & Appl. Chem.*, 1989, **61**, 717.
332. D. D. M. Wayner and A. Houmam, *Acta Chem. Scand.*, 1998, **52**, 377.
333. K. Daasbjerg, S. U. Pedersen and H. Lund, in *General Aspects of the Chemistry of Radicals*, ed. Z. B. Alfassi, Wiley, Chichester, Editon edn., 1999, pp. 385–427.
334. C. Y. Lin, M. L. Coote, A. Gennaro and K. Matyaszewski, *J. Am. Chem. Soc.*, 2008, **130**, 12762–12774.
335. J. M. Kern and P. Federlin, *Tetrahedron*, 1978, **34**, 661.
336. M. E. Niyazymbetov, Z. Rongfeng and D. H. Evans, *J. Chem. Soc., Perkin Trans.*, 1996, **2**, 1957.
337. A. A. Isse and A. Gennaro, *J. Phys. Chem. A*, 2004, **108**, 4180.
338. D. D. M. Wayner, D. J. McPhee and D. Griller, *J. Am. Chem. Soc.*, 1988, **110**, 132.
339. K. Wada and H. Hashimoto, *Bull. Chem. Soc. Jpn.*, 1968, **41**, 3001.
340. R. R. Jacobson, Z. Tyeklar and K. D. Karlin, *Inorg. Chim. Acta*, 1991, **181**, 111.
341. K. Matyjaszewski, S. M. Jo, H.-j. Paik and S. G. Gaynor, *Macromolecules*, 1997, **30**, 6398.
342. J. K. Kochi, *Science*, 1967, **155**, 415.
343. J. K. Kochi, *Acc. Chem. Res.*, 1974, **7**, 351.
344. C. L. Jenkins and J. K. Kochi, *J. Am. Chem. Soc.*, 1972, **94**, 843.
345. C. L. Jenkins and J. K. Kochi, *J. Am. Chem. Soc.*, 1972, **94**, 856.
346. K. Wada, J. Yamashita and H. Hashimoto, *Bull. Chem. Soc. Jpn.*, 1967, **40**, 2410.

347. K. Matyjaszewski, K. Davis, T. E. Patten and M. Wei, *Tetrahedron*, 1997, **53**, 15321–15329.

348. J.-F. Lutz and K. Matyjaszewski, *Macromol. Chem. Phys.*, 2002, **203**, 1385.

349. J.-F. Lutz and K. Matyjaszewski, *J. Polym. Sci.: Part A: Polym. Chem.*, 2005, **43**, 897.

350. A. Fuerstner and N. Shi, *J. Am. Chem. Soc.*, 1996, **118**, 12349.

351. M. Freiberg and D. Meyerstein, *J. Chem. Soc., Chem. Comm.*, 1977, 127.

352. G. Ferraudi, *Inorg. Chem.*, 1978, **17**, 2506.

353. K. Matyjaszewski and B. E. Woodworth, *Macromolecules*, 1998, **31**, 4718.

354. R. Poli, F. Stoffelbach and S. Maria, *Polym. Prepr.*, 2005, **46**(2), 305.

Subject Index